# APPLIED CIRCUIT THEORY:
## Matrix and Computer Methods

## ELLIS HORWOOD SERIES IN ELECTRICAL AND ELECTRONIC ENGINEERING

*Editor:* P. S. Brandon, Professor of Electrical Engineering, University of Cambridge

**APPLIED CIRCUIT THEORY: MATRIX AND COMPUTER METHODS**
P. R. Adby, University of London King's College

**NOISE IN SOLID STATE DEVICES**
D. A. Bell, University of Hull and
M. J. Buckingham, Royal Aircraft Establishment, Farnborough, Hampshire

**DIFFRACTION THEORY AND ANTENNAS**
John Brown and R. H. Clarke, Imperial College of Science and Technology, University of London

**INTEGRATED CIRCUIT TECHNOLOGY OF COMMUNICATION-BASED COMPUTERS**
C. Moir, Ministry of Defence, Royal Signals and Radar Establishment, Worcestershire

**HANDBOOK OF RECTIFIER CIRCUITS**
Graham J. Scoles, English Electric Valve Company, Chelmsford, Essex

**PRINCIPLES OF COMPUTER COMMUNICATION NETWORK DESIGN**
J. Seidler, Institute of Information Science, Technical University of Gdansk, Poland

**CONTROL SYSTEMS WITH TIME DELAYS**
A. Olbrot, Institute of Automatic Control, Technical University of Warsaw

# APPLIED CIRCUIT THEORY:
## Matrix and Computer Methods

P. R. ADBY, B.Sc., D.Phil.
Department of Electronic and Electrical Engineering
King's College, University of London

**ELLIS HORWOOD LIMITED**
Publishers Chichester

Halsted Press: a division of
**JOHN WILEY & SONS**
New York - Chichester - Brisbane - Toronto

First published in 1980 by

**ELLIS HORWOOD LIMITED**
Market Cross House, Cooper Street, Chichester, West Sussex, PO19 1EB, England

*The publisher's colophon is reproduced from James Gillison's drawing of the ancient Market Cross, Chichester.*

**Distributors:**

*Australia, New Zealand, South-east Asia:*
Jacaranda-Wiley Ltd., Jacaranda Press,
JOHN WILEY & SONS INC.,
G.P.O. Box 859, Brisbane, Queensland 40001, Australia.

*Canada:*
JOHN WILEY & SONS CANADA LIMITED
22 Worcester Road, Rexdale, Ontario, Canada.

*Europe, Africa:*
JOHN WILEY & SONS LIMITED
Baffins Lane, Chichester, West Sussex, England.

*North and South America and the rest of the world:*
Halsted Press, a division of
JOHN WILEY & SONS
605 Third Avenue, New York, N.Y. 10016, U.S.A.

**British Library Cataloguing in Publication Data**
Adby, P. R.
Applied circuit theory: Matrix and computer methods.
(Ellis Horwood series in electrical and electronic engineering).
1. Electric circuits
2. Matrices
I. Title
621.319'2'01512943 TK454 79–41458

ISBN 0–85312–071–4 (Ellis Horwood Ltd., Publishers, Library Edition)
ISBN 0–470–26908–1 (Halsted Press)

Type set in Press Roman by Ellis Horwood Ltd.
Printed in Great Britain by W. & J. Mackay Ltd., Chatham

# Table of Contents

## Chapter 2 MATRIX ANALYSIS OF NETWORKS



## Chapter 3 NON-LINEAR D.C. ANALYSIS

*To my wife Ann*

# Preface

During the last ten years the computer has become an indispensable tool of electrical and electronic circuit analysis and design. Every engineer must expect to use a computer at all stages of circuit design, from specification through to manufacture, with the familiarity that was formerly accorded to the slide rule and calculator. One recent trend in the growth of computation is the increasing availability of computers away from computer centres. Minicomputers and microcomputers now provide 32K of memory for less than £1000 ($2000), and every design laboratory, however small, can run significant analysis programmes at very low cost. Also, the programmes themselves have become more accessible to individuals for adaptation and improvement. Graphic outputs can easily be incorporated, and specialised programmes can be more readily written.

A modern circuit analysis programme running on a large computer achieves accuracy and efficiency by use of sophisticated numerical techniques. This book establishes the circuit theory on which such programmes are founded, and it provides an introduction to recent research in computer-aided circuit design without becoming overburdened with numerical methods. Emphasis throughout the text is on matrix methods and computation. Engineers in industry, and postgraduate students in engineering and physics, will therefore find the book particularly suitable if they have not previously followed this approach to circuit analysis. The book also fills the gap between fundamental circuit theory and computer-aided circuit design with an intermediate treatment, and is therefore also suitable for second and third year undergraduate courses in electronics and electrical engineering in English universities and polytechnics, and equivalent courses elsewhere. Knowledge of matrix algebra and basic circuit theory is assumed, including circuit theorems, sinusoidal steady-state analysis, and the Laplace transform.

Undergraduate courses in circuit theory have undergone a radical change to the computational approach in response to the widespread availability of computers and to the increasing complexity of circuits. One important consequence is that implementation in conjunction with laboratory work can be achieved early in the course. The student quickly becomes concerned with transistors, operational amplifiers, and other active devices both in practice and through circuit theory. This process motivates him with an increased awareness of electronic circuit design, and a familiarization with devices and their circuit models, which is of immense unifying benefit.

The treatment of circuit analysis in this book is predominantly theoretical. To retain adequate coverage of the mass of available material, examples have been restricted to relatively simple circuits. In undergraduate courses it is essential to balance theory with extensive application to practical circuits. Motivation to apply theoretical techniques through laboratory work and design projects rests with the instructor. He is in the best position to provide more extensive and realistic circuit examples which are suited to analysis as well as fitting in with the remainder of the course.

The role of the computer in undergraduate courses has been much debated. The view taken at King's College is that the computer should be used to achieve experience with all possible aspects of circuit behaviour in as wide a range of circuits as possible. Experimentation by computer simulation is encouraged throughout the course, although the underlying objective is the comparison of computed results with design and laboratory measurements. Students are not expected to write analysis programmes, but they do have to write short subroutines, and they frequently have to determine why analysis does not work. Computing problems are included at the end of each chapter, and students become involved in some of them. Problems vary from straightforward circuit investigations to requests for programmes. The use of time-shared BASIC for large analysis programmes, even if restricted to small circuits when implemented on a minicomputer, has been more than justified by the integration of computation into the course and by enthusiastic student participation.

Five main computer programmes are included in the text: they cover d.c. analysis, a.c. analysis, two-port analysis, transfer function analysis, and random simulation. The programmes may be freely reproduced and used by instructors, and ASCII coded paper tapes are available on request. It is inevitable that some errors remain in the programmes; a regularly updated list of errors and programme improvements will be available from the author in the form of a users' newsletter. Many of the programmes requested in the problems should also become available. More extensively modified and improved versions of these programmes adapted to other computer systems and different implementations of BASIC will be available from Dayton Electronics Ltd, South Way, Newhaven, Sussex, England.

This book has had the benefit of use in manuscript form by engineering students at the University of London King's College. I would like to thank them for their helpful comments and criticism. I would also like to thank Miss Rosemary Ainsworth, Mrs. Margaret Richards and Mrs. Jean Hynes for typing the text. Finally, I would like to acknowledge my debt to the pioneers of computer-aided circuit analysis and design, whose work I have freely used and adapted to produce a coherent subject for study. Above all else this book rests on their research.

PAUL R. ADBY
King's College, London, 1979

CHAPTER 1

# Basic Circuit Theory

Circuit and system design is the creative task of the design engineer, but not that of the computer. However, the computer is an almost indispensable tool in the design process as an analyser, a mathematical modeller, a simulator, and an optimiser of electric networks. An essential part of many systems is the integrated circuit which today is ten to ten thousand times more complex than the discrete circuit of just a few years ago. Analysis, which is still the most important design step, must be systematic, accurate, and economic if it is to be applied to such large circuits. The application of network theory through matrix methods provides an approach which is suited to digital computation and is also capable of considerable sophistication under the broad heading of computer-aided design.

Many readers will be familiar with some of the network theory to be studied, particularly in the first two chapters of this book. Chapter 1 briefly reviews basic circuit theory, and Chapter 2 presents the basic methods of circuit analysis. These are expressed in matrix form for the development in the remainder of the book of methods for d.c., a.c., transient, sensitivity, and tolerance analysis. Chapter 2 should therefore be regarded as important even if the work is familiar, as it orients analysis towards matrix and computer methods.

## 1.1 LUMPED CIRCUITS

Electromagnetic theory tells us that energy is radiated whenever an electron is accelerated. In any circuit, current varies with time, and energy is therefore radiated and lost. However, the wavelength of the generated electromagnetic wave is usually large in comparison with the physical dimensions of the circuit, and the energy loss is therefore negligible.

*Circuits in which radiated energy may be assumed negligible are called* **lumped circuits,** *and they are obtained by interconnecting* **lumped circuit elements.**

Of course, lumped elements and circuits are idealised models of physical elements and circuits. A lumped circuit is assumed to have negligible physical dimensions, and may be analysed exactly. The corresponding physical circuit, depending on its frequency of operation, may or may not behave as predicted by the analysis. The circuit analysed, therefore, is not the physical circuit but a model drawn in the form of a circuit diagram and subject to a number of assumptions. Within these assumptions, the analysis is simplified at the expense of accuracy in modelling the behaviour of physical circuits. The most important assumption throughout this book is that circuits are *lumped* and therefore obey Kirchhoff' laws.

*Circuits which are not lumped are called* **distributed.**

Usually the dimensions of a distributed circuit are specifically designed to utilise the electromagnetic wave properties at their frequency of operation. A television antenna, for example, is designed to receive some of the energy radiated from another antenna. The shape and size of the receiving antenna are designed to selectively receive the required band of frequencies from a particular direction. A less obvious example of a distributed circuit is the electricity supply system. The wavelength corresponding to 50Hz is 6000 kM. The supply system is of comparable size, and must therefore be considered as a distributed circuit.

The most common circuit elements such as resistors, capacitors, inductors, transistors, voltage sources, should be familiar to the student from first courses in electric circuits and electronics. The lumped versions of these elements are assumed to have zero physical dimensions, and therefore interact with the circuit voltage or current at a point. Each element has two or more **terminals** by which it is connected to other elements to form a circuit.

*A point in a circuit where two or more terminals are connected is called a* **node.** *A two-terminal element connected between two nodes in a circuit is called a* **branch.**

The circuit shown in Fig. 1.1(a) has four nodes numbered ⓪, ①, ②, and ③, and six branches labelled $i_1, i_2, \ldots i_6$ with arrows to show current direction.

*The* **branch voltage** *is the voltage across a branch, and the* **branch current** *is the current through a branch.*

Voltage polarity is illustrated in Figs. 1.1(b) and 1.1(c), where the branch voltage is, by convention, in a direction opposite to the branch current. Indication of polarity, either by means of an arrow or by + and − signs, is redundant on the circuit diagram, since the polarity is fixed by the choice of current direction and by the sign convention. In circuits with sinusoidal or other time-dependent sources polarity signifies the direction of positive variation.

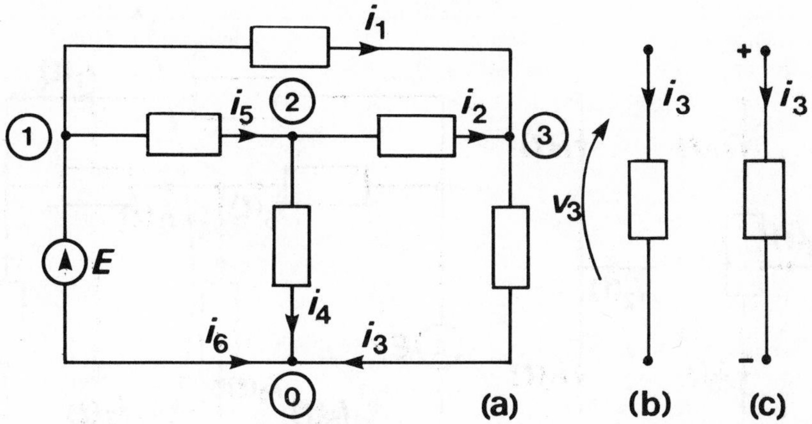


Fig 1.1(a) – a bridged-T circuit; (b), (c) the branch sign convention.

Not all textbooks utilise the same convention for indicating voltage polarity, and often arrows similar to the one in Fig. 1.1(b) refer to voltage drops. The convention used here ensures that measured voltages and currents in actual circuits have the same directions as those on the circuit diagram. Another consequence of the convention is that the product, branch voltage × branch current, represents the power delivered to the branch element.

## 1.2 KIRCHHOFF'S LAWS

Kirchhoff's current law (KCL) and voltage law (KVL) are the two fundamental laws which govern the voltages and currents in a circuit. They may be derived from Maxwell's equations under the lumped circuit assumption.

**Kirchhoff's current law** *states that the sum of the currents entering a lumped circuit element is zero at all times.*

**Kirchhoff's voltage law** *states that the sum of the voltages round a closed path in a lumped circuit is equal to zero at all times.*

KCL may be generalised by regarding a node as a circuit element, so that the sum of the currents entering a circuit node is zero. Also a circuit may be divided into two halves, each half being regarded as single elements with a larger number of terminals than usual. The sum of currents entering one half of the divided circuit is zero. KCL applied to a circuit element, circuit node, and a divided circuit is illustrated by Fig. 1.2. In each case KCL states that

$$i_1(t) + i_2(t) - i_3(t) = 0 \quad .$$

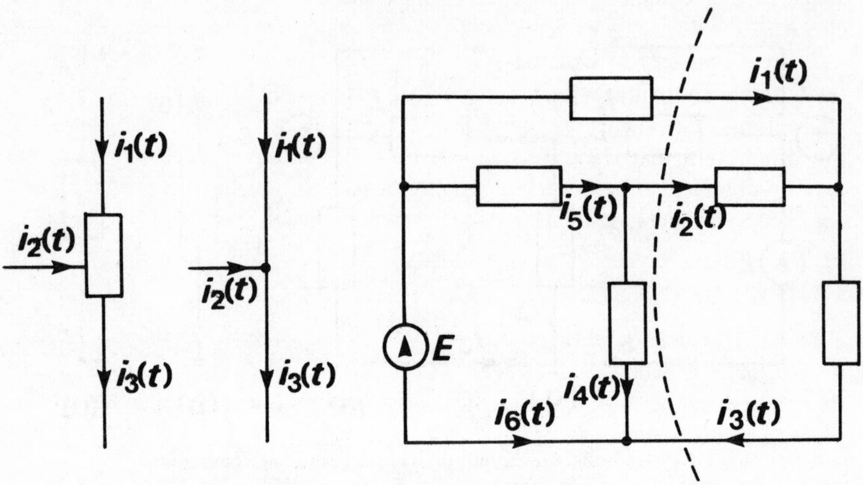


Fig. 1.2 – Kirchhoff's current law.

The circuit division shown in Fig. 1.2 as a broken line will later be defined precisely and called a *cut*.

The *node equations* for the circuit in Fig. 1.2 are obtained by application of KCL at each node, giving

$$-i_1 - i_5 - i_6 = 0$$

$$i_5 - i_2 - i_4 = 0$$

$$i_1 + i_2 - i_3 = 0$$

$$i_3 + i_4 + i_6 = 0 \ .$$

The *loop equations* for the circuit are obtained by application of KVL round the closed paths, or **loops**, as they are usually called. Using the three obvious loops and the outer loop we obtain

$$E - v_5 - v_4 = 0$$

$$-v_1 + v_2 + v_5 = 0$$

$$v_4 - v_2 - v_3 = 0$$

$$E - v_1 - v_3 = 0 \ .$$

Three further equations can be derived for the more complicated loops in the circuit.

These two sets of equations are linear; they express linear constraints on the currents and voltages in a circuit which exist solely as a result of the circuit connections. The elements themselves have not yet been considered except for the assumption that they are lumped. Both the node and loop equation sets may by observed to be *linearly dependent* since it is possible to express any one equation as a linear combination of the remaining three.

## 1.3 TELLEGEN'S THEOREM

Tellegen's theorem applies to any network which obeys Kirchhoff's laws; it is essentially a statement that energy is conserved in a lumped network. Consider a network with $n$ branches with branch currents $i_1, i_2, \ldots i_n$ and corresponding branch voltages $v_1, v_2, \ldots v_n$.

*Tellegen's theorem states that the sum* $\Sigma_{k=1}^{n} v_k i_k$ *equals zero at all times.*

We have already noted that the product of branch voltage and branch current gives the power delivered to the branch. The total power delivered to the circuit by a branch containing a source must therefore be absorbed by the remaining branches so that energy is conserved.

**Proof**

Consider the branch in a network connected between node ⓡ and node ⓢ. Introduce a reference point in the circuit so that node voltages $e_r$ and $e_s$ may be defined relative to it as shown in Fig. 1.3.

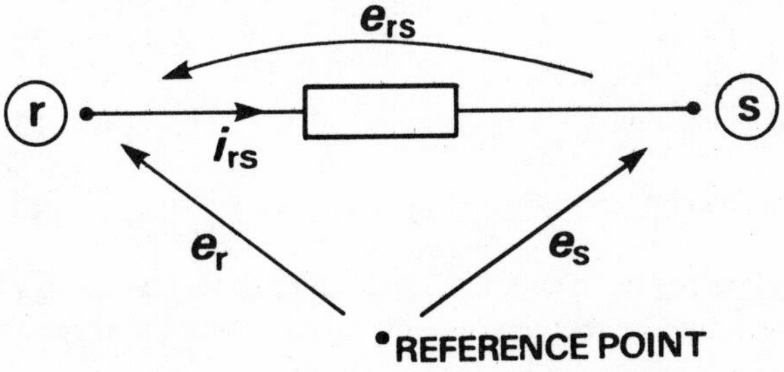


Fig. 1.3 – Tellegen's theorem.

Then

$$e_{rs} = e_r - e_s \ ,$$

and the power delivered to the branch is given by

$$e_{rs} i_{rs} = e_r i_{rs} - e_s i_{rs} \ .$$

The total delivered branch power for all the branches in the network is therefore given as the sum

$$\Sigma_{r=1}^{n} \Sigma_{s=1}^{n} e_{rs}i_{rs} = \Sigma_{r=1}^{n} \Sigma_{s=1}^{n} (e_r i_{rs} - e_s i_{rs}) .$$

This may be rewritten as

$$\Sigma_{r=1}^{n} e_r \Sigma_{s=1}^{n} i_{rs} - \Sigma_{s=1}^{n} e_s \Sigma_{r=1}^{n} i_{rs} .$$

The term $\Sigma_{s=1}^{n} i_{rs}$ can be recognised as the sum of all currents entering node ⓡ, which equals zero by KCL. Similarly $\Sigma_{r=1}^{n} i_{rs}$ can be recognised as the sum of all currents leaving node ⓢ, and is therefore also zero. The total of power delivered to the network branches is therefore equal to zero.

The only assumption made is that the circuit is lumped and obeys Kirchhoff's laws. No reference has been made to specific element values or types, and it is necessary only that the branch currents obey KCL and the branch voltages obey KVL. The actual values are arbitrary.

*Example* 1.1

In the bridge circuit (Fig. 1.4) let the branch voltages and currents take arbitrary values, and check Tellegen's theorem.

Assigning arbitrary branch voltages and currents which obey Kirchhoff's laws we obtain,

| | | |
|---|---|---|
| $i_1 = 2$ | $v_1 = -12$ | $i_1 v_1 = -24$ |
| $i_2 = 1$ | $v_2 = 2$ | $i_2 v_2 = 2$ |
| $i_3 = 1$ | $v_3 = 13$ | $i_3 v_3 = 13$ |
| $i_4 = 3$ | $v_4 = 1$ | $i_4 v_4 = 3$ |
| $i_5 = 4$ | $v_5 = -11$ | $i_5 v_5 = -44$ |
| $i_6 = 5$ | $v_6 = 10$ | $i_6 v_6 = 50$ |
| | | $\Sigma_{k=1}^{6} i_k v_k = 0$ |

The sets of branch voltages and currents should be checked to see that Kirchhoff's laws are satisfied, and other sets of values should be tried to check the validity of Tellegen's theorem.

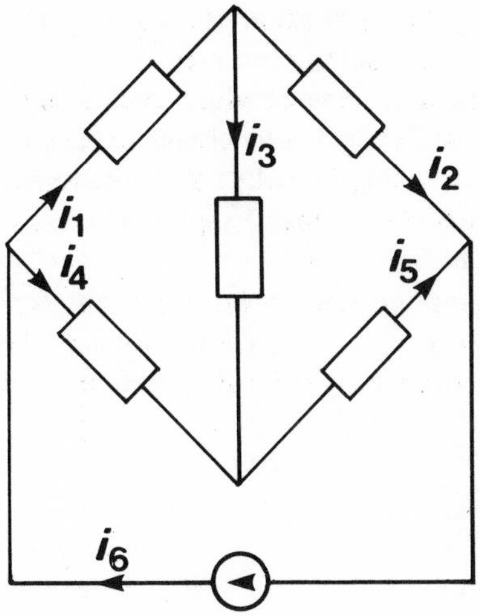


Fig. 1.4 – Bridge circuit.

## 1.4 TIME-INVARIANT CIRCUITS

Kirchhoff's laws and Tellegen's theorem apply to any lumped circuit at any instant of time. In most circuits the elements are normally fixed in value and vary only over a long period because of temperature changes and ageing.

*A circuit with an excitation voltage or current $x$ and an output voltage of current $y$ with the characteristic $y = f(x)$ is* **time invariant** *if the function $f$ is independent of time.*

Note that the function $f$ depends on the circuit element values such as resistance and capacitance, and it is these values which must remain constant with time. The circuit variables $x$ and $y$ are of course expected to vary with time. All circuits considered in this book are assumed to be time-invariant.

## 1.5 LINEAR CIRCUITS

*A circuit with an excitation voltage or current $x$ and an output voltage or current $y$ is* **linear** *if $y$ is proportional to $x$; that is, if $y = f(x)$ then $ky = f(kx)$.*

Circuit elements are included within this definition of linearity since they may be regarded as circuits. Resistors, capacitors, and inductors are normally close to linear within the range of operation specified by the manufacturers. In general, however, all physical components are *non-linear* owing to effects such as internal heating, voltage breakdown, and magnetic core saturation, which alter the component value and depend on the applied voltage.

As we shall see, the linearity assumption greatly simplifies circuit analysis. It is important to make this assumption even if circuits are only approximately linear. Analytical techniques, some of which exclusively apply to linear circuits, may then be used to gain insight into circuit behaviour even though approximation is involved in assuming linearity. The disadvantage in using non-linear analysis, which for accuracy is normally implemented as a computer programme, is that this human insight is then lost.

The linearity assumption has several important consequences. If we consider a change $\Delta x$ in the excitation $x$, a corresponding change $\Delta y$ is produced in $y$. In a linear circuit these are proportional, therefore

$$\frac{\Delta y}{y} = \frac{\Delta x}{x} = k$$

$$\text{and} \qquad \Delta y = f(\Delta x) \ . \qquad (1.1)$$

The changes in the circuit voltages and currents produced by variation of the fixed sources are therefore independent of the nominal values which exist in the circuit. Also, since

$$1 + \frac{\Delta y}{y} = 1 + \frac{\Delta x}{x} = 1 + k \ ,$$

$$(y + \Delta y) = f(x + \Delta x) \ , \qquad (1.2)$$

the effects of source variation are additive to the nominal values.

If the variations $\Delta x$ and $\Delta y$ are considered small compared with the nominal values $x$ and $y$, the analysis for variation only is called **small signal analysis.** It is performed separately from the **large signal analysis** used to determine nominal values. Usually, large signal analysis is meant to imply non-linear analysis since, as we have seen, large signals and small signals have proportional effects in a linear circuit.

**Superposition**

*The* **principle of superposition** *states that a linear circuit excited by n sources* $x_1, x_2, \ldots x_n$ *which when applied individually give outputs* $y_1, y_2 \cdots y_n$, *will give the output* $\sum_{k=1}^{n} y_k$ *when the sources are applied collectively.*

Superposition extends the properties of Eqs. (1.1) and (1.2) to linear circuits with more than one source by stating that the sources do not interact.

Application of this principle means that one complete analysis consists of separate analyses for each source in the circuit. All other independent voltage sources are short circuit, and all other independent current sources are open circuit, but any dependent sources, for example transistor $g_m$, must be left operative.

*Example* 1.2

Determine the voltage across the resistor $R_2$ in the circuit shown in Fig. 1.5.

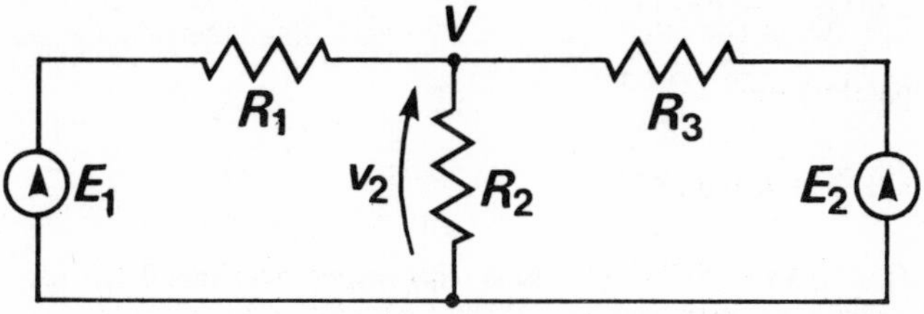


**Fig. 1.5 – Superposition example.**

With $E_2$ short circuit,

$$v_2 = E_1\left(\frac{R_2R_3}{R_2+R_3}\right)\Big/\left(R_1 + \frac{R_2R_3}{R_2+R_3}\right)$$

$$= \frac{R_2R_3E_1}{(R_1R_2+R_1R_3+R_2R_3)} \ .$$

With $E_1$ short circuit,

$$v_2 = \frac{E_2R_1R_2}{(R_1R_2+R_1R_3+R_2R_3)} \ ,$$

$$\text{therefore} \quad V = \frac{R_2R_3E_1+R_1R_2E_2}{R_1R_2+R_1R_3+R_2R_3} \ .$$

It is instructive to obtain this result by mesh analysis, since the linear form of the mesh equations and solution by Cramer's rule demonstrates that a solution *must* be obtained with separate factors for each of the sources present.

When the sources are sinusoidal, the principle of superposition implies that sinusoids are added in a linear circuit. Product terms, which give rise to harmonics and distortion, cannot arise. The property of linearity is therefore essential for sinusoidal steady-state analysis to be valid.

Finally, it should perhaps be noted that fixed sources are non-linear by our definition of linearity, because their voltage and current are independent. Also, it is reasonable to expect that a circuit is linear if it is composed only of interconnected linear circuit elements.

## 1.6 SINUSOIDAL STEADY-STATE ANALYSIS

The characteristic equation relating voltage and current is given for inductors by

$$v(t) = L\mathrm{d}i(t)/\mathrm{d}t \ , \tag{1.3}$$

and for capacitors by

$$i(t) = C\mathrm{d}v(t)/\mathrm{d}t \ . \tag{1.4}$$

Both equations involve differentiations with respect to time. Their use in analysis therefore produces a differential equation which can be solved to give both the transient and steady-state solutions. These are added to obtain the total response as a function of time. When the circuit input is sinusoidal and when the circuit is assumed to have reached the steady state, analysis for transient response may be replaced by sinusoidal steady-state analysis for frequency response. Here we will briefly recall the main steps in the transition from analysis in the time domain to analysis in the frequency domain through rotating vectors, phasors, and complex numbers.

A rotating vector of length $V$ is shown in Fig. 1.6 rotating about the origin of rectangular coordinate axes at a rate of $\omega$, radians/second in the anticlockwise direction. The projection on to the horizontal axis represents its instantaneous value as a function of time.

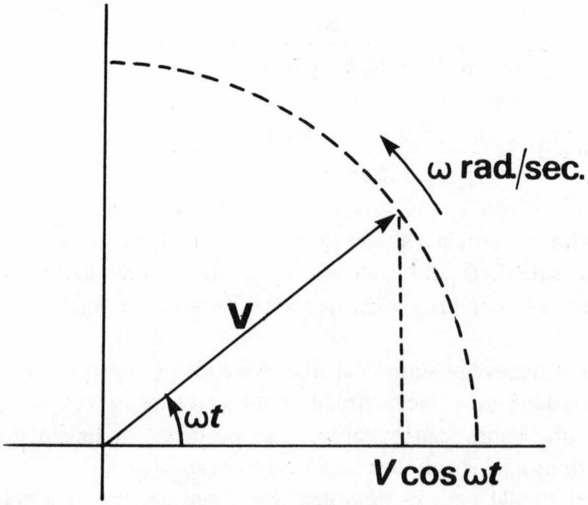


Fig. 1.6 – A rotating vector.

In a linear circuit excited by sinusoidally varying sources, all voltages and currents may be represented by rotating vectors. However, linearity ensures that all vectors rotate at the same angular frequency $\omega$, and it is more convenient to consider the vector stationary with the axes rotating at $\omega$ rad/sec in the opposite direction. The vectors are then called **phasors** and the diagram on which they are plotted is called a **phasor diagram**. A phasor $\mathbf{V}$ is plotted as a straight line of length $|\mathbf{V}|$ at an angle $\theta$ to the horizontal, where $|\mathbf{V}|$ and $\theta$ represent magnitude and phase of the phasor respectively. A phasor diagram portrays the relative amplitudes and phases of the circuit voltages and currents at a given frequency. Since sufficient information about a network in the steady state can be obtained from the phasor diagram, projection of the phasors on to the rotating axes is no longer necessary. When required, however, the rotational term can be reinserted, giving $v(t) = |\mathbf{V}| \cos(\omega t + \theta)$ for the phasor $\mathbf{V}$.

Analysis proceeds by addition, subtraction, integration, and differentiation of phasors, and is limited in practice because the technique, either graphically or algebraically, is cumbersome, the more so as networks become bigger and have large phase shifts. Also, a phasor diagram relates only to one frequency, whereas in most cases a range of frequencies are of interest and a **frequency response** is required.

Another representation of a phasor is derived through Euler's identity

$$e^{j\theta} = \cos\theta + j\sin\theta \ , \tag{1.5}$$

where $j = \sqrt{-1}$ .

$e^{j\theta}$ may be plotted on the Argand diagram as a vector of unit length at an angle $\theta$ to the horizontal axis, as in Fig. 1.7. An Argand diagram may be identi-

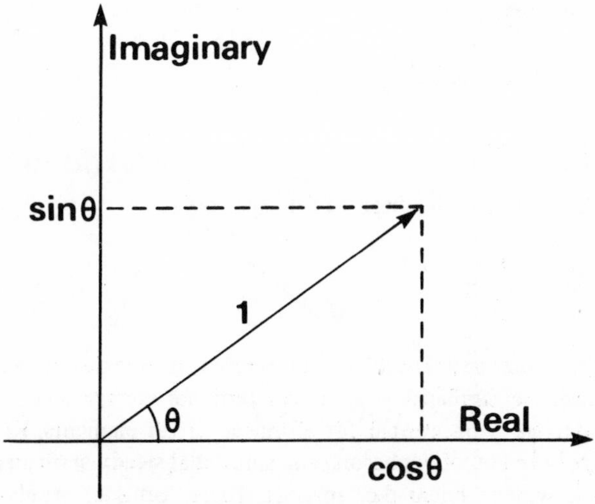


Fig. 1.7 – Argand diagram of $e^{j\theta}$.

fied with a phasor diagram, and the phasor $\mathbf{V} = V\angle\theta$ may be represented by the complex quantity $Ve^{j\theta}$ since it has magnitude $V$ and phase $\theta$.

If we excite a network by a current $i(t) = I\cos(\omega t + \phi)$ and obtain a response $v(t) = V\cos(\omega t + \theta)$ we can represent the phasors $\mathbf{I}$ and $\mathbf{V}$ as $Ie^{j\phi}$ and $Ve^{j\theta}$ respectively. The rotational term $e^{j\omega t}$ is omitted from these expressions because phasors are stationary, and it is applied to the rotating axes of the phasor diagram instead. To recover the time domain behaviour the rotation term $e^{j\omega t}$ must be reinserted, and then the real part taken to recover the cosine; that is,

$$i(t) = \mathrm{Re}\{\,|\mathbf{I}|\,e^{j\phi}e^{j\omega t}\} = I\cos(\omega t+\phi)$$

$$v(t) = \mathrm{Re}\{\,|\mathbf{V}|\,e^{j\theta}e^{j\omega t}\} = V\cos(\omega t+\theta)\ .$$

It is this cancellation of $e^{j\omega t}$ which simplifies analysis in the sinusoidal steady state by removal of trigonometric terms and conversion of differential equations to algebraic equations.

Substitution of the excitation current $I\cos(\omega t+\phi)$ into the inductor characteristic. Eq. (1.3) yields

$$v(t) = L\frac{d}{dt}\{I\cos(\omega t+\phi)\}$$

$$= \omega L\,I\cos(\omega t+\phi+\pi/2)\ .$$

In terms of phasors,

$$\mathbf{V} = \omega L\,I\,e^{j\phi}e^{j\pi/2}$$
$$= j\omega L\,\mathbf{I}\ . \qquad (1.6)$$

A similar result can be obtained for capcitors as

$$\mathbf{V} = \frac{1}{j\omega C}\mathbf{I}\ .$$

The quantities $j\omega L$ and $1/j\omega C$ are interpreted as complex impedance. Mutual inductors may be similarly treated and terminal voltage and current phasors thus shown to be proportional for all linear circuit elements. Kirchhoff's laws therefore apply to complex phasors, and sinusoidal steady-state analysis proceeds in the same way as linear d.c. analysis. Three forms of steady-state analysis are illustrated by the following example.

*Example* 1.3

Determine the output voltage of the filter circuit shown in Fig. 1.8.

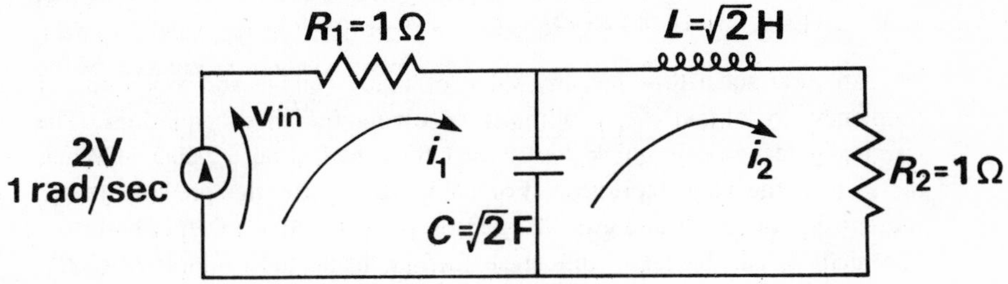


Fig. 1.8 – Low-pass filter.

The impedance of the two reactive components can be written in three forms.

| | Numerical | Function of $\omega$ | Function of $\omega$ and $L$ or $C$ |
|---|---|---|---|
| C | $1/\sqrt{2}\,j$ | $1/j\omega\sqrt{2}$ | $1/j\omega C$ |
| L | $\sqrt{2}\,j$ | $j\omega\sqrt{2}$ | $j\omega L$ |

Each of these forms can be used in analysis, giving three solutions of increasing usefulness in the steady state.;

If we use the numerical form of impedance, the mesh equations for the filter are

$$(1 - 0.7j)\,i_1 + 0.7j\,i_2 = 2$$

$$0.7j\;i_1 + (1+0.7j)i_2 = 0\;.$$

Solving for $i_2$ gives $i_2 = -0.7j$, hence

$$v_{OUT} = 0.7\,\angle{-90^\circ}\;.$$

In order to analyse the circuit at other frequencies the mesh equations must be re-formed and re-solved at each frequency required.

If impedance is a complex function of $\omega$ the mesh equations become

$$(1 + 0.7/j\omega)i_1 - (0.7/j\omega)i_2 = 2$$

$$-(0.7/j\omega)i_1 + (1+1.4j\omega + 0.7/j\omega)i_2 = 0\;.$$

Solution for $i_2$ gives $i_2 = 1/\{(j\omega)^2 + 1.4j\omega + 1\}$, and the voltage gain can be written

$$\frac{v_{OUT}}{v_{IN}} = \frac{0.5}{(j\omega)^2 + 1.4(j\omega) + 1} . \tag{1.8}$$

We can now substitute for any value of input voltage and any value of frequency to obtain $v_{OUT}$ without re-solving the mesh equations. The frequency response is quickly evaluated by substituting a range of values for $\omega$ into the gain expression. For comparison with the previous result, substitution of $\omega = 1$ and $v_{IN} = 2$ yields $v_{OUT} = 1/1.4j = -0.7j$ as before. The analysis can be taken one stage further, using the symbols $L$, $C$, $R_1$, and $R_2$ in place of numerical values. The mesh equations become

$$(R_1 + 1/j\omega C)i_1 - i_2/j\omega C = v_{IN}$$

$$(-1/j\omega C)i_1 + (1/j\omega C + j\omega L + R_2)i_2 = 0 .$$

This can be solved, noting that $v_{OUT} = i_2 R_2$, giving

$$\frac{v_{OUT}}{v_{IN}} = \frac{R_2}{(j\omega)^2 R_1 LC + j\omega(L+CR_2) + (1+R_2)} . \tag{1.9}$$

Numerical values for $R_1$, $R_2$, $L$, and $C$ can be substituted, giving gain as a function of $\omega$ identical to the previous results in Eq. (1.8) above. The additional advantage of this symbolic solution is that circuits with different component values may be analysed without re-solving the mesh equations. Also, the expression for gain can be differentiated with respect to the circuit components to predict the variation in gain as component values are varied.

The two expressions (1.8) and (1.9) are called **transfer functions.** In general, they take the form of a ratio of two polynomials in the variable $j\omega$. In order to investigate the properties of transfer functions we must generalise the frequency variable $j\omega$ to the complex frequency variables $s = \sigma + j\omega$. Where $j\omega$ has appeared we now have $s$. For example, the transfer function in Eq. (1.9) becomes

$$F(s) = \frac{R_2}{R_1 LCs^2 + (L+CR_2)s + (1+R_2)} ,$$

and the frequency response is evaluated by the substitution $s = j\omega$. The impedance of the inductor and capacitor become $sL$ and $1/sC$ respectively. Analysis to obtain transfer functions with numerical coefficients is called **transfer function analysis** or **$s$-domain analysis.** When coefficients are symbolic functions of the components, the analysis is called **symbolic analysis.** Transfer function analysis will be developed in Chapter 6, but for convenience we shall occasionally use $s$ in place of $j\omega$ in earlier chapters.

## 1.7 NON-LINEAR CIRCUITS

*A circuit which is not linear is non-linear*

As we have already noted, all practical circuits are non-linear. Fortunately, many circuits operate in regions of near linearity, and distortion due to non-linearity is usually small enough to ignore. Circuits in which this cannot be assumed cause major errors when linear analysis is employed, and special analysis techniques are necessary.

### Linearisation

The pn junction diode is a non-linear element which is governed at room temperture by the equation

$$i = I_s(e^{40v}-1) \ . \qquad (1.10)$$

Clearly, circuit analysis involving this relationship between voltage and current is not easy; but when the diode is biased to a known operating current, and signal currents in the diode are small in comparison, the circuit can be linearised.

The graph of the diode characteristic is shown in Fig. 1.9. Its gradient gives the value of the diode incremental or slope resistance. At any bias current

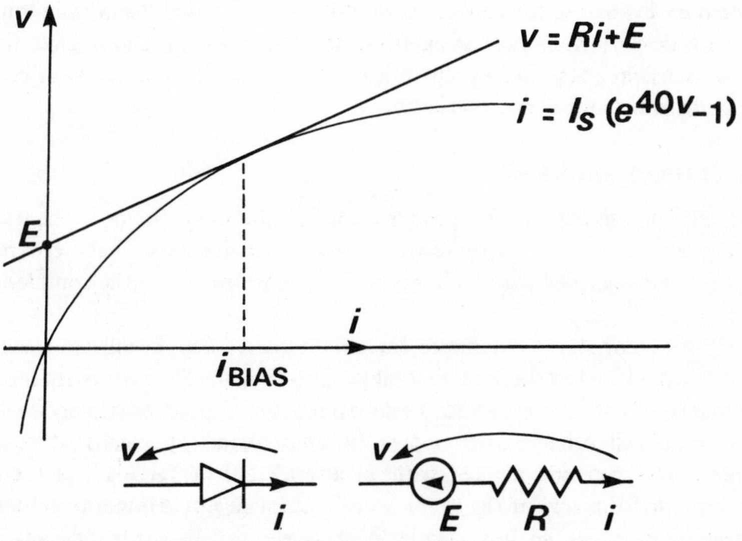


Fig. 1.9 – Diode linearisation.

$i_{BIAS}$ the tangent to the characteristic curve is an approximation to the diode behaviour in the region of $i_{BIAS}$. The intercept $E$ on the voltage axis represents a fixed voltage which appears in series with the diode slope resistance to correct the d.c. voltage across the diode.

The replacement of non-linear elements by a combination of sources and linear elements is called **linearisation**. The resulting linear circuit is valid, provided that small signals are assumed which do not deviate significantly from the bias (or nominal) values.

**Non-linear d.c. Analysis**

When transistors, diodes, and other non-linear elements are present in a circuit, analysis to determine d.c. voltages and currents is complicated by the variation of the element resistances. Until a solution is obtained it is not possible to assign the circuit resistance values necessary to obtain a solution. Some form of iterative solution is therefore required which adjusts the resistances in accordance with their characteristic equations as the solution is approached. This type of analysis is called **non-linear d.c. analysis.**

If the circuit then operates under small signal conditions it may be linearised, and linear analysis is used to find frequency or transient response. However, a large class of circuits, for example all digital switching circuits, operate under large-signal non-linear conditions. For these circuits we must employ non-linear transient analysis.

**Non-linear transient analysis**

Non-linear circuits under large-signal conditions are the most difficult to analyse since they involve simultaneous solution of differential and algebraic equations. The solution by numerical integration of these *state* equations gives the non-linear transient response. In this book we shall do little more than introduce the basic solution for linear networks.

## 1.8 CIRCUIT ELEMENTS

This section summarises the properties of all the basic circuit elements. Many readers will find some of the elements have not previously been covered, but they are included here to emphasise the relationships within the complete set of elements.

Many computer programmes for circuit analysis apply only to a subset of the elements listed in the summary table, even though all elements are important in modern circuit design. Various minimum subsets of elements may be defined from which all other elements may be constructed by means of equivalent circuits. The elements marked with an asterisk (*) in Table 1.1 at the end of this section form one of the more useful, although not minimum, subsets. For convenience, it should be possible to analyse a circuit containing any of the basic circuit elements without the need for equivalent circuits.

Resistors, capacitors, and inductors should need no further explanation beyond noting that they are assumed to be lumped, linear, and time-invariant, and that both positive and negative values are allowed.

**Dependent sources**

Dependent sources differ from the independent sources by being within a circuit. Independent sources are the inputs to a circuit, and they correspond to the signal generator or power source in practice. Dependent sources are idealised models of electronic devices such as transistors and operational amplifiers which form part of the circuit. The transconductance, or $g_m$, of a transistor is an example of a controlled source in which a current generator in the collector circuit is controlled by the voltage developed across the base emitter junction (neglecting the base resistance $r_b$ for a moment). The hybrid $\pi$ equivalent circuit of a transistor is shown in Fig. 1.10 with the voltage-controlled current source indicated.

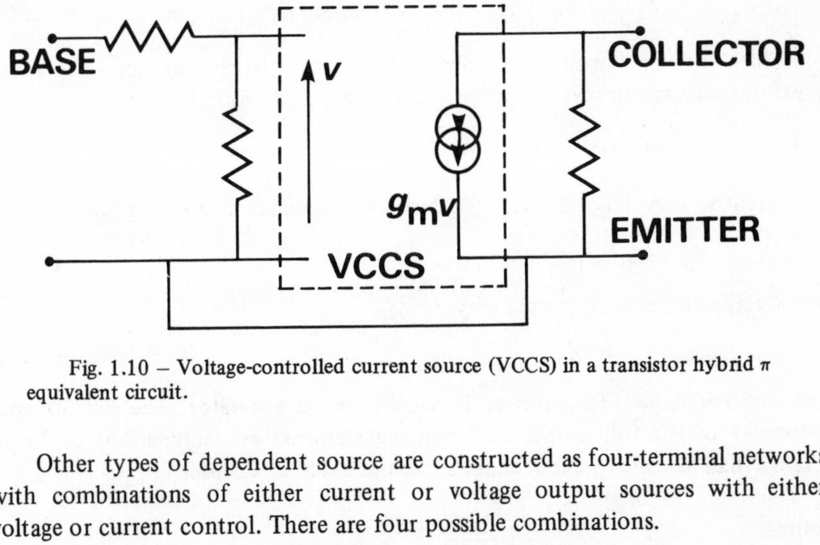


Fig. 1.10 – Voltage-controlled current source (VCCS) in a transistor hybrid $\pi$ equivalent circuit.

Other types of dependent source are constructed as four-terminal networks with combinations of either current or voltage output sources with either voltage or current control. There are four possible combinations.

| | |
|---|---|
| VCVS | Voltage Controlled Voltage Source |
| VCCS | Voltage Controlled Current Source |
| CCVS | Current Controlled Voltage Source |
| CCCS | Current Controlled Current Source |

The dependent sources are assumed to be linear and time-invariant, and the ratio of output/input is normally assumed to be real. None of these assumptions is necessarily true in practice. The circuits and properties of the four dependent sources are summarised in Table 1.1.

**Impedance converters and inverters**

The ideal transformer shown in Fig. 1.11 is an impedance converter since the impedance $Z_{in}$ is given by

$$Z_{in} = n^2 Z_L \tag{1.11}$$

The ideal transformer is assumed to be linear and time-invariant, and the factor $n$ is assumed to be real. The transformer polarity dots in Fig. 1.11 will be omitted in future. Instead, the value of $n$ will take a positive value when the transformer is connected in phase and a negative value when connected in anti-phase. The same convention will apply to other impedance converters and inverters and also to mutual inductors.

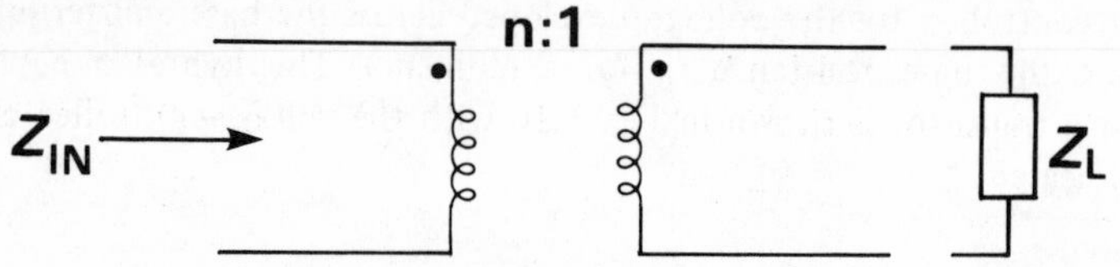


Fig. 1.11 – The ideal transformer.

Three similar impedance conversions are also of importance. The ideal transformer gives a positive impedance conversion. The equation

$$Z_{in} = -n^2 Z_L \tag{1.12}$$

gives negative impedance conversion. Positive impedance inversion is given by

$$Z_{in} = n^2/Z_L \ , \tag{1.13}$$

and negative impedance inversion by

$$Z_{in} = -n^2/Z_L \ . \tag{1.14}$$

The positive impedance inverter is usually called a **gyrator**. The circuits and properties of the four impedance-changing elements are summarized in Table 1.1. We shall investigate these circuits in more detail in Chapter 2.

**Sources**

Voltage and current sources should by now be known well enough. We should note that they are non-linear elements which may be either real or complex in value. The equations describing these sources are:

$$\mathbf{v} = V\angle\phi, \ \mathbf{i} = \text{arbitrary}$$

for a voltage source, and

$$\mathbf{I} = I\angle\phi, \ \mathbf{v} = \text{arbitrary}$$

for a current source.

Two further types of element are suggested by these equations in which **v** and **i** are either both fixed or both arbitrary. We may therefore define the **nullator** by the equation

$$\mathbf{v} = \mathbf{i} = 0$$

and the **norator** by the equations

$$\mathbf{v} = \text{arbitrary}, \quad \mathbf{i} = \text{arbitrary} \ .$$

We shall return to these elements in Chapter 4 where they will be found useful in reducing three-terminal and four-terminal models to circuits with two-terminal branches only.

**Table 1.1** – Basic circuit elements

| LINEAR CIRCUIT ELEMENTS | | |
|---|---|---|
| *RESISTOR |   | $v = Ri$ |
| *CAPACITOR | 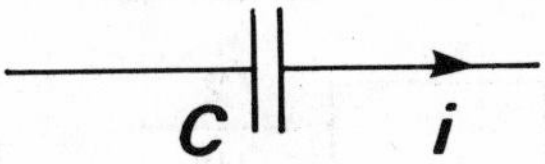  | $i = j\omega Cv$ |
| *INDUCTOR |   | $v = j\omega Li$ |
| MUTUAL INDUCTOR | 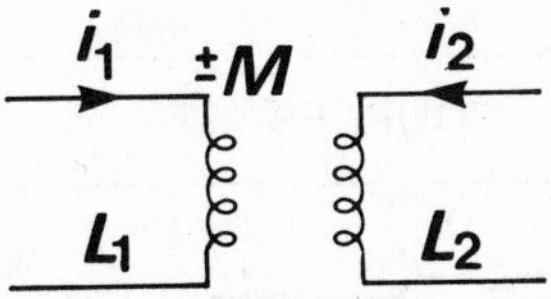  | $v_2 = j\omega L_2 i_2 \pm j\omega M i_1$ <br> $v_1 = j\omega L_1 i_1 \pm j\omega M i_2$ |
| **DEPENDENT SOURCES** | | |
| VCVS |   | $v_2 = \mu v_1$ |
| *VCCS | 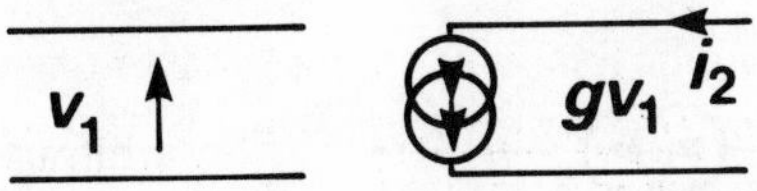  | $i_2 = gv_1$ |
| CCVS |   | $v_2 = ri_1$ |
| CCCS | 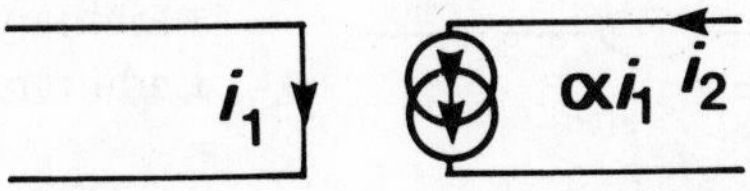  | $i_2 = \alpha i_1$ |

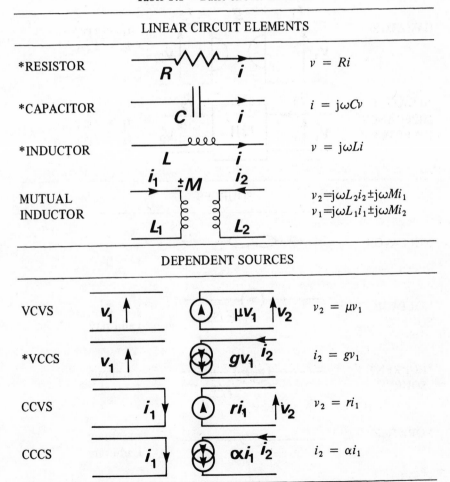

## IMPEDANCE CONVERTERS AND INVERTERS

IDEAL TRANSFORMER

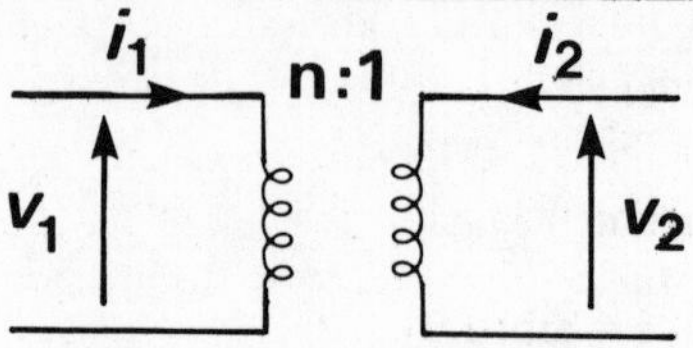


$$v_1 = \pm n v_2$$
$$-i_2 = \pm n i_1$$

NEGATIVE IMPEDANCE CONVERTER

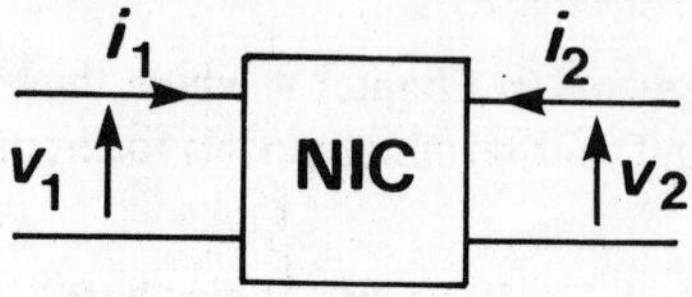


$$v_1 = \pm n v_2$$
$$-i_2 = \mp n i_1$$

GYRATOR

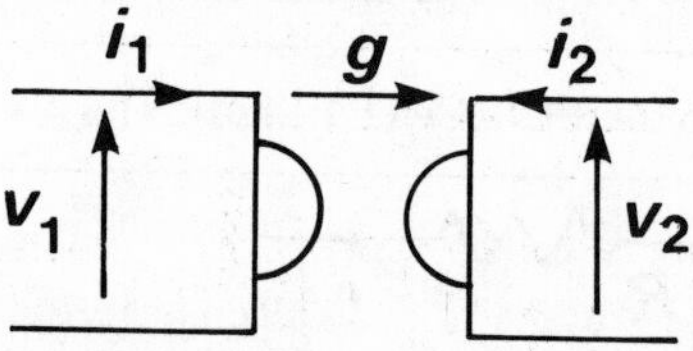


$$i_1 = \pm g v_2$$
$$-i_2 = \pm g v_1$$

NEGATIVE IMPEDANCE INVERTER

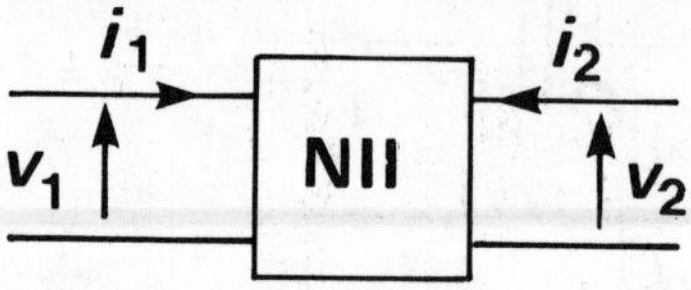


$$i_1 = \pm g v_2$$
$$-i_2 = \mp g v_1$$

## SOURCES

NULLATOR

v = 0
i = 0

VOLTAGE SOURCE

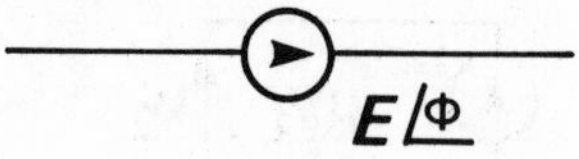


v = $E\angle\phi$
i arbitrary

*CURRENT SOURCE

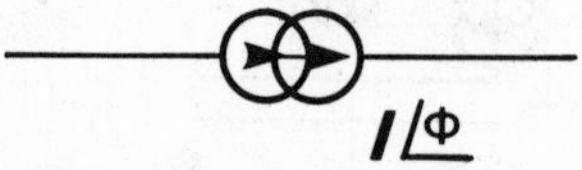


v arbitrary
i = $I\angle\phi$

NORATOR

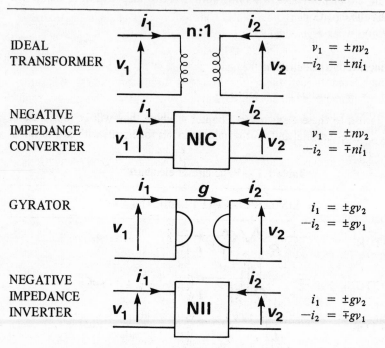

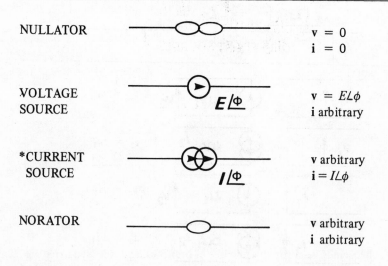

v arbitrary
i arbitrary

## 1.9 EQUIVALENT CIRCUITS

*Two circuits are* **equivalent** *if they have identical terminal voltages, currents, and waveforms when connected to any other network.*

The two circuits are therefore indistinguishable when the terminal behaviour is considered, even though they probably have different numbers of internal nodes and branches. Equivalent circuits are primarily used to simplify a complicated network, replacing it for analysis by a network with few elements and sources. Other equivalents arise when devices such as the transistor, which is neither lumped nor linear, is modelled by a lumped linear circuit for small-signal analysis. A number of important theorems dealing with equivalent circuits will be briefly reviewed here.

### Rosen's theorem

*The mesh network equivalent to a star network with n connected admittances $Y_1, Y_2, \ldots Y_n$ has $n(n-1)/2$ elements with admittances given by*

$$Y_{ij} = Y_i Y_j / \Sigma_{k=1}^{n} Y_k \quad i = 1, \ldots, n, \quad j = i+1, \ldots, n \ .$$

The three-branch ($n$=3) case is illustrated in Fig. 1.12.

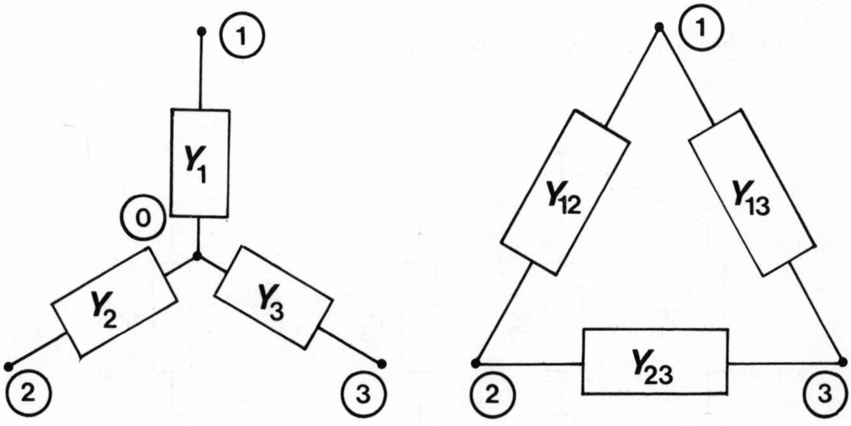


Fig. 1.12 – Rosen's theorem.

Rosen's theorem is valid for any number of branches, and it is used to reduce the number of nodes in a circuit by one.

**The star-delta transformation**

Rosen's theorem in reverse can only be used to convert a mesh of three branches (a delta) into a star.

*The star network equivalent to a mesh of three branches with impedances $Z_{12}$, $Z_{23}$, $Z_{31}$ has impedances given by*

$$Z_1 = Z_{12}\, Z_{31}/\, Z$$

$$Z_2 = Z_{12}\, Z_{23}/\, Z$$

$$Z_3 = Z_{23}\, Z_{31}/\, Z\ ,$$

$$\text{where } Z = Z_{12} + Z_{23} + Z_{31}\ .$$

Fig. 1.12 applies to the star-delta transformation, with the same node and branch numbering as before. The transformation is used to reduce the number of loops in a network by one.

**Millman's theorem**

*When n voltage generators of output voltage $E_1, E_2, \ldots E_n$ and output admittance $Y_1, Y_2, \ldots Y_n$ are connected in parallel, the equivalent single voltage generator has output voltage $E = \Sigma_{i=1}^{n} E_i Y_i / \Sigma_{i=1}^{n} Y_i$ and output admittance $Y = \Sigma_{i=1}^{n} Y_i$.*

The three-generator case is shown in Fig. 1.13.

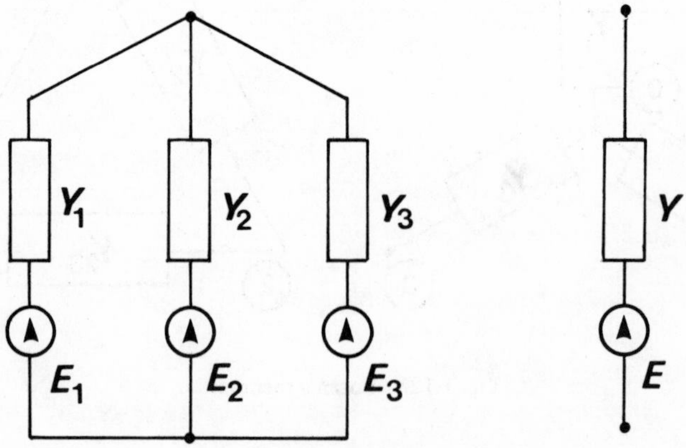


Fig. 1.13 – Millman's theorem.

**The Thévénin-Norton theorem**

Consider the linear network shown in Fig. 1.14 loaded at two output terminals by an arbitrary impedance $Z_L$.

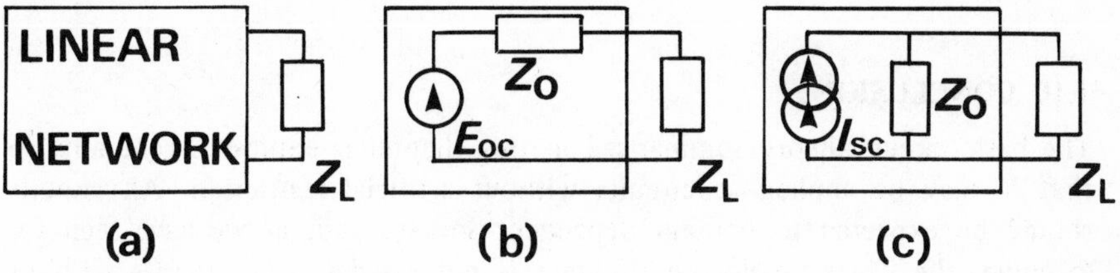


Fig. 1.14 – Thévénin-Norton theorem.
(a) loaded network; (b) Thévénin equivalent circuit; (c) Norton equivalent circuit.

*The* **Thévénin-Norton theroem** *states that a linear network loaded by an arbitrary impedance can be replaced by either the Thévènin equivalent circuit or the Norton equivalent circuit without affecting the terminal voltages, currents, or waveforms.*

No assumptions have been made about the load impedance $Z_L$ and it may be distributed, non-linear, time-varying as required. The network, however, is linear. The **Thévénin equivalent circuit** is a voltage source $E_{oc}$ with series impedance $Z_o$, and the **Norton equivalent circuit** is a current source $I_{sc}$ with parallel impedance $Z_o$.

The impedance $Z_o$ is obtained from the original network by replacing all independent voltage sources by short circuits, replacing all independent current sources by open circuits, and reducing the resulting network to a single impedance across the network terminals. Note that dependent sources are left operative. Also, in networks with sinusoidal sources, inductors, and capacitors, reduction to a single complex impedance is valid only at the frequency considered.

The voltage generator $E_{oc}$ in the Thévénin equivalent circuit is equal to the output voltage of the unloaded network with all fixed sources operative. The current generator $I_{sc}$ in the Norton equivalent circuit is equal to the output current of the network with short-circuited load and with all fixed sources operative. As before, in networks with sinusoidal sources, $E_{oc}$ and $I_{sc}$ are complex and valid only at the frequency considered.

**The substitution theorem**

*In an arbitrary network any uncoupled branch may be replaced either by an independent voltage source or by an independent current source with the same voltage or current waveform respectively as the branch, without affecting the branch voltages, currents, or waveforms in the remainder of the network.*

The definition of a network branch given in Section 1.1 was restricted to two-terminal elements. The substitution theorem is therefore also limited, and in particular does not apply to three-terminal or four-terminal elements involving coupling such as mutual inductors or dependent sources. It is also necessary that the network has a unique solution for the branch voltages and currents.

## 1.10 CONCLUSION

The basic circuit theory summarised in this chapter is expressed generally, so that it may be applied to circuits without artificial restriction. All circuits should be expected to contain dependent sources and, as we have seen for example, the Thévénin-Norton theorem is not restricted to circuits without them. *Passive* circuits which do not usually contain dependent sources are still important, but in circuit analysis using computers they are not distinguished from circuits in general.

In using computer programs for network analysis the basic circuit laws and theorems should not be overlooked. They are the most important part of circuit theory, giving the circuit designer insight into the behaviour of both practical circuits and computer simulations.

## FURTHER READING

Desoer, C. A. and Kuh, E. S., (1969), *Basic Circuit Theory,* McGraw-Hill, New York, Chapters 1 to 9 and 16.

(Ivison, J. M., (1977), *Electric Circuit Theory,* Van Nostrand Reinhold, Wokingham.

Williams, G., (1973), *An Introduction to Electrical Circuit Theory,* Macmillan, London.

## PROBLEMS

1.1 Determine the Thévénin and Norton equivalent networks for the one-port circuits in Fig. 1.15.

1.2 Determine the resistance or capacitance between the points A and B of the circuits in Fig. 1.16.

1.3 What are the voltage gain, input impedance, and output impedance of a transistor amplifier modelled by the small signal equivalent T network in Fig. 1.17 with common-emitter connection and 1kΩ load?

1.4 Determine the component values of the hybrid parameter network of Fig. 1.18 equivalent to the T-network of Fig. 1.17.

1.5 Determine the input admittance and input impedance of the circuit in Fig. 1.19 when the network $\mathcal{N}$ is assumed in turn to be each of the four impedance converters (with $n$=4) or inverters (with $g$=4).

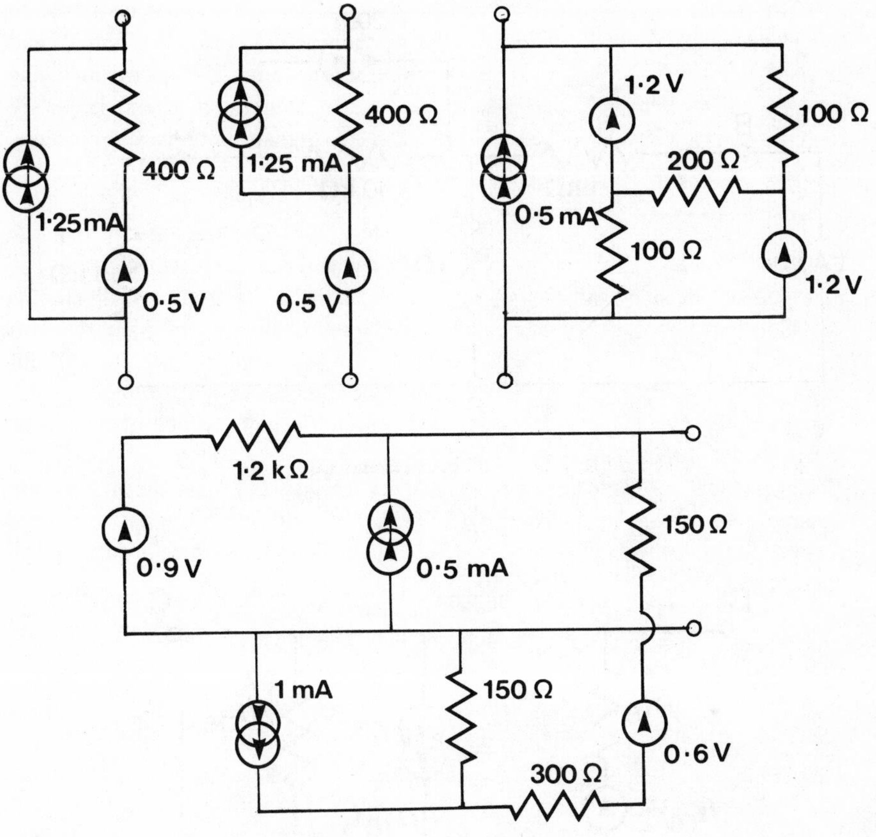


Fig. 1.15 – Problem 1.1.

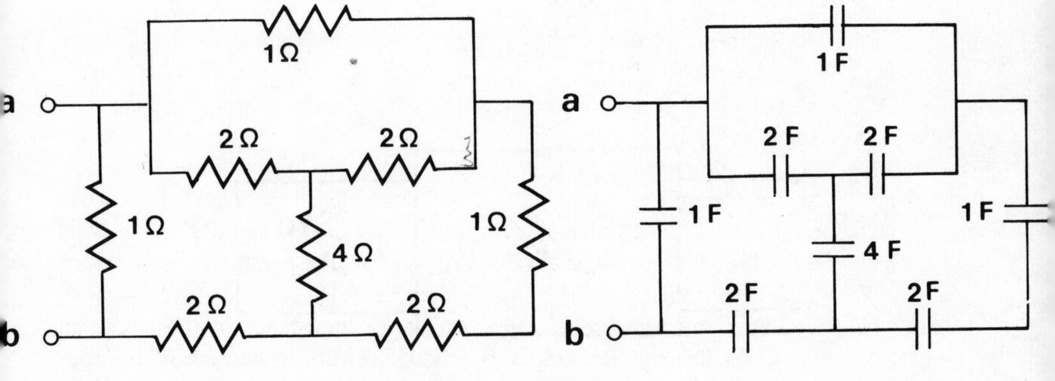


Fig. 1.16 – Problem 1.2.

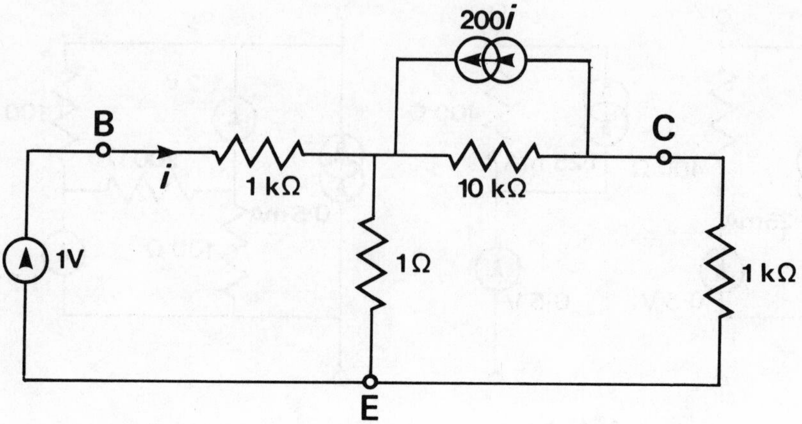


Fig. 1.17 – Problems 1.3 and 1.6.

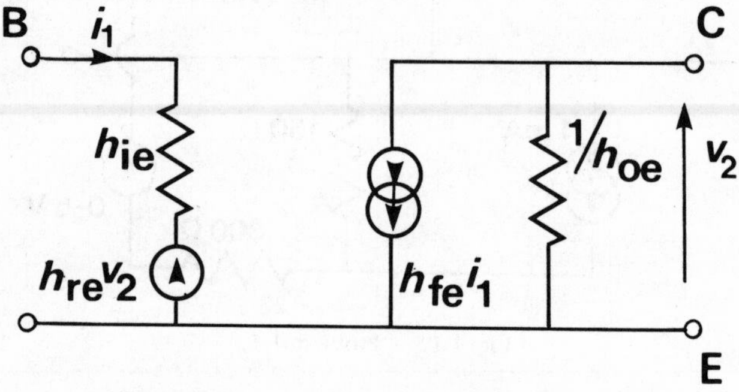


Fig. 1.18 – Problem 1.4.

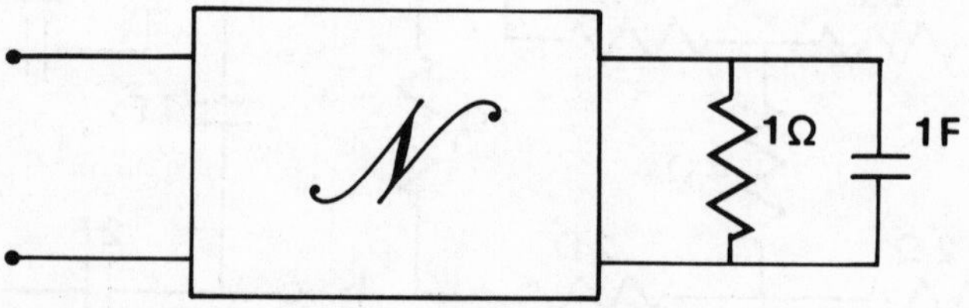


Fig. 1.19 – Problem 1.5.

1.6 Construct the duals of the circuits in Figs. 1.17 and 1.20.

1.7 Draw a phasor diagram showing all voltages and currents for the circuit in Fig. 1.21 with $M = 0.75$ H.

1.8 Determine the voltage gain transfer function of the circuit in Fig. 1.21 with $M$=0.75 H. Evaluate the gain in dB and phase in degrees at $\omega$=1 and 10 rad/sec.

1.9 What is the value of mutual inductance required for maximum power transfer to the load resistor $R_2$ in Fig. 1.21 at $\omega$=1 rad/sec?

1.10 Determine the input impedance of the circuit in Fig. 1.21 as seen by the voltage source with $M$=0.75 H. at $\omega$=1 rad/sec.

1.11 Solve for all voltages and currents in the networks of Figs. 1.22 and 1.23, using several different methods of analysis, for example by superposition, use of the Thévénin-Norton theorem, use of star-delta transformation, nodal analysis, or mesh analysis.

1.12 Plot the frequency response of the filter-circuit in Fig. 1.20 over the frequency range 0.01 Hz to 100 Hz.

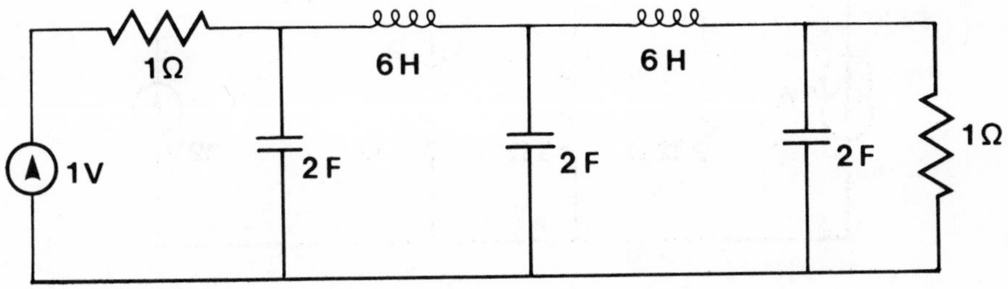


Fig. 1.20 – Problems 1.6 and 1.12.

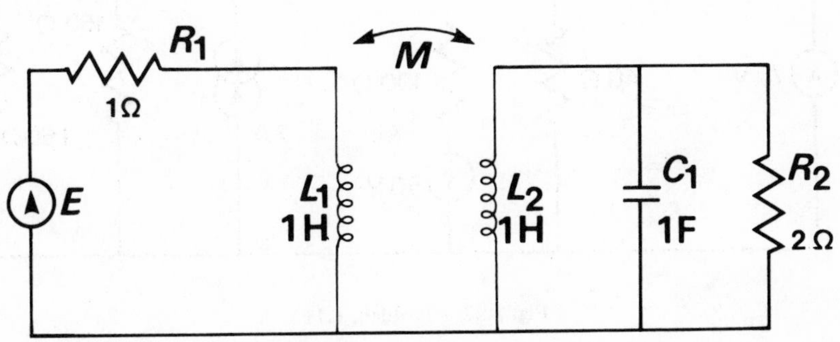


Fig. 1.21 – Problems 1.7 to 1.10.

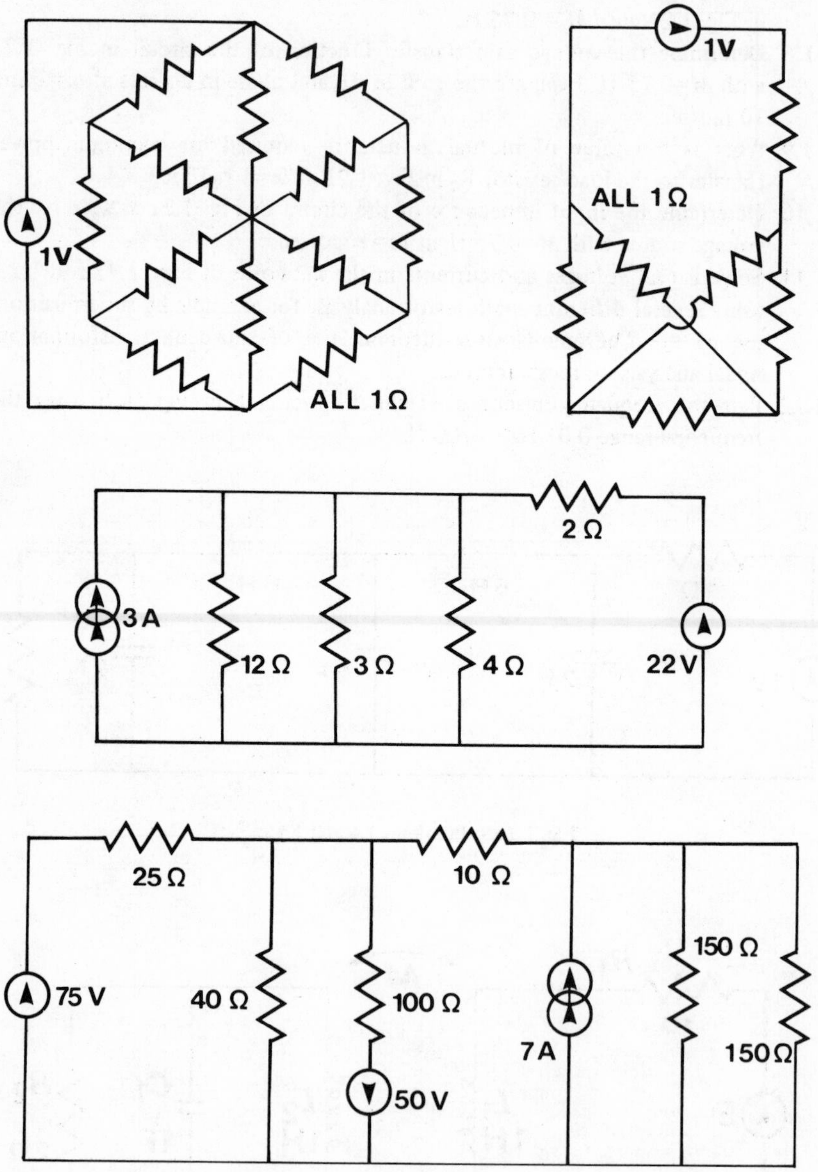


Fig. 1.22 – Problem 1.11.

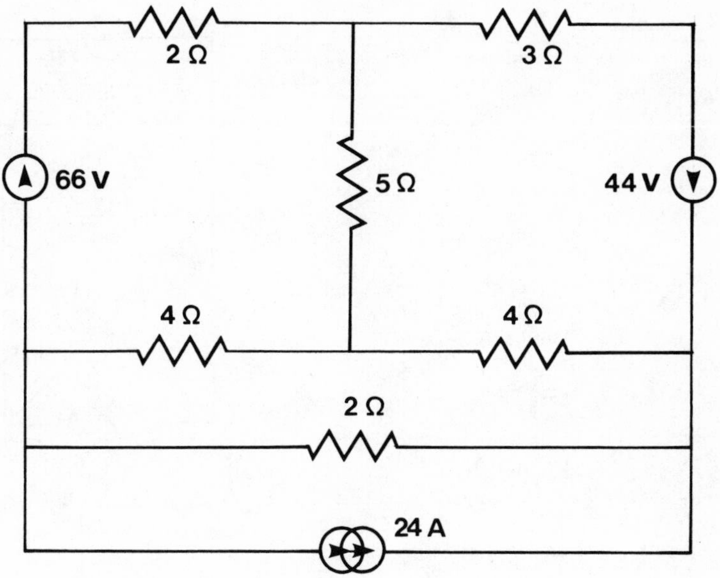


Fig. 1.23 – Problem 1.11.

CHAPTER 2

# Matrix Analysis of Networks

The methods employed for circuit analysis when computation is unnecessary or not available, are easily adapted to suit the network so that the quickest solution can be chosen, for example, in working the problems to the previous chapter, the labour of analysis can be much reduced by use of the Thévénin-Norton theorem or star-delta transformation, followed by combination of series and parallel components. Unfortunately the computer cannot be so readily programmed for such efficient adaptability, and all circuits must be solved by a sufficiently general predetermined method. Additionally, the input of circuit description and output of results must be formalised. These requirements for a general method of analysis are achieved through the use of matrices, and the conversion of a circuit diagram into a matrix description is carried out by using the concept of topology.

## 2.1 NETWORK TOPOLOGY

### Graphs

The first stage in converting a network description into matrix form is to draw the *circuit graph*. Consider the circuit shown in Fig. 2.1, which is adapted from Problem 1.11 in Chapter 1 (Fig. 1.23). Its graph is drawn by retaining the circuit nodes and replacing all branches by lines. The graph is *directed* by inserting arrows which indicate the current directions in the corresponding circuit branches. The **directed graph** of the circuit in Fig. 2.1 is shown in Fig. 2.2. It shows only the nodes and the branch connections with current direction; it gives no indication of the circuit elements.

As we have noted in Chapter 1, Kirchhoff's Laws and Tellegen's theorem are expressions of the constraints on the network voltages and currents solely due to the circuit connections. The directed graph is a topological description of the network and therefore gives sufficient information to apply the fundamental circuit laws. For the moment we will restrict our discussion and examples to networks containing two-terminal elements only. Coupling between branches, through, for example, dependent sources, will be discussed in Section 2.3.

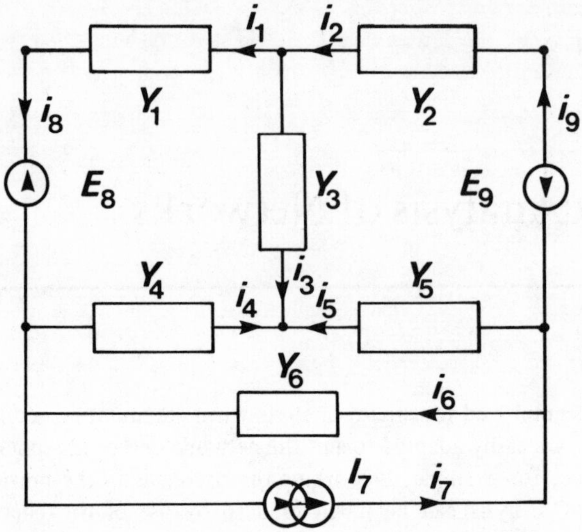


Fig. 2.1 – Circuit example.

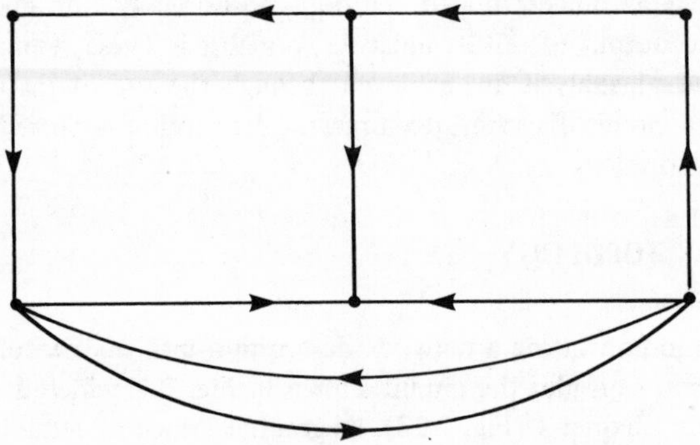

Fig. 2.2 – Directed graph of circuit in Fig. 2.1.

**Trees**

The second stage of converting the network description into matrix form involves selecting a *tree.*

*A* **tree** *is a set of branches which connects to all the nodes of a graph without containing any loops.*

*Branches forming part of the tree are called* **tree branches.**

*Branches not forming part of the tree are called* **links.**

In most networks there are a number of possible trees. Four trees for the graph in Fig. 2.2 are shown in Fig. 2.3. Tree branches are shown in solid lines, and links have broken lines.

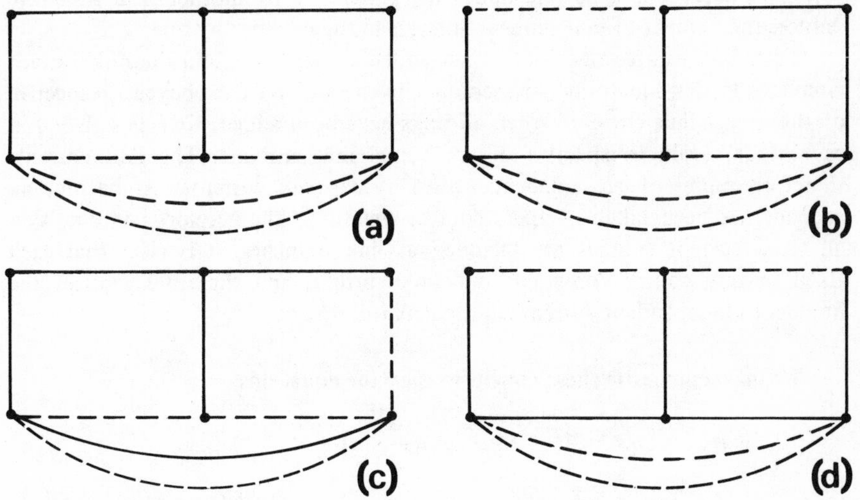


Fig. 2.3 – Trees of the graph in Fig. 2.2.

Let the number of nodes and the number of branches in a graph be given by $n_N$ and $n_B$ respectively. It is clear from Fig. 2.3 that the number of tree branches $n_T$ needed to connect the nodes is $n_N-1$. The remaining $n_L$ branches are links. Therefore,

$$n_T = n_N - 1$$

$$n_L = n_B - n_N + 1 \ ,$$

where $n_T$ and $n_L$ are the number of tree branches and the number of links respectively.

The tree branches form a network without loops, by definition. Each link when added to the tree forms just one loop consisting of the link and a number of tree branches. KVL around the loop ensures that the link branch voltage is defined as the negative of the sum of the tree branch voltages in the loop. All links can be similarly treated, and the set of link voltages is fully defined by the set of three branch voltages.

Since there are no loops in the tree, KVL cannot be invoked in order to relate the tree branch voltages together. The tree branch voltages are therefore independent. We can conclude that the number of independent voltages in a network is equal to the number of tree branches. However, no account has been taken of any fixed voltage sources present in the network. Assume that all fixed voltage sources have been included in the tree. It is clear that each fixed voltage source constrains one node voltage relative to another, and therefore reduces the number of independent voltages in the network by one.

Each link creates one loop when added to the trace, and the link current circulates in that loop only. At each node in the loop KCL is obeyed irrespective of the actual link current. When all links have been added, KCL is satisfied at each node, again irrespective of the actual link currents. The link currents therefore represent an independent set of network currents. As before, no account has been taken of fixed sources present in the network. Assume that all fixed current sources are included as link branches. It is clear that each fixed current source constrains one link current, and therefore reduces the number of independent currents in the network by one.

We may summarise these conclusions by the equations

$$n_V = n_N - 1 - n_{CV} = n_T - n_{CV}$$

$$n_I = n_B - n_N + 1 - n_{CI} = n_L - n_{CI} ,$$

where $n_V$ and $n_I$ are the number of independent voltages and currents in a network respectively, and $n_{CV}$ and $n_{CI}$ are the number of fixed voltage and fixed current sources respectively.

The selection of a network tree provides the means for selecting a minimum set of independent variables from which all the network voltages and currents can be found. Obviously, both the tree branch voltages and the link currents are suitable sets of variables for network analysis. Nodal analysis is based on the relationship $n_T = n_N - 1$. If one node is chosen as a reference, the $n_N-1$ remaining nodes can be shown to be independent. Similarly, in mesh analysis the number of meshes in a planar network is equal to $n_L$, since addition of each link to the network, beginning with the tree, creates one mesh. The mesh currents can also be shown to be independent.

**Cut-sets**

*A* **cut** *is a closed surface which divides a network into two parts.*

In Section 1.2 we noted that KCL could be generalised to apply to a cut, since the part of the circuit within the cut could be regarded as a lumped circuit element. A typical cut is shown in Fig. 2.4(a).

*A* **basic cut** *is a cut which cuts just one tree branch.*

A typical basic cut is shown in Fig. 2.4(b). For clarity on this and all future figures, cuts are not shown as closed surfaces. A convention is therefore required to define the interior of the cut surface. In this book the tree branch current is assumed to leave the cut.

*A* **cut-set** *is a set of basic cuts, one of which cuts each tree branch.*

A cut-set is shown in Fig. 2.4(c). The cuts are numbered with the tree branch number; for example, cut ② cuts tree branch 2.

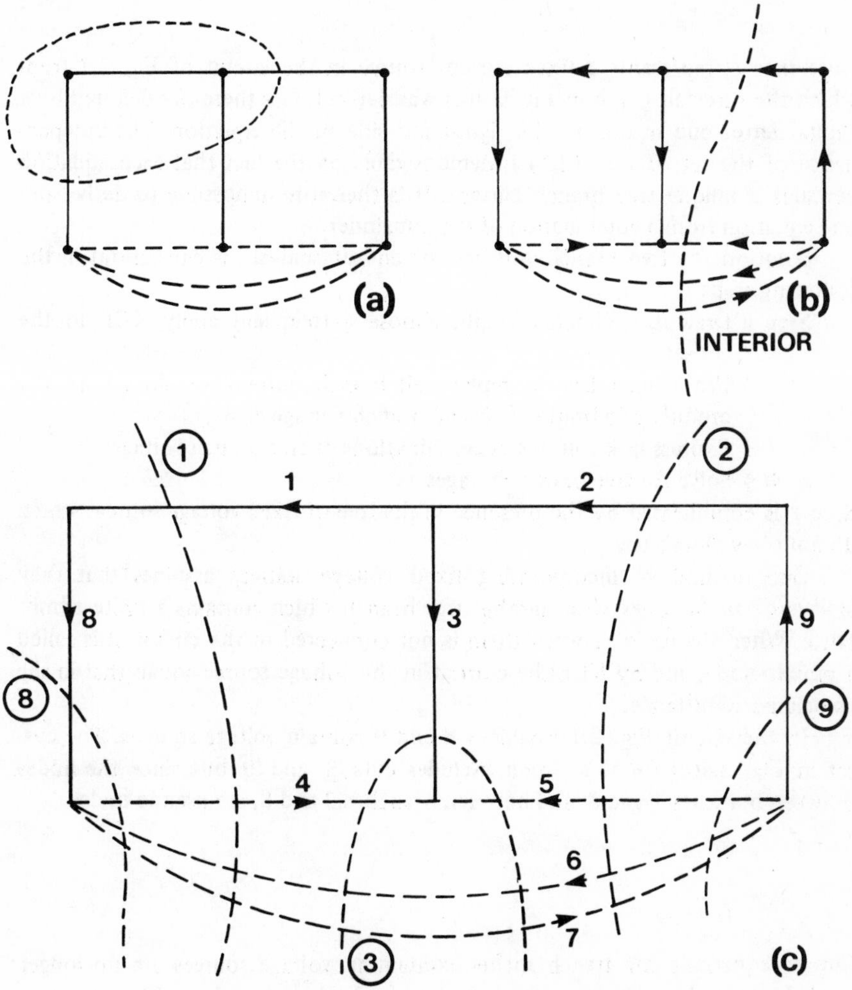


Fig. 2.4 – (a) a cut; (b) a basic cut; (c) a cut-set.

The cut-set provides a routine method for deriving the required independent set of $n_T$ equations which involve tree-branch voltages. KCL applied to each cut in the cut-set of Fig. 2.4(c) yields the equations.

$$\begin{aligned}
i_1 - i_4 + i_6 &= I_7 \\
i_2 + i_5 + i_6 &= I_7 \\
i_3 + i_4 + i_5 &= 0 \\
i_8 - i_4 + i_6 &= I_7 \\
i_9 + i_5 + i_6 &= I_7 \; .
\end{aligned} \tag{2.1}$$

Note that $I_7$ represents a fixed current source in the circuit of Fig. 2.1 from which the directed graph in Fig. 2.4(c) was derived. $I_7$ is therefore denoted by a capital letter and moved to the right-hand side of the equation. The independence of the set of Eqs. (2.1) is demonstrated by the fact that each equation contains a unique tree branch current. It is therefore impossible to derive any one equation from a combination of the remainder.

Solution for tree branch voltages, or cut-set snalysis, is carried out in the following steps.

Step 1 Draw the directed graph, choose a tree, and apply KCL to the cut-set.

2 Use Ohm's law to replace all branch current variables $i$ by the product of admittance $Y$ and branch voltage $e$.

3 Express link voltages as combinations of tree branch voltages.

4 Solve for tree branch voltages.

Step 2 is complicated by the presence in the tree of fixed voltage sources which do not obey Ohm's law.

One method of incorporating fixed voltage sources assumes that they always occur in series with another tree branch which contains a finite admittance. When the node between them is not connected to the circuit, it is called a pseudo-node, and by KCL the current in the voltage source equals that in the associated admittance.

In the circuit Fig. 2.1 branches 8 and 9 contain voltage sources. The cut-set in Fig. 2.4(c) for this circuit includes cuts ⑧ and ⑨ but since the nodes between branches 1 and 8, and between branches 2 and 8, are pseudo-nodes,

$$\begin{aligned}
i_1 &= i_8 \\
i_2 &= i_9 \; .
\end{aligned}$$

Cut-set equations for tree branches containing voltage sources are no longer needed, since the current constraints above reduce the number of independent voltages in the network by two to $n_V$ as required.

*Example* 2.1

Determine the tree branch voltages in the circuit shown in Fig. 2.5.

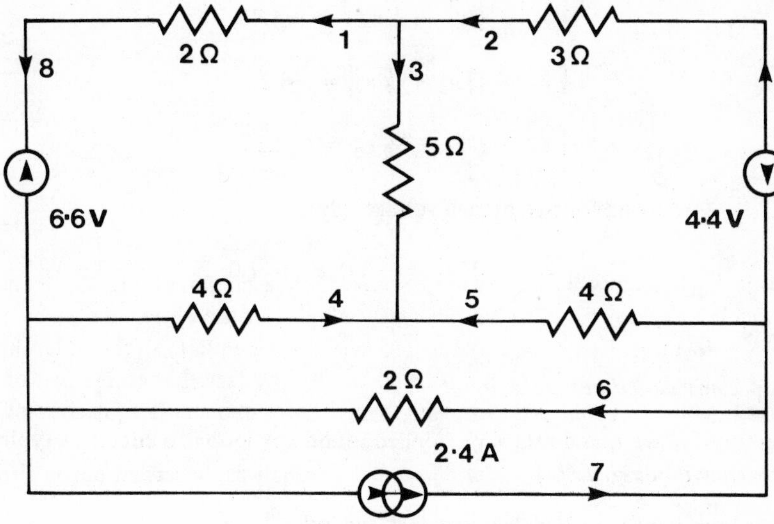


Fig. 2.5 – Analysis example.

Step 1. The tree and cut-set in Fig. 2.4(c) apply to this example. Ignoring cuts ⑧ and ⑨ since these are voltage source branches, and applying KCL yields

$$i_1 - i_4 + i_6 = 2.4$$

$$i_2 + i_5 + i_6 = 2.4$$

$$i_3 + i_4 + i_5 = 0\ .$$

Step 2. Apply Ohm's law to branches

Step 3. Express link votages in terms of tree branch voltages

$$i_1 = \tfrac{1}{2}\,v_1$$

$$i_2 = \tfrac{1}{3}\,v_2$$

$$i_3 = \tfrac{1}{5}\,v_3$$

$$i_4 = \tfrac{1}{4}\,v_4 \qquad = \tfrac{1}{4}(-v_1 + v_3 - 6.6)$$

$$i_5 = \tfrac{1}{4}\,v_5 \qquad = \tfrac{1}{4}(v_2 + v_3 + 4.4)$$

$$i_6 = \tfrac{1}{2}\,v_6 \qquad = \tfrac{1}{2}(v_2 + v_1 + 6.6 + 4.4).$$

Step 4. Substituting from Steps 2 and 3 into the current equations of Step 1 gives

$$\tfrac{5}{4} v_1 + \tfrac{1}{2} v_2 - \tfrac{1}{4} v_3 = -4.75$$

$$\tfrac{1}{2} v_1 + \tfrac{13}{12} v_2 + \tfrac{1}{4} v_3 = -4.2$$

$$-\tfrac{1}{4} v_1 + \tfrac{1}{4} v_2 + \tfrac{7}{10} v_3 = 0.55$$

Solving for tree branch voltages gives

$$v_1 = -2.4, v_2 = -3.0, v_3 = 1.0 \ .$$

**Tie-sets**

*A* **loop** *is any closed path in a circuit*

In Section 1.2 we noted that KVL applied round any loop in a circuit. A typical loop is shown in Fig. 2.6(a).

*A* **basic loop** *is a loop which traverses just one link.*

A typical basic loop is shown in Fig. 2.6(b). By convention the loop is traversed in the direction of the link voltage.

*A* **tie-set** *is a set of basic loops one of which traverses each link.*

A tie-set is shown in Fig. 2.6(c). The loops are numbered with the link branch number, for example loop ④ traverses link 4.

The tie-set provides a routine method for deriving the required independent set of $n_L$ equations which involve link voltages. KVL applied to each loop in the tie-set of Fig. 2.6(c) yields the equations

$$v_4 - v_3 + v_1 = -E_8$$

$$v_5 - v_3 - v_2 = E_9$$

$$v_6 - v_2 - v_1 = E_8 + E_9$$

$$v_7 + v_2 + v_1 = -E_8 - E_9 \ . \qquad (2.2)$$

As before, fixed sources have been removed to the right-hand side of the equations and denoted by capital letters. The independence of Eqs. (2.2) is demonstrated by the fact that each question contains a unique link voltage, and it is therefore impossible to derive any equation as a combination of the remainder.

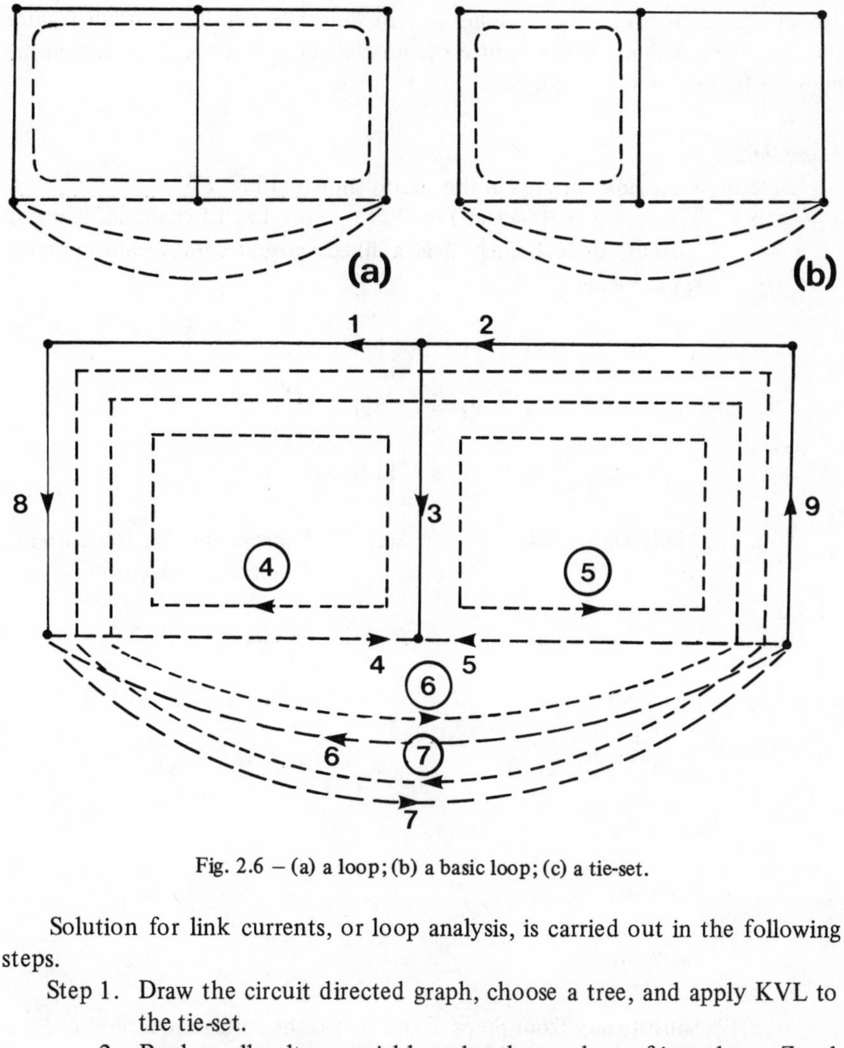


Fig. 2.6 – (a) a loop; (b) a basic loop; (c) a tie-set.

Solution for link currents, or loop analysis, is carried out in the following steps.

Step 1. Draw the circuit directed graph, choose a tree, and apply KVL to the tie-set.

2. Replace all voltage variables $v$ by the product of impedance $Z$ and branch current $i$.

3. Express tree-branch currents as combinations of link currents.

4. Solve for link currents.

As before in cut-set analysis, Step 2 is complicated by the presence of a fixed source. All fixed current sources are assumed to occur in parallel with a link containing a finite impedance. The voltage across the current source is then constrained and equals that across the associated impedance. In this circuit the fixed current source in branch 7 is in parallel with branch 6, and

$$v_7 = -v_6 \ .$$

Tie-set equations for links containing current sources are no longer needed, since the voltage constraints above reduce the number of independent currents in the network by one to $n_{\text{I}}$ as required.

*Example* 2.2

Determine the link currents in the circuit shown in Fig. 2.5.

Step 1. The tree and tie-set in Fig. 2.6(c) apply to this example. Ignoring loop ⑦ since branch 7 is a fixed current source, and applying KVL, yields

$$v_4 - v_3 + v_1 = -6.6$$

$$v_5 - v_3 - v_2 = 4.4$$

$$v_6 - v_2 - v_1 = 11.0\ .$$

Step 2. Apply Ohm's law to branches

Step 3. Express tree branch currents in terms of link currents

$$v_1 = 2i_1 = 2(i_4 - i_6 + 2.4)$$

$$v_2 = 3i_2 = 3(-i_5 - i_6 + 2.4)$$

$$v_3 = 5i_3 = 5(-i_4 - i_5)\ .$$

$$v_4 = 4i_4$$

$$v_5 = 4i_5$$

$$v_6 = 2i_6$$

Step 4. Substituting from Steps 2 and 3 into the voltage equations of Step 1 gives

$$11i_4 + 5i_5 - 2i_6 = -11.4$$

$$5i_4 + 12i_5 + 3i_6 = 11.6$$

$$-2i_4 + 3i_5 + 7i_6 = 23\ .$$

Solving for link currents gives

$$i_4 = -0.8, i_5 = 0.6, i_6 = 2.8\ .$$

**Duality**

*Two networks are said to be* **dual** *if the tree-branch voltage equations of one network are identical to the link current equations of the other network.*

This definition uses the duality of cut-set analysis and loop analysis to avoid the artificial restriction to planar networks which arises when using mesh analysis in the definition. Of course, generalisation to non-planar networks makes it difficult to construct duals.

Possibly we should not be concerned with duality in anything more than very simple networks. Any insight we might gain into basic circuit theory applied to simple networks is lost in most practical analysis situations. A large proportion of circuits are non-planar, and analysis methods used must apply to any general network. Duality and dual networks are therefore of little use in analysis. This is not to say that the concept is unimportant, and indeed, duality is an important stimulus to thought in circuit theory and design.

## 2.2 MATRIX ANALYSIS METHODS

### The Standard circuit branch

In both tree-branch-voltage and link-current methods of analysis, fixed sources have been incorporated by assuming them to appear either in series or in parallel with finite impedances. This is usually true in practice, although there are exceptions which must be handled by means of equivalent circuits. The requirements of both analytical methods can be met by defining the standard circuit branch shown in Fig. 2.7. In preparation for the matrix analysis to follow, suffix $r$ indicates the $r^{\text{th}}$. branch in the circuit $Z_{rr}$ and $Y_{rr}$ represent the self-impedance or admittance of the $r^{\text{th}}$ branch.

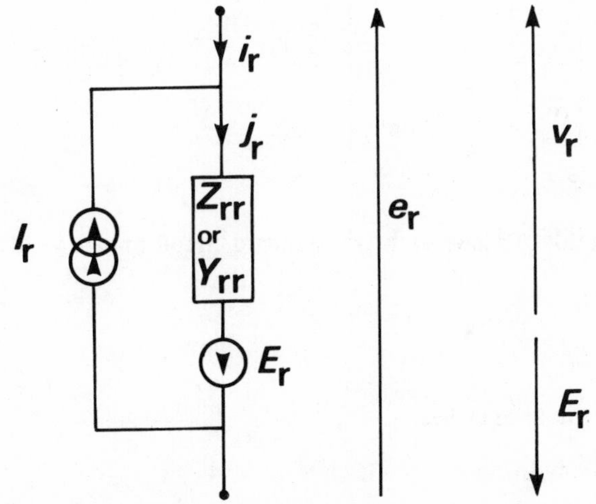


Fig. 2.7 – The standard circuit branch.

The fixed current source appears across the branch terminals as required for link current analysis, and the fixed voltage source appears in series with the impedance $Z_{rr}$ as required for tree-branch analysis. Thévénin's theorem may of course be used to convert the standard branch to the single equivalent current source $(I_r - Y_{rr}E_r)$ or the single equivalent voltage source $(E_r - Z_{rr}I_r)$ for the analysis method in use. The pseudo-node is no longer shown.

Ohm's law for the $r^{th}$ branch states that

$$v_r = Z_{rr}j_r \ ,$$

For all branches simultaneously we may use the matrix equation

$$\mathbf{v} = \mathbf{Zj} \ , \tag{2.3}$$

where

$$\mathbf{v}^t = [v_1, v_2, \ldots v_{n_B}]$$

$$\mathbf{j}^t = [j_1, j_2, \ldots j_{n_B}]$$

$$\mathbf{Z} = \begin{bmatrix} Z_{11} & & 0 \\ & Z_{12} & \\ & & \\ 0 & & Z_{n_Bn_B} \end{bmatrix} .$$

(Note that $\mathbf{v}^t$ is the transpose of $\mathbf{v}$.)

Also,

$$\mathbf{j} = \mathbf{Y}\mathbf{v} \ , \tag{2.4}$$

where,

$$\mathbf{Y} = \begin{bmatrix} Y_{11} & & 0 \\ & Y_{22} & \\ & & \\ 0 & & Y_{n_Bn_B} \end{bmatrix} .$$

$$= \mathbf{Z}^{-1}$$

Application of Kirchhoff's laws to the $r^{th}$ standard circuit branch yields

$$j_r = I_r + i_r$$

$$v_r = E_r + e_r \ ,$$

or in matrix form for all branches,

$$\mathbf{j} = \mathbf{I} + \mathbf{i} \tag{2.5}$$

$$\mathbf{v} = \mathbf{E} + \mathbf{e} \ , \tag{2.6}$$

where $\mathbf{E}^t = [E_1, E_2, \ldots E_{n_B}]$

$\mathbf{I}^t = [I_1, I_2, \ldots I_{n_B}]$

$\mathbf{e}^t = [e_1, e_2, \ldots e_{n_B}]$

$\mathbf{i}^t = [i_1, i_2, \ldots i_{n_B}]$ .

Substitution into Eqs. (2.3) and (2.4) and rearranging yields

$$(\mathbf{E} - \mathbf{ZI}) + \mathbf{e} = \mathbf{Zi} \tag{2.7}$$

$$(\mathbf{I} - \mathbf{YE}) + \mathbf{i} = \mathbf{Ye} \ . \tag{2.8}$$

These equations apply to the branches individually and contain the information obtained by application of KCL, KVL, and Ohm's law to the standard circuit branch. The matrix methods of analysis which follow utilise these equations and consider the constraints on the branch voltage and currents caused by the circuit connections.

**The cut-set matrix**

The directed graph and cut-set shown in Fig. 2.4(c) for the circuit in Figs. 2.1 and 2.5 must now be modified to take the standard circuit branch into account. The circuit is redrawn in Fig. 2.8. Note particularly that the voltage sources are now negative in order that their directions comply with the assumed current direction in branches 1 and 2 on the directed graph.

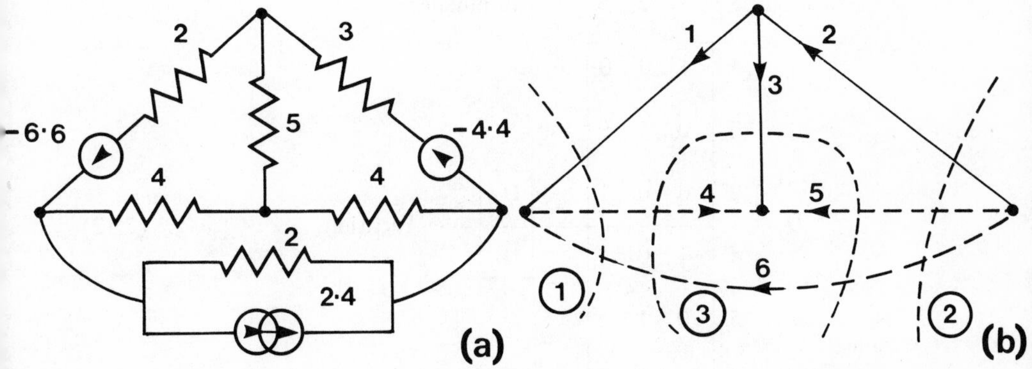


Fig. 2.8 – Redrawn versions of Figs. 2.5 and 2.4(c).
(a) circuit example; (b) the cut-set.

Since fixed sources no longer appear in separate branches, the cut-set Eqs. (2.1) are modified to

$$i_1 - i_4 + i_6 = 0$$

$$i_2 + i_5 + i_6 = 0$$

$$i_3 + i_4 + i_5 = 0 \,. \tag{2.9}$$

In matrix form these may be rewritten

$$\begin{bmatrix} 1 & 0 & 0 & -1 & 0 & 1 \\ 0 & 1 & 0 & 0 & 1 & 1 \\ 0 & 0 & 1 & 1 & 1 & 0 \end{bmatrix} \begin{bmatrix} i_1 \\ i_2 \\ \cdot \\ \cdot \\ \cdot \\ i_6 \end{bmatrix} = \begin{bmatrix} 0 \\ 0 \\ \cdot \\ \cdot \\ \cdot \\ 0 \end{bmatrix} , \tag{2.10}$$

or $\mathbf{D}^t\mathbf{i} = \mathbf{0}$.

The matrix **D** composed of elements with values 1, 0, or −1, is called the **cut-set matrix**, and Eq. (2.11) expresses KCL applied to the cut-set in matrix form. The cut-set matrix **D** for the circuit graph in Fig. 2.8(b) is repeated in Eq. (2.12) to illustrate its derivation and partition.

$$\begin{array}{c} \\ \\ \mathbf{D} = \end{array} \begin{array}{c|ccc|} & 1 & 2 & 3 \\ \hline 1 & 1 & 0 & 0 \\ 2 & 0 & 1 & 0 \\ 3 & 0 & 0 & 1 \\ \hline 4 & -1 & 0 & 1 \\ 5 & 0 & 1 & 1 \\ 6 & 1 & 1 & 0 \end{array} = \begin{bmatrix} \mathbf{D}_T \\ \hline \mathbf{D}_L \end{bmatrix} \text{--- Partition.} \tag{2.12}$$

(Columns: Cut number 1 2 3; rows: Branch number 1–6.)

By convention the tree-branches are numbered first, beginning from unity. The matrix row number corresponds to the circuit branch with the same number. Similarly, each matrix column corresponds to a cut. The partition separates the upper section of the matrix, denoted by $\mathbf{D}_T$, which corresponds to tree branches, from the lower section of the matrix $\mathbf{D}_L$, which corresponds to the links. Since the number of cuts in the cut-set equals the number of tree branches, and since a basic cut can cut only one tree branch, $\mathbf{D}_T$ must be a square matrix equal to the unit matrix.

A simple set of rules for filling the cut-set matrix is easily derived by inspection of this example.

Matrix entries are +1 if the branch identified by the row number is cut by the cut identified by the column number, and if the branch current is in the same relative direction as that of the tree branch associated with the cut. The matrix entry is −1 if the current is in the reserve direction, and the entry is zero if the branch is not cut.

A further property of the cut-set matrix is illustrated by considering the expression $\mathbf{D}\mathbf{e}_T$, where $\mathbf{e}_T$ is the part of $\mathbf{e}$ relating to tree branches. Expanding fully

$$\begin{bmatrix} 1 & 0 & 0 \\ 0 & 1 & 0 \\ 0 & 0 & 1 \\ -1 & 0 & 1 \\ 0 & 1 & 1 \\ 1 & 1 & 0 \end{bmatrix} \begin{bmatrix} e_1 \\ e_2 \\ e_3 \end{bmatrix} = \begin{bmatrix} e_1 \\ e_2 \\ e_3 \\ -e_1+e_3 \\ e_2+e_3 \\ e_1+e_2 \end{bmatrix} = \begin{bmatrix} e_1 \\ e_2 \\ e_3 \\ e_4 \\ e_5 \\ e_6 \end{bmatrix}, \tag{2.13}$$

that is, $$\mathbf{D}\,\mathbf{e}_T = \mathbf{e}\ . \tag{2.14}$$

The relationship in Eq. (2.13) between tree-branch voltages and link voltages are identical to those in Eq. (2.2), remembering that the fixed voltage sources are included through the expression $\mathbf{v} = \mathbf{E} + \mathbf{e}$ from Eq. (2.6).

The two major steps in cut-set analysis are the application of KCL to the cut-set and the expression of the link voltages in terms of tree-branch voltages. Both have been accomplished by the use of the cut-set matrix, and the analysis can therefore be expressed in matrix form. The branch relationships from Eq. (2.8) are

$$(\mathbf{I} - \mathbf{Y}\mathbf{E}) + \mathbf{i} = \mathbf{Y}\mathbf{e}\ . \tag{2.15}$$

Multiplication by $\mathbf{D}^t$ yields

$$\mathbf{D}^t(\mathbf{I} - \mathbf{YE}) + \mathbf{D}^t\mathbf{i} = \mathbf{D}^t\mathbf{Ye} . \tag{2.16}$$

As we have seen, $\mathbf{D}^t\mathbf{i}$ is equal to zero from KCL. Substitution for $\mathbf{e}$ from Eq. (2.14) yields

$$\mathbf{D}^t(\mathbf{I} - \mathbf{Ye}) = \mathbf{D}^t\mathbf{Y}\,\mathbf{D}\,\mathbf{e}_T . \tag{2.17}$$

Since we have taken great care to formulate a solution in terms of the independent set of tree-branch voltages, the matrix $\mathbf{D}^t\mathbf{YD}$ can be inverted, giving

$$\mathbf{e}_T = (\mathbf{D}^t\mathbf{YD})^{-1}\,\mathbf{D}^t(\mathbf{I}-\mathbf{YE}) . \tag{2.18}$$

All other voltages and currents in the network may be derived by substitution into Eqs. (2.5), (2.6), (2.14), and (2.15).

Branch voltages $\mathbf{e} = \mathbf{D}\,\mathbf{e}_T$

Branch currents $\mathbf{i} = \mathbf{Ye} - (\mathbf{I}-\mathbf{YE})$

Element voltages $\mathbf{v} = \mathbf{e} + \mathbf{E}$

Element currents $\mathbf{j} = \mathbf{i} + \mathbf{I}$ .

**The tie-set matrix**

The tie set for the Fig. 2.8(a) circuit, shown in Fig. 2.9, is a redrawn version of Fig. 2.6(c) in which fixed sources have been incorporated into the circuit branches.

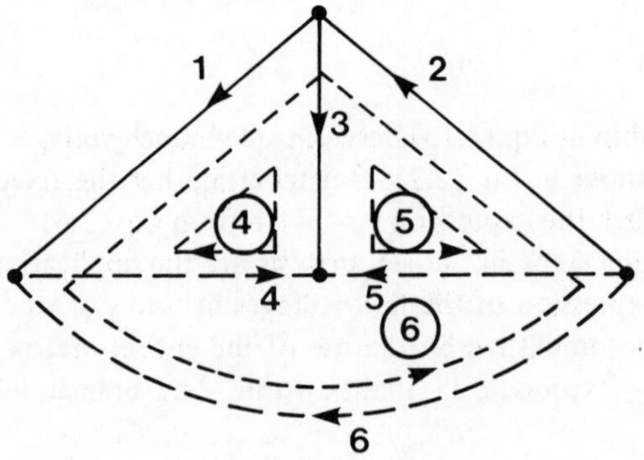


Fig. 2.9 – Tie-set for the circuit in Fig. 2.8(a).

KVL applied to each loop in the tie-set gives the equations

$$\begin{aligned} e_4 - e_3 + e_1 &= 0 \\ e_5 - e_3 - e_2 &= 0 \\ e_6 - e_1 - e_2 &= 0\ , \end{aligned} \tag{2.19}$$

or in matrix form

$$\begin{bmatrix} 1 & 0 & -1 & 1 & 0 & 0 \\ 0 & -1 & 1 & 0 & 1 & 0 \\ -1 & -1 & 0 & 0 & 0 & 1 \end{bmatrix} \begin{bmatrix} e_1 \\ e_2 \\ \cdot \\ \cdot \\ \cdot \\ e_6 \end{bmatrix} = \begin{bmatrix} 0 \\ 0 \\ \cdot \\ \cdot \\ \cdot \\ 0 \end{bmatrix}, \tag{2.20}$$

that is, $$\mathbf{C}^t\mathbf{e} = \mathbf{0}\ . \tag{2.21}$$

The matrix **C** is called the **tie-set matrix**, and Eq. (2.21) is the matrix form of KVL applied to the tie-set. The matrix is repeated below with added row and column numbering to illustrate its derivation from the directed graph.

$$\mathbf{C} = \begin{array}{cl} & \begin{array}{ccc} 4 & 5 & 6 \end{array} \quad \text{Loop number} \\ \begin{array}{c} 1 \\ 2 \\ 3 \\ 4 \\ 5 \\ 6 \end{array} & \left[\begin{array}{rrr} 1 & 0 & -1 \\ 0 & -1 & -1 \\ -1 & -1 & 0 \\ \hline 1 & 0 & 0 \\ 0 & 1 & 0 \\ 0 & 0 & 1 \end{array}\right] = \left[\begin{array}{c} \mathbf{C}_T \\ \hline \mathbf{C}_L \end{array}\right] \text{— Partition}\ . \\ \begin{array}{c} \text{Branch} \\ \text{number} \end{array} & \end{array} \tag{2.22}$$

The matrix is partitioned so that $\mathbf{C}_T$ and $\mathbf{C}_L$ refer to the tree branches and the links respectively.

Entries are made into the tie-set matrix according to the following rules. Entries are +1 if the branch identified by the row number of the matrix is traversed by the loop identified by $n_T$ + the column number in the same direction as the associated link. The entry is −1 if the branch is traversed in the opposite direction, and 0 if the branch is not traversed. Since each loop in the tie-set can traverse only one link, $\mathbf{C}_L$ is equal to the unit matrix.

On the grounds of duality between cut-set analysis and loop analysis $\mathbf{C}\mathbf{i}_L$ should express the branch currents in terms of the link currents. Expanding fully for the cut-set matrix in Eq. (2.20) gives

$$\begin{bmatrix} 1 & 0 & -1 \\ 0 & -1 & -1 \\ -1 & -1 & 0 \\ 1 & 0 & 0 \\ 0 & 1 & 0 \\ 0 & 0 & 1 \end{bmatrix} \begin{bmatrix} i_4 \\ i_5 \\ i_6 \end{bmatrix} = \begin{bmatrix} i_4 - i_6 \\ -i_5 - i_6 \\ -i_4 - i_5 \\ i_4 \\ i_5 \\ i_6 \end{bmatrix} = \begin{bmatrix} i_1 \\ i_2 \\ i_3 \\ i_4 \\ i_5 \\ i_6 \end{bmatrix}, \tag{2.23}$$

that is, $$\mathbf{C}\mathbf{i}_L = \mathbf{i} \ . \tag{2.24}$$

The relationships between link currents and tree-branch currents in Eqs. (2.23) correspond to those in Eq. (2.1), remembering that current sources have now been incorporated into the standard branch.

Loop analysis is based on the application of KVL to the tie-set and on the expression of tree-branch currents in terms of link currents. Both of these have been achieved through the tie-set matrix. The branch relationships from Eq. (2.7) are

$$(\mathbf{E} - \mathbf{Z}\mathbf{I}) + \mathbf{e} = \mathbf{Z}\mathbf{i} \ . \tag{2.25}$$

Multiplication by $\mathbf{C}^t$ yields

$$\mathbf{C}^t(\mathbf{E} - \mathbf{Z}\mathbf{I}) + \mathbf{C}^t\mathbf{e} = \mathbf{C}^t\mathbf{Z}\mathbf{i} \ . \tag{2.26}$$

From KVL applied to the tie-set $\mathbf{C}^t\mathbf{e}$ is zero. Substitution for $\mathbf{i}$ from Eq. (2.24) gives

$$\mathbf{C}^t(\mathbf{E} - \mathbf{Z}\mathbf{I}) = \mathbf{C}^t\mathbf{Z}\mathbf{C}\,\mathbf{i}_L \ , \tag{2.27}$$

and taking the inverse of $\mathbf{C}^t\mathbf{Z}\mathbf{C}$ gives the required solution for link currents

$$\mathbf{i}_L = (\mathbf{C}^t\mathbf{Z}\mathbf{C})^{-1}\,\mathbf{C}^t(\mathbf{E} - \mathbf{Z}\mathbf{I}) \ . \tag{2.28}$$

The solution for the remaining network voltages and currents is completed by substitution into Eqs. (2.5), (2.6), (2.24), and (2.25).

Branch currents $\mathbf{i} = \mathbf{C}\mathbf{i}_L$

Branch voltages $\mathbf{e} = \mathbf{Z}\mathbf{i} - (\mathbf{E} - \mathbf{Z}\mathbf{I})$

Element voltages $\mathbf{v} = \mathbf{e} + \mathbf{E}$

Element currents $\mathbf{j} = \mathbf{i} + \mathbf{I}$ .

Cut-set and loop analysis are the fundamental methods of analysis based on topological methods for the determination of an independent set of equations for solution. Node and mesh analysis utilise more convenient sets of variables which may be easily identified without the trouble of drawing a tree. The number of nodes $n_N$ in a network was derived as $n_T + 1$. If one node is eliminated by designating it as the reference node, the number of nodes remaining $n_N - 1$ is equal to the number of independent voltages in the network $n_T$. Similarly the number of meshes in a planar network $n_m$ can be observed to equal the number of links $n_L + 1$. The number of mesh currents is therefore equal to the number of independent currents in the network if the outer mesh is omitted. Node and mesh analysis methods are based on these observations coupled with demonstration of the fact that the node voltages and the mesh currents are sets of independent variables.

**The incidence matrix**

The incidence matrix is obtained by application of KCL at the circuit nodes. The directed graph of the circuit in Fig. 2.8(a) is redrawn in Fig. 2.10 to identify the circuit nodes.

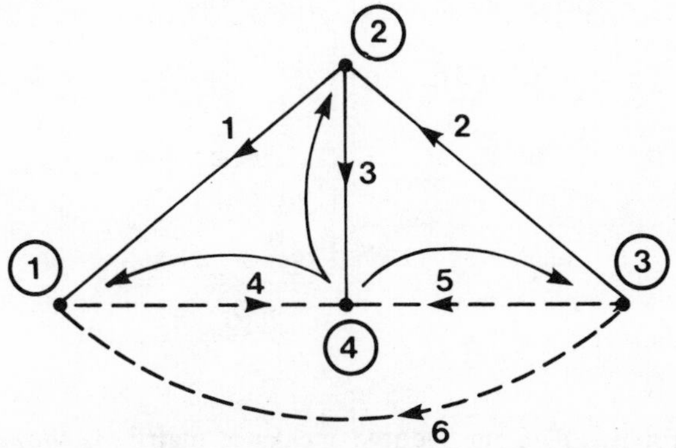


Fig. 2.10 – Directed graph of circuit in Fig. 2.8(a) with nodes identified by circled numbers

Application of KCL at the circuit nodes yields

$$-i_1 + i_4 + i_6 = 0$$

$$i_1 - i_2 + i_3 = 0$$

$$i_2 + i_5 + i_6 = 0$$

$$-i_3 - i_4 - i_5 = 0 \ , \tag{2.29}$$

or in matrix form

$$\begin{bmatrix} -1 & 0 & 0 & 1 & 0 & -1 \\ 1 & -1 & 1 & 0 & 0 & 0 \\ 0 & 1 & 0 & 0 & 1 & 1 \\ 0 & 0 & -1 & -1 & -1 & 0 \end{bmatrix} \begin{bmatrix} i_1 \\ i_2 \\ \cdot \\ \cdot \\ \cdot \\ i_6 \end{bmatrix} = \mathbf{0} \ . \tag{2.30}$$

As we have noted, the number of nodes $n_N$ is greater than the number of independent voltages in the network. In Eq. (2.30) the matrix is linearly dependent since all columns sum to give zero. This matrix, called the **incidence matrix**, can be made linearly independent by deleting one row. Choosing node ④ as a reference node and removing row 4 gives

$$\begin{bmatrix} -1 & 0 & 0 & 1 & 0 & -1 \\ 1 & -1 & 1 & 0 & 0 & 0 \\ 0 & 1 & 0 & 0 & 1 & 1 \end{bmatrix} \begin{bmatrix} i_1 \\ i_2 \\ \cdot \\ \cdot \\ \cdot \\ i_6 \end{bmatrix} = \mathbf{0} \ , \tag{2.31}$$

or $$\mathbf{A}^t\mathbf{i} = \mathbf{i} \ . \tag{2.32}$$

The matrix **A**, called the **reduced incidence matrix**, is shown below with branches and nodes indicated by row and column numbers respectively.

$$
\begin{array}{cccc}
 & 1 & 2 & 3 \quad \text{Node number}
\end{array}
$$

$$
\mathbf{A} = 
\begin{array}{c}
1 \\
2 \\
3 \\
4 \\
5 \\
6
\end{array}
\begin{bmatrix}
-1 & 1 & 0 \\
0 & -1 & 1 \\
0 & 1 & 0 \\
1 & 0 & 0 \\
0 & 0 & 1 \\
-1 & 0 & 1
\end{bmatrix}.
$$

Branch
number

Each element of the reduced incidence matrix is filled with $+1$ if the branch identified by the row number begins at the node identified by the column number. The element is $-1$ if the branch terminates at the node, and is zero if the branch is not connected to the node.

The node voltages $\mathbf{e}'$ are defined as the set of voltages from each circuit node in turn referred to the reference node; that is

$$
\mathbf{e}' = 
\begin{bmatrix}
e_1 - e_4 \\
e_2 - e_4 \\
e_3 - e_4
\end{bmatrix}
=
\begin{bmatrix}
e_1 \\
e_2 \\
e_3
\end{bmatrix},
\tag{2.34}
$$

(assuming $e_4$ is the circuit ground connection).

The expression $\mathbf{Ae}'$ is, one hopes, significant to nodal analysis. For the circuit graph in Fig. 2.10

$$
\mathbf{Ae}' = 
\begin{bmatrix}
-1 & 1 & 0 \\
0 & -1 & 1 \\
0 & 1 & 0 \\
1 & 0 & 0 \\
0 & 0 & 1 \\
-1 & 0 & 1
\end{bmatrix}
\begin{bmatrix}
e_1 \\
e_2 \\
e_3
\end{bmatrix}
=
\begin{bmatrix}
e_2 - e_1 \\
e_3 - e_2 \\
e_2 \\
e_1 \\
e_3 \\
e_3 - e_1
\end{bmatrix}
=
\begin{bmatrix}
e_1 \\
e_2 \\
e_3 \\
e_4 \\
e_5 \\
e_6
\end{bmatrix}
$$

$$
= \mathbf{e} .
$$

Clearly $\mathbf{Ae}'$ expresses the branch voltages in terms of the node voltages.

Nodal analysis proceeds from Eq. (2.8),

$$(\mathbf{I} - \mathbf{YE}) + \mathbf{i} = \mathbf{Ye}\ . \tag{2.36}$$

Multiplication $\mathbf{A}^t$ yields

$$\mathbf{A}^t(\mathbf{I} - \mathbf{YE}) + \mathbf{A}^t\mathbf{i} = \mathbf{A}^t\mathbf{Ye}\ , \tag{2.37}$$

and since $\mathbf{A}^t\mathbf{i} = \mathbf{0}$ from KCL and $\mathbf{Ae}' = \mathbf{e}$,

$$\mathbf{A}^t(\mathbf{I} - \mathbf{YE}) = \mathbf{A}^t\mathbf{Y}\,\mathbf{Ae}'\ , \tag{2.38}$$

Solving for node voltages yields

$$\mathbf{e}' = (\mathbf{A}^t\mathbf{YA})^{-1}\,\mathbf{A}^t(\mathbf{I}-\mathbf{YE})\ . \tag{2.39}$$

The solution obtained is similar to that for the tree-branch voltages except that the reduced incidence matrix has replaced the cut-set matrix. As before, all other circuit voltages and currents can be obtained by substitution.

Branch voltages $\mathbf{e} = \mathbf{Ae}'$

Branch currents $\mathbf{i} = \mathbf{Ye} - (\mathbf{I}-\mathbf{YE})$

Element voltages $\mathbf{v} = \mathbf{e} + \mathbf{E}$

Element currents $\mathbf{j} = \mathbf{i} + \mathbf{I}\ .$

**Mesh analysis**

The development of mesh analysis proceeds in the same way as nodal analysis, but since the analysis is applicable only to planar networks it will be described only briefly.

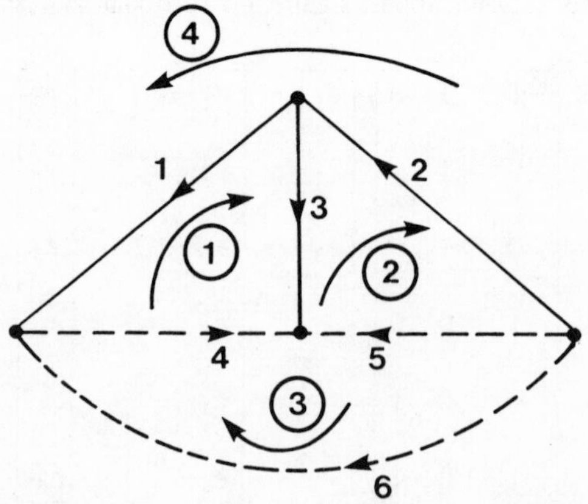


Fig. 2.11 – Directed graph of circuit in Fig. 2.8(a) with meshes identified by circled numbers.

Application of KVL to each mesh in Fig. 2.11, including the outer mesh, gives

$$-e_1 + e_3 - e_4 = 0$$

$$-e_3 - e_2 + e_5 = 0$$

$$e_4 - e_5 + e_6 = 0$$

$$e_1 + e_2 - e_6 = 0\ , \tag{2.40}$$

or in matrix form

$$\begin{bmatrix} -1 & 0 & 1 & -1 & 0 & 0 \\ 0 & -1 & -1 & 0 & 1 & 0 \\ 0 & 0 & 0 & 1 & -1 & 1 \\ 1 & 1 & 0 & 0 & 0 & -1 \end{bmatrix} \begin{bmatrix} e_2 \\ e_2 \\ \vdots \\ e_6 \end{bmatrix} = \begin{bmatrix} 0 \\ 0 \\ \vdots \\ 0 \end{bmatrix} . \tag{2.41}$$

These equations are linearly dependent since columns in the matrix sum to zero. As in nodal analysis, one row of the matrix, corresponding to the outer mesh, is deleted, leaving the reduced matrix; that is,

$$\begin{bmatrix} -1 & 0 & 1 & -1 & 0 & 0 \\ 0 & -1 & -1 & 0 & 1 & 0 \\ 0 & 0 & 0 & 1 & -1 & 1 \end{bmatrix} \begin{bmatrix} e_1 \\ e_2 \\ \vdots \\ e_6 \end{bmatrix} = \begin{bmatrix} 0 \\ 0 \\ \vdots \\ 0 \end{bmatrix} , \tag{2.42}$$

or

$$\mathbf{M}^t \mathbf{e} = \mathbf{0}\ . \tag{2.43}$$

Matrix **M**, called the **reduced mesh matrix**, also relates the mesh currents to the branch currents. If

$$\mathbf{i}' = \begin{bmatrix} i_1 \\ i_2 \\ i_3 \end{bmatrix} ,$$

where $\mathbf{i}'$ is the column vector of mesh currents, then check that

$$\mathbf{M}\,\mathbf{i}' = \mathbf{i}\ . \tag{2.44}$$

**Mesh analysis** is based on the two relationships (2.43) and (2.44), and substitution into Eq. (2.7) yields the solution for mesh currents

$$\mathbf{i}' = (\mathbf{M}^t\,\mathbf{Z}\,\mathbf{M})^{-1}\,\mathbf{M}^t\,(\mathbf{E} - \mathbf{Z}\mathbf{I})\ , \tag{2.45}$$

from which all other circuit voltages and currents can be determined.

Planar networks are usually fairly obvious when drawn on paper, and mesh analysis by hand is then straightforward, particularly as it is not necessary to draw a tree. However, the computer usually works from a network description which lists the circuit components and their node connections. It is difficult for the computer to identify meshes and generate the mesh matrix. Even if it could do so, the programme would be limited to planar networks. Computer methods of analysis must therefore be based on nodal, tree-branch, or loop analysis.

*Example* 2.3

Complete the analysis of Examples 2.1 and 2.2 by each of the matrix methods.

The matrices **A, C, D,** and **M** in the text were derived for the circuit in this analysis example under the four matrix analysis headings in this section. Additionally we require the **I, E, Y,** and **Z** matrices. With reference to the circuit and tree, redrawn in Fig. 2.12, these matrices are easily filled.

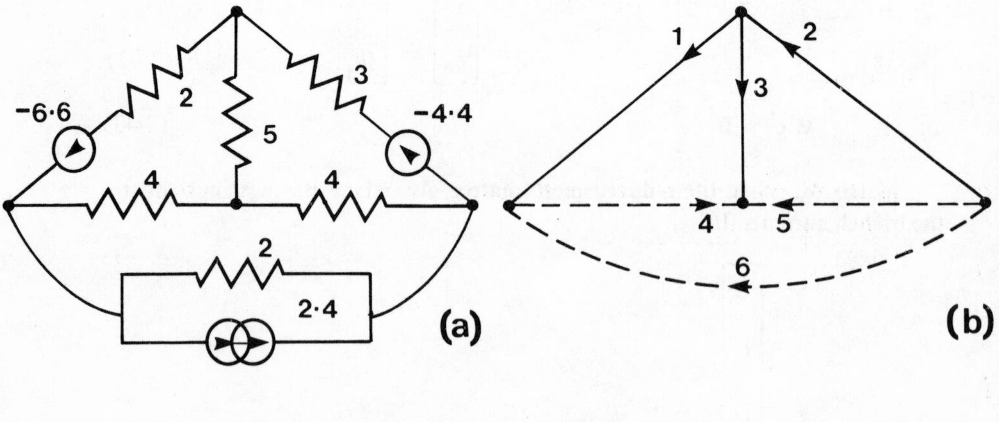


Fig. 2.12 – (a) circuit example; (b) tree.

$$\mathbf{I}^t = [\,0 \quad 0 \quad 0 \quad 0 \quad 0 \quad 2.4\,]$$

$$\mathbf{E}^t = [\,-6.6 \quad -4.4 \quad 0 \quad 0 \quad 0 \quad 0\,]$$

$$\mathbf{Y} = \begin{bmatrix} 1/2 & 0 & 0 & 0 & 0 & 0 \\ 0 & 1/3 & 0 & 0 & 0 & 0 \\ 0 & 0 & 1/5 & 0 & 0 & 0 \\ 0 & 0 & 0 & 1/4 & 0 & 0 \\ 0 & 0 & 0 & 0 & 1/4 & 0 \\ 0 & 0 & 0 & 0 & 0 & 1/2 \end{bmatrix}$$

$$\mathbf{Z} = \begin{bmatrix} 2 & 0 & 0 & 0 & 0 & 0 \\ 0 & 3 & 0 & 0 & 0 & 0 \\ 0 & 0 & 5 & 0 & 0 & 0 \\ 0 & 0 & 0 & 4 & 0 & 0 \\ 0 & 0 & 0 & 0 & 4 & 0 \\ 0 & 0 & 0 & 0 & 0 & 2 \end{bmatrix}.$$

*Cut-set analysis*

The cut-set is shown in Fig. 2.8(b), and the cut-set matrix from Eq. (2.12) is

$$\mathbf{D}^t = \begin{bmatrix} 1 & 0 & 0 & -1 & 0 & 1 \\ 0 & 1 & 0 & 0 & 1 & 1 \\ 0 & 0 & 1 & 1 & 1 & 0 \end{bmatrix}.$$

Tree-branch voltages are given by

$$\mathbf{e}_T = (\mathbf{D}^t\,\mathbf{Y}\,\mathbf{D})^{-1}\,\mathbf{D}^t\,(\mathbf{I} - \mathbf{YE}) .$$

$(\mathbf{I} - \mathbf{YE})$ is equivalent to Thévénin conversion of voltage sources to current sources. $\mathbf{D}^t(\mathbf{I} - \mathbf{YE})$ converts branch current sources into an equivalent set of tree-branch current sources.

By substitution,

$$(\mathbf{I} - \mathbf{YE}) = \begin{bmatrix} 0 \\ 0 \\ 0 \\ 0 \\ 0 \\ 2.4 \end{bmatrix} - \begin{bmatrix} 1/2 & & & & & 0 \\ & 1/3 & & & & \\ & & 1/5 & & & \\ & & & 1/4 & & \\ & & & & 1/4 & \\ 0 & & & & & 1/2 \end{bmatrix} \begin{bmatrix} -6.6 \\ -4.4 \\ 0 \\ 0 \\ 0 \\ 0 \end{bmatrix},$$

$$= [3.3 \quad 1.46667 \quad 0 \quad 0 \quad 0 \quad 2.4]^t ,$$

$$\mathbf{D}^t(\mathbf{I}-\mathbf{YE}) = \begin{bmatrix} 1 & 0 & 0 & -1 & 0 & 1 \\ 0 & 1 & 0 & 0 & 1 & 1 \\ 0 & 0 & 1 & 1 & 1 & 0 \end{bmatrix} \begin{bmatrix} 3.3 \\ 1.46667 \\ 0 \\ 0 \\ 0 \\ 2.4 \end{bmatrix},$$

$$= [5.7 \quad 3.86667 \quad 0]^t ,$$

$$\mathbf{D}^t\mathbf{Y} = \begin{bmatrix} 1/2 & 0 & 0 & -1/4 & 0 & 1/2 \\ 0 & 1/3 & 0 & 0 & 1/4 & 1/2 \\ 0 & 0 & 1/5 & 1/4 & 1/4 & 0 \end{bmatrix},$$

$$\mathbf{D}^t\mathbf{YD} = \begin{bmatrix} 5/4 & 1/2 & -1/4 \\ 1/2 & 13/12 & 1/4 \\ -1/4 & 1/4 & 7/10 \end{bmatrix},$$

that is,

$$\mathbf{e}_T = \begin{bmatrix} e_1 \\ e_2 \\ e_3 \end{bmatrix} = \begin{bmatrix} 5/4 & 1/2 & -1/4 \\ 1/2 & 13/12 & 1/4 \\ -1/4 & 1/4 & 7/10 \end{bmatrix}^{-1} \begin{bmatrix} 5.7 \\ 3.86667 \\ 0 \end{bmatrix},$$

$$= [4.2 \quad 1.4 \quad 1.0]^t .$$

Since

$$\mathbf{v} = \mathbf{e} + \mathbf{E} ,$$

$$\mathbf{v}_T = [-2.4 \quad -3 \quad 1]^t .$$

This agrees with the answer to Example 2.1.

*Loop analysis*

The tie-set is shown in Fig. 2.9, and the tie-set matrix from Eq. (2.22) is

$$\mathbf{C}^t = \begin{bmatrix} 1 & 0 & -1 & 1 & 0 & 0 \\ 0 & -1 & -1 & 0 & 1 & 0 \\ -1 & -1 & 0 & 0 & 0 & 1 \end{bmatrix} .$$

By substitution

$$\mathbf{C}^t(\mathbf{E} - \mathbf{Z}\mathbf{I}) = [-6.6 \quad 4.4 \quad 6.2]^t ,$$

$$\mathbf{C}^t\mathbf{Z}\,\mathbf{C} = \begin{bmatrix} 11 & 5 & -2 \\ 5 & 12 & 3 \\ -2 & 3 & 7 \end{bmatrix} .$$

Therefore

$$\mathbf{i}_L = \begin{bmatrix} i_4 \\ i_5 \\ i_6 \end{bmatrix} = \begin{bmatrix} 11 & 5 & -2 \\ 5 & 12 & 3 \\ -2 & 3 & 7 \end{bmatrix}^{-1} \begin{bmatrix} -6.6 \\ 4.4 \\ 6.2 \end{bmatrix},$$

$$= [-0.8 \quad 0.6 \quad 0.4]^t .$$

Since $\mathbf{j} = \mathbf{i} + \mathbf{I}$

$$\mathbf{j}_L = [-0.8 \quad 0.6 \quad 2.8]^t .$$

This agrees with the answer to Example 2.2.

*Nodal analysis*

Nodes are defined in Fig. 2.10, and the reduced incidence matrix from Eq. (2.31) is

$$\mathbf{A}^t = \begin{bmatrix} -1 & 0 & 0 & 1 & 0 & -1 \\ 1 & -1 & 1 & 0 & 0 & 0 \\ 0 & 1 & 0 & 0 & 1 & 1 \end{bmatrix} .$$

By substitution,

$$\mathbf{A}^t(\mathbf{I} - \mathbf{YE}) = [-5.7 \quad 1.8333 \quad 3.8667]^t ,$$

$$(\mathbf{A}^t\mathbf{Y}\,\mathbf{A}) = \begin{bmatrix} 5/4 & -1/2 & -1/2 \\ -1/2 & 31/30 & -1/3 \\ -1/2 & -1/3 & 13/12 \end{bmatrix} .$$

Therefore

$$\mathbf{e}' = \begin{bmatrix} e_1 \\ e_2 \\ e_3 \end{bmatrix} = (\mathbf{A}^t\mathbf{YA})^{-1}\,\mathbf{A}^t(\mathbf{I} - \mathbf{YE}),$$

$$= [-3.2 \quad 1.0 \quad 2.4]^t .$$

Also $\mathbf{e} = \mathbf{Ae}'$

$$= [4.2 \quad 1.4 \quad 1.0 \quad -3.2 \quad 2.4 \quad 5.6]^t .$$

and $\mathbf{i} = \mathbf{Ye} - (\mathbf{I} - \mathbf{YE})$

$$= [-1.2 \quad -1.0 \quad 0.2 \quad -0.8 \quad 0.6 \quad 0.4]^t .$$

These answers agree with those from cut-set and loop analysis.

*Mesh analysis*

Meshes are defined in Fig. 2.11, and the reduced mesh matrix from Eq. (2.42) is

$$\mathbf{M}^t = \begin{bmatrix} -1 & 0 & 1 & -1 & 0 & 0 \\ 0 & -1 & -1 & 0 & 1 & 0 \\ 0 & 0 & 0 & 1 & -1 & 1 \end{bmatrix} .$$

By substitution,

$$\mathbf{M}^t(\mathbf{E} - \mathbf{ZI}) = [6.6 \quad 4.4 \quad -4.8]^t$$

$$(\mathbf{M}^t\,\mathbf{Z}\,\mathbf{M}) = \begin{bmatrix} 11 & -5 & -4 \\ -5 & 12 & -4 \\ -4 & -4 & 10 \end{bmatrix} .$$

Therefore

$$\mathbf{i}' = \begin{bmatrix} i_1 \\ i_2 \\ i_3 \end{bmatrix} = (\mathbf{M}^t\,\mathbf{Z}\,\mathbf{M})^{-1}\,\mathbf{M}^t\,(\mathbf{E} - \mathbf{ZI})$$

$$= [1.2 \quad 1.0 \quad 0.4]^t$$

Also $\mathbf{i} = \mathbf{Mi}'$

$$= [-1.2 \quad -1 \quad 0.2 \quad -0.8 \quad 0.6 \quad 0.4]^t ,$$

and $\mathbf{e} = \mathbf{Zi} - (\mathbf{E} - \mathbf{ZI})$ ,

$$= [4.2 \quad 1.4 \quad 1.0 \quad -3.2 \quad 2.4 \quad 5.6]^t .$$

These answers agree with those from cut-set, loop, and nodal analysis. Solutions are summarised in Fig. 2.13.

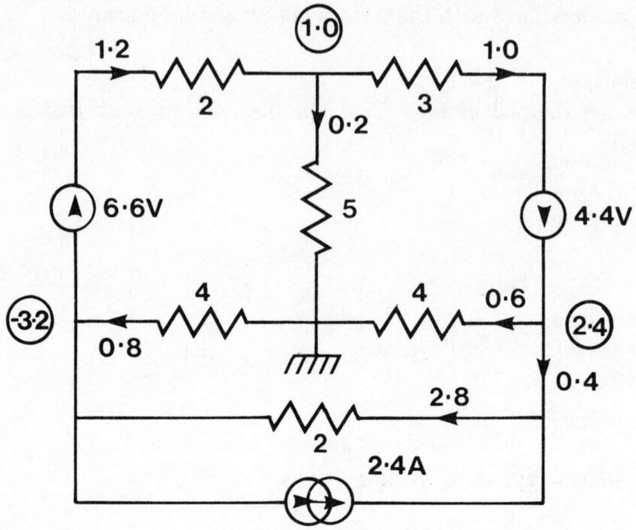


Fig. 2.13 – Analysis solution.

## 2.3 ACTIVE CIRCUITS

The four fundamental matrix analysis methods derived in the previous section apply to circuits consisting of resistors, inductors, capacitors, and fixed voltage or current sources. Each solution is in terms of a minimum set of voltages or currents from which all voltages and currents in the network may be determined.

Cut-set and nodal analysis are both based on the admittance matrix **Y** with diagonal element $Y_{jj}$, $j = 1, n_B$, equal to the admittance of branch $j$. In d.c. analysis the elements of **Y** are real conductances of the form $1/R$. Also all energy sources are d.c., and the elements of **E** and **I** are real. The solutions therefore involve the multiplication and inversion of matrices with real elements. Tie-set and mesh analysis both use the impedance matrix **Z** which is the inverse of the admittance matrix. The elements of **Z** are dual to the elements of **Y**, and analysis in terms of impedance is dual to analysis in terms of admittance. Chapter 3 will explore d.c. analysis in detail, concentrating on the non-linear circuit where the resistors are functions of the voltages or currents in the network.

In a.c. analysis the admittances $Y_{jj}$ are of the form $1/R$, $1/\text{j}\omega L$, or $\text{j}\omega C$ and are complex. Energy sources are also complex, representing sinusoidal excitations of the same frequency but of different phases. The impedances $Z_{jj}$ of the form $R$, $\text{j}\omega L$, and $1/\text{j}\omega C$ are of course also complex. Therefore the solutions required are complex and involve multiplication and inversion of complex matrices. A.C. analysis will be considered in Chapter 4 and developed further in Chapter 6 to obtain transfer functions.

Cut-set analysis and its dual, tie-set analysis, require identification of a tree. This is not difficult, but in numerically sophisticated computer programs it is found necessary to choose a tree so that matrix inversion can be performed with maximum efficiency. This does not concern us here, and for the present the labour of choosing a tree is unnecessary since nodal analysis may be used. In electronic circuits, where the number of nodes is normally less than the number of links, and where mutual inductors are relatively uncommon, the nodal analysis method is predominant. This is also true in the remainder of this book, since important practical analysis techniques, such as symbolic analysis and sensitivity analysis, are derived from nodal analysis.

Almost all circuits are active and include one or more dependent sources, each of which may be one of the four types given in Table 1.1. All other four-terminal elements can be modelled as a combination of two dependent sources. It is therefore essential to generalize the matrix methods from the previous section to include dependent sources. All future development of the basic matrix methods will then apply to active circuits. To achieve the generalisation without destroying the two-terminal branch structure of the circuits considered, dependent sources will be incorporated in the form of energy coupling between standard branches. Four-terminal networks which cannot be dealt with in this way will be incorporated by means of a simple equivalent circuits which approximate their behaviour.

**Transfer impedance coupling**

The current-controlled voltage source (CCVS) from Table 1.1 is governed by the equation

$$v_2 = ri_1$$

where $r$ is known as the **forward transfer resistance**. The CCVS in Fig. 2.14 is connected between two standard circuit branches k and $l$ such that the voltage $r_{l\mathrm{k}}j_{\mathrm{k}}$ is injected in series with branch $l$. Application of Kirchhoff's laws and Ohm's law to the modified circuit branch $l$ yields

$$(E_l - Z_{ll}I_l) + e_l - r_{l\mathrm{k}}\, j_{\mathrm{k}} = Z_{ll}\, i_l \ .$$

Since $j_{\mathrm{k}} = i_{\mathrm{k}} + I_{\mathrm{k}}$,

$$(E_l - Z_{ll}I_l - r_{l\mathrm{k}}I_{\mathrm{k}}) + e_l = Z_{ll}i_l + r_{l\mathrm{k}}i_{\mathrm{k}} \ . \qquad (2.46)$$

In general, $r_{l\mathrm{k}}$ is replaced by the impedance $Z_{l\mathrm{k}}$, since the transfer impedance may be resistive, inductive, or capacitive. The impedance $Z_{l\mathrm{k}}$ may be identified as an element of **Z** appearing in column k because it must multiply with the $k^{\mathrm{th}}$ element in both **i** and **I**. $Z_{l\mathrm{k}}$ appears in row $l$ because Eq. (2.46) refers to branch

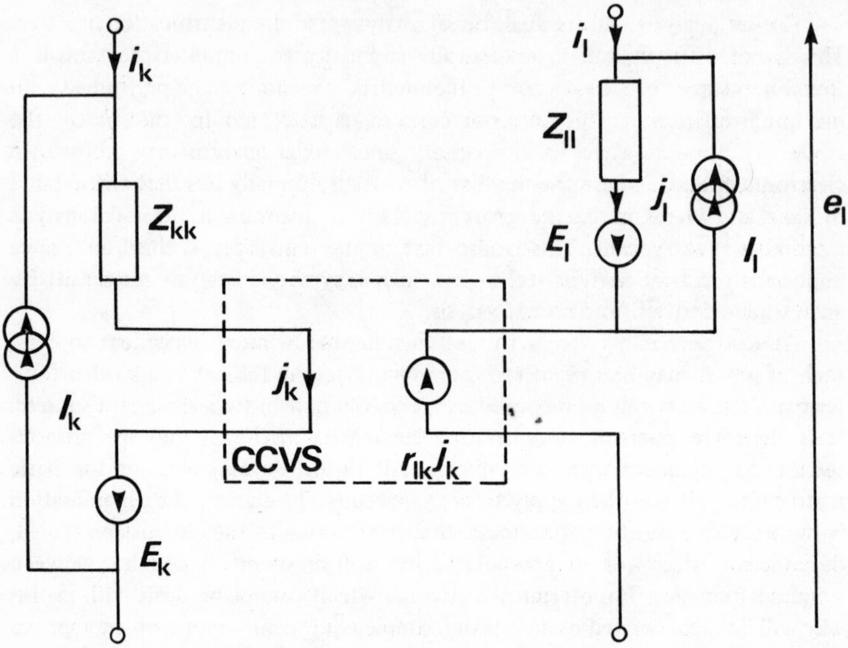


Fig. 2.14 – Transfer impedance coupling.

*l*. The choice of suffixes is therefore confirmed. The matrix $\mathbf{Z_m}$ is a modified version of $\mathbf{Z}$ which includes all transfer impedance terms and is in general neither diagonal or symmetric. Converting Eq. (2.46) into matrix form for all branches yields

$$(\mathbf{E} - \mathbf{Z_m I}) + \mathbf{e} = \mathbf{Z_m i} \ . \tag{2.47}$$

The tie-set solution for link currents proceeds as before, giving the modified version of Eq. (2.28).

$$\mathbf{i_L} = (\mathbf{C^t Z_m C})^{-1}\, \mathbf{C^t}(\mathbf{E} - \mathbf{Z_m I}) \ . \tag{2.48}$$

Solving for other currents and voltages in the circuit,

$$\mathbf{i} = \mathbf{C i_L} \ ,$$

$$\mathbf{j} = \mathbf{i} + \mathbf{I} \ ,$$

and $$\mathbf{e} = \mathbf{Z_m i} - (\mathbf{E} - \mathbf{Z_m I}) \ .$$

The solution for element voltages must be corrected to take the dependent sources into consideration giving

$$\mathbf{v} = \mathbf{e} + \mathbf{E} - (\mathbf{Z_m} - \mathbf{Z})(\mathbf{i} + \mathbf{I}) \tag{2.49}$$

$$= \mathbf{Z}(\mathbf{i} + \mathbf{I}) = \mathbf{Zj}\ .$$

Loop analysis with transfer impedance elements in **Z** is identical to that for passive circuits derived in Section 2.2, except for the correction term in Eq. (2.49). The voltage controlled voltage source is incorporated into $\mathbf{Z_m}$ using Ohm's law, since $v_k = Z_{kk}j_k$. When a branch is coupled by a dependent current-source instead of a voltage source, the Thévenin-Norton theorem may be used for conversion provided that the branch is not involved in any other coupling.

**Mutual inductance**

The mutual inductor from Table 1.1 is redrawn in Fig. 2.15 as two standard circuit branches mutually coupled by two CCVS-dependent sources. Transfer impedances are inductive and equal $j\omega M$. By inspection, the circuit equations are

$$e_k = j\omega L_1 i_k + j\omega M i_l$$

$$e_l = j\omega M i_k + j\omega L_2 i_l$$

where $L_1 = L_{kk}$, $L_2 = L_{ll}$, and $M = M_{lk} = M_{kl}$.

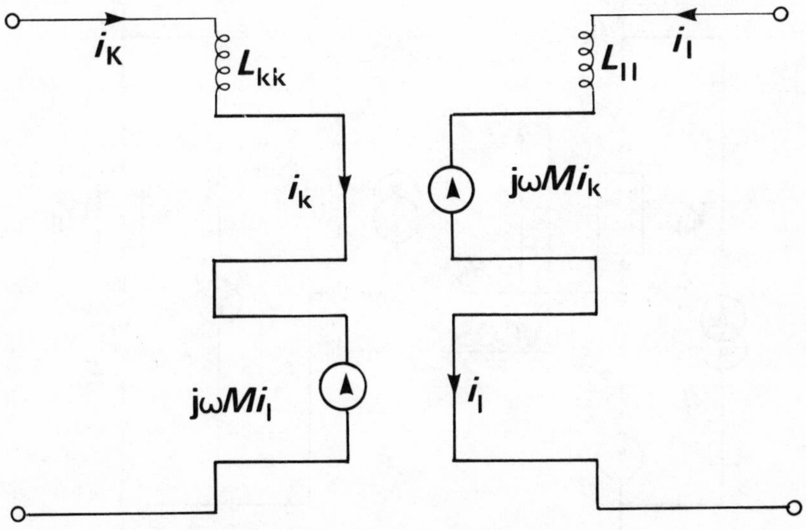


Fig. 2.15 – Dependent source model of mutual inductor.

The elements of the **Z** matrix are clearly

$$\begin{bmatrix} \ddots & & & & \\ & j\omega L_{kk} & - & - & j\omega M_{kl} \\ & & & & \\ & j\omega M_{lk} & - & - & j\omega L_{ll} \\ & & & & \ddots \end{bmatrix} .$$

When the determinant of the mutual inductor terms in **Z** is zero, the coefficient of coupling $M/\sqrt{L_1L_2}$ is unity. In this case neither $\mathbf{Z}_m$ nor $(\mathbf{C}^t\mathbf{Z}_m\mathbf{C})$ can be inverted. Both the matrix $\mathbf{Y}_m$ and the solution for link currents are therefore unobtainable. We will return to this problem when equivalent circuits are discussed later in this section.

**Transfer admittance coupling**

The voltage-controlled current source (VCCS) from Table 1.1 is defined by the equation

$$i_2 = gv_1$$

where $g$ is the **forward transfer admittance.**

The VCCS connection between two standard circuit branches is shown in Fig. 2.16. The circuit in this figure is the dual of that in Fig. 2.14, and the corresponding equation dual to Eq. (2.46) is

$$(I_l - Y_{ll}E_l - g_{lk}E_k) + i_l = Y_{ll}e_l + g_{lk}e_k . \tag{2.50}$$

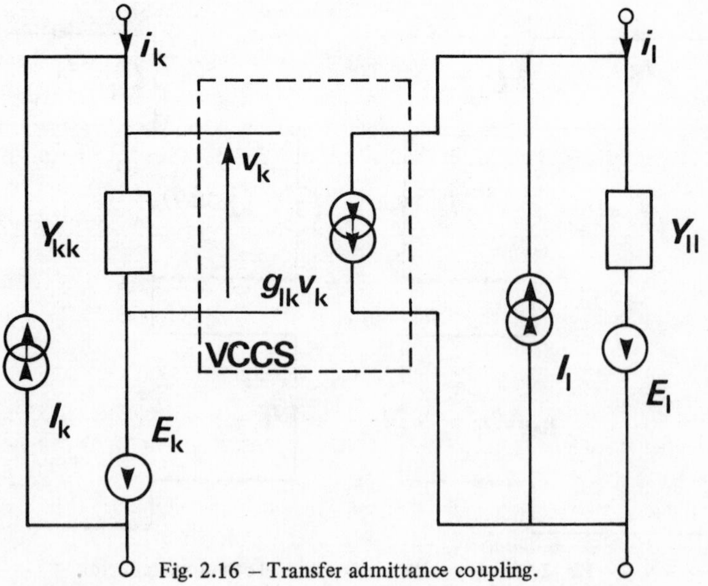


Fig. 2.16 – Transfer admittance coupling.

The term $g_{lk}$, replaced by the admittance $Y_{lk}$, is inserted in row $l$ column k of the **Y** matrix, giving the modified form $\mathbf{Y}_m$. The matrix version of Eq. (2.50) for all network branches is then

$$(\mathbf{I} - \mathbf{Y}_m\mathbf{E}) + \mathbf{i} = \mathbf{Y}_m\mathbf{e} \ . \tag{2.51}$$

The solution for tree-branch voltages is obtained as the modified form of Eq. (2.18),

$$\mathbf{e}_T = (\mathbf{D}^t\mathbf{Y}_m\mathbf{D})^{-1}\,\mathbf{D}^t(\mathbf{I}-\mathbf{Y}_m\mathbf{E}) \ . \tag{2.52}$$

Solving for remaining voltages and currents,

$$\mathbf{e} = \mathbf{D}\mathbf{e}_T \ , \tag{2.53}$$

$$\mathbf{v} = \mathbf{e} + \mathbf{E} \ ,$$

$$\mathbf{i} = \mathbf{Y}_m\mathbf{e} - (\mathbf{I} - \mathbf{Y}_m\mathbf{E}) \ ,$$

$$\mathbf{j} = \mathbf{i} + \mathbf{I} - (\mathbf{Y}_m - \mathbf{Y})(\mathbf{e}+\mathbf{E}) \ , \tag{2.54}$$

$$= \mathbf{Y}(\mathbf{e}+\mathbf{E}) = \mathbf{Y}\mathbf{v} \ .$$

The element current solution contains the correction term $(\mathbf{Y}_m-\mathbf{Y})(\mathbf{e}+\mathbf{E})$ to take dependent sources into account. Nodal analysis is similar, requiring use of the incidence matrix **A** in place of the cut-set matrix **D** in Eqs. (2.52) and (2.53).

The current-controlled current source is easily incorporated into $\mathbf{Y}_m$ using Ohm's law since $j_k = Y_{kk}\, v_k$. When dependent voltage sources are present the Thévénin-Norton theorem can be used to convert them to the current sources required, provided that the branch is not involved in any other coupling. In cases where $\mathbf{Z}_m$ is available when $\mathbf{Y}_m$ is required, the relationship $\mathbf{Y}_m = \mathbf{Z}_m{}^{-1}$ is used. This may also be invoked for part of a network which contains transfer impedance coupling only (for example the mutual inductor) and the inverse taken for that part of the network only. In more complicated cases where several dependent sources of different types interact with a single branch, the generalised relationships developed in the following section must be used.

*Example* 2.4

Find an admittance matrix for the long-tail pair amplifier shown in Fig. 2.17. Assume small-signal nodal analysis using the transistor hybrid parameters $h_{ie} = 3.6\text{k}$, $h_{re} = 0$, $h_{fe} = 280$, $h_{oe} = 25\mu\text{s}$.

Transistor models have not yet been considered, and the simplified model shown in Fig. 2.17 will be used. This model is based on the hybrid parameters quoted for the BC107 transistor at 2mA collector current.

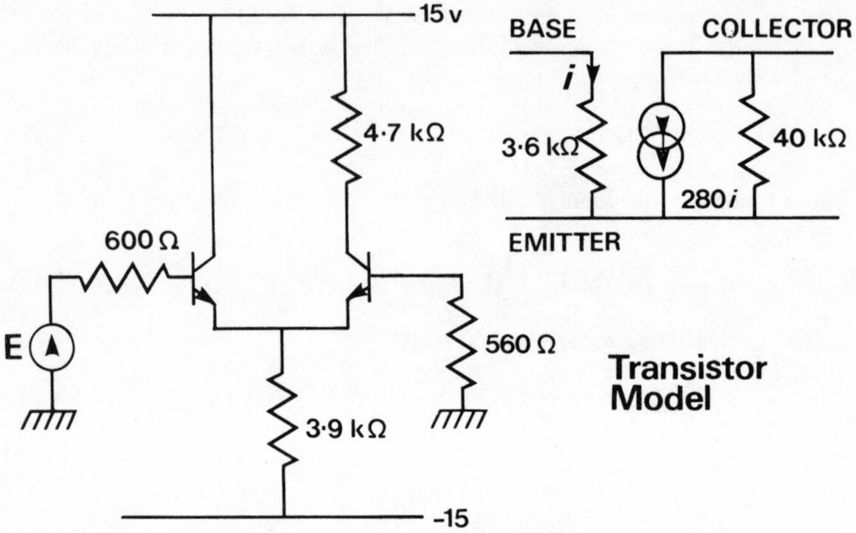


Fig. 2.17 – Amplifier for Example 2.4.

The first step in preparing a circuit for analysis is to redraw the circuit in a form showing two-terminal branch elements and, for nodal analysis, transfer admittance coupling. Since we are concerned here with small-signal analysis, the power supply voltage sources are set to zero, effectively grounding the supply rails. When the transistor model is inserted, the circuit diagram for analysis in Fig. 2.18 is obtained. The CCCS in the transistor has been converted to voltage control, using Ohm's law in the base to emitter resistor. Nodes numbered ① to ③, and branch currents and directions, are shown on the figure.

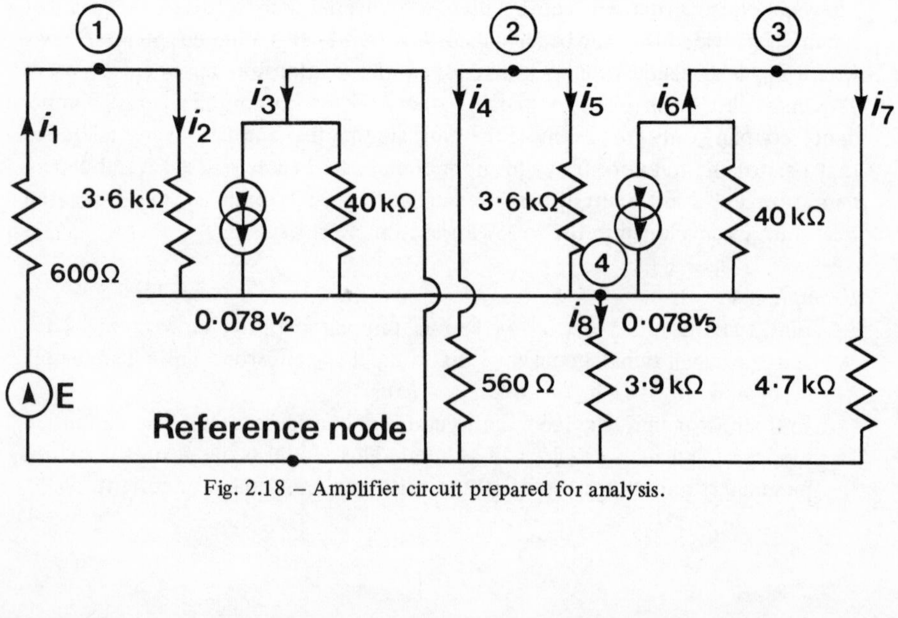


Fig. 2.18 – Amplifier circuit prepared for analysis.

The **Y** matrix can now be filled by inspection. The transfer admittance term of value 0.078 appears in the **Y** matrix in row 3 column 2 for the VCCS between branches 2 and 3. The second transfer admittance appears in row 6 column 5 but with a negative sign, since the assumed direction of branch 6 is opposite to that of the current source. All other off-diagonal terms in the admittance matrix are zero.

$$\mathbf{Y} = \begin{bmatrix} 1/600 & & & & & & & 0 \\ & 1/3600 & & & & & & \\ & 0.078 & 1/40000 & & & & & \\ & & & 1/560 & & & & \\ & & & & 1/3600 & & & \\ & & & & -0.078 & 1/40000 & & \\ & & & & & & 1/4700 & \\ 0 & & & & & & & 1/3900 \end{bmatrix}.$$

**Generalization**

The two previous subsections have separately treated analysis in terms of the impedance and admittance matrices. The CCVS and VCVS have been included by modification to **Z** , and the VCCS and CCCS by modification to **Y**. This process can be generalised to include all dependent source types into both **Y** and **Z**.

Consider the circuit branch in Fig. 2.19. The location and direction of all dependent source types are defined on the figure. The source $r_{l\mathrm{p}} j_{\mathrm{p}}$ is due to a CCVS with transfer resistance $r_{l\mathrm{p}}$ from branch p. Similarly, $m_{l\mathrm{q}} v_{\mathrm{q}}$ is due to a VCVS with voltage gain $m_{l\mathrm{q}}$ from branch q;$g_{l\mathrm{r}}\, v_{\mathrm{r}}$is due to a VCCS with transfer admittance $g_{l\mathrm{r}}$ from branch r; and $n_{l\mathrm{s}}$ is due to a CCCS with current gain $n_{l\mathrm{s}}$ from branch s.

From Kirchhoff's laws.

$$j_l = i_l + I_l - g_{l\mathrm{r}} v_{\mathrm{r}} - n_{l\mathrm{s}} j_{\mathrm{s}}$$

$$v_l = e_l + E_l - m_{l\mathrm{q}} v_{\mathrm{q}} - r_{l\mathrm{p}} j_{\mathrm{p}} \;.$$

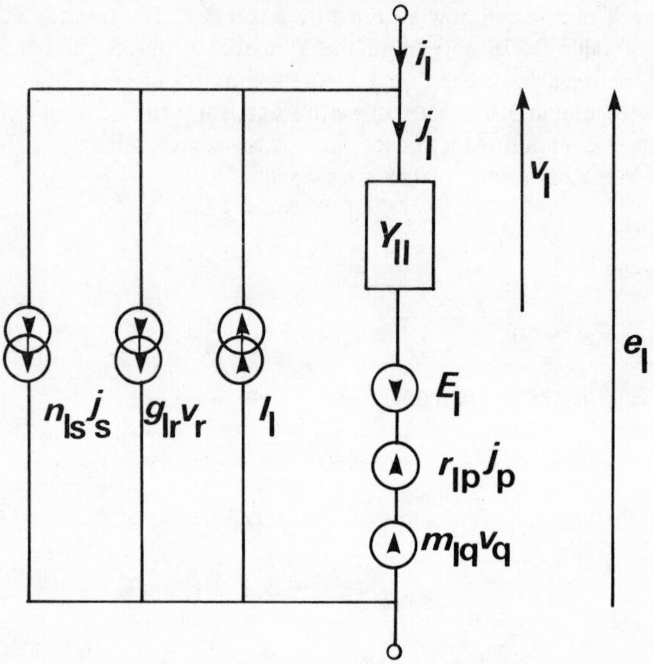


Fig. 2.19 – The generalised circuit branch.

In matrix form for all branches,

$$\mathbf{j} = \mathbf{i} + \mathbf{I} - \mathbf{G}\mathbf{v} - \mathbf{N}\mathbf{j} \; , \tag{2.55}$$

$$\mathbf{v} = \mathbf{e} + \mathbf{E} - \mathbf{M}\mathbf{v} - \mathbf{R}\mathbf{j} \; , \tag{2.56}$$

where **G**, **N**, **M**, and **R** are matrices containing the dependent source gains. Rearranging yields

$$(\mathbf{Y} + \mathbf{N}\mathbf{Y} + \mathbf{G})\mathbf{v} = \mathbf{i} + \mathbf{I} \; ,$$

$$(\mathbf{U} + \mathbf{M} + \mathbf{R}\mathbf{Y})\mathbf{v} = \mathbf{e} + \mathbf{E} \; ,$$

where **U** is the unit matrix. Taking the inverse and substituting,

$$(\mathbf{Y} + \mathbf{N}\mathbf{Y} + \mathbf{G})(\mathbf{U} + \mathbf{M} + \mathbf{R}\mathbf{Y})^{-1}(\mathbf{e} + \mathbf{E}) = (\mathbf{i} + \mathbf{I}) \; . \tag{2.57}$$

The dual expression may be similarly derived as

$$(\mathbf{Z} + \mathbf{M}\mathbf{Z} + \mathbf{R})(\mathbf{U} + \mathbf{N} + \mathbf{G}\mathbf{Z})^{-1}(\mathbf{i} + \mathbf{I}) = (\mathbf{e} + \mathbf{E}) \; . \tag{2.58}$$

The modified admittance matrix $\mathbf{Y}_m$ in Eq. (2.51) can be identified with $(\mathbf{Y} + \mathbf{NY} + \mathbf{G})$ in Eq. (2.57). Similarly $\mathbf{Z}_m$ from Eq. (2.47) may be identified with $(\mathbf{Z} + \mathbf{MZ} + \mathbf{R})$ in Eq. (2.58). The generalised modification to $\mathbf{Y}$ and $\mathbf{Z}$ to incoprporate all dependent sources in the fully modified matrices $\mathbf{Y}_c$ and $\mathbf{Z}_c$ are given by

$$\mathbf{Y}_c = (\mathbf{Y} + \mathbf{NY} + \mathbf{G})(\mathbf{U} + \mathbf{M} + \mathbf{RY})^{-1} , \tag{2.59}$$

$$\mathbf{Z}_c = (\mathbf{Z} + \mathbf{MZ} + \mathbf{R})(\mathbf{U} + \mathbf{N} + \mathbf{GZ})^{-1} . \tag{2.60}$$

Analysis proceeds in terms of admittance or impedance matrices as before, but with $\mathbf{Y}_c$ or $\mathbf{Z}_c$ in place of $\mathbf{Y}_m$ or $\mathbf{Z}_m$ in Eqs. (2.48) and (2.52).

Expressions (2.59) and (2.60) are not of direct practical use, since the matrices involved in computing $\mathbf{Y}_c$ or $\mathbf{Z}_c$ are large and very sparse; that is, they contain very few non-zero elements. It is not therefore efficient to store the matrices separately. Nevertheless, the matrix expressions are relvant to the standard methods of circuit analysis and must be used when branches are no longer coupled in isolation.

## 2.4 IMPEDANCE CONVERTERS AND INVERTERS

The four-terminal network elements from Table 1.1 which involve mutual coupling between two branches are shown in Table 2.1 connected between two branches of finite impedance. Each element is modelled as a pair of dependent sources identified directly from the element equations.

The **Y**,**Z**,**N**,**M**,**G**, and **R** matrices of the previous section may be derived by inspection of Table 2.1 for each element in turn, and new models may be developed which are suitable for direct inclusion in the $\mathbf{Y}_m$ and $\mathbf{Z}_m$ matrices. Use of the more complicated expressions for $\mathbf{Y}_c$ and $\mathbf{Z}_c$ can then be avoided within the assumption that the four-terminal elements are connected between branches in which no other coupling is present. The majority of practical cases are then analysed without recourse to the generalised formulae.

### The Ideal Transformer

The modified admittance matrix $\mathbf{Y}_c$ is obtained from Eq. (2.59), assuming in this case the presence of two branches only. Filling each matrix yields

$$\mathbf{Y} = \begin{bmatrix} Y_{11} & 0 \\ 0 & Y_{22} \end{bmatrix}, \quad \mathbf{N} = \begin{bmatrix} 0 & 0 \\ -n & 0 \end{bmatrix}, \quad \mathbf{M} = \begin{bmatrix} 0 & n \\ 0 & 0 \end{bmatrix},$$

$$\mathbf{Z} = \begin{bmatrix} Z_{11} & 0 \\ 0 & Z_{22} \end{bmatrix}, \text{ and } \mathbf{G} = \mathbf{R} = \mathbf{0} .$$

| EQUATIONS | BRANCH CONNECTIONS AND DEPENDENT SOURCE MODEL |
|---|---|
| Ideal transformer<br>$v_1 = nv_2$<br>$-i_2 = ni_1$ | 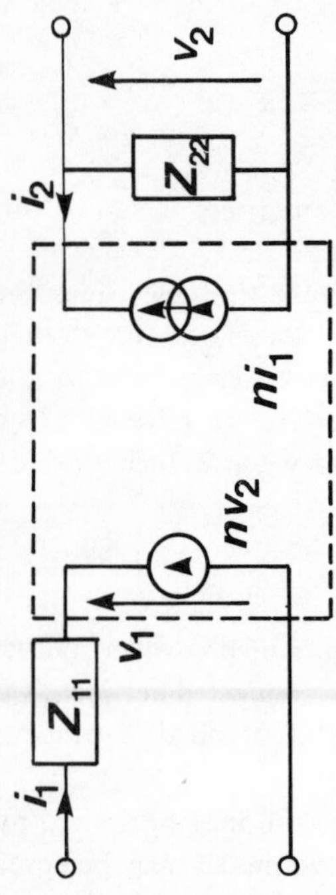 |
| Negative impedance converter<br>$v_1 = nv_2$<br>$-i_2 = -ni_1$ | 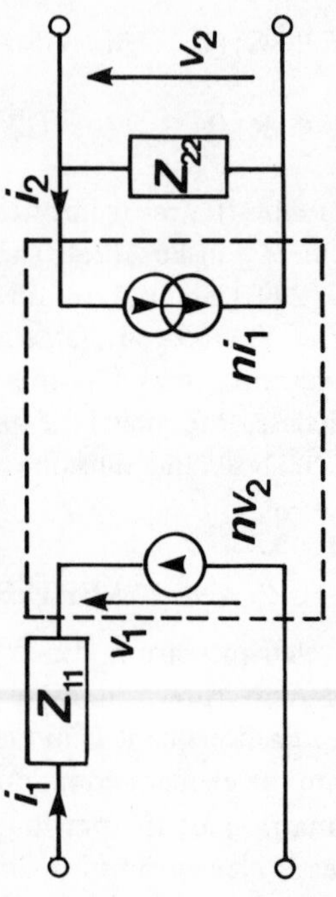 |
| Gyrator<br>$i_1 = gv_2$<br>$-i_2 = gv_1$ | 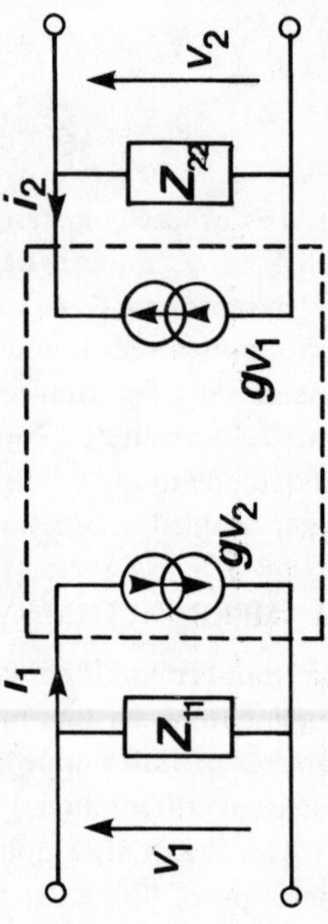 |

| | |
|---|---|
| Negative impedance inverter<br>$i_1 = gv_2$<br>$-i_2 = -gv_1$ | 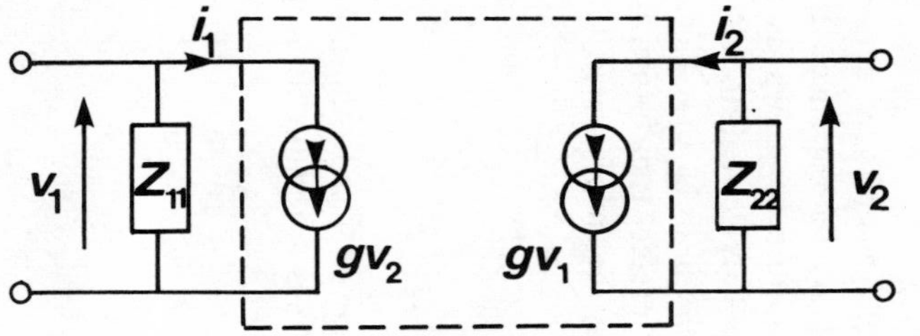 |
| Mutual inductance<br>$L_1L_2 \neq M^2$<br>$v_2 = j\omega Mi_1 + j\omega L_2i_2$<br>$v_1 = j\omega L_1i_1 + j\omega Mi_2$ | 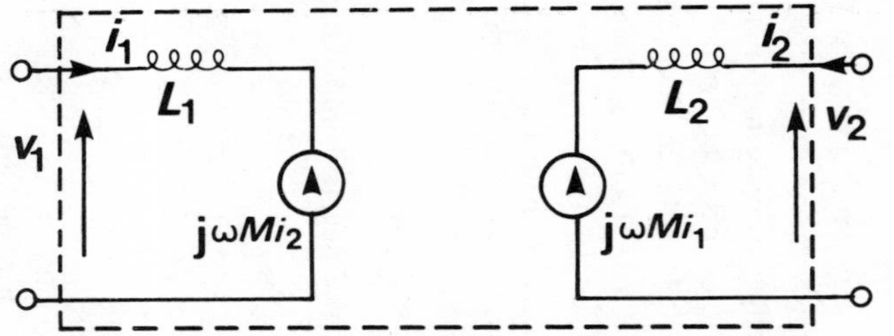 |
| Mutual inductance<br>$L_1L_2 = M^2$<br>$v_2 = j\omega Mi_1 + j\omega L_2i_2$<br>$v_1 = j\omega L_1i_1 + j\omega Mi_2$ | 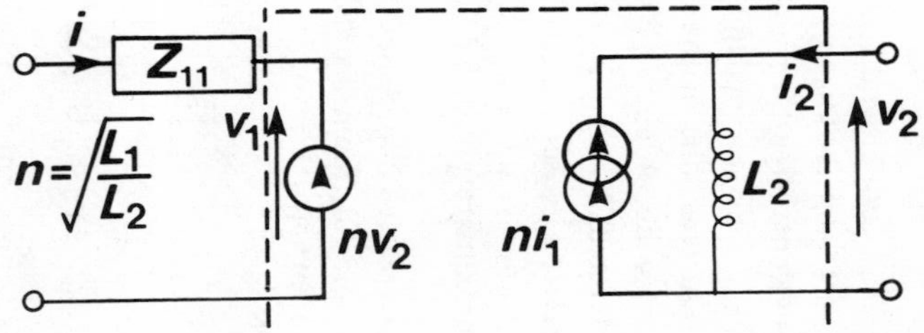 |

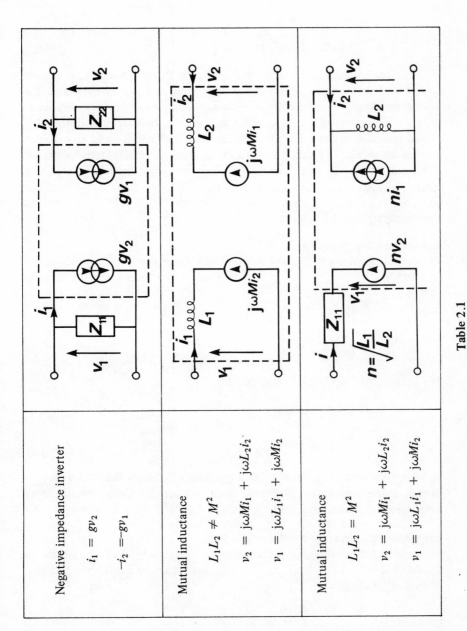

**Table 2.1**
Four-terminal network elements

By substitution

$$\mathbf{Y}_c = \begin{bmatrix} Y_{11} & -nY_{11} \\ -nY_{11} & n^2Y_{11} + Y_{22} \end{bmatrix}, \quad \mathbf{Z}_c = \begin{bmatrix} Z_{11} + n^2Z_{22} & nZ_{22} \\ nZ_{22} & Z_{22} \end{bmatrix}.$$

The terms $-nY_{11}$ in $\mathbf{Y}_c$ may be identified with voltage-controlled current sources, and the diagonal terms may be interpreted as branch admittances. The equivalent circuit of the perfect transformer may therefore be drawn as in Table 2.2. The impedance form of this circuit may be similarly derived from $\mathbf{Z}_c$.

At least one of the branch impedances must be finite in order that $\mathbf{Y}_c$ and $\mathbf{Z}_c$ may be obtained. When $Z_{11} = 0$ and $Z_{22} = \infty$, the prefect transformer has undefined $\mathbf{Y}_c$ and $\mathbf{Z}_c$ matrices. In this case they may be incorporated into analysis by inserting either a small resistor $R_{11}$ or a large transistor $R_{22}$. When the resistance value chosen is suitable it will affect the circuit only slightly, and an approximate solution is found.

**The negative impedance converter**

The only difference between the negative impedance converter and the perfect transformer is the sign of the current generator which leads to the modification of **N** to

$$\mathbf{N} = \begin{bmatrix} 0 & 0 \\ n & 0 \end{bmatrix}.$$

The expressions for $\mathbf{Y}_c$ and $\mathbf{Z}_c$ may be derived as

$$\mathbf{Y}_c = \begin{bmatrix} Y_{11} & -nY_{11} \\ nY_{11} & -n^2Y_{11} + Y_{22} \end{bmatrix}, \quad \mathbf{Z}_c = \begin{bmatrix} Z_{11} - n^2Z_{22} & nZ_{22} \\ -nZ_{22} & Z_{22} \end{bmatrix}$$

yielding the equivalent circuits in Table 2.2. As in the case of the ideal transformer, at least one branch impedance must be finite, and similar remarks apply when solutions are required with $Z_{11} = 0$ and $Z_{22} = \infty$.

**The gyrator and negative impedance inverter**

The gyrator and negative impedance converter are already in a form suitable for derivation of $\mathbf{Y}_m$ since their models incorporate transfer admittance coupling. Therefore for the gyrator

$$\mathbf{Y}_c = \mathbf{Y}_m = \begin{bmatrix} Y_{11} & g \\ -g & Y_{22} \end{bmatrix}$$

and, for the negative impedance inverter,

$$\mathbf{Y}_c = \mathbf{Y}_m = \begin{bmatrix} Y_{11} & g \\ g & Y_{22} \end{bmatrix}.$$

The gyrator impedance matrix is given by the matrix inversion $\mathbf{Y}_m^{-1}$, that is,

$$\mathbf{Z}_m = \begin{bmatrix} Z_{11}/(1+g^2Z_{11}Z_{22}) & -Z_{11}Z_{22}g/(1+g^2Z_{11}Z_{22}) \\ Z_{11}Z_{22}g/(1+g^2Z_{11}Z_{22}) & Z_{22}/(1+g^2Z_{11}Z_{22}) \end{bmatrix}.$$

When branch impedances $Z_{11}$ and $Z_{22}$ become infinite, $\mathbf{Z}_m$ reduces to the more usual form

$$\mathbf{Z}_m = \begin{bmatrix} 0 & -1/g \\ 1/g & 0 \end{bmatrix}.$$

Equivalent circuits for the gyrator are shown in Table 2.2.

The negative impedance inverter may be similarly treated to yield

$$\mathbf{Z}_m = \begin{bmatrix} Z_{11}/(1-g^2Z_{11}Z_{22}) & -Z_{11}Z_{22}g/(1-g^2Z_{11}Z_{22}) \\ -Z_{11}Z_{22}g/(1-g^2Z_{11}Z_{22}) & Z_{22}/(1-g^2Z_{11}Z_{22}) \end{bmatrix},$$

which reduces for $Z_{11}$ and $Z_{22}$ infinite to

$$\mathbf{Z}_m = \begin{bmatrix} 0 & 1/g \\ 1/g & 0 \end{bmatrix}.$$

Equivalent circuits are shown in Table 2.2.

**Mutual inductance**

We have already noted that the $\mathbf{Z}_m$ matrix for the mutual inductor is given by

$$\mathbf{Z}_m = \begin{bmatrix} \mathrm{j}\omega L_1 & \mathrm{j}\omega M \\ \mathrm{j}\omega M & \mathrm{j}\omega L_2 \end{bmatrix}.$$

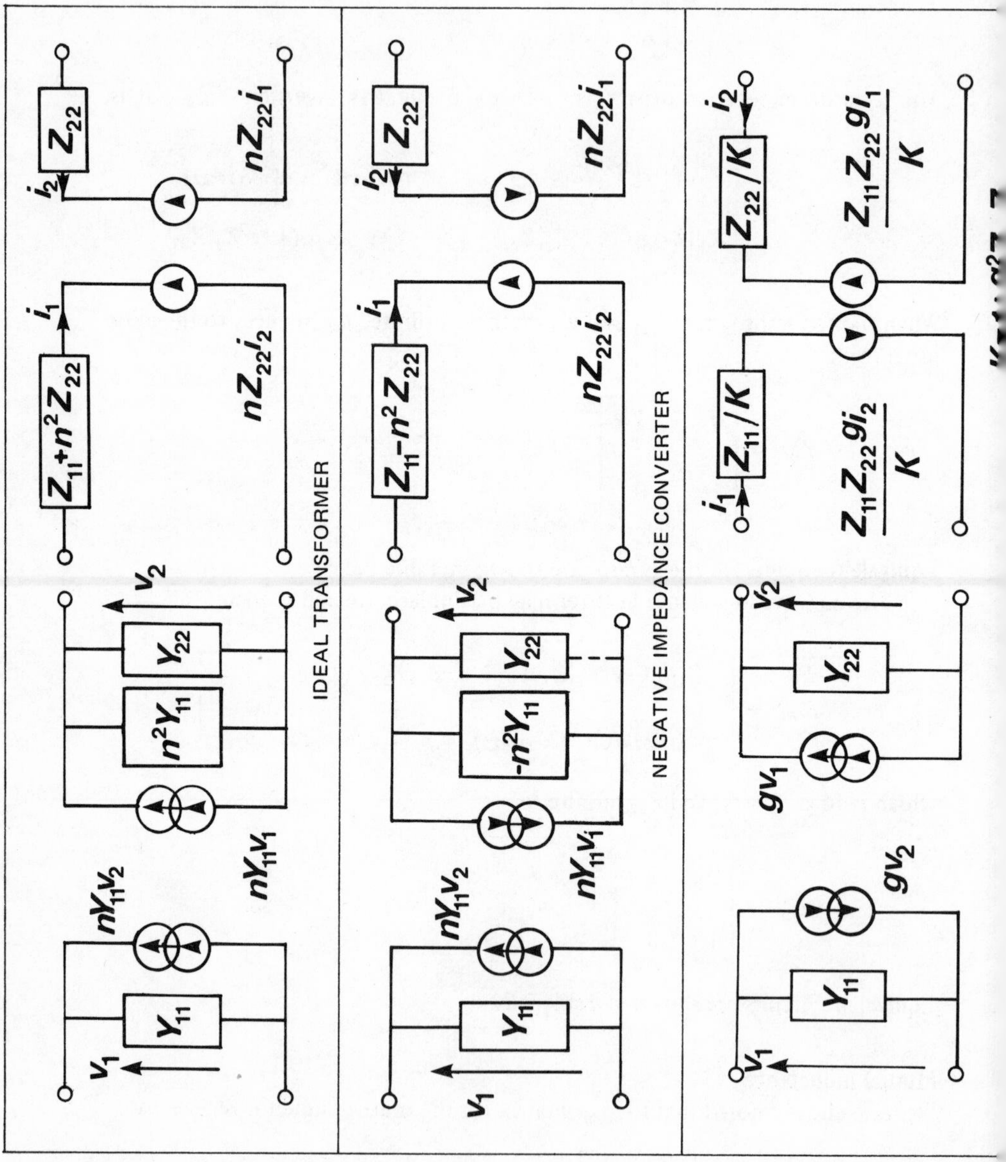
IDEAL TRANSFORMER
NEGATIVE IMPEDANCE CONVERTER
$v_1$
$Y_{11}$
$nY_{11}v_2$
$nY_{11}v_1$
$n^2Y_{11}$
$Y_{22}$
$v_2$
$Z_{11}+n^2Z_{22}$
$i_1$
$nZ_{22}i_2$
$Z_{22}$
$i_2$
$nZ_{22}i_1$
$-n^2Y_{11}$
$Z_{11}-n^2Z_{22}$
$gv_2$
$gv_1$
$Z_{11}/K$
$Z_{22}/K$
$\frac{Z_{11}Z_{22}gi_2}{K}$
$\frac{Z_{11}Z_{22}gi_1}{K}$

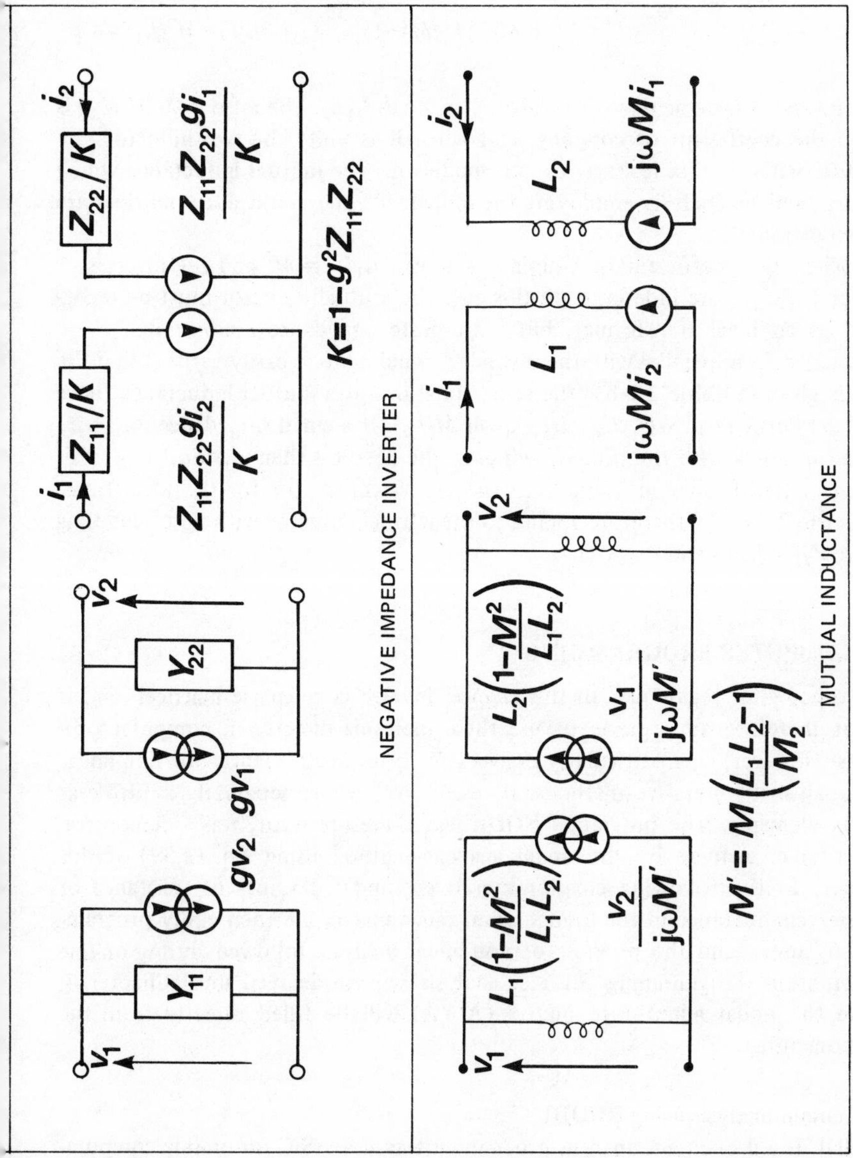


**Table 2.2**

**Y** and **Z** matrix equivalent circuits for four-terminal elements

Therefore,

$$\mathrm{Y_m} = \mathrm{Z_m^{-1}} = \begin{bmatrix} 1/\{j\omega L_1(1-M^2/L_1L_2)\} & 1/\{j\omega M(L_1L_2/M^2-1)\} \\ -1/\{j\omega M(L_1L_2/M^2-1)\} & 1/\{j\omega L_2(1-M^2/L_1L_2)\} \end{bmatrix}.$$

This inverse can be achieved provided that $M^2 \neq L_1L_2$. The ratio $\sqrt{(M^2/L_1L_2)}$ is called the **coefficient of coupling**, and when it is unity the two inductors are coupled without flux leakage. Circuit models for the mutual inductance which can be used in analysis employing the impedance and admittance matrices are shown in Table 2.2.

When the coefficient of coupling is unity, $L_1L_2 = M^2$ and the inverses of $Z_m$ and $C^tZ_mC$ are indefinite. In this case the mutual inductor must be recognised as an ideal transformer, but with finite impedance incorporated as an inductance $L_2$ in parallel with the secondary winding. It is easily verified that the circuit given in Table 2.1 has the same equations as a mutual inductance (note that the turns ratio $\sqrt{(L_1L_2)}$ also equals $M/L_2$). If a small impedance $Z_{11}$ is included in series with the primary winding, the circuit is then identical to that of the ideal transformer in Table 2.1. The equivalent circuit for the ideal transformer in Table 2.2 therefore applies to the mutual inductor with unity coupling where $Z_{22} = j\omega L_2$ and $n = \sqrt{(L_1L_2)}$.

## 2.5 COMPUTER PROGRAMMING

The topological techniques of this chapter involve large sparse matrices, and it is not therefore appropriate to use these methods directly in computer programs. However, it is particularly convenient in the next chapter for non-linear d.c. analysis by iterative methods, if resistor values are separately available as matrix elements. The program RNODE was therefore written as a demonstration of d.c. analysis by the nodal analysis method using Eq. (2.39) of this chapter. In the following chapter, iteratively updated values of resistance or transfer conductance in the branch admittance matrix are then easily programmed by users, and the progress of non-linear analysis followed during on-line computation. Programming for a.c. analysis will be delayed until Chapter 4, where the nodal admittance matrix ($\mathbf{A^tYA}$) will be filled directly from the component list.

### D.C. circuit analysis using RNODE

RNODE is a d.c. circuit analysis program written in BASIC for on-line computation from time-shared terminals. A 10-node, 20-branch circuit will run in approximately 10K bytes of memory, using Data General Basic 3.8 on the NOVA 1220 computer. The program listing is given in the Appendix to this chapter.

The matrix equation required for solution by the nodal analysis method and summarized here are based on Eq. (2.39) in Section 2.2 and on the active circuit version given in Eqs. (2.52) and (2.54) in Section 2.3. RNODE uses the following equations:

$$e' = (A^t Y_m A)^{-1} A^t (I - Y_m E)$$

$$e = Ae'$$

$$v = e + E$$

$$i = Y_m e - (I - Y_m E)$$

$$j = Yv \ .$$

One further facility provided by RNODE, useful, for example, for common-base input amplifiers, is for the VCCS transfer admittances to be dependent on branch voltage instead of resistor voltage. The modification required in the system is that $(\mathbf{I}-\mathbf{Y}_m\mathbf{E})$ is replaced by $(\mathbf{I}-\mathbf{YE})$.

*Circuit input*

Program statements 10 to 400 are concerned with the input of circuit component data and filling of the matrices $\mathbf{Y}_m$, **E**, **I**, and **A** in preparation for analysis. Corresponding RNODE arrays are Y(B,B), E(B,1), I(B,1), and A(B,N) respectively. Circuit nodes must be numbered 1 to N with reference node 0. Resistors must be numbered 1 to B, and VCCS transfer admittances numbered 1 to S.

Only two types of component input are allowed. Each resistor is entered in the form of the standard branch (Fig. 2.7) together with associated fixed voltage and current sources. On each request R? the required input is Resistor number $r$, value $R_r$, +ve node, −ve node, $E_r$, $I_r$.

Each VCCS is entered on request GM? and the required input is VCCS number, value $GM_{ij}$, send resistor number $j$, receive resistor number $i$.

Both resistors and VCCS can be modified after analysis on command MC by input of the component number followed by the new value, and for resistors, new source values also. Component connections must remain unchanged. Component modification is programmed at statement numbers 970 to 1140.

One restriction, which applies only when sensitivity is required (see Chapter 8, Section 8.6), is that the number of VCCS for which sensitivity can be determined is limited to a maximum of B. This could quite easily be programmed differently, but usually S<<B.

*Analysis and output*

Program statements 410 to 570 utilise the matrix statements available in BASIC to implement the equations previously summarised. On completion of the solution

Z(N,N) contains the inverse of the nodal admittance matrix $(\mathbf{A}^t\mathbf{Y}\mathbf{A})^{-1}$

L(N,1) contains the node voltages **e**,

K(B,1) contains the branch voltages **e**,

M(B,1) contains the branch currents **i**, and

J(B,1) contains the resistor voltages **v**.

These matrices are available for use in non-linear iteration or sensitivity subroutines. However, we are concerned at the moment only with the outputs provided. Program statements 580 to 760 tabulate all the available circuit voltages and currents under suitable headings.

*Commands*

Program statements 770 to 890 organise the action of the program after the completion of analysis. The following commands contained in M$(8) are provided at present

ER End the computer run.

MC Modify the component values of the circuit and repeat the analysis.

SE Call the sensitivity subroutine. For details see Chapter 8.

NL Call for non-linear iteration.

Program statements 900 to 960 organise the non-linear iteration by setting up N4 calls of subroutine 5000 which linearises a non-linear d.c. circuit, followed by linear d.c. analysis. This procedure will be discussed fully in Chapter 3, and circuit examples will be programmed for linearisation.

*Other programme variables*
C8, I9 are control variables.
B1,B2,B9,E1,G1,I1,I2,I3,N1,N2,R,S9,N$(3),D(N,B) are used for calculation space. Z(N,B) is also used for calculation space but it eventually contains the inverse of the nodal admittance matrix $(\mathbf{A}^t\mathbf{Y}_m\mathbf{A})^{-1}$. The present sensitivity subroutine also uses Z(N,B) and leaves it filled with the VCCS sensitivity, thus destroying the inverse matrix.

*Example* 2.5

Use RNODE to carry out d.c. analysis of the transistor amplifier in Fig. 2.17, assuming the transistor model to contain an additional fixed source of 0.5 V in series with the base-to-emitter resistor.

The circuit was prepared for small-signal analysis in Fig. 2.18, and must be slightly modified to include the fixed sources in the power supply as well as those required in the transistor model. Normally this analysis should be treated as non-linear. However, we have assumed that the base-emitter diode and $r_{BB'}$ for the transistor can be modelled by the linear combination of a 3.6 kΩ resistor in series with a 0.5 V fixed source.

The modifications to Fig. 2.18 necessary to incorporate the 0.5 V sources and the 15 V supplies may be summarised as follows:

Branch 2 Add $E_2 = -0.5$ V

5 Add $E_5 = -0.5$ V

3 Add $E_3 = 15$ V (See Problem 1.1).

7 Add $E_7 = -15$ V.

8 Add $E_8 = 15$ V.

Recall that the signs of these added sources depend on the assumed direction of branch currents. Note also that it will now be necessary to specify the VCCS as resistor-dependent, since the original CCCS current control was converted to voltage control, using the branch resistor.

Analysis with $E_1$=0 will yield the quiescent operating point of the amplifier. The change in node ③ voltage when $E_1$ is modified to 0.01 V, will yield the value of the small-signal d.c. voltage gain of the amplifier. The RNODE computer run for this example is reproduced in Fig. 2.20. The input data was prepared from Fig. 2.18 with the addition of the listed sources. The quiescent d.c. voltages are largely as expected, and the d.c. gain is clearly 147. The sensitivity of this amplifier is determined in Chapter 8.

```
RNODE     DC CIRCUIT ANALYSIS
NO. OF NODES, NO.OF RESISTORS, NO.OF DEPENDENT SOURCES ? 4,8,2
RESISTOR NO.,VALUE,+VE NODE NO.,-VE NODE NO.,SOURCE E,SOURCE I
R ? 1,680,0,1,0,0
R ? 2,3600,1,4,-.5,0
R ? 3,40E3,0,4,15,0
R ? 4,560,2,0,0,0
R ? 5,3600,2,4,-.5,0
R ? 6,40E3,4,3,0,0
R ? 7,4700,3,0,-15,0
R ? 8,3900,4,0,15,0
GM BRANCH OR RESISTOR DEPENDENT: INPUT B OR R ? R
GM NO.,VALUE,SEND R NO.,RECEIVE R NO.
GM ? 1,.078,2,3
GM ? 2,-.078,5,6
NODE          VOLTAGE
 1            -3.73882E-03
 2            -3.16793E-03
 3             6.68695
 4            -.523533
BRANCH OUTPUT YES OR NO ? YES
               RESISTOR        RESISTOR        BRANCH          BRANCH
R              VOLTAGE         CURRENT         VOLTAGE         CURRENT
 1              3.73882E-03     5.49826E-06     3.73882E-03     5.49826E-06
 2              1.97944E-02     5.49844E-06     .519794         5.49844E-06
 3              15.5235         3.88088E-04     .523533         1.93205E-03
 4             -3.16793E-03    -5.65701E-06    -3.16793E-03    -5.65701E-06
 5              2.03653E-02     5.65702E-06     .520365         5.65702E-06
 6             -7.21048        -1.80262E-04    -7.21048        -1.76876E-03
 7             -8.31305        -1.76873E-03     6.68695        -1.76873E-03
 8              14.4765         3.71192E-03    -.523533         3.71192E-03
NEXT COMMAND: END RUN      ER, MODIFY CIRCUIT          MC
            : SENSITIVITY SE, NON-LINEAR ITERATION NL
NEXT ? MC
NUMBER OF ALTERED BRANCHES, AND DEPENDENT SOURCES ? 1,0
R NO.,NEW VALUE,NEW E SOURCE,NEW I SOURCE
R ? 1,680,.01,0
NODE          VOLTAGE
 1             5.49746E-03
 2            -2.46888E-03
 3             8.16176
 4            -.51834
BRANCH OUTPUT YES OR NO ? NO
NEXT ? SE
```

Fig. 2.20 – RNODE computation for Example 2.5.

## 2.6 CONCLUSION

It is not good practice to directly apply the matrix methods of circuit analysis derived in this chapter in the form of computer programs. Nevertheless, the topological matrices and basic analytical methods are fundamental to most circuit analysis programs in current use. Complete familiarity with the material of this chapter is therefore an essential prerequisite to further study.

Much of the sophistication of circuit analysis computer programs, and indeed much of the remainder of this book, is derived from modification of the basic methods, with the primary objectives of reduction in computation time and memory, and of increased numerical accuracy. Specialised matrix inversion techniques are used in nodal admittance matrix analysis in Chapter 4, and through transfer functions in Chapter 6, so that these objectives can be realised. In two-port analysis in Chapter 5 and state-space analysis in Chapter 7 more convenient solution variables are chosen to meet the same objectives.

## FURTHER READING

Desoer, C. A. and Kuh, E. S., (1969), *Basic Circuit Theory*, McGraw-Hill, New York, Chapters 10 and 11.

Tropper, A. M., (1962), *Matrix Theory for Electrical Engineering Students*, Harrap, London.

## PROBLEMS

2.1 Draw a directed graph of a network with the reduced incidence matrix

$$\mathbf{A} = \begin{bmatrix} 1 & 1 & 0 & 0 & 0 & -1 & 0 & 0 \\ 0 & 0 & 1 & 0 & 0 & 0 & 1 & -1 \\ 0 & 0 & 0 & -1 & -1 & 1 & 0 & 1 \\ 0 & -1 & -1 & 1 & 0 & 0 & 0 & 0 \end{bmatrix}^t .$$

Determine the cut-set and tie-set matrices from the graph. Prove and verify the following relationships between the sub-matrices $\mathbf{A}_T$, $\mathbf{A}_L$, $\mathbf{C}_T$, and $\mathbf{D}_L$.

$$\mathbf{C}_T = -\mathbf{D}_L^t = -\ [\mathbf{A}_T^t]^{-1}\,\mathbf{A}_L^t\ .$$

2.2 Draw directed graphs for each circuit in Fig. 1.22. In each case indicate a tree and draw a cut-set and a tie-set. Fill the circuit matrices **A**, **C**, **D**, **Y**, **Z**, **I**, and **E** and solve for all network voltages and currents, using node or tie-set analysis. Verify KCL and KVL for the networks, using the solutions obtained.

2.3 Assemble the complete set of KVL and KCL equations for the circuit in Fig. 2.5. Select one tree and extract the KVL equations for the corresponding tie-set. Show that all KVL equations can be expressed as linear combinations of these equations for the basic loops. Extract the KCL equations for the corresponding cut-set, and show that all KCL equations can be similarly expressed.

2.4 Assign arbitrary branch voltages and currents to the network in Fig. 2.5, and verify Tellegen's theorem.

2.5 Determine the dual of the standard resistive circuit branch from Fig. 2.7, and transform it back into the same form as the original, using the circuit theorems. What are the new component and source values?

2.6 Assemble the full set of matrix equations required for network analysis with branch-dependent VCCS.

2.7 Determine node voltages and voltage gain for each circuit in Fig. 2.21.

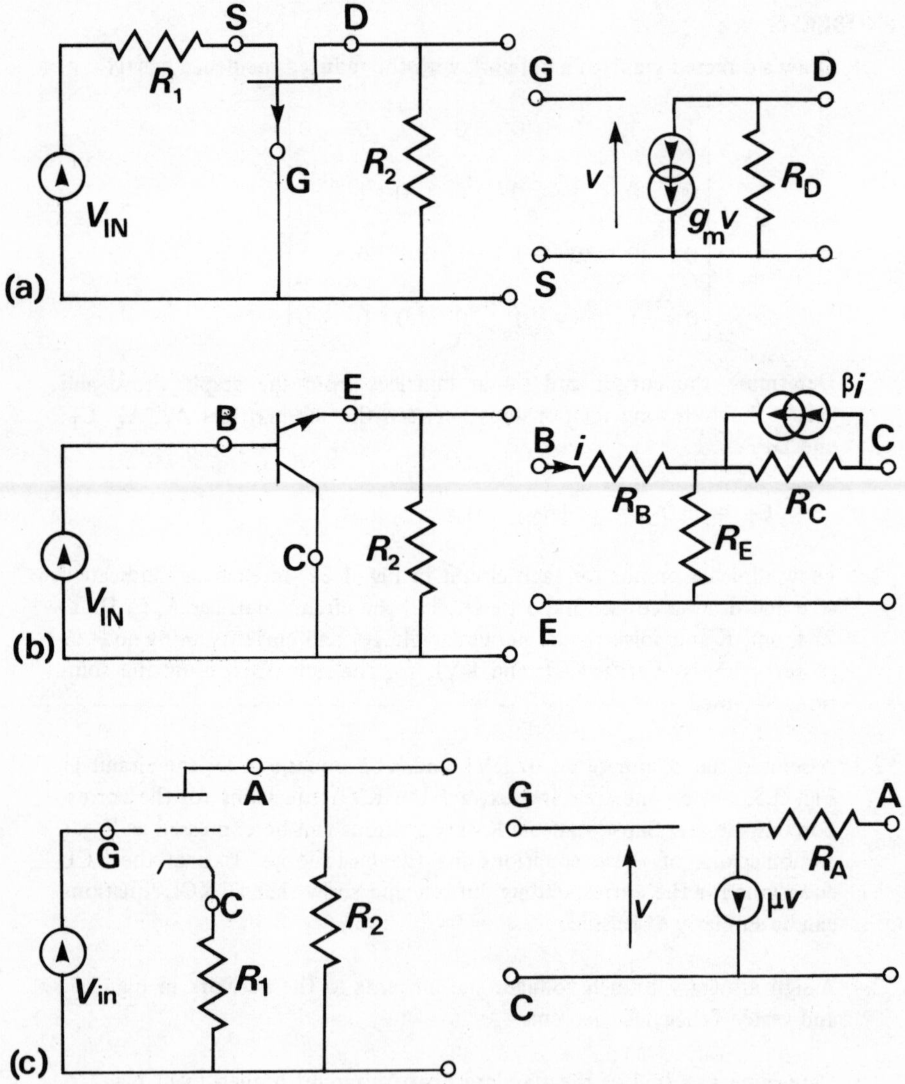


Fig. 2.21 – Problem 2.7.

2.8 Solve for branch currents, by mesh analysis of the network in Fig. 1.17.

2.9 Determine the elements of the **Y** and **Z** matrices for the network in Fig. 1.21 with $M = 0.75$ and $M = 1$.

2.10 Show that the network in Fig. 2.22 is a negative-impedance inverter, and that two of them connected in cascade are equivalent to an ideal transformer with unity turns ratio.

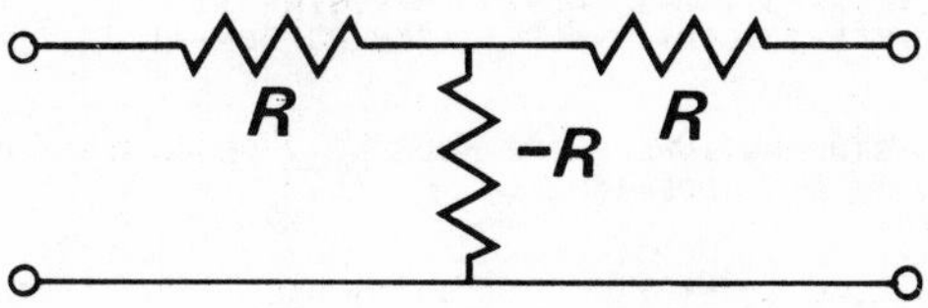


Fig. 2.22 – Problems 2.10 and 2.12.

2.11 Show that both networks in Fig. 2.23 are equivalent to negative-impedance converters. What is the difference between the two circuit configurations?

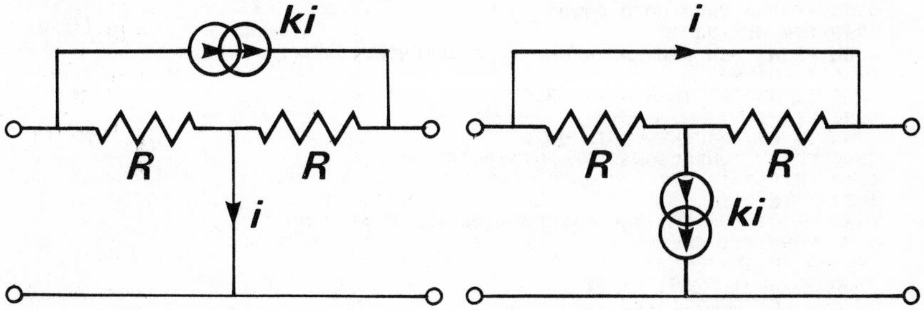


Fig. 2.23 – Problem 2.11.

2.12 Show that a gyrator is obtained if a negative-impedance converter is cascaded with a negative-impedance inverter.

*Computing*

2.13 Utilise program RNODE by solving for all network voltages and currents in familiar networks, for example those in Fig.s 1.15, 1.17, 1.22, 2.5, 2.17.

2.14 Replace the BASIC statements involving matrix inverse by writing an inversion subroutine which does not fail because of a zero on the diagonal. Error messages due to zero determinant can also be handled more conveniently with return to the NEXT command. This can be regarded as the first step in converting RNODE to run on computers with 8K BASIC.

2.15 Write a program to find a network tree. This is one of the operations required for tie-set or cut-set analysis.

## APPENDIX

```
0010 PRINT 'RNODE    DC CIRCUIT ANALYSIS'
0020 PRINT 'NO. OF NODES, NO.OF RESISTORS, NO.OF DEPENDENT SOURCES';
0030 INPUT N,B,S
0040 IF S>B THEN PRINT 'SENSITIVITY TO FIRST ';B;' GM ONLY'
0050 DIM Y[B,B],E[B,1],I[B,1],A[B,N],M$[8],N$[3]
0060 DIM J[B,1],K[B,1],L[N,1],M[B,1],D[N,B],Z[N,B]
0070 LET C8=1
0080 LET N4=1
0090 PRINT 'RESISTOR NO.,VALUE,+VE NODE NO.,-VE NODE NO.,';
0100 PRINT 'SOURCE E,SOURCE I'
0110 FOR I1=1 TO B
0120   PRINT 'R';
0130   INPUT B1,R,N1,N2,E1,I2
0140   LET E[B1,1]=E1
0150   LET I[B1,1]=I2
0160   LET Y[B1,B1]=1/R
0170   IF N1=0 THEN GOTO 0190
0180   LET A[B1,N1]=1
0190   IF N2=0 THEN GOTO 0210
0200   LET A[B1,N2]=-1
0210 NEXT I1
0220 LET I9=1
0230 IF S=0 THEN GOTO 0410
0240 DIM G[S,2]
0250 PRINT 'GM BRANCH OR RESISTOR DEPENDENT: INPUT B OR R';
0260 INPUT N$
0270 IF N$='B' THEN GOTO 0330
0280 LET I9=2
0290 IF N$='R' THEN GOTO 0330
0300 PRINT 'UNRECOGNISED - RETYPE';
0310 LET I9=1
0320 GOTO 0260
0330 PRINT 'GM NO.,VALUE,SEND R NO.,RECEIVE R NO.'
0340 FOR I1=1 TO S
0350   PRINT 'GM';
0360   INPUT S9,G1,B1,B2
0370   LET G[S9,1]=B2
0380   LET G[S9,2]=B1
0390   LET Y[B2,B1]=G1
0400 NEXT I1
0410 ON I9 THEN GOTO 0420, 0460
0420 FOR I1=1 TO B
0430   LET J[I1,1]=I[I1,1]-Y[I1,I1]*E[I1,1]
0440 NEXT I1
0450 GOTO 0480
0460 MAT J=Y*E
0470 MAT J=I-J
0480 MAT D=TRN(A)
0490 MAT K=D*J
0500 MAT Z=D*Y
0510 MAT D=Z*A
0520 MAT Z=INV(D)
0530 MAT L=Z*K
0540 MAT K=A*L
0550 MAT M=Y*K
0560 MAT M=M-J
0570 MAT J=K+E
0580 IF N4=1 THEN GOTO 0650
0590 PRINT 1+N5-N4,
```

*

```
0600 FOR I1=1 TO N
0610   PRINT L[I1,1],
0620 NEXT I1
0630 PRINT
0640 GOTO 0940
0650 PRINT "NODE","VOLTAGE"
0660 FOR I1=1 TO N
0670   PRINT I1,L[I1,1]
0680 NEXT I1
0690 PRINT "BRANCH OUTPUT YES OR NO";
0700 INPUT N$
0710 IF N$="NO" THEN GOTO 0770
0720 PRINT " ","RESISTOR","RESISTOR","BRANCH","BRANCH"
0730 PRINT "R","VOLTAGE","CURRENT","VOLTAGE","CURRENT"
0740 FOR I1=1 TO B
0750   PRINT I1,J[I1,1],J[I1,1]*Y[I1,I1],K[I1,1],M[I1,1]
0760 NEXT I1
0770 IF C8>1 THEN GOTO 0820
0780 LET M$="ERMCNLSE"
0790 PRINT "NEXT COMMAND: END RUN      ER, MODIFY CIRCUIT       MC"
0800 PRINT "              : SENSITIVITY SE, NON-LINEAR ITERATION NL"
0810 LET C8=2
0820 PRINT "NEXT";
0830 INPUT N$
0840 FOR I1=1 TO 4
0850   IF N$=M$[2*I1-1,2*I1] THEN GOTO 0890
0860 NEXT I1
0870 PRINT "UNRECOGNISED COMMAND - RETYPE";
0880 GOTO 0830
0890 ON I1 THEN GOTO 1160, 0970, 0900, 1170
0900 PRINT "NO. ITERATIONS";
0910 INPUT N5
0920 LET N4=N5+1
0930 PRINT "ITERATION","NODE VOLTAGES 1 TO ";N
0940 GOSUB 5000
0950 LET N4=N4-1
0960 GOTO 0410
0970 PRINT "NUMBER OF ALTERED BRANCHES, AND DEPENDENT SOURCES";
0980 INPUT B9,S9
0990 IF B9=0 THEN GOTO 1080
1000 PRINT "R NO.,NEW VALUE,NEW E SOURCE,NEW I SOURCE"
1010 FOR I1=1 TO B9
1020   PRINT "R";
1030   INPUT B1,R,E1,I2
1040   LET E[B1,1]=E1
1050   LET I[B1,1]=I2
1060   LET Y[B1,B1]=1/R
1070 NEXT I1
1080 IF S9=0 THEN GOTO 0410
1090 PRINT "GM NO., NEW VALUE"
1100 FOR I1=1 TO S9
1110   PRINT "GM";
1120   INPUT G1,E1
1130   LET Y[G[G1,1],G[G1,2]]=E1
1140 NEXT I1
1150 GOTO 0410
1160 STOP
1170 GOSUB 2000
1180 GOTO 0820

*
```

Fig. 2.24 – Program RNODE.

CHAPTER 3

# Non-Linear D.C. Analysis

We noted in Chapter 1 that all circuit elements are non-linear under some conditions of operation. However, most circuits are designed to operate within regions where non-linearity is controlled, so that a linear approximation is almost valid. Such circuits can therefore be analysed by linear methods, assuming that signals are sufficiently small in comparison with the limits of non-linearity. The matrix methods of analysis in Chapter 2 are limited to lumped circuit elements by Kirchhoff's laws, and more strictly to linear elements by the use of Ohm's law and by the definition of idealised circuit elements.

In this chapter the nodal method for d.c. analysis will be extended to include non-linear resistive circuit elements. A solution will then be sought for the d.c. node voltages when the circuit is biased to the steady-state d.c. operating point. The analysis is large-signal and non-linear, since the fixed sources in this case are the power supply voltages, and the resulting bias voltages are large in comparison to the regions of linear operation of most active devices in use.

The dynamic or transient behaviour of a circuit in reaching ist steady-state point or points (for there may be more than one) is another important area of analysis which will be discussed briefly in Chapter 7. Non-linear d.c. analysis can be regarded as a by-product of non-linear transient analysis at infinite time. Alternatively, it may be regarded as necessary preliminary analysis giveing the final value of transients. Several solutions are often obtained, indicating the existence of various wanted and unwanted stable d.c. states of a network. Unwanted d.c. states may cause a circuit to *latch* into a condition from which it cannot be released, and correct operation of the circuit is inhibited. The identification of latch conditions is one further important application of non-linear analysis.

## 3.1 CIRCUITS WITH ONE NON-LINEAR RESISTOR

### Newton's iteration in one dimension

Newton's method for finding the root of a non-linear equation is based on the Taylor expansion of a function $f(x)$ about a point $x_0$, given by

$$f(x_0 + \Delta x_0) = f(x_0) + \frac{\mathrm{d}f}{\mathrm{d}x}\Delta x_0 + \frac{1}{2!}\frac{\mathrm{d}^2 f}{\mathrm{d}x^2}\Delta x_0^2 + \ldots,$$

where all derivatives are evaluated at $x_0$. Assuming $\Delta x_0$ is small, the first-order approximation to the behaviour of the non-linear function is described by the linear equation

$$f(x_0 + \Delta x_0) = f(x_0) + \frac{\mathrm{d}f}{\mathrm{d}x}\Delta x_0 \ .$$

If the point $f(x_0 + \Delta x_0)$ is the required root, then

$$f(x_0 + \Delta x_0) = 0 = f(x_0) + \frac{\mathrm{d}f}{\mathrm{d}x} \Delta x_0$$

and
$$\Delta x_0 = -f(x_0) \Big/ \frac{\mathrm{d}f}{\mathrm{d}x} \ . \tag{3.1}$$

The Newton iteration is based on this expression.

Since the non-linear function has been approximated by a linear function, expression (3.1) is not exact. Let the new point generated be $x_1 = x_0 + \Delta x_0$, then

$$x_1 = x_0 - f(x_0) \Big/ \left(\frac{\mathrm{d}f}{\mathrm{d}x}\right)_{x\,=\,x_0} \ . \tag{3.2}$$

At each new point generated by Eq. (3.2) both the function and its gradient are computed, and the process is repeated iteratively. At the $n^{\text{th}}$ point, the $(n+1)^{\text{th}}$ point is generated by the general expression

$$x_{n+1} = x_n - f(x_n)/f'(x_n) \ . \tag{3.3}$$

In this section we will show how this simple iterative technique can be interpreted in terms of circuit linearisation, and can be used for non-linear circuit analysis.

**Non-linear resistor networks**

The matrix analysis methods developed in Chapter 2 are based on the standard two-terminal branch. Modifications to the admittance or impedance matrices are necessary for networks incoprporating three-terminal or four-terminal elements which couple energy between branches. Non-linear elements are therefore considered to be two-terminal. However, the methods described in this chapter can be extended to the more general case of $n$-terminal non-linearities.

A linear resistor obeys the Ohm's law relationship $v=Rj$, and the resistance $R$ is constant for all values of the voltage $v$ and current $j$. A non-linear resistor is defined by a more general functional relationship between voltage and current of the form $f(v,j)=0$. We shall have to limit discussion in an introductory text such as this to relationships of the forms

$$v=f(j) \text{ and } j=f(v) \ .$$

This is taken to imply that the graph of the characteristics of the non-linear resistor is single-valued in the dependent variable. Consider, for example, the tunnel diode characteristic shown in Fig. 3.1. For each value $v'$ of the independent variable $v$ there is a unique value of diode current $j'$. For a given value of $j'$, however, there are in some cases three possible values of diode voltage $v$. It is therfore necessary for the diode to exhibit the feature of memory so that the part of the characteristic on which it is operating is defined. The tunnel diode equation in the general form $j=f(v)$ falls within our definition of acceptable non-linear elements.

The normal diode characteristic, shown for comparison in Fig. 3.1, represents the most common form of non-linear resistor appearing in networks either directly as a diode or indirectly through transistor and FET equivalent circuits. Diode-resistor networks are therefore selected for illustration throughout this chapter. The methods developed are general and apply to most of the non-linear resistive elements commonly encountered. Devices such as thermistors, lamps, and various other types of diodes are treated by identical techniques to those given, with the substitution of the appropriate characteristic equation.

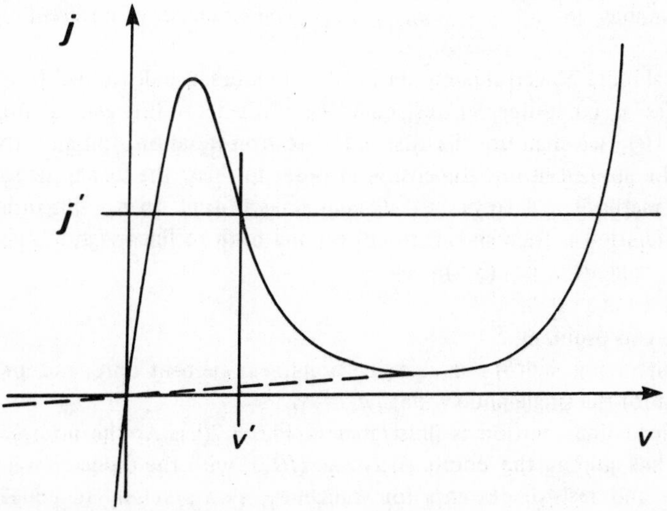


Fig. 3.1 – Tunnel Diode characteristic ———, Normal diode characteristic – – – –.

The predominant non-linear equations encountered in practical circuits are of the form $j=f(v)$ suggesting the use of the admittance matrix. This is adopted here, and the nodal method of analysis is generalized. While convenient, use of admittance matrix methods is not essential, and the following treatment can be applied, using any other analysis technique.

The solution for node voltages in Section 2.2 was given in Eq. (2.39) as

$$\mathbf{e}' = (\mathbf{A}^t\,\mathbf{Y}\mathbf{A})^{-1}\,\mathbf{A}^t\,(\mathbf{I} - \mathbf{Y}\mathbf{E}) \;, \tag{3.4}$$

where the elements of the admittance matrix **Y** in d.c. analysis are the conductances of the network resistors. In a non-linear circuit the values of conductance are dependent on the set of network voltages and current obtained from the solution (3.4). As we shall see later in this section, the fixed sources **I** or **E** also depend on the solution for $\mathbf{e}'$ through linearization of the non-linear resistors. Nevertheless, the solution (3.4) is still valid provided that the conductances and sources are simultaneously correct.

The mathematical problem can be formulated in general terms by rewriting Eq. (3.4) to show the functional dependence of **Y**, **I**, and **E** on the solution $\mathbf{e}'$. Removing the matrix inverse and factorising (3.4) yields

$$\mathbf{A}^t\,\mathbf{Y}(\mathbf{e}')\,[\mathbf{A}\mathbf{e}' + \mathbf{E}(\mathbf{e}')] \;-\; \mathbf{A}^t\,\mathbf{I}(\mathbf{e}') = 0 \tag{3.5}$$

or

$$\mathbf{f}(\mathbf{e}') = 0 \tag{3.6}$$

Equations (3.5) and (3.6) represent a set of non-linear simultaneous equations in the $n_N$ variables $[e_1'\; e_2' - - - e_{n_N}']$. In general a number of different solutions will exist.

Several quite powerful computational techniques, usually available as library subroutines at computer centres, could be utilised for this general non-linear problem. Here we shall use the first-order Newton iteration. Initially, the iteration will be applied in one dimension in order to relate the technique to simple graphical methods and to circuit element linearization. In $n_N$ dimensions the Newton iteration is then shown to correspond both to linearization and to the taking of gradients in Eq. (3.5).

**The diode bias problem**

The circuit in Fig. 3.2(a) contains one non-linear element only, and a solution is required for the single node voltage $e_1' = v_D$.

The load line solution is illustrated in Fig. 3.2(b). At the intersection of the load line joining the points $(0,I)$ and $(IR,0)$ with the diode characteristic, the diode and resistor currents for voltage $v_D$ sum precisely to equal $I$. The intersection defines the point on the diode characteristic where KCL at the circuit node is satisfied.

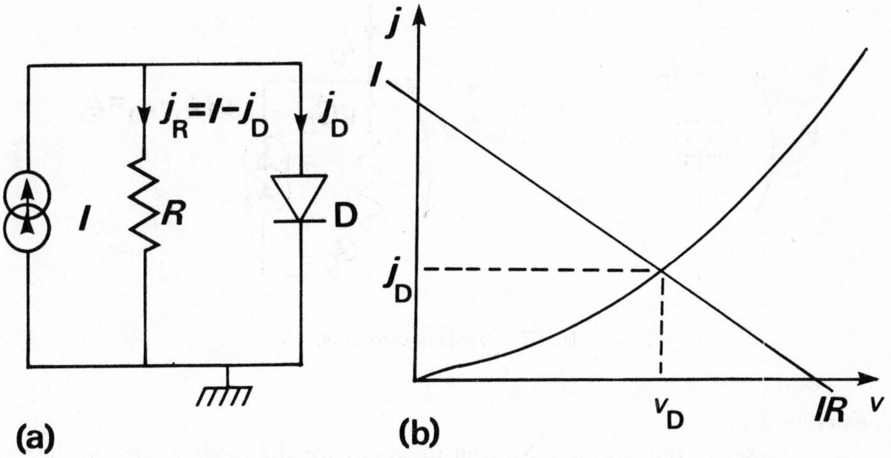


Fig. 3.2(a) – single diode circuit; (b) load line solution.

The diode characteristic at room temperature is given by

$$j = I_s\,(e^{40v} - 1)$$

and the load line by

$$j = I - v/R\ .$$

At the point of intersection therefore,

$$v/R - I + I_s(e^{40v} - 1) = 0\ . \tag{3.7}$$

Equation (3.7) corresponds to the general form of Eq. (3.6) and must be solved to find $v_D$.

The same equation can be derived from Eq. (3.6). Substitution yields

$$[1 \quad 1] \begin{bmatrix} 1/R & 0 \\ 0 & G_D \end{bmatrix} \begin{bmatrix} 1 \\ 1 \end{bmatrix} [v] - [1 \quad 1] \begin{bmatrix} I \\ I_D \end{bmatrix} = 0$$

where $G_D$ and $I_D$ are the conductance and current source from the diode linearisation shown in Fig. 3.3. Note also that the node voltage $e_1'$ equals the element voltage $v$. Expanding the matrices yields Eq. (3.7) as expected.

The one-dimensional Newton iteration from Eq. (3.3) may be applied directly to Eq (3.7) in order to find the node voltage $v_D$. The progress of the iteration is best followed in a numerical example.

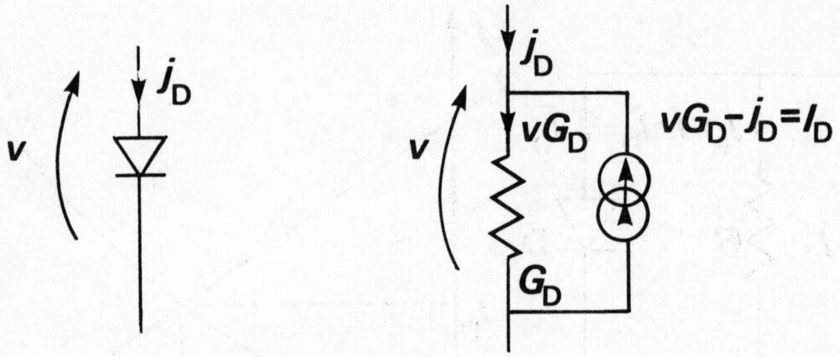


Fig. 3.3 – Diode linearization.

*Example* 3.1

Compute the first few Newton iterations for the diode circuit in Fig. 3.2(a) assuming $I = 10^{-2}$ A, $I_s = 10^{-11}$ A, and $R = 1$ kΩ.

From Eq. (3.7) neglecting $I_s << I$,

$$f(v) = 10^{-11}\, e^{40v} + 10^{-3}v - 10^{-2} .$$

Differentiating with respect to $v$,

$$f'(v) = 4 \times 10^{-10} e^{40v} + 10^{-3} .$$

$$\text{Therefore, } f(v)/f'(v) = \frac{10^{-8}e^{40v} + v - 10}{40 \times 10^{-8}e^{40v} + 1} .$$

The progress of the iteration is now easily tabulated, assuming a first approximation $v_0 = 0.5$.

| $n$ | $v_n$ | $f(v_n)/f'(v_n)$ | $v_n + 1$ |
|---|---|---|---|
| 0 | 0.5 | –0.0238 | 0.5238 |
| 1 | 0.5238 | $0.6141 \times 10^{-2}$ | 0.5177 |
| 2 | 0.5177 | $0.9272 \times 10^{-3}$ | 0.5168 |
| 3 | 0.5168 | $0.4478 \times 10^{-4}$ | 0.5167 |

In order to show that the iteration can become slow and involve a large overshoot of the required answer, assume $v_0$=0.4. The first two iterations are tabulated below.

| $n$ | $v_n$ | $f(v_n)/f'(v_n)$ | $v_n + 1$ |
|---|---|---|---|
| 0 | 0.4 | −2.088 | 2.488 |
| 1 | 2.488 | 0.025 | 2.463 |
| 2 | 2.463 | 0.025 | 2.438 . |

For all $v_n$ >0.55 the value of $f(v)/f'(v)$ approaches $1/40 = 0.025$ since $e^{40v} \gg 10$. Further iteration down from $v_2 = 2.438$ will proceed in steps of about 0.025 until $v_n$ regains a value less than 0.55.

### Geometrical interpretation

The first order Newton iteration for finding the zero of a non-linear one-dimensional function has a simple geometric interpretation. Rewriting Eq. (3.3),

$$f'(x_n)\,[x_n - x_{n+1}] = f(x_n) \ . \tag{3.8}$$

Fig. 3.4 shows the first few iterations of a solution similar to that of the

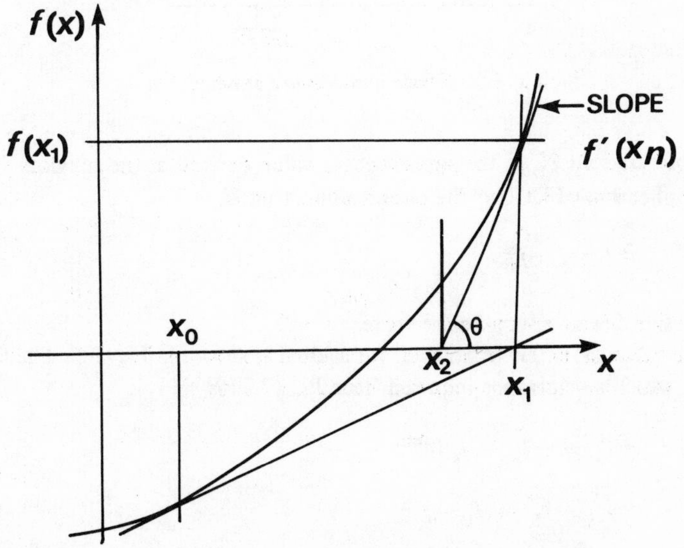


Fig. 3.4 – Geometric interpretation of Newton iteration.

diode bias network. At the point $x_1$, the tangent to the curve has slope $f'(x_1)$. If the tangent intersects the $x$-axis at $x_2$ then

$$(x_1-x_2)\tan\theta = (x_1-x_2)f'(x_1) = f(x_1)$$

and Eq. (3.8) is verified.

The successive values $x_1, x_2, x_3 \ldots$ are the $x$-axis intercepts of the tangent to the function at $x_0$, $x_1$, $x_2$ ... respectively. The iteration is clearly terminated when $|x_{n+1} - x_n|$ is sufficiently small.

**Circuit interpretation**

The diode bias network in Fig. 3.2(a) is shown in Fig. 3.5 at iteration $n$. The superscript $n$, as for example in $j_D^n$ identifies the value of variables at iteration $n$, and $G_D^n$ is the diode conductance.

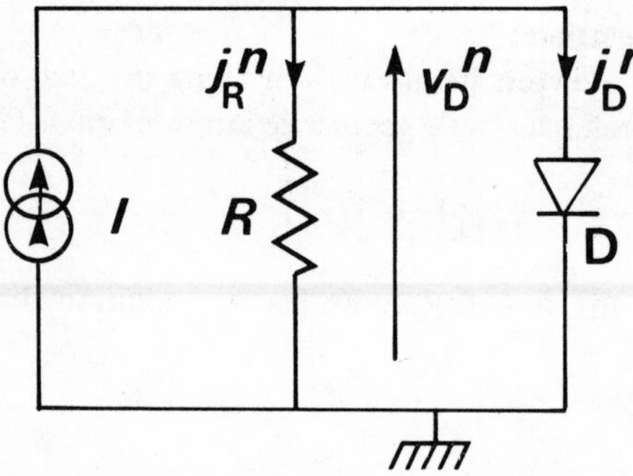


Fig. 3.5 – Diode bias network at iteration $n$.

The diode current $\bar{j}_D^n$ is the approximate value derived at the previous iteration from application of KCL at the circuit node, that is,

$$\bar{j}_D^n = I - j_R^n \ . \qquad (3.9)$$

The bar signifies an approximate value.

The graphical interpretation of iteration $n$ is shown in Fig. 3.6. The function $f(v)$ for which a solution is required from Eq. (3.7) is

$$f(v) = v/R + I_s\,(e^{40v}-1) - I \qquad (3.10)$$

$$= j_R + j_D - I \ . \qquad (3.11)$$

The vertical axis of the graph in Fig. 3.6 therefore corresponds to the error between the actual diode current $j_D$ and the approximate value $\bar{j}_D$Differentiation of Eq. (3.10) yields

$$f'(v) = 1/R + 40I_s e^{40v} \qquad (3.12)$$

$$= 1/R + G_D \; . \qquad (3.13)$$

The slope of the tangent to the function $f(v)$ is therefore equal to the combined conductance of the diode and resistor in parallel.

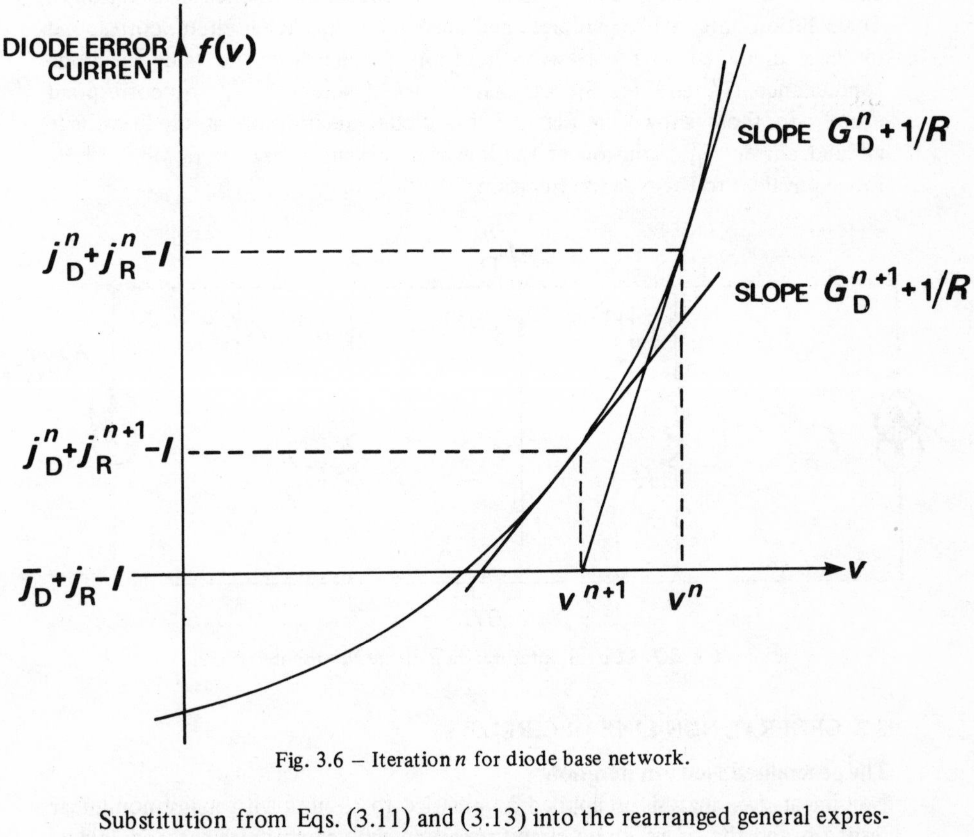


Fig. 3.6 – Iteration $n$ for diode base network.

Substitution from Eqs. (3.11) and (3.13) into the rearranged general expression for the Newton iteration in Eq. (3.8) yields

$$(v^n - v^{n+1})(G_D^n + 1/R) = j_D^n + j_R^n - I \; .$$

Expanding the brackets,

$$v^n G_D^n - v^{n+1} G_D^n = j_D^n + j_R^n - I + v^{n+1}/R - v^n/R \; .$$

Now, $v^n/R = j_R^n$,

$$v^{n+1}/R = j_R^{n+1} = I - \bar{j}_D^{n+1} ,$$

therefore, $v^n G_D^n - v^{n+1} G_D^n = j_D^n - \bar{j}_D^{n+1}$

$$\text{or,} \qquad \bar{j}_D^{n+1} = j_D^n - v^n G_D^n + v^{n+1} G_D^n . \qquad (3.14)$$

The circuit model corresponding to Eq. (3.14) is shown in Fig. 3.7. It is clear that all sources and components of the circuit are defined at iteration $n$. The solution obtained for voltages and currents in this linear circuit correspond to those predicted by the Newton iteration. Furthermore the value for diode conductance $G_D^n$ and for the associated current source $v^n G_D^n - j_D^n$ correspond exactly to those shown in Fig. 3.3 for diode linearisation at diode voltage $v_D^n$ and current $j_D^n$. Solution of the linearised circuit at each iteration is therefore equivalent to the Newton iteration.

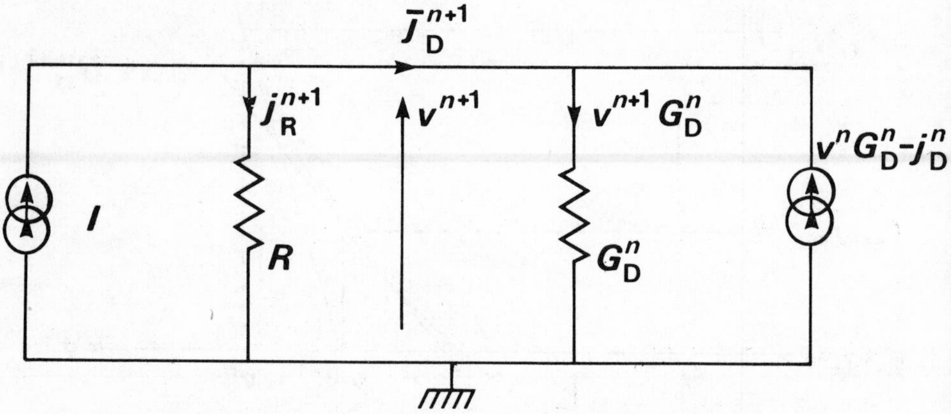


Fig. 3.7 – Circuit interpretation of the Newton iteration.

## 3.2 GENERAL NON-LINEAR CIRCUITS

**The generalised Newton iteration**

Non-linear d.c. analysis is applied in practice to circuits with many non-linear resistors, and the single diode circuit was used only to establish the principle of the Newton iteration and its relationship to linearisation. The method must now be generalised to $n_N$-dimensions.

The Taylor expansion of a function of the $n_N$ variables $\mathbf{x} = [x_1, x_2 \ldots x_{n_N}]^t$ is given by

$$f(\mathbf{x} + \Delta\mathbf{x}) = f(\mathbf{x}) + \mathbf{g}\,\Delta\mathbf{x} + \frac{1}{2}\Delta\mathbf{x}^t\,\mathbf{H}\,\Delta\mathbf{x}$$

where $\Delta \mathbf{x} = [\Delta x_1 \, \Delta x_2 \ldots \Delta x_{n_N}]^t$ ,

$$\mathbf{g} = \begin{bmatrix} \frac{\partial f(x)}{\partial x_1} & \cdots & \frac{\partial f(x)}{\partial x_{n_N}} \end{bmatrix},$$

and

$$\mathbf{H} = \begin{bmatrix} \frac{\partial^2 f(x)}{\partial x_1^2} & \cdots & \frac{\partial^2 f(x)}{\partial x_1 \, \partial x_{n_N}} \\ , & & , \\ , & & , \\ , & & , \\ \frac{\partial^2 f(x)}{\partial x_{n_N} \, \partial x_1} & & \frac{\partial^2 f(x)}{\partial x_{n_N}^2} \end{bmatrix}.$$

**g** is known as the gradient vector and **H** is known as the Hessian matrix. The $n_N$ functions which are to be simultaneously set to zero are defined in the Newton iteration by

$$f_1(\mathbf{x} + \Delta\mathbf{x}) = f_1(\mathbf{x}) + \mathbf{g}_1 \Delta\mathbf{x}$$

$$f_2(\mathbf{x} + \Delta\mathbf{x}) = f_2(\mathbf{x}) + \mathbf{g}_2 \Delta\mathbf{x}$$

$$,$$

$$,$$

$$,$$

$$f_{n_N}(\mathbf{x} + \Delta\mathbf{x}) = f_{n_N}(\mathbf{x}) + g_{n_N} \Delta x \ .$$

Gradient terms involving derivatives of second and higher orders have been neglected. Solving simultaneously for $f_1(\mathbf{x} + \Delta\mathbf{x}) = f_2(\mathbf{x} + \Delta\mathbf{x}) = \ldots = f_{n_N}(\mathbf{x} + \Delta\mathbf{x}) = 0$, and expanding the gradient vectors $\mathbf{g}_1, \ldots \mathbf{g}_{n_N}$ for clarity.

$$\frac{\partial f_1(x)}{\partial x_1} \Delta x_1 + \frac{\partial f_1(x) \Delta x_2}{\partial x_2} \cdots \frac{\partial f_1(x) \Delta x_{n_N}}{\partial x_{n_N}} = -f_1(x);$$

$$\frac{\partial f_2(x)}{\partial x_1} \Delta x_1 \ \ldots\ldots \qquad = -f_2(x)$$

$$,$$

$$,$$

$$,$$

$$\frac{\partial f_{n_N}(x)}{\partial x_1} \Delta x_1 \ \ldots. \qquad \frac{\partial f_{n_N}(x)}{\partial x_{n_N}} \Delta x_{n_N} = -f_{n_N}(x)$$

In matrix form this may be written;

$$\mathbf{G}\,\Delta\mathbf{x} = -\mathbf{f} \tag{3.15}$$

where $\mathbf{G} = [\mathbf{g}_1\mathbf{g}_2 \ldots \mathbf{g}_{n_N}]^t$.

and $\mathbf{f} = [f_1(\mathbf{x})\ f_2(\mathbf{x})\ \ldots\ f_{n_N}(\mathbf{x})]^t$ .

Eq. (3.15) is the generalised form of the Newton iteration.

## Analysis by Newton's iteration

In a circuit with $n_N$ nodes the functions to be reduced to zero are those defined by Eq. (3.5). Expansion of the matrices in that equation results in the required column vector $\mathbf{f}$ when the variables $\mathbf{x}$ in the mathematical derivation of the Newton iteration correspond to the node voltages $\mathbf{e}' = [e_1'\ e_2' \ldots e_{n_N}]^t$. From Eq. (3.5),

$$\begin{bmatrix} f_1(\mathbf{e}') \\ f_2(\mathbf{e}') \\ \cdot \\ \cdot \\ f_{n_N}(\mathbf{e}') \end{bmatrix} = \mathbf{A}^t\ \mathbf{Y}(\mathbf{e}')\ [\mathbf{A}\mathbf{e}' + \mathbf{E}] - \mathbf{A}^t\mathbf{I}\ . \tag{3.16}$$

The functions $\mathbf{E}(\mathbf{e}')$ and $\mathbf{I}(\mathbf{e}')$ have been made constants equal to the fixed sources in the circuit on the basis of the following assumed form of non-linear resistor specification. The characteristic equation $j = f(v)$ is in terms of circuit element voltage and currents. In order to incorporate the expression into Eq. (3.16) we must recall from Chapter 2 that

$$\mathbf{j} = \mathbf{I} + \mathbf{i} = \mathbf{Y}(\mathbf{e} + \mathbf{E}) = \mathbf{Y}(\mathbf{A}\mathbf{e}' + \mathbf{E})\ .$$

The terms in $\mathbf{Y}(\mathbf{e}')\ [\mathbf{A}\mathbf{e}' + \mathbf{E}]$ relating to non-linear resistors must therefore be interpreted simply as the relevant terms of $\mathbf{j}$. Differentiation with respect to $\mathbf{v} = \mathbf{e} + \mathbf{E}$ will then result in the linearised conductance as expected. Since the non-linear resistor currents are incorporated exactly by this interpretation, the source vectors $\mathbf{E}$ and $\mathbf{I}$ do not contain correction terms arising from linearisation.

Differentiation of Eq. (3.16) with respect to the elements of $\mathbf{e}'$ to obtain the partial differential coefficient in $\mathbf{G}$ for the Newton iteration must be carried out element by element while incorporating the characteristic equations of the non-linear resistors. This will be done in Example 3.2 which follows this section.

Here we will abbreviate the treatment by defining the following matrices,

$$\mathbf{dv} = [\mathrm{d}v_1 \ldots \mathrm{d}v_{n_\mathrm{B}}]^\mathrm{t}$$

$$\mathbf{de} = [\mathrm{d}e_1 \ldots \mathrm{d}e_{n_\mathrm{B}}]^\mathrm{t}$$

$$\mathbf{de}' = [\mathrm{d}e_1' \ldots \mathrm{d}e_{n_\mathrm{N}}']^\mathrm{t}$$

The proper mathematical interpretation of these matrices is in terms of finite increments in all the variables, for example $\delta v_1$ and $\delta e_1$. Then $\mathrm{d}v_1/\mathrm{d}e_1$ equals the $\lim_{\delta e_1 \to 0} (\delta v_1/\delta e_1)$.

Since $\mathbf{v} = \mathbf{e} + \mathbf{E} = \mathbf{A}\mathbf{e}' + \mathbf{E}$,

$$\begin{bmatrix} v_1 \\ \\ v_2 \\ , \\ , \\ v_{n_\mathrm{B}} \end{bmatrix} = \mathbf{A} \begin{bmatrix} e_1' \\ , \\ , \\ e_{n_\mathrm{N}} \end{bmatrix} + \begin{bmatrix} E_1 \\ , \\ , \\ E_{n_\mathrm{N}} \end{bmatrix} .$$

Matrix $\mathbf{A}$ consists of elements which may only equal +1, −1, and 0. $\mathbf{A}$ is therefore given by

$$\mathbf{A} = \frac{\mathbf{dv}}{\mathbf{de}'} = \begin{bmatrix} \dfrac{\mathrm{d}v_1}{\mathrm{d}e_1'} & \dfrac{\mathrm{d}v_1}{\mathrm{d}e_2'} & \cdots & \dfrac{\mathrm{d}v_1}{\mathrm{d}e_{n_\mathrm{N}}'} \\ , & & & \\ , & & & \\ , & & & \\ \dfrac{\mathrm{d}v_{n_\mathrm{B}}}{\mathrm{d}e_1'} & \cdots\cdots & & \dfrac{\mathrm{d}v_{n_\mathrm{B}}}{\mathrm{d}e_{n_\mathrm{N}}'} \end{bmatrix}$$

This is easily seen if all the differentiations are carried out individually. We can now differentiate Eq. (3.16) simultaneously for all variables in $\mathbf{f}$ with respect to all node voltages $\mathbf{e}'$ in order to obtain the elements of $\mathbf{G}$. Therefore,

$$\mathbf{G} = \frac{\mathrm{d}}{\mathbf{de}'}\left\{\mathbf{A}^\mathrm{t}\mathbf{Y}[\mathbf{A}\mathbf{e}' + \mathbf{E}]\right\}$$

$$= \mathbf{A}^\mathrm{t} \frac{\mathrm{d}}{\mathbf{dv}}\left\{\mathbf{Y}\mathbf{v}\right\}\frac{\mathbf{dv}}{\mathbf{de}'}$$

$$= \mathbf{A}^{t} \frac{\mathrm{d}}{\mathbf{dv}} \left\{ \mathbf{Yv} \right\} \mathbf{A}$$

$$= \mathbf{A}^{t}\overline{\mathbf{Y}}\mathbf{A} \ , \tag{3.17}$$

where $\overline{\mathbf{Y}}$ is the linearised admittance matrix. The elements of $\mathbf{Yv}$ relating to non-linear resistors are interpreted as currents of the form $j_i = f(v_i)$ for the element i. When differentiated with respect to $v_i$ the linearised conductance is obtained. For linear resistors the conductance is, of course, obtained directly. The elements of the resulting matrix $\overline{\mathbf{Y}}$ are therefore the linearised set of conductance.

The matrix differentiation implied by

$$\frac{\mathrm{d}}{\mathbf{dv}} \left\{ \mathbf{Yv} \right\}, \text{ where } \mathbf{Yv} \text{ is a column vector,}$$

is given in full by

$$\left[ \frac{\mathrm{d}}{\mathrm{d}v_1} \begin{bmatrix} Y_1 v_1 \\ Y_2 v_2 \\ , \\ , \\ , \\ Y_{n_\mathrm{B}} v_{n_\mathrm{B}} \end{bmatrix} \quad \frac{\mathrm{d}}{\mathrm{d}v_2} \begin{bmatrix} Y_1 v_1 \\ Y_2 v_2 \\ , \\ , \\ , \\ Y_{n_\mathrm{B}} v_{n_\mathrm{B}} \end{bmatrix} \quad \cdots \quad \frac{\mathrm{d}}{\mathrm{d}v_{n_\mathrm{B}}} \begin{bmatrix} Y_1 v_1 \\ Y_2 v_2 \\ , \\ , \\ , \\ Y_{n_\mathrm{B}} v_{n_\mathrm{B}} \end{bmatrix} \right]$$

The matrix $\overline{\mathbf{Y}}$ obtained is therefore square and diagonal.

Substitution of the expression for $\mathbf{f}$, (3.16), and the gradient, (3.17), into the general expression for the Newton iteration (3.15) yields

$$(\mathbf{A}^{t}\overline{\mathbf{Y}}_n\mathbf{A})(\mathbf{e}'_{n+1} - \mathbf{e}'_n) = \mathbf{A}^{t}\mathbf{I} - \mathbf{A}^{t}\mathbf{Y}(\mathbf{A}\mathbf{e}' + \mathbf{E}) \ . \tag{3.18}$$

Solution of the linearised circuit at iteration $n$ was based on the equality $(\mathbf{A}^{t}\mathbf{Y}\mathbf{A})\mathbf{e}'_n = \mathbf{A}^{t}(\mathbf{I}-\overline{\mathbf{Y}}\mathbf{E})$. Also we must once again interpret $\mathbf{Y}(\mathbf{A}\mathbf{e}'+\mathbf{E})$ on the right-hand side of the equation as the circuit element current vector $\mathbf{j}_n$ which is evaluated exactly. Therefore,

$$(\mathbf{A}^{t}\overline{\mathbf{Y}}_n\mathbf{A})\mathbf{e}'_{n+1} = \mathbf{A}^{t}(\mathbf{I}-\overline{\mathbf{Y}}\mathbf{E}) + \mathbf{A}^{t}(\mathbf{I}-\mathbf{j}_n) \ , \tag{3.19}$$

where $\mathbf{j}_n$ is the vector of exact element currents computed from the characteristic equations for the non-linear resistors and from $\mathbf{Y}(\mathbf{A}\mathbf{e}'+\mathbf{E})$ for linear resistors.

Since $\quad \mathbf{j} = \mathbf{i} + \mathbf{I}$ ,

$$\mathbf{A}^t(\mathbf{I} - \mathbf{j}_n) = -\mathbf{A}^t\mathbf{i}_n \ . \qquad (3.20)$$

In order to demonstrate the equivalence of this expression to that of the current generators introduced during linearisation we must incorporate the branch currents from the linearised solution at iteration $n$. In a linear circuit $\mathbf{A}^t\mathbf{i}$ is an expression of KCL at the nodes. Let the branch current vector of the linearised solution be $\bar{\mathbf{i}}_n$ at iteration $n$, then $\mathbf{A}^t\bar{\mathbf{i}}_n = 0$. Adding this to the right-hand side of Eq. (3.20),

$$\mathbf{A}^t(\mathbf{I}-\mathbf{j}_n) = \mathbf{A}^t(\bar{\mathbf{i}}_n-\mathbf{i}_n)$$

Since $\quad \bar{\mathbf{i}}_n = \bar{\mathbf{j}}_n - \mathbf{I} = \bar{\mathbf{Y}}_n\mathbf{v}_n - \mathbf{I}$ ,

$$\mathbf{A}^t(\mathbf{I}-\mathbf{j}_n) = \mathbf{A}^t(\bar{\mathbf{Y}}_n\mathbf{v}_n - \mathbf{I} - \mathbf{i}_n)$$

$$= \mathbf{A}^t(\bar{\mathbf{Y}}_n\mathbf{v}_n - \mathbf{j}_n) \ .$$

Therefore, from (3.19),

$$\mathbf{e}'_{n+1} = (\mathbf{A}^t\bar{\mathbf{Y}}_n\mathbf{A})^{-1}\mathbf{A}^t(\mathbf{I}+\bar{\mathbf{I}}_n-\bar{\mathbf{Y}}_n\mathbf{E}) \qquad (3.21)$$

where $\quad \bar{\mathbf{I}}_n = (\bar{\mathbf{Y}}_n\mathbf{v}_n - \mathbf{j}_n)$ .

The elements of $\bar{\mathbf{I}}_n$ correspond to the current sources introduced during linearisation of non-linear resistors as shown in Fig. 3.3.

The solution for node voltages in Eq. (3.21) was derived from substitution into the generalised Newton iteration. Since Eq. (3.21) is also the solution for node voltages in the linearised circuit, direct equivalence of the Newton iteration to analysis of the linearised circuit has been demonstrated. It is therefore possible to iteratively analyse a non-linear circuit by component linearisation. The importance of this equivalence is that non-linear components may be modelled and linearised individually, and it is not necessary to explicitly formulate the non-linear equations for solution. Also, since the iterative solution for node voltages has been shown to be dependent only on linearisation, the actual method used for analysis is no longer relevant and any analysis technique can be used.

*Example* 3.2

Carry out the first iteration for solution of the circuit in Fig. 3.8 by Newton's iteration and by linearisation. Assume $I_s = 10^{-11}$ A for both diodes and $\mathbf{e}'_0 = [1 \quad \tfrac{1}{2}]^t$.

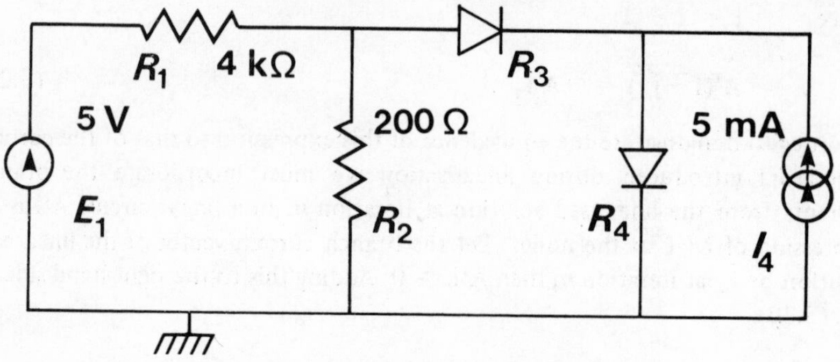


Fig. 3.8 – Circuit for Example 3.2.

In order to illustrate the differentiations for **f**′(e′) in Newton's iteration the circuit will be analysed in more detail than necessary for numerical solution. Exapnding all matrices in $\mathbf{f}(\mathbf{e}') = \mathbf{A}^t[\mathbf{Y}(\mathbf{A}\mathbf{e}'+\mathbf{E}) - \mathbf{I}]$ yields,

$$\begin{bmatrix} f_1(e') \\ f_2(e') \end{bmatrix} = \begin{bmatrix} -1 & 1 & 1 & 0 \\ 0 & 0 & -1 & 1 \end{bmatrix} \left[ \begin{bmatrix} Y_1 & & & 0 \\ & Y_2 & & \\ & & Y_3 & \\ 0 & & & Y_4 \end{bmatrix} \left[ \begin{bmatrix} -1 & 0 \\ 1 & 0 \\ 1 & -1 \\ 0 & 1 \end{bmatrix} \begin{bmatrix} e'_1 \\ e'_2 \end{bmatrix} + \begin{bmatrix} E_1 \\ 0 \\ 0 \\ 0 \end{bmatrix} \right] - \begin{bmatrix} 0 \\ 0 \\ 0 \\ I_4 \end{bmatrix} \right]$$

$$= \begin{bmatrix} Y_1(e'_1 - E_1) + Y_2 e'_1 + Y_3(e'_1 - e'_2) \\ Y_3(e'_2 - e'_1) + Y_4 e'_2 - I_4 \end{bmatrix}.$$

Directly differentiating,

$$\mathbf{G} = \begin{bmatrix} \dfrac{\partial f_1(\mathrm{e}')}{\partial e'_1} & \dfrac{\partial f_1(\mathrm{e}')}{\partial e'_2} \\[2ex] \dfrac{\partial f_2(\mathrm{e}')}{\partial e'_1} & \dfrac{\partial f_2(\mathrm{e}')}{\partial e'_2} \end{bmatrix} = \begin{bmatrix} Y_1 + Y_2 + \bar{Y}_3 & -\bar{Y}_3 \\ -\bar{Y}_3 & \bar{Y}_3 + \bar{Y}_4 \end{bmatrix},$$

where $\bar{Y}_3 = \mathrm{d}j_3/\mathrm{d}v_3$ and $\bar{Y}_4 = \mathrm{d}j_4/\mathrm{d}v_4$.

Alternatively we may differentiate with respect to **v** as was done in obtaining Eq. (3.17). In terms of **v** therefore,

$$\begin{bmatrix} f_1(\mathbf{e}') \\ f_2(\mathbf{e}') \end{bmatrix} = \begin{bmatrix} -1 & 1 & 1 & 0 \\ 0 & 0 & -1 & 1 \end{bmatrix} \begin{bmatrix} Y_1 v_1 \\ Y_2 v_2 \\ j_3 \\ j_4 - I_4 \end{bmatrix},$$

and

$$\mathbf{Y} = \frac{\mathrm{d}}{\mathrm{d}\mathbf{v}} \begin{bmatrix} Y_1 v_1 \\ Y_2 v_2 \\ j_3 \\ j_4 - I_4 \end{bmatrix} = \begin{bmatrix} Y_1 & & & 0 \\ & Y_2 & & \\ & & \bar{Y}_3 & \\ 0 & & & \bar{Y}_4 \end{bmatrix}.$$

Therefore $\mathbf{G} = \mathbf{A}^t \bar{\mathbf{Y}} \mathbf{A}$

$$= \begin{bmatrix} Y_1 + Y_2 + \bar{Y}_3 & -\bar{Y}_3 \\ -\bar{Y}_3 & \bar{Y}_3 + \bar{Y}_4 \end{bmatrix}.$$

The Newton iteration can now be applied numerically to the circuit in Fig. 3.8. The diode equation $j = 10^{-11}(e^{40v} - 1)$ must first be linearised at the estimated values $\mathbf{v}_0 = [\tfrac{1}{2}\ \ 1]^t$, that is, at $v = 0.5$ V in both cases,

$$j = 10^{-11}(e^{20} - 1) = 4.852 \text{ mA}$$

$$\bar{Y} = \mathrm{d}j/\mathrm{d}v = 40\,(j + 10^{-11}) = 0.194 \text{ S}$$

$$\bar{I} = \bar{Y}v - j = 92.2 \text{ mA}\ .$$

Therefore

$$\mathbf{G} = \begin{bmatrix} 0.19925 & -0.194 \\ -0.194 & 0.388 \end{bmatrix}$$

$$\mathbf{f} = \begin{bmatrix} 0.00873 \\ -0.005 \end{bmatrix}.$$

By direct application of the Newton iteration,

$$\mathbf{e}'_1 = -\mathbf{G}^{-1}\mathbf{f} + \mathbf{e}'_0$$

$$= -25.2 \begin{bmatrix} 0.388 & 0.194 \\ 0.194 & 0.19925 \end{bmatrix} \begin{bmatrix} 0.00873 \\ -0.005 \end{bmatrix} + \begin{bmatrix} 1.0 \\ 0.5 \end{bmatrix}$$

$$= \begin{bmatrix} 0.939 \\ 0.4825 \end{bmatrix} .$$

Solution by linearisation is shown in Fig. 3.9.

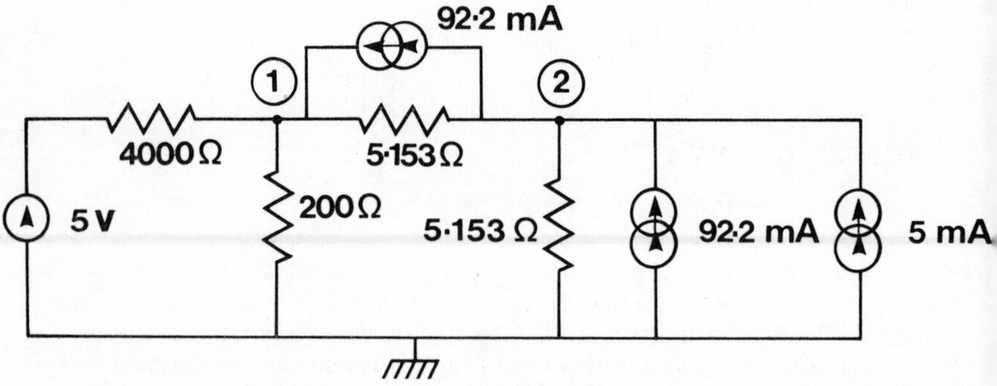


Fig. 3.9 – Linearised circuit for Example 3.2.

This circuit is easily solved by application of Thévénin's theorem to create a single loop of resistors and voltage sources. The node voltages obtained are of course the same as those found by Newton's iteration, that is, $[0.939\ 0.482]^t$.

**Non-linear d.c Nodal Analysis**

The solution of the linearised circuit was shown in the previous section to correspond to application of the generalised Newton iteration for determination of the zero of a function vector. The equivalence was demonstrated in terms of the matrix equation for nodal analysis while omitting any mention of transconductance terms in the admittance matrix. It is to be expected that linear transconductances do not affect the discussion since terms such as $g_m v$ arising in $\mathbf{Y}\ \mathbf{v}$ differentiate readily to $g_m$ for entry into $\bar{\mathbf{Y}}$. The generalisation of the analysis to incorporate non-linear transconductance has not been carried out here although updating the value of $g_m$ at each iteration is the obvious simple method.

The steps in the solution of non-linear d.c circuits by nodal analysis can be formulated as follows.

Step 1. Input the circuit description including the characteristic equations or other data defining the non-linear resistors together with initial estimates of the non-linear branch voltages.

2. Determine the exact currents **j** in the non-linear elements from their equations and the latest available set of terminal voltages **v**.
3. Linearise the non-linear resistors to obtain $\bar{\mathbf{Y}}$ and $\bar{\mathbf{I}}$.
4. Solve for node voltages from the matrix equation

$$\mathbf{e}' = (\mathbf{A}^t\,\bar{\mathbf{Y}}\,\mathbf{A})^{-1}\,\mathbf{A}^t\,(\mathbf{I} + \bar{\mathbf{I}} - \bar{\mathbf{Y}}\,\mathbf{E})$$

   and obtain **e**, **v**, $\bar{\mathbf{i}}$ and $\bar{\mathbf{j}}$.
5. Iteratively repeat Steps 2 to 4 until convergence is obtained, for example when the node voltages change by sufficiently small voltages.
6. Output the required circuit voltages and currents (and compute and output the sensitivity of the linearised circuit; see Chapter 8).

The BASIC computer programme RNODE is applied to non-linear analysis in Section 3.5. The nodal analysis used also illustrates the matrix methods developed in Chapter 2 as applied to d.c. circuits. The use of separate matrices **A** and **Y** is of course wasteful in terms of storage and computation time. Nevertheless their use is retained in this chapter for the purpose of simplified programming and direct comparison with matrices in the theoretical treatment. In practice the nodal admittance matrix $\mathbf{A}^t\mathbf{Y}\mathbf{A}$ is generated directly and then inverted by numerical techniques optimised to take advantage of the character of matrices obtained in circuit analysis. These modifications, made in the interest of efficiency, do not alter the Newton iteration. Neither do they overcome the practical problems arising from the use of iterative methods. Some of these problems and steps taken to overcome them are outlined in the next section.

## 3.3 COMPUTATIONAL DIFFICULTIES

### Scaling

It is good numerical analysis practice to invert matrices with element magnitudes scaled to a range centred on unity in order to avoid round-off errors or overflow during the computation. Admittances in **Y** in typical electronic circuits are predominantly in the range $10^{-2}$ to $10^{-5}$, and solutions in **i** are similarly small. Extreme values of either voltage or current are best avoided by scaling. This particularly applies to iterative non-linear analysis since errors once introduced are difficult to remove by iteration.

Current scaling is carried out by multiplication of all network currents by a factor $k$, that is, the scaled currents $\mathbf{j}_s = k\mathbf{j}$. In order to retain voltages $v = jR$ unchanged, impedances must be scaled by $1/k$ or admittance by $k$. In non-linear resistors current scaling of admittances involving $j = f(v)$ are similarly scaled since $j_s = kj = kf(v)$. In the case of expressions of the form $v = f(j)$, $f(kj) \neq k\{f(j)\}$ and the differentiation of impedance must be repeated in the current scaled condition. An obvious example of scaling is to work all computations in the units kΩ and mA for nodal analysis of low power electronic circuits.

Voltage scaling is similar except that all voltages are scaled by $k$, that is, $\mathbf{v}_s = k\mathbf{v}$. Resistors are then scaled by $k$ and admittances by $1/k$. Special treatment is now required for non-linear resistors with characteristic equations of the form $j = f(v)$.

*Example* 3.3

Current scale the diode with characteristic $j = 10^{-11}(e^{40v}-1)$ by $k = 10^{-3}$ at $v = 0.5$ V. Also voltage scale the same diode by $k = 10$ at $j = 4.85 \times 10^{-3}$ A.

Current scaled units $j_s$ mA, $R_s$ kΩ, $Y_s$ mS.

At $\quad v = \frac{1}{2}$ V,

$$j = 4.85 \times 10^{-3}\text{ A}, R = 5.15\Omega, Y = 0.194\text{ S}.$$

In scaled units

$$j_s = 4.85\text{ mA}, \; R_s = 00515\text{ k}\Omega, \; Y_s = 194\text{ mS}.$$

Voltage scaled units $v_s$ 100 mV, $R_s$ $10^{-1}$ Ω, $Y_s$ 10 S.

At $\quad j = 4.85 \times 10^{-3}$ A,

$$v = 0.5\text{ V}, \; R = 5.15\;\Omega, \; Y = 0.194\text{ S}.$$

In scaled units

$$v_s = 5\,(100\text{ mV}), \; R_s = 51.5\,(0.1\;\Omega), \; Y_s = 0.0194\,(10\text{ S}).$$

However, these values cannot be obtained directly from the diode characteristic equation by insertion of $v = 5$. In unscaled units,

$$Y = \frac{dj}{dv} = 4 \times 10^{-10}\,e^{40v}.$$

Therefore

$$Y_s = \lambda\left(\frac{dj}{dv}\right)_{v=5} = 0.0194$$

$$= \lambda \times 2.89 \times 10^{77} ,$$

and $\lambda = 6.712 \times 10^{-80}$ .

Unfortunately $\lambda$ depends on the evaluation voltage $v$, and voltage scaling must be performed after linearisation of the diode equations in the form $j = f(v)$.

**Convergence**

The Newton iteration is not guaranteed to converge to the expected solution; indeed it may not converge to any solution. Three simple difficulties arising from non-monatonic characteristics in one-dimension are illustrated in Fig. 3.10. In Fig. 3.10(a) several solutions are available, and convergence may be to any one of them dependent on the initial approximation to $v$. A first approximation at either the maximum or minimum on the graph results in a gradient $df(v)/dv$ equal to zero, and the iteration cannot proceed since the tangent does not intersect the $v$-axis. Additionally, the very small gradients obtained near maxima and minima will result in the prediction of a value $v$ for the next iteration far removed from the region of solution.

In Fig. 3.10(b) an oscillation is set up where each successive iteration predicts a new point equal to that obtained at the previous iteration. In the generalised iteration in many dimensions much more complicated situations occur. The problem in overcoming such difficulties often lies in recognition of the type of situation that has arisen. The only real guard against such behaviour is to make sure that the first estimate of non-linear branch voltages or currents is as accurate as possible.

Experimental diode characteristics suffer from extreme non-linearity, usually in the region of 0.3 to 0.7 V, where the range of validity of the linearisation is far exceeded by the predicted voltage for the following iteration. At voltages lower than the knee in the characteristic, small gradients cause a large overshoot of the normal operationing region for a diode. For example, the diode in Example 3.3 linearises to a resistance of $1.062 \times 10^{-8}\ \Omega$ in parallel with a current source of $9.18 \times 10^{7}$ A at $v = 1$ V, $j = 2.35 \times 10^{6}$ A.

Extreme values of resistance and current such as these introduce numerical inaccuracy and, as we saw in Example 3.1, cause slow convergence. Overshoot can be restricted by limiting the maximum forward voltage and current to the device rating. However, for diodes with positive voltage $v$ it is usually preferable to use the inverse characteristic $v = 0.025 \log_n(1 + j/I_s)$. It is then necessary to solve for the diode current $\bar{j}$ and update the linearisation at the exact diode

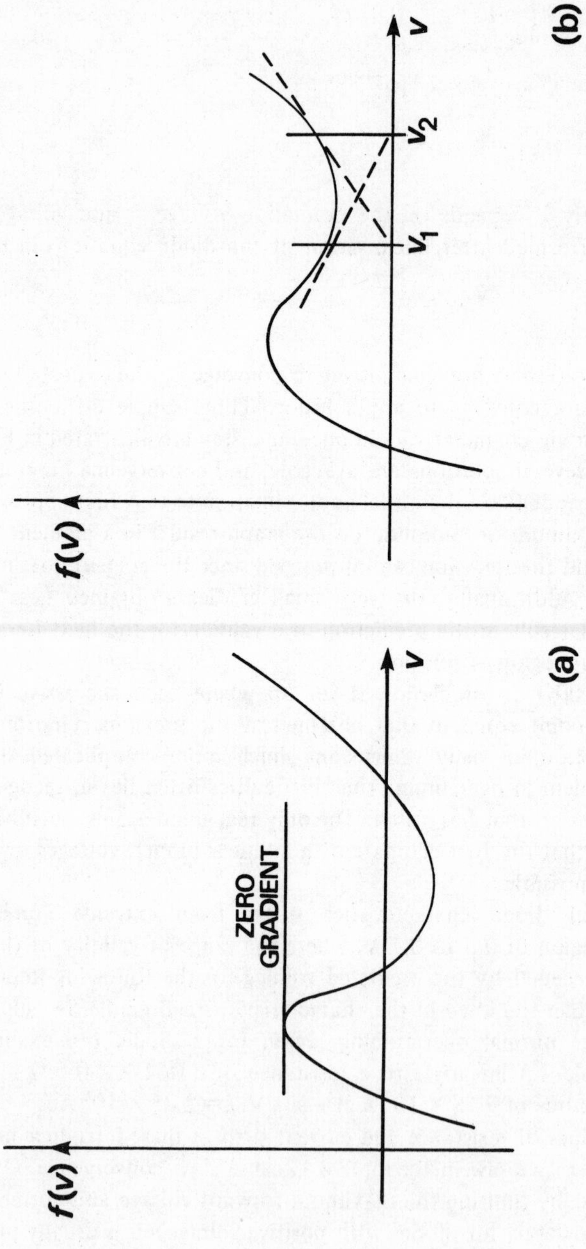


Fig. 3.10 – Non-converging examples of Newton iteration.
(a) Zero gradient (and multiple solutions).
(b) Oscillating iterations.

voltage $v$. This modified Newton iteration converges significantly faster than the original iteration.

The modified Newton iteration is used for all non-linearities which cannot be expressed in the standard form $j = f(v)$ either because the equation cannot be solved or for the reason of convergence.

**General non-linearities**

Non-linear resistors in active device models are frequently defined by measurement or approximation instead of by systematic equations. Returning to the tunnel diode characteristic from Fig. 3.1, a set of measurements for the diode could be taken resulting in the graph shown in Fig. 3.11. Alternatively a coarse approximation to the shape of the graph could be obtained by means of a *piecewise linear* approximation which is also shown.

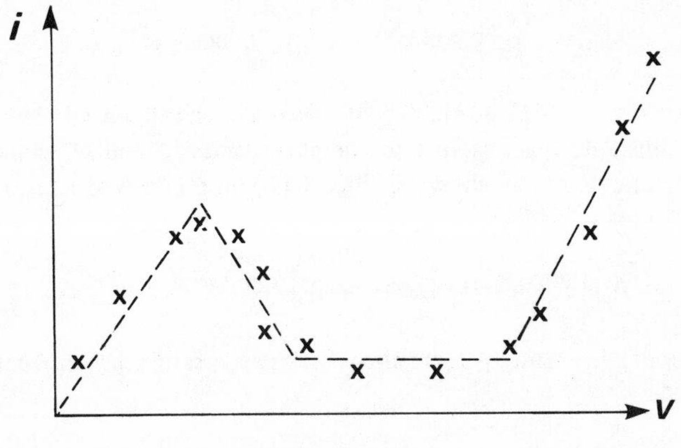


Fig. 3.11 – Tunnel diode characteristic. $x$ Measurements.
----- Piecewise linear approximation.

Both types of specification can be incorporated into non-linear d.c. analysis. Measurements, which are more precise than piecewise linear approximation but otherwise similar, can be directly incorporated into the Newton iteration by fitting the measured points by a set of cubic polynomials. These are defined by the amplitude and gradient at the two end points of any section of the graph. The cubic polynomials or *splines* are obtained by means of a computer subroutine which assigns gradients at each datum point such that the graph and its first derivatives are continuous. The cubics, however, differ at each segment.

Piecewise linear approximations can be directly incorporated into some versions of circuit analysis with specially adapted matrix inversion procedures.

## 3.4 ACTIVE DEVICE MODELLING

Non-linear d.c. analysis computer programmes are of only limited use without adequate models of the most common active devices. The process of modelling their behaviour from the laws of solid state physics is outside the scope of this book. However, we are concerned with using models and with measuring the model parameters. To illustrate the use of active device models in non-linear d.c. analysis this section will concentrate on the Ebers-Moll model of the bipolar transistor. The techniques developed can be applied to other active devices, the most important of which are field effect and MOS transistors.

**Bipolar transistors**

The Ebers-Moll injection current model of the d.c. behaviour of an npn bipolar transistor is defined by the two equations,

$$i_E = I_{EF}\,(e^{\theta_E v_{BE}}-1) - \alpha_R I_{CR}(e^{\theta_C v_{BC}}-1) \tag{3.22}$$

$$i_C = -\alpha_F I_{EF}\,(e^{\theta_E v_{BE}}-1) + I_{CR}(e^{\theta_C v_{BC}}-1) \tag{3.23}$$

where $\theta = q/kT$, $\theta_E = \theta/M_E$, $\theta_C = \theta/M_C$, $q$ is the charge on an electron, $k$ is Boltzmann's constant, $T$ is absolute temperature, and $M_E$ and $M_C$ are emission constants. The circuit model shown in Fig. 3.12 can be derived from the equations by identifying

$$j_1 = I_{EF}(e^{\theta_E v_{BE}}-1) \text{ and } j_2 = I_{CR}(e^{\theta_C v_{BC}}-1) \tag{3.24}$$

as diode currents. $I_{EF}$ and $I_{CR}$ are then the reverse saturation currents of the diodes.

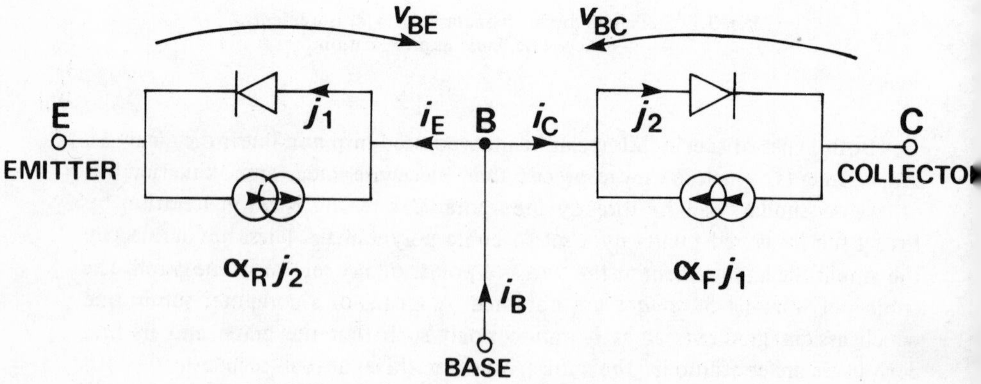


Fig. 3.12 – Ebers-Moll simplified bipolar npn transistor model

Linearisation of the transistor model is identical to that for non-linear resistors since Eqs. (3.22) and (3.23) have been interpreted with two-terminal diode non-linearities. Slight complication arises from the CCCS dependent sources which are controlled by currents in the non-linear elements. Also in practice the values of $\alpha_R$ and $\alpha_F$ are themselves current dependent. The linearised circuit in Fig. 3.13(a) is obtained from Fig. 3.12 by separately linearising the diodes. New variables $v_1$, $v_2$, $i_1$, and $i_2$ equal to $v_{BE}$, $v_{BC}$, $i_E$, and $i_C$ respectively, have been introduced to conform with our branch numbering convention. Superscripts $n$ and $n+1$ indicate the iteration number as in Fig. 3.7 and throughout this chapter.

Solution of the linearised circuit has been shown to be equivalent to the Newton iteration, and the current and voltage in the non-linear resistors determined by linear circuit analysis correspond to the approximate values, $\bar{\mathbf{j}}^{n+1}$ and $\bar{\mathbf{v}}^{n+1}$, for the next iteration. These values are therefore applicable to Fig. 3.13(a). The appropriate branch currents must be labelled $\bar{\mathbf{i}}^{n+1}$ since they contain $\bar{\mathbf{j}}^{n+1}$. The controlling currents of the dependent sources enter the solution as $\bar{\mathbf{j}}^{n+1}$ since they can then be converted by using Ohm's law into solution variables $\mathbf{v}^{n+1}$ through the relationships,

$$\alpha_R \bar{j}_2^{n+1} = \alpha_R(v_1^{n+1} G_2^n - \bar{I}_2^n) \tag{3.25}$$

$$\alpha_F \bar{j}_1^{n+1} = \alpha_F(v_1^{n+1} G_1^n - \bar{I}_1^n) \; . \tag{3.26}$$

Eqs. (3.25) and (3.26) convert each CCCS into a VCCS which can be incorporated directly in nodal analysis. The additional terms $\alpha_R \; \bar{I}_2^n$ and $\alpha_F \; \bar{I}_1^n$ convert the controlled source dependence from non-standard branch dependence to element dependence as required in Section 2.3. A version of the linearised circuit for nodal analysis is shown in Fig. 3.13(b).

The circuit for analysis contains only components known at iteration $n$, and solution via the node voltages is in terms of branch voltages and currents which are the estimated values for the following iteration. In order to linearise the diode equations the exact diode currents $j_1$ and $j_2$ must be obtained from the diode Eqs. (3.24) and $\alpha_F$ and $\alpha_R$ must be updated. The solution is applicable to all regions of transistor operation. Convergence is, however, more difficult to achieve than in the simple diode circuits since the transistor model is essentially an interacting pair of diodes. Two modifications to the solution are of considerable benefit, possibly even essential.

The first and simplest modification is to limit the forward voltage of all diodes to a specified maximum value which is used whenever a larger value is predicted for the next iteration. Silicon diodes could, for example, be limited to a maximum forward voltage of 0.8 V.

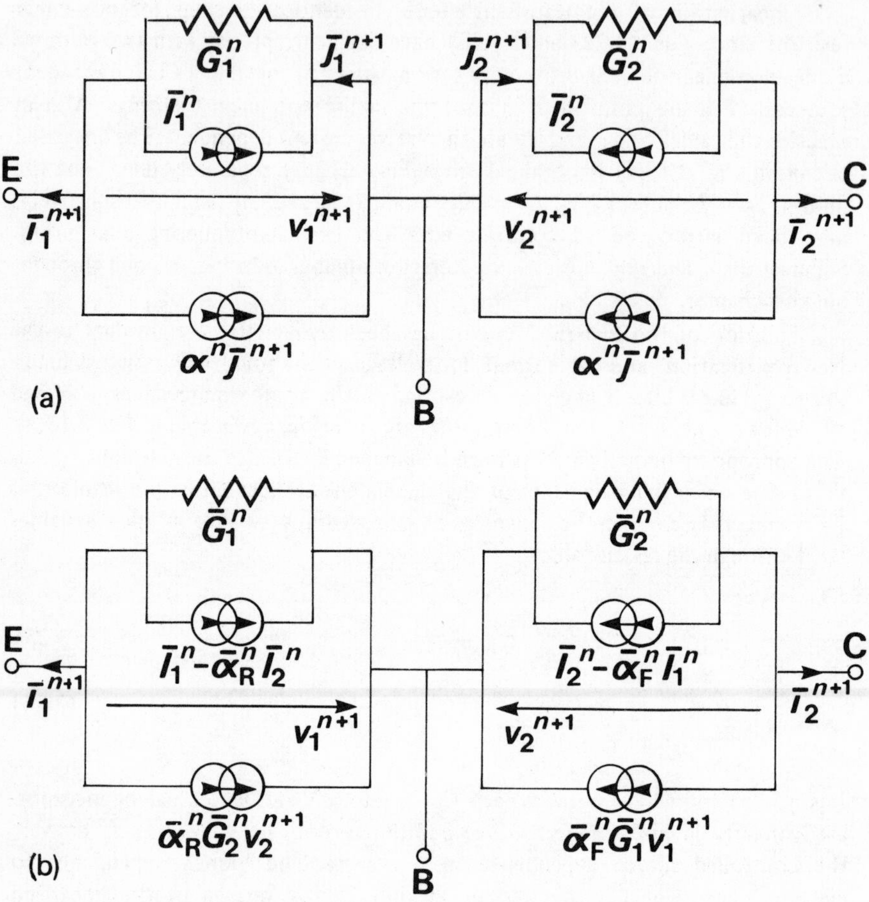


Fig. 3.13 – (a) linearised Ebers-Moll transistor model; (b) version for nodal analysis.

$$\bar{G}_1^n = \theta_E(j_1^n - I_{EF}) \qquad \bar{G}_2^n = \theta_C(j_2^n - I_{CR})$$

$$\bar{I}_1^n = v_1^n \bar{G}_1^n - j_1^n \qquad \bar{I}_2^n = v_2^n \bar{G}_2^n - j_2^n$$

The second modification involves recognition of the various operating regions of a transistor so that a simplified model can be used for any diode operating with reverse bias. In the normal operating region, for example, the base-collector diode is reverse biased, and in the cut-off region both diodes are reverse biased. A simple and often adequate model of a reverse biased diode consists of a fixed resistance equal to the resistance of the diode at zero bias. Application to the base collector diode begins with the diode equation,

$$j_2 = I_{CR}\,(e^{\theta_C v_{BC}} - 1) \ .$$

Differentiation yields,

$$G_2 = \frac{\partial j_2}{\partial v_{BC}} = \theta_C I_{CR} e^{\theta v_{BC}} \tag{3.27}$$

$$= \theta_C I_{CR} \ ,$$

at $v_{BC} = 0$. Also, of course, the parallel current generator $I_2 = 0$ at zero bias.

At large values of reverse bias ($-v_{BC} \gg 1/\theta_C$) the diode conductance is much greater than the value derived for zero bias. At $v_{BC} = -0.1$ V for the diode in Example 3.3, $G_2$ is reduced by a factor of 50, and at $-1$ V by a factor $> 10^{17}$. Leakage must be considered in order to predict behaviour more accurately, and it is usual to include fixed resistors in the transistor model to take it into account. Leakage resistors $R_{EL}$ and $R_{CL}$ are shown in the practical Ebers-Moll simplified transistor model of Fig. 3.14. The base resistance $R_{BB'}$ is also important and is included in the model. Collector and emitter bulk resistors are also shown. The additional resistors are considered fixed and may be treated as separate branches in the analysis. $R_{EL}$ and $R_{CL}$ may also be combined with $G_1^n$ and $G_2^n$ if this is convenient. Further modification to take the Early effect into account involves another adjustment of $\alpha_F$.

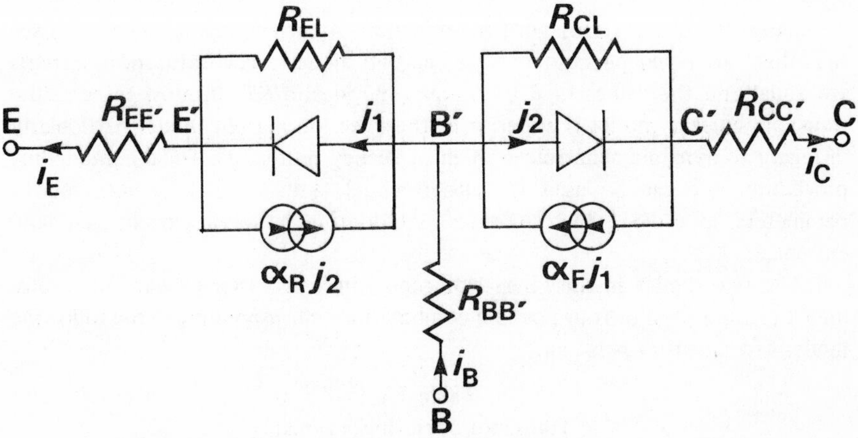


Fig. 3.14 – Practical Ebers-Moll model showing bulk and leakage resistors.

**Determination of the Ebers-Moll model parameters**

The determination of model parameters can become time consuming and expensive if the ultimate objective is full information including the statistical variation of parameters. Automated testing backed by a large computer programme becomes a necessity when accurate information on a complete transistor model is required for use up to high frequencies. Here our need is for d.c. model determination from a few simple measurements with sufficient accuracy for circuit design.

The linearisation procedure from the previous section may be summarised in four steps.

Step 1. Solve for $v^n_{B'E}$ and $v^n_{B'C}$ by circuit analysis.

2. Use the diode Eqs. (3.24) to determine the exact values of current $j^n_1$ and $j^n_2$.
3. Calculate new values $\bar{\alpha}^n_F$ $\bar{\alpha}^n_R$
4. Linearise to obtain $\bar{G}^n_1$ $\bar{G}^n_2$ $\bar{I}^n_1$ $\bar{I}^n_2$.

In order to carry out Step 2 the constants $I_{EF}$, $I_{CR}$, $M_E$, $M_C$, and the temperature $T$ must be available as well as the voltages from Step 1. In Step 3 the updated values of current gain must be evaluated from theoretical or tabulated variation against transistor voltages or currents. Two further constants are involved.

Before we begin evaluating the transistor model parameters it is useful to introduce a second equivalent set of Ebers-Moll equations based on the transport currents $\alpha_N j_1$ and $\alpha_I j_2$. They are defined by

$$\alpha_N j_1 = I_0(e^{\theta v_{B'E}}-1) \;.$$

$$\alpha_I j_2 = I_0(e^{\theta v_{B'C}}-1) \;. \qquad (3.28)$$

Comparing these equations with the injection of current Eqs. (3.4) we can see that there are fewer parameters since the two diode reverse saturation currents are equal and the values of $\theta$ both equal the ideal $q/KT$. In most respects the transport current model is superior to the injection current model particularly in regard to dynamic behaviour of high frequency transistors. The d.c. modelling procedure here can be used for either model, and in fact uses both sets of parameters to provide the functional relationship between current gain and current.

The two diodes in the Ebers-Moll model may be either forward or reverse biased giving a total of four possible combinations corresponding to the following modes of transistor operation.

**Table 3.1**

Transistor operating regions

| | BASE-EMITTER DIODE | BASE-COLLECTOR DIODE |
|---|---|---|
| NORMAL ACTIVE REGION | Forward bias | Reverse bias |
| SATURATION | Forward bias | Forward bias |
| CUT-OFF | Reverse bias | Reverse bias |
| INVERSE ACTIVE REGION | Reverse bias | Forward bias |

The normal active and the inverse regions of operation are the most suited for measurement since in each case one diode conducts while the other is reverse biased. The characteristics of the two diodes are therefore separated and may be evaluated individually.

*Normal active transistor*

At low to medium currents in the active region the bulk resistances, in particular $R_{BB'}$, can be neglected. The base-collector diode is reverse biassed and $j_2$ is assumed negligible. Sufficient collector voltage maintains the reverse bias but the voltage is kept as small as possible to reduce power dissipation at the junction and to avoid the Early effect. $v_{B'E'}$ is assumed sufficient for the $-1$ to be neglected in the diode equation, and the measured terminal voltage $v_{BE}$ is assumed equal to $v_{B'E'}$.

The emitter current $i_E$ corresponds to $i_1 = j_1$, and from Eqs. (3.24) and (3.28) complies with both

$$i_E = I_{EF} e^{\theta v_B / M_E}$$

$$\alpha_N i_E = I_0 e^{\theta v_{BE}} \tag{3.29}$$

where $\alpha_N = 1 - i_B / i_E$ .

Measurements of $v_{BE}$, $i_B$, and $i_E$ for constant $v_{BC}$ will yield values for $I_0$, $I_{EF}$, $\theta$, $M_E$, and tabulated values of $\alpha_N$ with diode current $j_1$. The experimental circuit is shown diagrammatically in Fig. 3.15(a). Taking logarithms to base e,

$$\log(i_E) = \log(I_{EF}) + (\theta / M_E) v_{BE}$$

$$\log(\alpha_N i_E) = \log(I_0) + \theta v_{BE} \ . \tag{3.30}$$

Graphical plots or least squares fitting to these linear relationships will give slope and intercept in both cases to give $\theta / M_E$, $\theta$, log $I_{EF}$, and log $I_0$. In theory only two measurements are necessary to fit the linear Eqs. (3.30), and the values can be obtained by exact solution. However, least squares will give more reliable answers.

At high base currents the graph deviates from a straight line owing to the volt drop in the bulk base resistance $R_{BB'}$. The value $v_{B'E'}$ which applies is significantly less than the measured value $v_{BE}$. At high base currents therefore

$$\log(i_E) = \log(I_{EF}) + (\theta / M_E)(v_{BE} - i_B R_{BB'}) \ . \tag{3.31}$$

Substitution of measured values for $i_E$, $v_{BE}$, and $i_B$ for a high base current, and $I_{EF}$, $\theta$, and $M_E$ determined from low base current measurements, gives a solution for $R_{BB'}$.

Similar extreme measurements can be used to determine $R_{CC'}$ and $R_{EE'}$. However, typical values for low current transistors (5 Ω and 0.5 Ω respectively) show that they can be ignored except at very high emitter currents. In fact default values in the absence of measurements could be obtained from $1/i_{EMAX}$ and $0.1/i_{EMAX}$ respectively.

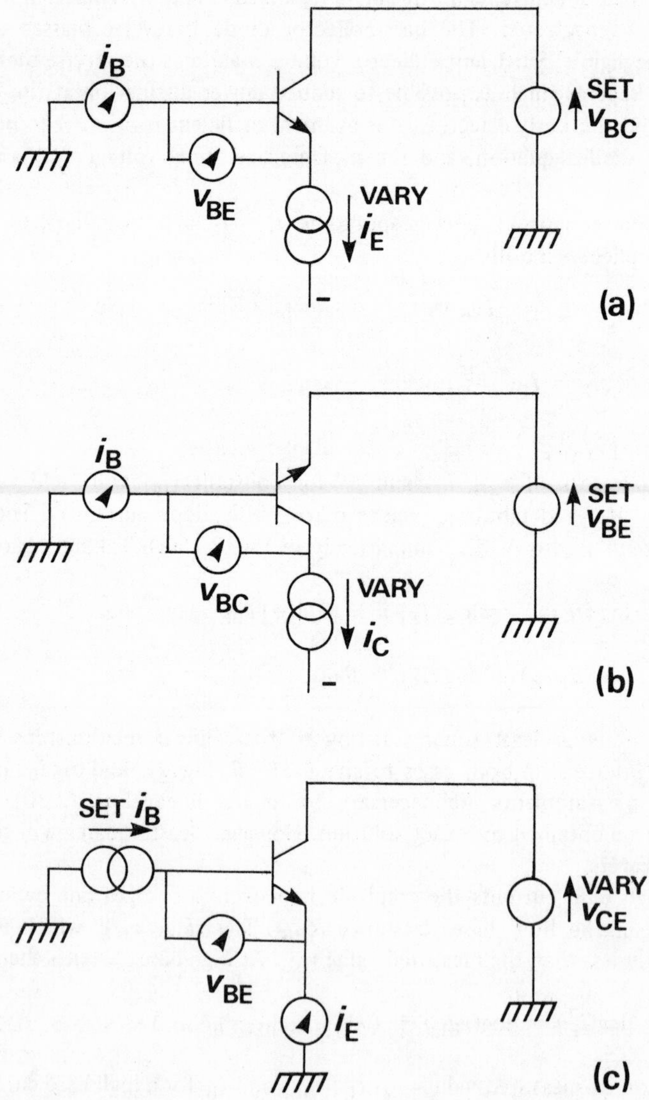


Fig. 3.15 – Measurement circuits for npn bipolar transistor modelling.
(a) normal active transistor; (b) inverse active transistor; (c) output characteristic.

*Inverse active transistor*

Exactly the same procedure can be followed in the inverse active region as in the normal active region of transistor operation. In practice the emitter and collector connections are merely interchanged. $I_{CR}$, $M_C$, $\theta$, and $I_0$ are then obtained by the same computational procedure as previously, and tabulated values of $\alpha_I$ with base-collector diode current $j_2$ are also available. The experimental circuit is shown diagrammatically in Fig. 3.15(b). Although the circuit is identical to that in Fig. 3.15(a), the voltage and current sources and the meters have been renamed to conform to the new transistor connection. Proceeding as before from Eqs. (3.24) and (3.28) and noting that $i_C$ corresponds to $i_2 = j_2$,

$$i_C = I_{CR}\, e^{\theta v_{BC}/M_C}$$

$$\alpha_I i_C = I_0\, e^{\theta v_{BC}} \qquad (3.32)$$

where $\alpha_1 = 1 - i_B/i_C$. Measurements of $v_{BC}$, $i_B$, and $i_C$ for a constant $v_{BE}$ will yield values for $I_0, I_{CR}, \theta$, and $M_C$.

*Current gain variation*

Variation in the values of normal and inverse current gains $\alpha_N$ and $\alpha_I$ as a function of diode currents $j_1$ and $j_2$ can be predicted by rearrangement of Eqs. (3.29) and (3.32) and taking logarithms,

$$\theta v_{B'E} = \log\left(\frac{\alpha_N j}{I_0}\right) = \log\left(\frac{j_1}{I_{EF}}\right)^{M_E}$$

$$\theta v_{B'C} = \log\left(\frac{\alpha_I j_2}{I_0}\right) = \log\left(\frac{j_2}{I_{CR}}\right)^{M_C} .$$

$$\text{Therefore,}\quad \alpha_N = \frac{I_0}{I_{EF}^{M_E}}\, j_1^{(M_E-1)} \qquad (3.33)$$

$$\alpha_I = \frac{I_0}{I_{CR}^{M_C}}\, j_2^{(M_C-1)} . \qquad (3.34)$$

These equations are valid over a large range and can be used in place of the tabulated values which must either be interpolated or approximated by a least squares polynomial before use.

A further correction to the value of the forward current gain $\alpha_N$ must be made to account for the effective increase in current gain due to the Early effect. In fact $\alpha_I$ can be similarly corrected. In order to achieve correction independent of collector current and temperature, the transistor common emitter current gain $\beta_N$ is assumed to vary with $v_{CB}$ as

$$\beta_N = \beta_{NM} + (v_{CB} - v_{CBM})/V_N \ . \tag{3.35}$$

$\beta_{NM}$ is the value of common emitter current gain at the measurement voltage $v_{CB} = v_{CBM}$ for a specified current and temperature. All measurements for the normal active transistor are taken at this fixed low value of collector-base voltage, but it is possible to avoid the use of $v_{CBM}$ as a model parameter provided that the values for $\alpha_N$ are corrected to $v_{CB} = 0$ before modelling.

From Eq. (3.35),

$$\beta_{NO} = \beta_{NM} - v_{CBM}/V_N \ ,$$

where $\beta_{NO}$ is the corrected value of $\beta_{NM}$ for $v_{CB} = 0$.

The tabulated data for the normal active transistor is corrected for $v_{CB} = 0$ by reducing $i_E$ to $i_{EO} = i_E - i_B \, v_{CBM}/V_N$. If $\alpha_{NO}$ is a value of common-base current gain referred to $v_{CB} = 0$ derived from the corrected measurements or Eq. (3.33), the value $\alpha_N$ for any collector-base voltage $v_{CB}$ can be obtained from

$$\alpha_N = \frac{\alpha_{NO}/(1-\alpha_{NO}) + v_{CB}/V_N}{1/(1-\alpha_{NO}) + v_{CB}/V_N} \ , \tag{3.36}$$

remembering that $\alpha_N = \beta_N/(1 + \beta_N)$ and $\beta_{NO} = \alpha_{NO}/(1 - \alpha_{NO})$. The value of the parameter $V_N$ is therefore necessary before modelling commences. Fortunately it can be derived from a single output characteristic since the Early effect is substantially independent of the temperature and transistor current.

The circuit for measurement of the output characteristic for a normal active transistor is shown in Fig. 3.15(c). Precautions are necessary to ensure that the junction temperature remains constant, and dissipation must be carefully minimised. At a chosen value of base current $i_B$, the emitter current $i_E$, the base-emitter voltage $v_{BE}$, and the collector emitter voltage $v_{CE}$, are recorded as $v_{CE}$ is varied. A graphical plot of $\beta_N$ against $v_{CB}$ or least squares fitting will yield a value of $V_N$.

*Model parameters*

The final product of transistor modelling using the Ebers-Moll equations is a list of parameter values. All of these are necessary in order to carry out the four steps of linearisation procedure summarised at the beginning of this section. A list of parameters together with typical values for a silicon planar transistor is given in the following table. When either the injection model or the transport model is used exclusively, or possibly for increased accuracy, the current gain data will also need to be available as a tabulation or a polynomial approximation.

**Table 3.2**

Ebers-Moll d.c. Parameters with typical values for a silicon planar transistor

| | | | |
|---|---|---|---|
| Base-Emitter | Diode inverse saturation current | $I_{EF}$ | $3.8 \times 10^{-14}$ A |
| | Diode emission constant | $M_E$ | 1.05 |
| | Leakage resistor | $R_{EL}$ | $10^7\ \Omega$ |
| Base-Collector | Diode inverse saturation current | $I_{CR}$ | $1.2 \times 10^{-12}$ A |
| | Diode emission constant | $M_C$ | 1.2 |
| | Leakage resistor | $R_{CL}$ | $10^8\ \Omega$ |
| Bulk Resistors | Base | $R_{BB'}$ | $100\ \Omega$ |
| | Collector | $R_{CC'}$ | $5\ \Omega$ |
| | Emitter | $R_{EE'}$ | $0.5\ \Omega$ |
| Transport Model | Inverse saturation current | $I_0$ | $10^{-14}$ A |
| Early Effect | Inverse slope | $V_N$ | 1 V |

**MOS transistors**

Applications of MOS transistors are generally in integrated circuit form and therefore involve inherently more complicated circuit models than discrete devices. The fixed potential of the integrated circuit substrate, for example, is common to a large number of otherwise separate devices and must be separately incorporated into the equation for each MOS transistor. Also, modelling is based on transistor geometry and physical constants from the manufacturing process, making the interpretation of the model for analysis in terms of circuit elements artificial. The scope of this section is therefore limited to one simple model from which various versions applicable to numerous types of MOS and field effect transistors may be developed.

The general equation for the N-channel MOS transistor in Fig. 3.16 in the non-saturation region is in the form

$$i_D = K[f(v_G, v_S) - f(v_G, v_D)] \tag{3.37}$$

where $f(v_G\ v) = (v_G - v_T - v)^2 + \dfrac{4k}{3}(v - v_b + 2\phi_f)^{3/2}$ ,

$$K = \mu C_{ox} W/2L \ ,$$

$$k = \sqrt{2q\epsilon_{si}N}/C_{ox} \ ,$$

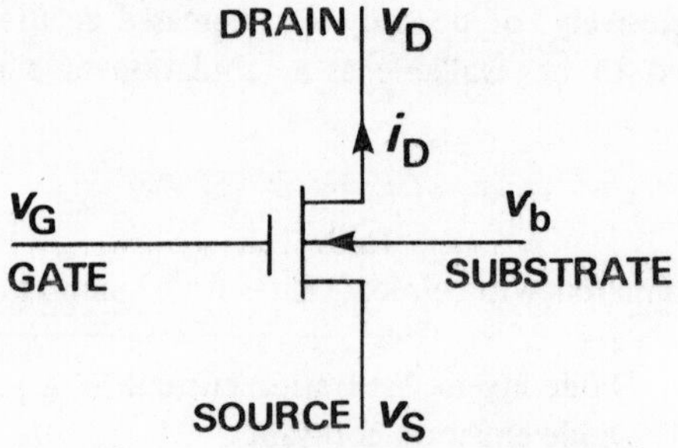


Fig. 3.16 – MOS transistor voltages.

$v_T$ is the threshold voltage, $\phi_f$ is the Fermi potential, $\mu$ is the surface hole mobility (negative for N-channel devices), $C_{ox}$ is the gate oxide capacitance per unit area, $W$ is the channel width, $L$ is the channel length, $N$ is the substrate doping concentration, $\epsilon_{Si}$ is the permittivity of silicon, and $q$ is the charge on an electron. Assuming the impurity density small and the depletion charge term negligible, Eq. (3.37) may be written in the simplified form

$$i_D = K[(v_G - v_T - v_S)^2 - (v_G - v_T - v_D)^2] \ , \tag{3.38}$$

and if $v_S = 0$, the well known Sah equation is obtained as

$$i_D = 2K[(v_G - v_T)v_D - v_D^2/2] \ . \tag{3.39}$$

Eq. (3.38) applies to the non-saturation region of operation defined by $v_G - v_D > v_T$. In saturation where $v_G - v_D \leqslant v_T$, the second term of the equation is zero and

$$i_D = K[(v_G - v_T - v_S)^2] \ . \tag{3.40}$$

The cut-off region of MOS transistor operation in which $i_D = 0$ is defined by the condition $v_G - v_S < v_T$.

In preparation for modelling in all regions of MOS transistor operation, the following functions for Eq. (3.27) may be identified.

$$\begin{aligned} j_1 = f(v_G, v_S) &= -K(v_G - v_T - v_S)^2, \ [v_G - v_S] \geqslant [v_T] \\ &= 0 \qquad , \ [v_G - v_S] < [v_T] \end{aligned} \tag{3.41}$$

$$\begin{aligned} j_2 = f(v_G, v_D) &= -K(v_G - v_T - v_D)^2, \ [v_G - v_S - v_T] > [v_D - v_S] \\ &= 0 \qquad , \ [v_G - v_S - v_T] \leqslant [v_D - v_S] \end{aligned} \tag{3.42}$$

The voltage conditions for cut-off and saturation in Eqs. (3.41) and (3.42) cater for N-channel devices and must be modified for P-channel. An Ebers-Moll type of model can be derived from these expressions by recognising the two functions $f(v_G, v_S)$ and $f(v_G, v_D)$ as currents $j_1$ and $j_2$ respectively which flow in MOS non-linear diodes replacing the junction diodes of the bipolar transistor model. The model which is shown in Fig. 3.17 is almost identical to Fig. 3.12. This is achieved because the form of Eq. (3.37) is similar to the Ebers-Moll equations (3.22) and (3.23). Only one equation is necessary for the MOS transistor since the gate current is zero.

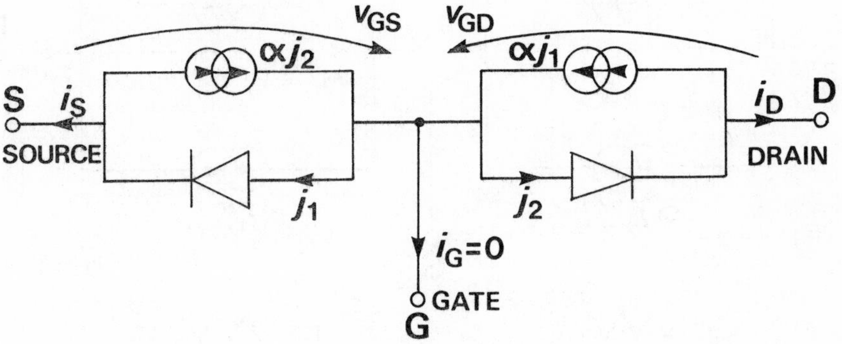


Fig. 3.17 – MOS transistor model of Ebers-Moll type.

The MOS diode characteristics are given by expressions (3.41) and (3.42) or by the full expression defined for Eq. (3.37). However, it must be emphasised that the MOS diodes defined here are purely for mathematical convenience and computation. They do not represent actual diodes which could be investigated experimentally. In fact they are not even necessary to the model except for the requirement in some computer programmes that a finite admittance exists in each circuit branch.

The MOS diode currents, $j_1 = -K(v_{GS} - v_T)^2$ and $j_2 = -K(v_{GD} - v_T)^2$, are the only differing parameters between the bipolar transistor model and the MOS transistor model. Linearisation of the N-channel MOS model proceeds in exactly the same way as before, and Fig. 3.18, which corresponds to Fig. 3.13 for bipolar transistors, is easily derived.

The model is modified in the saturation and cut-off regions of operation since either $j_1$ or $j_2$ or both $j_1$ and $j_2$ are zero. In saturation, therefore, from Eq. (3.42), linearisation yields $I_2 = G_2 = 0$. Similarly from Eq. (3.41) in cut-off $I_1 = G_1 = 0$. Normally in cut-off $I_2 = G_2 = 0$ as well. Further modification to the model in the saturation region is necessary in order to account for the finite drain conductance of a typical MOS device which arises from various second order effects not included in the analysis. It is usually sufficient to account for these effects by modification of $K$ in Eqs. (3.41) and (3.42) to a value given by $K(1 + \gamma[v_1 - v_2])$, noting that $(v_1 - v_2) = (v_D - v_S)$. The drain current is therefore increased by a small increment due to both the drain sources voltage and the drain current itself. Finally, leakage resistors can be added to the model from drain to gate, and from source to gate.

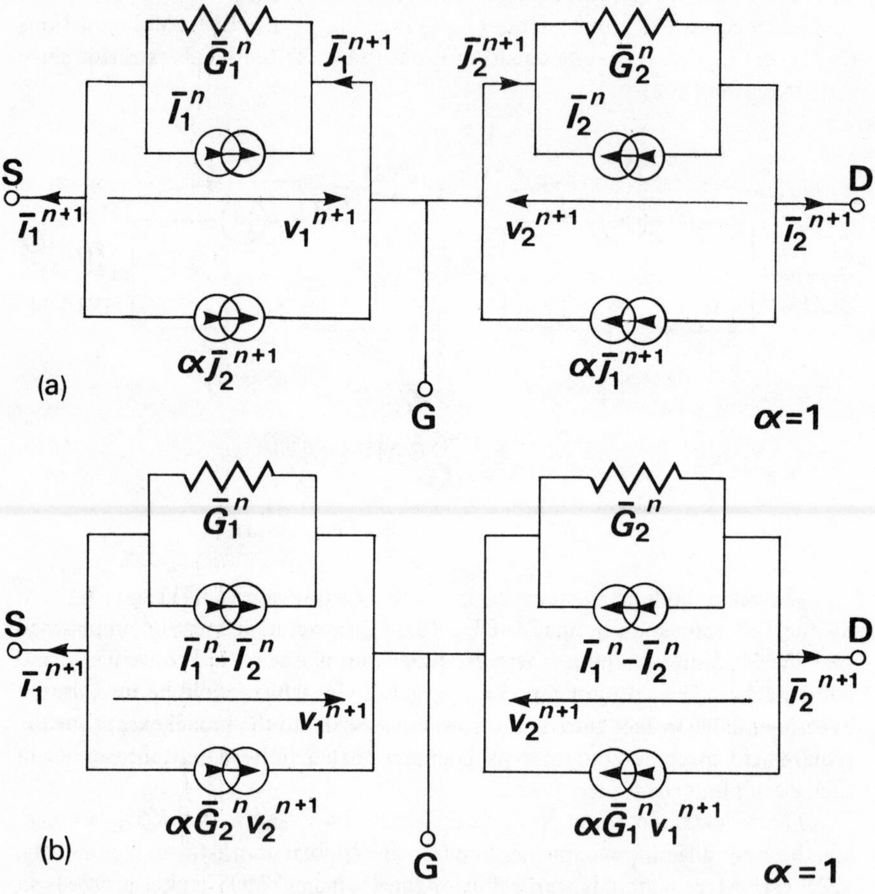


Fig. 3.18 – (a) linearised N-channel MOS transistor model; (b) version for nodal analyses.

$$G_1^n = -2K(v_1^n - V_T) \qquad G_2^n = -2K(v_2^n - v_T)$$

$$I_1^n = -K(v_1^{n^2} - v_T^2) \qquad I_2^n = -K(v_2^{n^2} - v_T^2)$$

The unknown model parameters, $K$, $v_T$, and $\gamma$, may easily be determined from three simple measurements made in the saturation region. From Eq. (3.40) and incorporating the modification to $K$,

$$i_D = -K(1 + \gamma[v_D - v_S])(v_G - v_S - v_T)^2 \ . \qquad (3.43)$$

If the transistor is operated with the gate connected to the drain to ensure saturation,

$$v_D - v_S = v_G - v_S = v_{GS}, \text{ and}$$

$$-i_D = \gamma K v_{GS}^3 + K(1 - 2\gamma v_T)\, v_{GS}^2 + K v_T(\gamma v_T - 2) v_{GS} + v_T^2 K \,. \tag{3.44}$$

The coefficients of Eq. (3.44) may be determined from a minimum of three measurements of the two values $i_D$ and $v_{GS}$ in the saturation region with $v_{GS} > v_T$. Substitution into Eq. (3.44) yields, after elimination of the constant term, three equations which must then be solved for the coefficients of the cubic equal to $K\gamma$, $K(1 - \gamma v_T)$, and $K v_T\ (\gamma v_T - 2)$. These must in turn be solved for $\gamma$, $K$, and $v_T$. When more than three measurements are made the coefficients of the cubic can be obtained more accurately by least squares.

Typical values of $K$, $\gamma$, and $V_T$ are $-2 \times 10^{-5}$ A/V$^2$, 0.01, and 4 V. It should be noted that these values are temperature dependent. Theoretically $K$ should vary according to a $-3/2$ power law with absolute temperature; that is, $(K/K_0) = (T/T_0)^{-3/2}$, where $T_0$ is the temperature of measurement. $v_T$ also varies by a few millivolts per degree, reducing with increased temperature.

## 3.5 COMPUTED EXAMPLES

The BASIC computer program RNODE from Chapter 2 contains facilities for non-linear iteration at the command NL. The program must initially complete a linear analysis, and it is therefore necessary to provide first estimates of non-linear resistor values together with parallel current sources arising from linearisation. Each subsequent iteration calls subroutine 5000 to update the linearisation, using the node and branch voltages and currents available from the linear RNODE solution. Node voltages are printed at each iteration so that convergence can be assessed.

Steps 2 and 3 of the non-linear analysis procedure summarised at the end of Section 3.2 are the only steps to differ from linear analysis. In Step 2, the exact currents in each non-linear resistor are determined from the solution for resistor voltages contained in the program matrix J(B,1). The matrix location of all RNODE solutions are listed in Section 2.5. When the modified form of the Newton iteration is to be used, exact voltages are determined from the computed resistor currents. In Step 3 the non-linear resistors are linearised and the resulting resistor and parallel current source are entered into the branch admittance matrix Y(B,B) and the fixed current source matrix I(B,1). Both steps are carried out in Subroutine 5000 which is separately programmed for each circuit. The procedure is illustrated in the two following examples. These are relatively straightforward and do not require safeguards to ensure convergence.

The use of RNODE in the analysis of non-linear circuits is mainly intended for demonstration of non-linear iteration and convergence in small circuits. However, there is considerable freedom to write more sophisticated linearisation subroutines with ease, owing to the structure of RNODE and use of the branch admittance matrix. When used on minicomputer systems the main limitation is accuracy and of course speed in slowly converging solutions. The program can also be transferred to a large computer and, if necessary, Input statements changed to Read for batch processing.

*Example* 3.4

Complete the iteration of Example 3.2, using the programme RNODE.

The linearisation subroutine written for this example is reproduced in Fig. 3.19. Statements 5010 to 5030 first calculate the exact diode current in $R_3$ from the diode equation and the resistor voltage stored in J(3,1) after each analysis. The linearised conductance is then entered into Y(3,3) and the parallel current source is entered into I(3,1).

```
5000 REM SUBROUTINE - EXAMPLE 3.2 DIODE LINEARISATION
5010 LET I1=1E-11*(EXP(40*J[3,1])-1)
5020 LET Y[3,3]=40*(I1+1E-11)
5030 LET I[3,1]=Y[3,3]*J[3,1]-I1
5040 LET I1=1E-11*(EXP(40*J[4,1])-1)
5050 LET Y[4,4]=40*(I1+1E-11)
5060 LET I[4,1]=Y[4,4]*J[4,1]-I1+.005
5070 RETURN

*
```

Fig. 3.19 – Diode linearisation subroutine for RNODE.

The second diode $R_4$ is similarly treated, noting that the fixed 5 mA current source must also be added into I(4,1). In this program, separate matrices for fixed current sources and linearisation sources have not been provided.

The RNODE run for this example is shown in Fig. 3.20. The first analysis utilises normal RNODE inputs, and an estimated linearised model for the two diodes was obtained as 5.123 Ω in parallel with 92.2 mA from the diode equation, assuming each diode forward voltage to be 0.5 V. An additional 5 mA source is required for $R_4$ input to account for the fixed source. After the first analysis, whch agrees with that in Example 3.2, command NL requests 8 iterations, and the solution quickly converges. It can be seen that diode $R_3$ is in fact reverse biased and non-conducting with branch current almost zero. If the program is re-run with the fixed voltage source equal to 20 V the solution converges even more quickly to a state with both diodes conducting.

```
RNODE     DC CIRCUIT ANALYSIS
NO. OF NODES, NO.OF RESISTORS, NO.OF DEPENDENT SOURCES ? 2,4,0
RESISTOR NO.,VALUE,+VE NODE NO.,-VE NODE NO.,SOURCE E,SOURCE I
R ? 1,4000,0,1,5,0
R ? 2,200,1,0,0,0
R ? 3,5.123,1,2,0,.0922
R ? 4,5.123,2,0,0,.0972
NODE           VOLTAGE
 1              .932921
 2              .479268
BRANCH OUTPUT YES OR NO ? YES
               RESISTOR       RESISTOR       BRANCH         BRANCH
R              VOLTAGE        CURRENT        VOLTAGE        CURRENT
 1              4.06708        1.01677E-03   -.932921        1.01677E-03
 2              .932921        4.6646E-03     .932921        4.6646E-03
 3              .453653        8.85522E-02    .453653       -3.6478E-03
 4              .479268        9.35522E-02    .479268       -3.64774E-03
NEXT COMMAND: END RUN      ER, MODIFY CIRCUIT          MC
            : SENSITIVITY SE, NON-LINEAR ITERATION NL
NEXT ? NL
NO. ITERATIONS ? 8
ITERATION      NODE VOLTAGES 1 TO  2
 1              .808161        .47797
 2              .261228        .51365
 3              .238095        .503574
 4              .238095        .500906
 5              .238095        .500754
 6              .238095        .500753
 7              .238095        .500753
NODE           VOLTAGE
 1              .238095
 2              .500753
BRANCH OUTPUT YES OR NO ? YES
               RESISTOR       RESISTOR       BRANCH         BRANCH
R              VOLTAGE        CURRENT        VOLTAGE        CURRENT
 1              4.7619         1.19048E-03   -.238095        1.19048E-03
 2              .238095        1.19048E-03    .238095        1.19048E-03
 3             -.262658       -2.87052E-15   -.262658       -9.99972E-12
 4              .500753        .100151        .500753        0
NEXT ? ER

STOP AT 1160
*
```

Fig. 3.20 – RNODE run for Example 3.4.

*Example* 3.5

Determine the output regulation of the zener diode stabilised power supply in Fig. 3.21(a) for various load resistors from 200 Ω to 2 Ω.

The zener diode and transistor models both incorporate diodes and for simplicity we will assume them to be identical with the equation $i = 10^{-11}(e^{40v}-1)$. The circuit is prepared for analysis in Fig. 3.22.

A linearisation subroutine for use in RNODE is given in Fig. 3.23. Diodes are treated as in the previous example, noting that the zener voltage can be included within the corresponding current source terminals and therefore complies with configuration of a standard branch. Transistor coupling to the collector circuit depends on the value of the linearised diode resistance, and transfer admittance in Y(4,5) is also updated together with the fixed source arising from the conversion of the transistor model to nodal analysis.

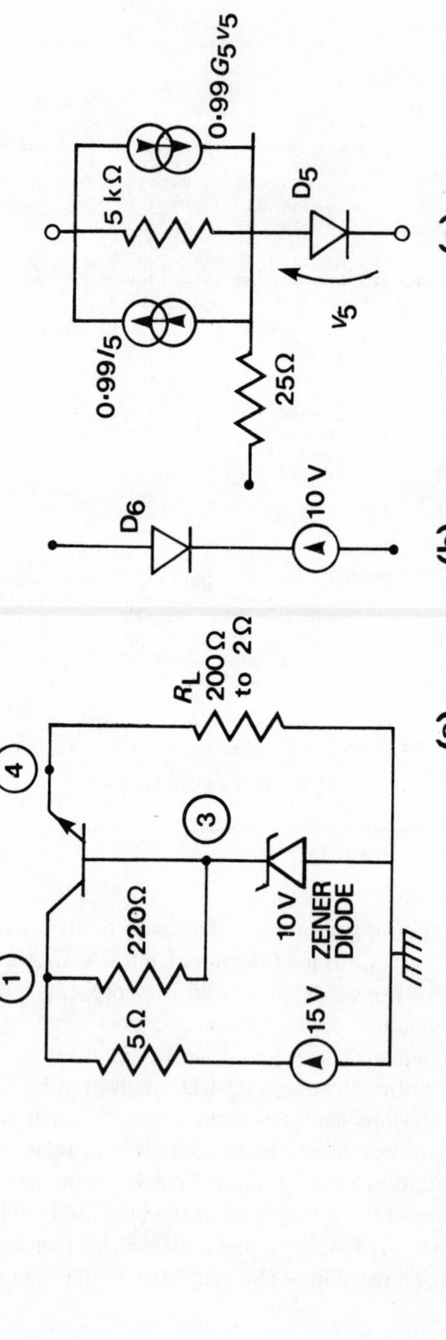


Fig. 3.21 – (a) voltage regulator for Example 3.5; (b) zener diode model; (c) transistor model.

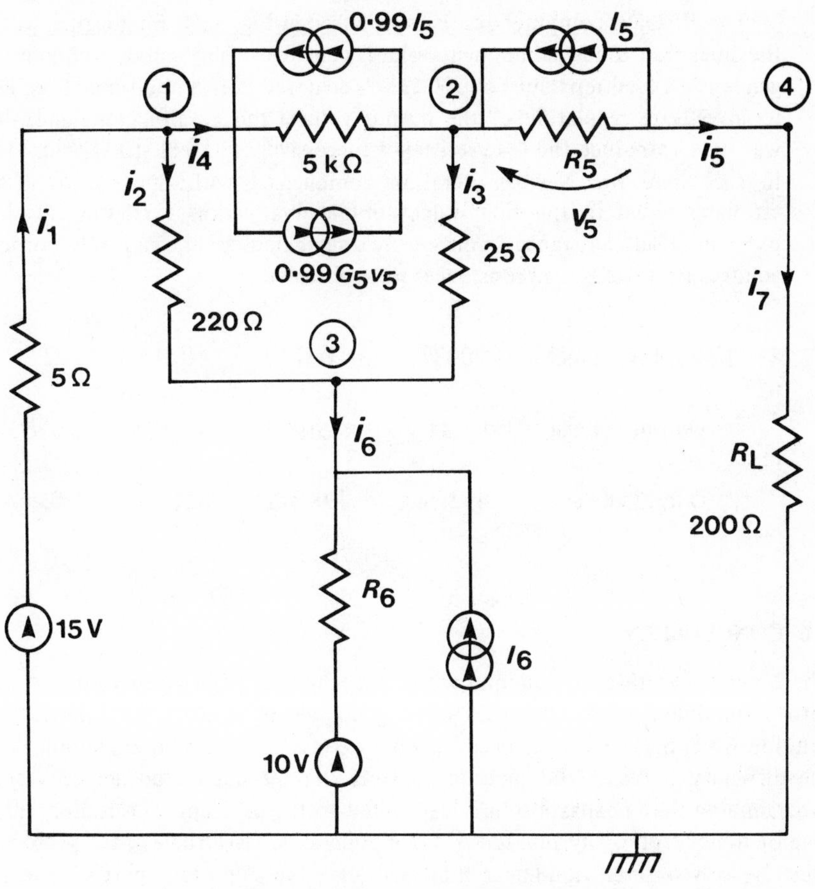


Fig. 3.22 – Voltage regulator prepared for analysis.
First estimates: $R_5 = 0.5\ \Omega, I_5 = 1.11$ A,
$R_6 = 5\ \Omega, I_6 = 0.1$ A

```
5000 REM SUBROUTINE - VOLTAGE REGULATOR
5010 LET I1=1E-11*(EXP(40*J[5,1])-1)
5020 LET Y[5,5]=40*(I1+1E-11)
5030 LET I[5,1]=Y[5,5]*J[5,1]-I1
5040 LET I[4,1]=.99*I[5,1]
5050 LET Y[4,5]=.99*Y[5,5]
5060 LET I1=1E-11*(EXP(40*J[6,1])-1)
5070 LET Y[6,6]=40*(I1+1E-11)
5080 LET I[6,1]=Y[6,6]*J[6,1]-I1
5090 RETURN

*
```

Fig. 3.23 – Voltage regulator linearisation subroutine.

The RNODE computer run is reproduced in Fig. 3.24. First estimates to the linearised diode component values were reasonably good, and convergence for a load resistance of 200 Ω was achieved in five iterations. In order to investigate regulation of the output voltage the modify command MC was used to reduce the load resistance successively to 50 Ω, 10 Ω, and 2 Ω. In each case the Newton iteration commenced with linearisation first estimates equal to the final values obtained at voltages relevant to the previous load resistance. Convergence remained good, and the output voltages for various load resistances were obtained as

| | | | | |
|---|---|---|---|---|
| Load resistance | 200 Ω | 50 Ω | 10 Ω | 2 Ω |
| Output voltage | 9.984 V | 9.902 V | 8.290 V | 3.076 V |
| Output current | 49.9 mA | 198 mA | 829 mA | 1.538 A |

## 3.6 CONCLUSION

The Newton iteration, in common with all methods of iterative solution for the zero or minimum of a function, cannot guarantee to converge to a particular solution when many exist, or even in some cases to any solution at all. Much of the difficulty in using the methods of this chapter lies in the necessity for programming the linearisation in a way suited to the problem. This is especially true of the exceptionally non-linear diode equations. Nevertheless, the problem must be solved since non-linear d.c. analysis is an important part of circuit design.

## FURTHER READING

Calahan, D. A., (1972), *Computer-Aided-Network Design,* Ch. 3, McGraw-Hill, New York.

Spence R., (1974), *Resistive Circuit Theory,* Ch. 11, McGraw-Hill, New York.

Wing, O., (1972), *Circuit Theory,* Ch. 5., Holt, Rinehart & Winston, New York.

*Device Models*

Herskowitz, G. J. and Schilling, R. B. (eds.), (1972), *Semiconductor Device Modeling for Computer-Aided Design,* McGraw-Hill, New York.

Penney, W. M. and Lau, L. (eds.), (1972), *MOS Intgrated Circuits,* Van Nostrand Reinhold, New York.

```
RNODE    DC CIRCUIT ANALYSIS
NO. OF NODES, NO.OF RESISTORS, NO.OF DEPENDENT SOURCES ? 4,7,1
RESISTOR NO.,VALUE,+VE NODE NO.,-VE NODE NO.,SOURCE E,SOURCE I
R ? 1,5,0,1,15,0
R ? 2,220,1,3,0,0
R ? 3,25,2,3,0,0
R ? 4,5000,1,2,0,1.099
R ? 5,.5,2,4,0,1.11
R ? 6,5,3,0,-10,.1
R ? 7,200,4,0,0,0
GM BRANCH OR RESISTOR DEPENDENT: INPUT B OR R ? B
GM NO.,VALUE,SEND R NO.,RECEIVE R NO.
GM ? 1,1.98,5,4
NODE           VOLTAGE
 1              14.6561
 2              10.5986
 3              10.5934
 4              10.0186
BRANCH OUTPUT YES OR NO ? NO
NEXT COMMAND: END RUN      ER, MODIFY CIRCUIT        MC
            : SENSITIVITY SE, NON-LINEAR ITERATION NL
NEXT ? NL
NO. ITERATIONS ? 5
ITERATION      NODE VOLTAGES 1 TO  4
 1              14.6552        10.5786        10.5707        10.013

 2              14.6552        10.5595        10.5515        10.0002

 3              14.6552        10.5471        10.539         9.98876

 4              14.6551        10.5427        10.5346        9.98443

NODE           VOLTAGE
 1              14.6551
 2              10.5423
 3              10.5342
 4              9.98401
BRANCH OUTPUT YES OR NO ? NO
NEXT ? MC
NUMBER OF ALTERED BRANCHES, AND DEPENDENT SOURCES ? 1,0
R NO.,NEW VALUE,NEW E SOURCE,NEW I SOURCE
R ? 7,50,0,0
NODE           VOLTAGE
 1              13.9424
 2              10.4959
 3              10.528
 4              9.86385
BRANCH OUTPUT YES OR NO ? NO
NEXT ? NL
NO. ITERATIONS ? 5
ITERATION      NODE VOLTAGES 1 TO  4
 1              13.9405        10.4947        10.5269        9.88248

 2              13.9389        10.4945        10.5269        9.89582

 3              13.9387        10.4946        10.5269        9.9012

 4              13.9387        10.4946        10.5269        9.90187

NODE           VOLTAGE
 1              13.9386
 2              10.4946
 3              10.5269
 4              9.90186
BRANCH OUTPUT YES OR NO ? NO
NEXT ? MC
```

```
NUMBER OF ALTERED BRANCHES, AND DEPENDENT SOURCES ? 1,0
R NO.,NEW VALUE,NEW E SOURCE,NEW I SOURCE
R ? 7,10,0.0
NODE            VOLTAGE
 1               10.2743
 2               10.2447
 3               10.4834
 4               9.55635
BRANCH OUTPUT YES OR NO ? NO
NEXT ? NL
NO. ITERATIONS ? 5
ITERATION      NODE VOLTAGES 1 TO  4
 1               10.3091          10.1242          10.3611          9.45827

 2               10.8132          9.02856          9.22949          8.38232

 3               10.8569          8.91916          9.11687          8.28561

 4               10.8546          8.91786          9.11548          8.28883

NODE            VOLTAGE
 1               10.8543
 2               8.91838
 3               9.11594
 4               8.28986
BRANCH OUTPUT YES OR NO ? NO
NEXT ? MC
NUMBER OF ALTERED BRANCHES, AND DEPENDENT SOURCES ? 1,0
R NO.,NEW VALUE,NEW E SOURCE,NEW I SOURCE
R ? 7,2,0,0
NODE            VOLTAGE
 1               7.31381
 2               3.72385
 3               4.09018
 4               3.0744
BRANCH OUTPUT YES OR NO ? NO
NEXT ? NL
NO. ITERATIONS ? 5
ITERATION      NODE VOLTAGES 1 TO  4
 1               7.3121           3.71989          4.08632          3.07536

 2               7.31158          3.71953          4.08598          3.07552

 3               7.31212          3.71933          4.08591          3.07535

 4               7.31102          3.71957          4.08601          3.07561

NODE            VOLTAGE
 1               7.31062
 2               3.71988
 3               4.08629
 4               3.07593
BRANCH OUTPUT YES OR NO ? NO
NEXT ? ER

STOP AT 1160
*
```

Fig. 3.24 – RNODE computer run for Example 3.5.

## PROBLEMS

**Note** Assume all diodes to be characterised by the equation $i = 10^{-11}(e^{40v}-1)$ at a temperature of 290°K. Assume all transistors to be defined by Table 3.2. Approximate solutions to Problems 3.1 to 3.10 should be obtained by hand. Accurate solutions can then be found by computer, using these approximations to derive first estimates for the non-linear elements.

3.1 Sketch the $i_1 - v_1$ characteristics of the diode networks in Fig. 3.25

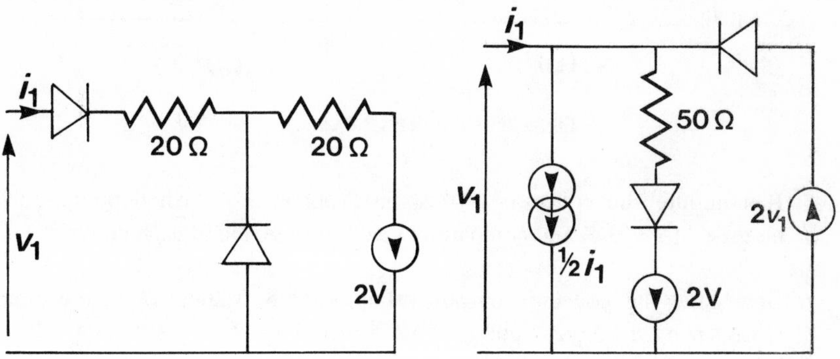


Fig. 3.25 – Problems 3.1 and 3.2.

3.2 Repeat Problem 3.1 at an input voltage of 2 V and vary the diode temperatures in the range 260°K to 320°K. Examine the variation of network voltages and currents.

3.3 Determine the values of the linearised Ebers-Moll bipolar transistor model defined by Table 3.2 at the following junction voltages.

| | | | | |
|---|---|---|---|---|
| $v_{B'C}$ | −5 V | 0.6 V | 0.6 V | −5 V |
| $v_{B'E}$ | 0.7 V | 0.7 V | −1 V | −1 V |

3.4 The transistor amplifiers in Fig. 3.26 all operate at a collector voltage of about 7 V. Determine stability factors for each circuit, assuming a transistor current gain of 80±40. Find the collector, base, and emitter voltages of each amplifier, using a simplified transistor model derived from Table 3.2 at 290°K with $\alpha_I$ negligible.

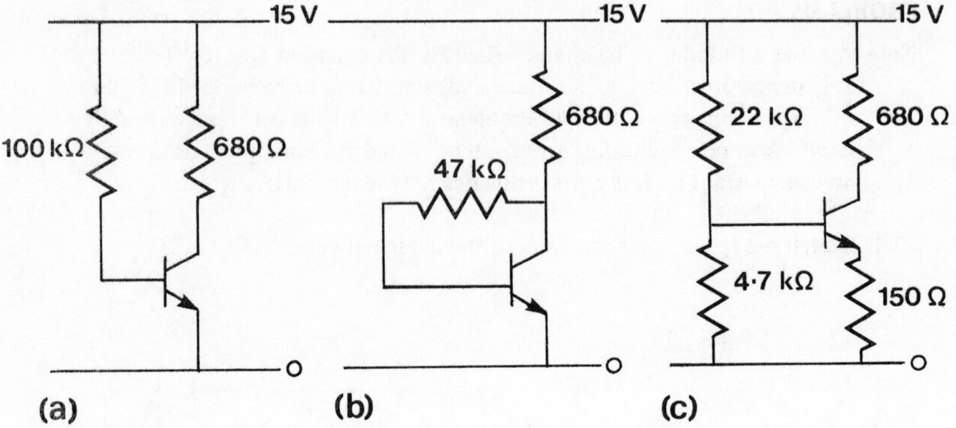


Fig. 3.26 – Problems 3.4 to 3.7.

3.5 Re-compute the collector voltages in Problem 3.4 when temperature is increased to 340°K and compare the change observed in each circuit.

3.6 Determine the collector output voltage in Fig. 3.26(a) when the input is driven by a voltage source of resistance 100 Ω as the source voltage is increased linearly from 0 V to 2 V.

3.7 What are the maximum and minimum values of collector voltage in Fig. 3.26(c) when resistors are subject to a tolerance of ±5%?

3.8 Show that the circuit in Fig. 3.27 provides an output voltage proportional to $\log_n v_1/R$ when the operational amplifier is ideal. Compute the output voltage for various input voltages between 1 mV and 20 V, using the amplifier model also given in Fig. 3.27.

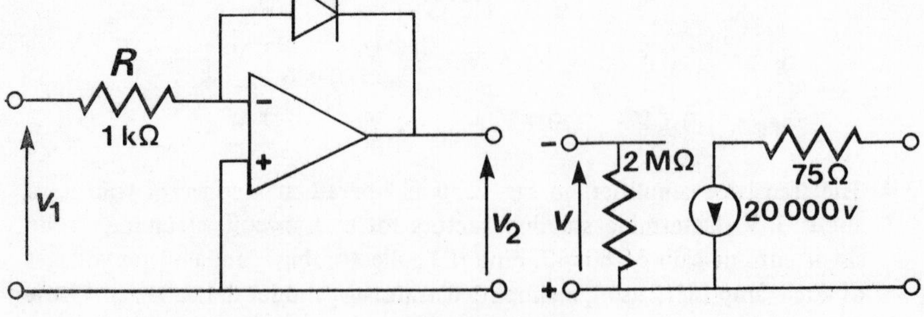


Fig. 3.27 – Problem 3.8.

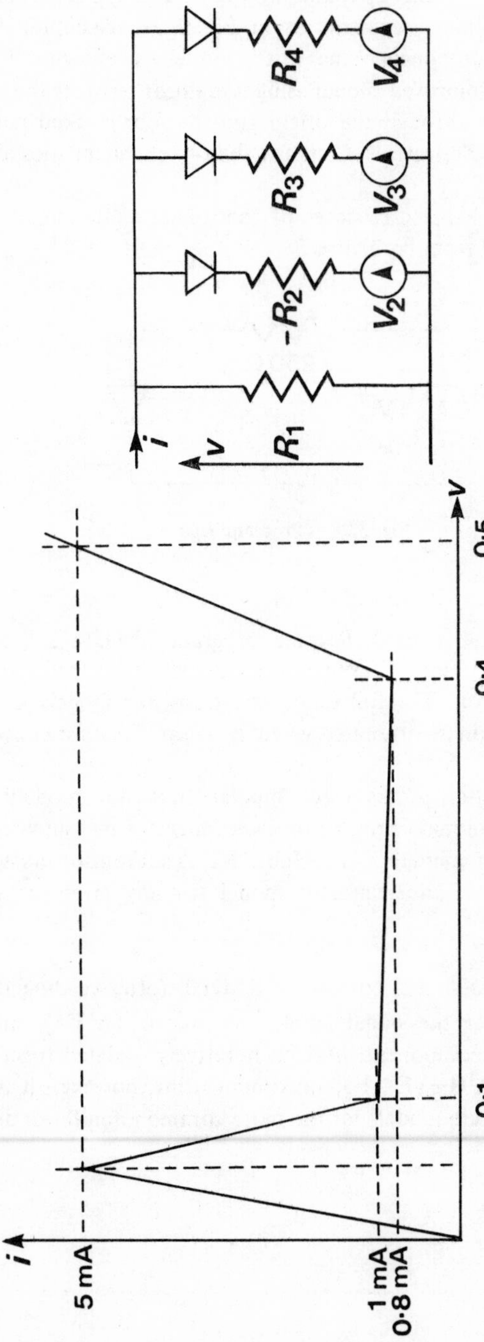


Fig. 3.28 – Problems 3.9 and 3.10.

3.9 The tunnel diode characteristic in Fig. 3.1 may be approximated by four piecewise-linear segments as in Fig. 3.28. Assuming ideal diodes, determine the component values of the model also shown in Fig. 3.28.

Develop an improved model using non-linear resistors the characteristic of which passes through the origin and the four marked points, and has continuous first derivatives. Compute the *i–v* characteristics of the model.

3.10 Determine the stable d.c. states of the tunnel diode circuit in Fig. 3.29, using the model from Problem 3.9.

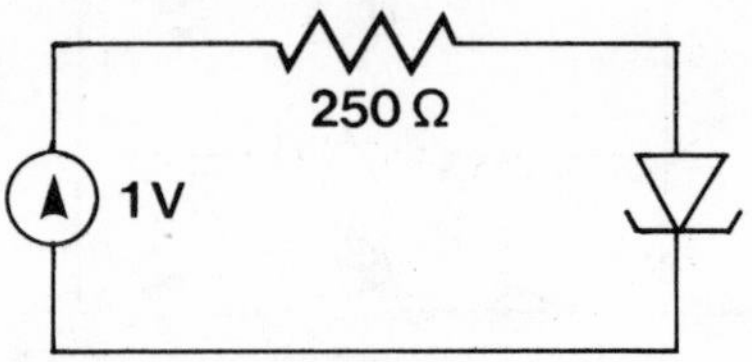


Fig. 3.29 – Problem 3.10.

*Computing*

3.11 Solve Problems 3.1 to 3.10 using program RNODE and compare convergence speeds obtained, using normal and modified forms of the Newton-Raphson iteration. The full diode and transistor models should be used and simplification incorporated where necessary to assist convergence.

3.12 Write a computer program for bipolar transistor modelling based on Section 3.4. The main program utilises transistor measurements to derive a set of model parameters as in Table 3.2. A subroutine must then be provided to linearise the transistor model for any given set of operating conditions.

3.13 It is probably desirable to rewrite RNODE (after reading the following Chapter) so that the nodal admittance matrix ($\mathbf{A}^t\mathbf{YA}$) can be directly filled from the component lists and iteratively updated together with the source vector $\mathbf{A}^t(\mathbf{I}-\mathbf{YE})$. For maximum utilit, however, it is also important to incorporate models for the more common non-linear devices.

CHAPTER 4

# Nodal Admittance Matrix Methods

Lumped linear time-invarient networks may contain any of the basic circuit elements tabulated in Chapter 1. When reactive elements are present in the network, linear circuit analysis in the steady state is generally the most common analysis requirement. We must assume that the circuit is linear or has been linearised, that voltages and currents are small in comparison with any non-linearity, and that the circuit is excited by independent sources which are sinusoidal and all of the same angular frequency. All voltages and currents are therefore phasors represented by complex numbers, and admittances or impedances are complex functions of frequency. Except for the use of complex arithmetic, however, no essential modification to the matrix methods of analysis in Chapter 2 is necessary.

A number of practical difficulties arise when using the basic matrix solutions for steady-state analysis. The first is that the sheer size of a practical circuit prevents the storage and manipulation of the incidence, cut-set, or tie-set matrices. A typical network might contain 100 nodes and 1000 branches and requires storage of 100 000 numbers, mostly equal to zero, in order to describe its topology. In nodal analysis it is necessary in practice to proceed directly to the **nodal admittance matrix** defined by the matrix expression $\mathbf{A}^t\mathbf{YA}$.

The second difficulty arises from the need to repeatedly analyse a circuit at a number of different frequencies so that the frequency response can be generated. Standard matrix inversion procedures are too slow for efficient use of the computer, and a single frequency response evaluation can become unnecessarily expensive. In large circuit analysis the programming effort in adapting matrix inversion to the form of the matrices arising from circuits is amply rewarded by improvements in speed of several orders of magnitude.

Another technique allied to the quest for efficiency is that of node suppression. Nodes of little or no interest to the circuit designer, for example those within a transistor or operational amplifier, may be removed from the analysis. More importantly the Y-parameters of a transistor or other device which may recur many times in a single circuit can be tabulated in advance of the main analysis and entered for solution only at terminal nodes. Five or more nodes per device can often be removed from the analysis, significantly reducing the order of the nodal admittance matrix.

## 4.1 THE NODAL ADMITTANCE MATRIX

The solution for node voltages in Chapter 2 was given in Eq. (2.39) as

$$\mathbf{e}' = (\mathbf{A}^t\mathbf{Y}\mathbf{A})^{-1}\,\mathbf{A}^t(\mathbf{I} - \mathbf{Y}\mathbf{E}) \tag{4.1}$$

where **A** was the reduced incidence matrix, **Y** was the branch admittance matrix modified if necessary to take dependent sources into account, and **I** and **E** were the independent source vectors. The choice of the set of $n_N-1$ node voltages, excluding the reference node, guarantees that the set of simultaneous equations obtained are independent and that the inverse $(\mathbf{A}^t\mathbf{Y}\mathbf{A})^{-1}$ exists. The matrix $\mathbf{Y}' = \mathbf{A}^t\mathbf{Y}\mathbf{A}$ is called the **nodal admittance matrix** (NAM).

The solution for node voltages is slightly more flexible if the equations are formed with the reference node included in the non-reduced incidence matrix $\mathbf{A}_I$. The resulting $n_N \times n_N$ admittance matrix $\mathbf{Y}_I' = \mathbf{A}_I^t\mathbf{Y}\mathbf{A}_I$ is called the **indefinite nodal admittance matrix** (INAM). Before $\mathbf{Y}_I'$ can be inverted its order must be reduced by one. Flexibility in the solution is achieved because nodes designated as input or output, and the reference node, can be changed at will.

### Filling the Nodal Admittance Matrix

Re-examination of an example from Chapter 2 will illustrate the three simple rules for writing down nodal admittance matrix elements by inspection for any circuit. The long-tailed pair amplifier in Example 2.4 was prepared for analysis in Fig. 2.18. This figure is repeated here in Fig. 4.1 with symbols in place of numerical values.

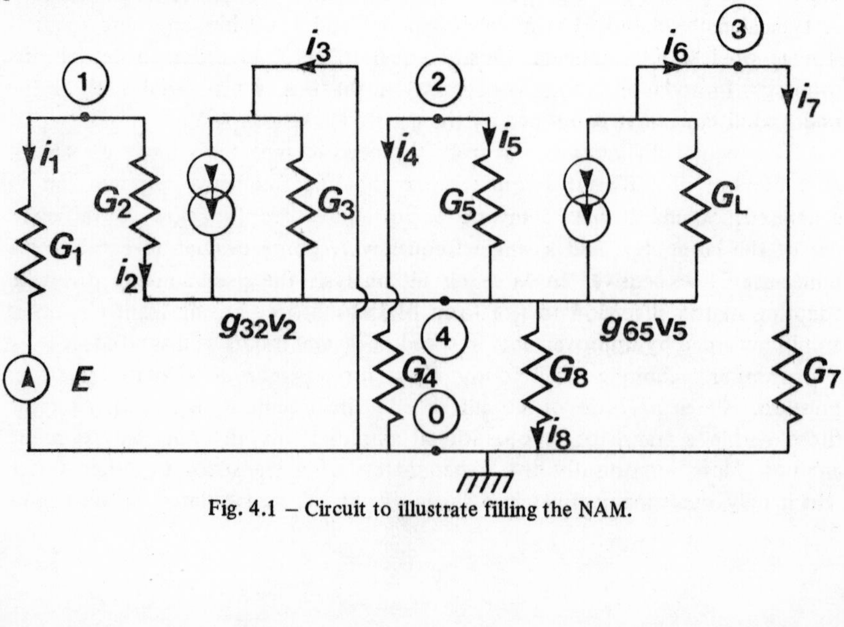


Fig. 4.1 – Circuit to illustrate filling the NAM.

The incidence matrix $\mathbf{A}_\mathrm{I}$ and modified admittance matrix $\mathbf{Y}_\mathrm{m}$ are given by

$$\mathbf{A}_\mathrm{I} = \begin{bmatrix} -1 & 1 & 0 & 0 & 0 \\ 0 & 1 & 0 & 0 & -1 \\ 1 & 0 & 0 & 0 & -1 \\ -1 & 0 & 1 & 0 & 0 \\ 0 & 0 & 1 & 0 & -1 \\ 0 & 0 & 0 & -1 & 1 \\ -1 & 0 & 0 & 1 & 0 \\ -1 & 0 & 0 & 0 & 1 \end{bmatrix}, \quad \mathbf{Y}_\mathrm{m} = \begin{bmatrix} G_1 & & & & & & & 0 \\ & G_2 & & & & & & \\ & g_{32} & G_3 & & & & & \\ & & & G_4 & & & & \\ & & & & G_5 & & & \\ & & & & & G_6 & & \\ & & & & & g_{65} & G_7 & \\ 0 & & & & & & & G_8 \end{bmatrix}$$

The indefinite nodal admittance matrix $\mathbf{A}_\mathrm{I}^\mathrm{t}\mathbf{Y}_\mathrm{m}\mathbf{A}_\mathrm{I}$ obtained by matrix multiplication is given by

$$\begin{bmatrix} G_1+G_3+G_4+G_7+G_8 & -G_1+g_{32} & -G_4 & -G_7 & -G_3-G_8-g_{32} \\ -G_1 & G_1+G_2 & 0 & 0 & -G_2 \\ -G_4 & 0 & G_4+G_5 & 0 & -G_5 \\ -G_7 & 0 & -g_{65} & G_6+G_7 & -G_6+g_{65} \\ -G_3-G_8 & -G_2-g_{32} & g_{65}-G_5 & -G_6 & G_2+G_3+G_5+G_6+G_8+g_{32}-g_{65} \end{bmatrix}$$

The following observations are readily apparent from this example.

1. In all rows and in all columns the sum of the elements is zero. The matrix cannot therefore be inverted since it is indefinite.
2. Considering only branch admittances, each diagonal element $y_{ii}$ of the matrix is the sum of branch admittances connected to the relevant node ⓘ.
3. Each off diagonal element $y_{ij}$ and $y_{ji}$ of the matrix is the negative of the sum of branch admittances connected directly from node ⓘ to node ⓙ.
4. Transfer admittances terms appear in groups of four with two positive and two negative elements.

The rules for filling the INAM are no more than a formal statement of these observations. Before formalising the rules we must introduce a consistent notation for us here and also later when sensitivity is discussed in Chapter 8. Let all circuit admittances including transfer admittances be represented by $y_{kl}^{ij}$. When $i=k$ and $j=l$ the admittance represented is a self admittance connected between nodes ⓘ and ⓙ. When $i \neq k$ or $j \neq l$ the admittance represents a transfer admittance (VCCS) as shown in Fig. 4.2(b).

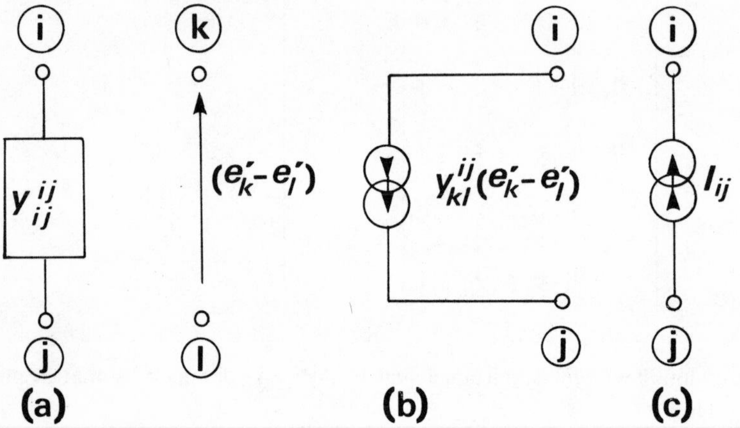


Fig. 4.2 – Self admittance $y_{ij}^{ij}$; (b) Transfer admittance $y_{kl}^{ij}$; (c) Fixed source $I_{ij}$.

A single rule can now be formulated for entering each admittance $y_{kl}^{ij}$ into the nodal admittance matrix. A second rule is required for filling $I'$.

RULE 1

*Each admittance $y_{kl}^{ij}$ is entered into four element positions in the INAM. In position (i,k) and (j,l) the value $y_{kl}^{ij}$ is added to the element, and in positions (j,k) and (i,l) the value $y_{kl}^{il}$ is subtracted from the element.*

RULE 2

*Each fixed source current $I_{ij}$ from node ⓙ to node ⓘ is entered into two element positions in $I'$. In position i the value $I_{ij}$ is added to the element, and in position j the value $I_{ij}$ is the subtracted from the element.*

Several minor limitations appear as a result of discarding the incidence matrix and using the INAM by direct entry. Use of $\mathbf{A}^t(\mathbf{I}-\mathbf{YE})$ is more flexible than filling $\mathbf{I}'$ by Rule 2. In particular the Thévenin conversion $\mathbf{YE}$ may result in frequency dependent elements in $\mathbf{I}'$ which must be then distinguished and entered separately at each frequency. Use of $\mathbf{I}'$ also implies that all VCCS are element dependent since branch or source dependence implies separate treatment of the modified branch admittance matrices in $(\mathbf{A}^t\mathbf{Y}_m\mathbf{A})$ and in $\mathbf{A}^t(\mathbf{I}-\mathbf{Y}_m\mathbf{E})$. $\mathbf{I}'$ can still be properly filled in the common cases where branch and element dependence are different, but care is needed.

Limitations with transfer impedances and voltage or current amplifiers are sometimes more difficult to overcome. Thévénin's theorem or Ohm's law may again be used to convert coupled branches to VCCS form. When reactive branches are involved, however, frequency dependent VCCS are obtained and it is also possible for $\omega^2$ dependence to arise. In fact it is also desirable for transfer admittances or impedances to be complex to allow for mutual inductor and Y-parameter models. In cases where the generalised analysis of Section 2.3 is required some difficulty may well arise in entering the resulting branch admittance elements into the INAM. In many computer programs generality is not attempted, and various restrictions are tolerated on the type of branch or VCCS allowed.

**Direct Solution for Node Voltages**

The indefinite $n_N \times n_N$ nodal admittance matrix $\mathbf{Y}'_I$ must be reduced to $(n_N-1) \times (n_N-1)$ before solution can commence. Usually a reference node is chosen common to both input and output of the circuit. If node ⓘ is the reference, the row $i$ and column $i$ must be deleted from $\mathbf{Y}'_I$ to obtain $\mathbf{Y}'$. Row $i$ must also be deleted in $\mathbf{I}'_I$. In situations where excitation is by one current source $I_{jk}$ only, it is often convenient to delete equation $k$ in $\mathbf{Y}'_I\mathbf{e}' = \mathbf{I}'_I$ leaving $n_{N-1}$ rows. This leaves only one non-zero element in the reduced source matrix $\mathbf{I}'$. Choice of voltage reference node ⓘ makes $e'_i = 0$, and column $i$ in $\mathbf{Y}'_I$ is deleted as before.

The solution for node voltages is given formally by

$$\mathbf{e}' = (\mathbf{Y}')^{-1}\mathbf{I}' \,, \tag{4.2}$$

$$= \begin{bmatrix} \dfrac{\Delta_{11}}{\Delta} & \cdots & \dfrac{\Delta_{(n_N-1)1}}{\Delta} \\ \vdots & & \vdots \\ \dfrac{\Delta_{1(n_N-1)}}{\Delta} & \cdots & \dfrac{\Delta_{(n_N-1)(n_N-1)}}{\Delta} \end{bmatrix} \mathbf{I}' \,, \tag{4.3}$$

where $\Delta_{ij}$ is the cofactor of the element $y_{ij}$ in $\mathbf{Y}'$ and $\Delta$ is equal to the determinant $|\mathbf{Y}'|$. In general the elements $\mathbf{Y}'$ and $\mathbf{I}'$ are complex and frequency dependent. Direct numerical solution for node voltages using Eq. (4.3) (or by Cramer's rule) implies evaluation of $(n-1)n!$ complex operations† and is out of the question for most practical circuits. The methods of solution described in the following sections all using $(n^3+3n^2-n)/3$ operations are still considered inefficient, and quite sophisticated numerical techniques have been developed to

† Multiplications and divisions.

improve computation speed, measured roughly as the number of operations, by another factor of 10; For a 10-node circuit Cramer's rule requires $>3.2 \times 10^7$ operations compared to 430 for the Gauss Matrix Reduction which follows.

**Gauss Matrix Reduction**

The objective of Gauss reduction is to process the matrix equation $\mathbf{Y}'\mathbf{e}' = \mathbf{I}'$ by row and column operations into an equivalent equation $\mathbf{U}\mathbf{e}' = \bar{\mathbf{I}}'$, where $\mathbf{U}$ is an upper triangular matrix (that is, having zero elements below leading diagonal elements equal to unity). The current source column matrix $\mathbf{I}'$ is simultaneously processed to yield $\bar{\mathbf{I}}'$. Solution for $\mathbf{e}'$ is then carried out by substitution instead of by matrix multiplication.

Consider the matrix equation for a general 3-node solution,

$$\begin{bmatrix} y_{11} & y_{12} & y_{13} \\ y_{21} & y_{22} & y_{23} \\ y_{31} & y_{32} & y_{33} \end{bmatrix} \begin{bmatrix} e_1' \\ e_2' \\ e_3' \end{bmatrix} = \begin{bmatrix} I_1' \\ I_2' \\ I_3' \end{bmatrix} . \tag{4.4}$$

Node voltage $e_1'$ can be eliminated from equations row 2 and row 3 of the matrix by generating new rows 2 and 3 in which $y_{21} = y_{31} = 0$. Row 1 requires only that element $u_{11}$ of $\mathbf{U}$ is unity, therefore row 1 is divided through by $y_{11}$ to yield the first row of the partly reduced $\mathbf{Y}'$ matrix in Eq. (4.5). A new row 2 in which the first element is zero is generated from the original row 2 minus $y_{21} \times$ the new row 1. Row 3 is similarly treated, giving finally,

$$\left[\begin{array}{c|cc} 1 & y_{12}/y_{11} & y_{13}/y_{11} \\ \hline 0 & y_{22}-y_{21}y_{12}/y_{11} & y_{23}-y_{21}y_{13}/y_{11} \\ 0 & y_{32}-y_{31}y_{12}/y_{11} & y_{33}-y_{31}y_{13}/y_{11} \end{array}\right] \begin{bmatrix} e_1' \\ e_2' \\ e_3' \end{bmatrix} = \left[\begin{array}{c} I_1'/y_{11} \\ \hline I_2'-I_1'y_{21}/y_{11} \\ I_3'-I_1'y_{31}/y_{11} \end{array}\right] \tag{4.5}$$

The general term $y_{ij}$ has been replaced by $y_{ij} - y_{i1}y_{1j}/y_{11}$ for $i \neq 1$. The process can clearly be repeated for the $2 \times 2$ matrix below the partition indicated by the broken line in Eq. (4.5) to yield a fully reduced matrix with unit terms on the leading diagonal and zero terms below.

Each reduction involves division by a diagonal element. In this example the first element used was $y_{11}$. At the second reduction the element $(y_{22}-y_{21}y_{12}/y_{11})$ will be used. In both cases the element must be non-zero. In general it is usually necessary to change the order of the equations in the matrix to avoid zero elements on the diagonal so that division yields finite elements.

For numerical accuracy it is additionally desirable that the non-zero element chosen (a *pivot*) should be the largest available. The reduction will fail if the equations are indefinite since an all zero row will be generated.

The final form of the equation derived by Gauss reduction is

$$\begin{bmatrix} 1 & u_{12} & u_{13} & u_{14} & \cdots & u_{1n_{N}-1} \\ 0 & 1 & u_{23} & u_{24} & & u_{2n_{N}-1} \\ 0 & 0 & 1 & u_{34} & & \\ \cdot & & & & & \\ \cdot & & & & & \\ \cdot & & & & & \\ 0 & 0 & 0 & & & 1 \end{bmatrix} \begin{bmatrix} e'_1 \\ \cdot \\ \cdot \\ \\ \\ \\ e'_{n_{N}-1} \end{bmatrix} = \begin{bmatrix} \bar{I}_1 \\ \cdot \\ \cdot \\ \\ \\ \\ \bar{I}_{n_{N}-1} \end{bmatrix} \tag{4.6}$$

Solution for node voltages is carried out by *back substitution* in which the final equation in the matrix is solved first, giving

$$\mathbf{e}'_{n_N-1} = \bar{I}_{n_N-1} \ .$$

Each solution is substituted back into the preceding equation in the matrix yielding in turn

$$\mathbf{e}'_{n_N-2}, \mathbf{e}'_{n_N-3}, \ldots, \mathbf{e}'_1 \ .$$

The Gauss reduction is inconvenient in circuit analysis because it is also applied to the column matrix of source currents. Repeated solution with different sets of driving currents is a frequent requirement, and the LU factorisation which preserves the matrix is preferred.

**LU Factorisation**

The LU factorisation is a numerical procedure for factorising a square matrix into the product of a lower triangular matrix **L** and an upper triangular matrix **U**. The upper matrix **U** was used in the previous section. The lower triangular matrix is similar but occupies the lower half of the matrix below the leading diagonal. All terms above the leading diagonal are zero. Terms on the diagonal are not necessarily equal to unity. The nodal equation $\mathbf{Y}'\mathbf{e}' = \mathbf{I}'$ is therefore factorised to give

$$\mathbf{LU}\,\mathbf{e}' = \mathbf{I}', \tag{4.7}$$

or in expanded form for a general 3-node solution

$$\begin{bmatrix} l_{11} & 0 & 0 \\ l_{21} & l_{22} & 0 \\ l_{31} & l_{32} & l_{33} \end{bmatrix} \begin{bmatrix} 1 & u_{12} & u_{13} \\ 0 & 1 & u_{23} \\ 0 & 0 & 1 \end{bmatrix} \begin{bmatrix} e_1' \\ e_2' \\ e_3' \end{bmatrix} = \begin{bmatrix} I_1' \\ I_2' \\ I_3' \end{bmatrix} . \tag{4.8}$$

The numerical factorisation procedure is quite straightforward and is easily observed by equating the elements of $\mathbf{Y}'$ and $\mathbf{LU}$, that is:

$$\begin{bmatrix} l_{11} & l_{11}u_{12} & l_{11}u_{13} \\ l_{21} & l_{21}u_{12}+l_{22} & l_{21}u_{13}+l_{22}u_{23} \\ l_{31} & l_{31}u_{12}+l_{32} & l_{31}u_{13}+l_{32}u_{23}+l_{33} \end{bmatrix} = \begin{bmatrix} y_{11} & y_{12} & y_{13} \\ y_{21} & y_{22} & y_{23} \\ y_{31} & y_{32} & y_{33} \end{bmatrix} .$$

Clearly the first column of each matrix directly gives $l_{11}$ $l_{21}$ $l_{31}$. Substitution down the second column yields $u_{12}$ $l_{22}$ and $l_{32}$ and down the third column yields $u_{13}$ $u_{23}$ and $l_{33}$. In practice the elements of **L** and **U** are stored in a single matrix in the form

$$\begin{bmatrix} l_{11} & u_{12} & u_{13} \\ l_{21} & l_{22} & u_{23} \\ l_{31} & l_{32} & l_{33} \end{bmatrix} ,$$

and since the elements of $\mathbf{Y}'$ used in the solution are not required further, once they have been used, they may be overwritten by the elements of **L** and **U**.

Solution of nodal equations which have been factorised into **L** and **U** proceeds in a similar manner to that for Gaussian reduction. The first step is to solve for $\mathbf{U}\,\mathbf{e}'$ by *forward substitution*. This is the same as back substitution but solves the equations in forward order. Back substitution is then used to obtain $\mathbf{e}'$ from $\mathbf{U}\,\mathbf{e}'$ as in Gaussian reduction. It should be noted that the matrix **U** in LU factorisation and in Gaussian reduction is the same matrix, and that $\bar{\mathbf{I}} = \mathbf{L}^{-1}\mathbf{I}'$. The advantage of LU factorisation is that the current source matrix $\mathbf{I}'$ is not changed and that repetitive solution with different source currents involve only the $n^2$ operations incurred by resubstitution.

**Network Reduction**

In Chapter 1 the star-delta transformation was defined. The effect of the transformation is to eliminate the node at the centre of the star, and it can quite easily be shown that Gaussian matrix reduction is equivalent to network reduction by star-delta transformation. The two numerical calculations, however, are not equivalent, and network reduction is inherently more accurate. The two methods can be illustrated for the simple T network composed of three conductances shown in Fig. 4.3.

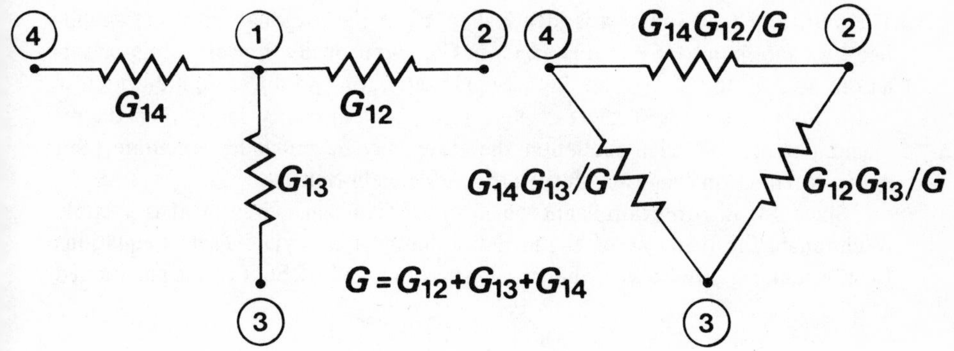


Fig. 4.3 – Star – delta transformation.

The INAM for both star and delta circuits are given by

$$\mathbf{Y}_* = \begin{bmatrix} G_{12}+G_{13}+G_{14} & -G_{12} & -G_{13} & -G_{14} \\ -G_{12} & G_{12} & 0 & 0 \\ -G_{13} & 0 & G_{13} & 0 \\ -G_{14} & 0 & 0 & G_{14} \end{bmatrix}$$

$$\mathbf{Y}_\Delta = \frac{1}{G}\begin{bmatrix} G_{12}G_{13}+G_{12}G_{14} & -G_{12}G_{13} & -G_{12}G_{14} \\ -G_{12}G_{13} & G_{12}G_{13}+G_{13}G_{14} & -G_{13}G_{14} \\ -G_{12}G_{14} & -G_{13}G_{14} & G_{12}G_{14}+G_{13}G_{14} \end{bmatrix}$$

where $G = G_{12} + G_{13} + G_{14}$. Gaussian reduction of the first column of $\mathbf{Y}_*$ results in the matrix

$$\left[\begin{array}{c|ccc} 1 & -G_{12}/G & -G_{13}/G & -G_{14}/G \\ \hline 0 & G_{12}-G_{12}^2/G & -G_{12}G_{13}/G & -G_{12}G_{14}/G \\ 0 & -G_{13}G_{12}/G & G_{13}-G_{13}^2/G & -G_{13}G_{14}/G \\ 0 & -G_{12}G_{14}/G & -G_{13}G_{14}/G & G_{14}-G_{14}^2/G \end{array}\right].$$

The partitioned 3X3 matrix is identical to $\mathbf{Y}_\Delta$ if the diagonal terms are simplified by cancellation of the squared term. Computationally, however, this cancellation, which does not arise in network reduction, produces round-off error which may be serious if the cancelled term is significantly larger than the remaining terms. Gaussian reduction therefore may be much less accurate than network reduction even though the processes are equivalent.

Since LU factorisation is another form of Gaussian reduction it is possible to eliminate the first row of **U** and first column of **L** to yield a set of equations for the reduced circuit with the first node eliminated. If Eq. (4.8) is partitioned as

$$\left[\begin{array}{c|c} l_{11} & \mathbf{0} \\ \hline \mathbf{l} & \mathbf{L}_R \end{array}\right] \left[\begin{array}{c|c} 1 & \mathbf{u} \\ \hline \mathbf{0} & \mathbf{U}_R \end{array}\right] \left[\begin{array}{c} e_1' \\ \hline \mathbf{e}_R' \end{array}\right] = \left[\begin{array}{c} I_1' \\ \hline \mathbf{I}_R' \end{array}\right],$$

then it can easily be shown that

$$[\mathbf{L}_R\,\mathbf{U}_R - \mathbf{l}\,\mathbf{u}]\,\mathbf{e}_R' = \mathbf{I}_R' - l_{11}^{-1}\mathbf{I}_1'\,\mathbf{l}\ .$$

This set of reduced equations is the same as those obtained by Gaussian reduction of the first column, and $[\mathbf{L}_R\mathbf{U}_R - \mathbf{l}\,\mathbf{u}] = \mathbf{Y}_\Delta$.

### Sparse Matrices

The importance of the LU factorisation is not readily apparent unless the special character of network equations is taken into account. Most practical circuits have a nodal admittance matrix with entries clustered near the leading diagonal and with a large number of zero entries elsewhere. This results from the numbering of nodes progressively through a circuit and the predominance of connections to nearby nodes. Obviously, particular circuits can be used as counter examples in which this is not the case. Also it is possible to number nodes in an arbitrary way filling the NAM more evenly. In general, however; the NAM is predominantly filled with zeros. Such matrices are called **sparse**. It is only by exploiting this characteristic feature of circuit matrices that efficient solutions for large networks are possible.

The first important consideration is to preserve sparsity in the factorised matrices. The inverse of the NAM is in general entirely filled, and solution of $(\mathbf{Y}')^{-1}\,\mathbf{I}'$ invariably involves $n^2$ operations. Forward and backward substitution in the solution $\mathbf{L}\,\mathbf{U}\,\mathbf{e}' = \mathbf{I}'$ similarly requires $n^2$ operations. If, however, sparsity can be exploited by avoiding zero elements the $n^2$ operations can be substantially reduced. It is therefore important to maintain sparsity in **L** and **U**.

The problem can be illustrated with the simple T network of Fig. 4.3. The non-zero elements in $\mathbf{Y}_*$ are marked x and zero elements are marked o. If we then carry out the LU factorisation and similarly mark the elements of **L U** we obtain the two matrix forms

$$\mathbf{Y}_*\begin{bmatrix} x & x & x & x \\ x & x & o & o \\ x & o & x & o \\ x & o & o & x \end{bmatrix} \qquad \mathbf{LU}\begin{bmatrix} x & x & x & x \\ x & x & x & x \\ x & x & x & x \\ x & x & x & x \end{bmatrix}.$$

The LU matrix is completely filled. However, if we renumber nodes by interchanging ① and ④ we then obtain

$$\mathbf{Y}_*\begin{bmatrix} x & o & o & x \\ o & x & o & x \\ o & o & x & x \\ x & x & x & x \end{bmatrix} \qquad \mathbf{LU}\begin{bmatrix} x & o & o & x \\ o & x & o & x \\ o & o & x & x \\ x & x & x & x \end{bmatrix}$$

and sparsity is fully preserved. In larger examples the correct order of node numbering is most important, and numbering algorithms have been developed to assign an optimal order. Solution time can be reduced by an order of magnitude with proper ordering.

The second problem is to avoid processing zero elements during factorisation and substitution. This becomes essential when the NAM matrix is too large to keep in computer memory. The first step in this direction is to store only non-zero elements.

The third problem arises in frequency response and non-linear computations where the elements of NAM change between solutions. Here it is important to avoid re-factoring the whole admittance matrix when only part of it changes. Various ordering and partition arrangements are used to reduce the number of operations required.

Most modern large analysis programmes incorporate features based on sparse matrix techniques. These topics and many others are discussed more fully in textbooks and research papers on computer-aided design. Here we have merely mentioned some of the problems.

*Example* 4.1

Fill the INAM for the filter circuit in Fig. 4.4(a) at 10 kHz normalised to 2 rad/sec.

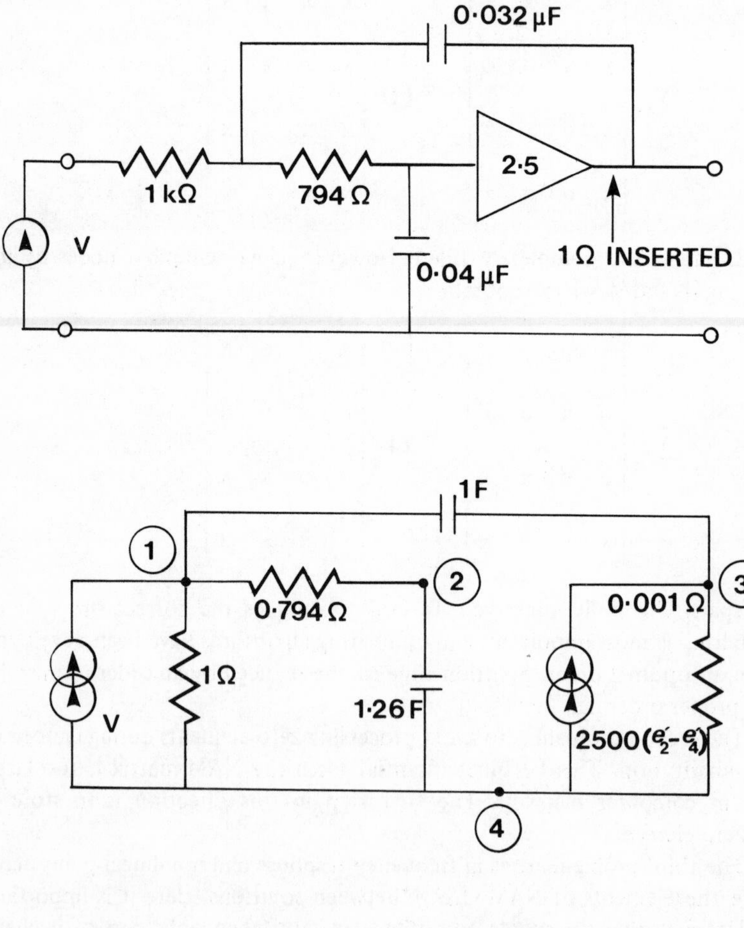


Fig. 4.4 – (a) Filter circuit for Example 4.1.; (b) Normalised circuit for analysis. Normalising Factors: Impedance $10^{-3}$, Frequency $1/10^4\,\pi$, Current $10^3$.

The circuite has been redrawn for analysis in Fig. 4.4(b). Three points about the redrawn circuit should be noted.

1. The amplifier is specified as a perfect VCVS with gain 2.5. In practice it is likely to be an operational amplifier with resistive feedback to the negative input, and the inserted 1 Ω output resistance is likely to exist.
2. The circuit is drawn as a voltage amplifier, but analysis by the NAM implies excitation by a current source. To avoid introduction of a source resistance the voltage source is converted to a current source using the 1 kΩ input resistance of the filter.
3. The reason for scaling the circuit for analysis is that solution is more accurate and computer overflow avoided if the NAM elements are as near to unity as possible. Two factors are used, and to obtain the scaled, or **normalised**, circuit elements both the impedance and the frequency have to be multiplied by their scaling factor. In this example the impedance scale factor is $10^{-3}$, therefore resistors and inductors are multiplied by $10^{-3}$ and capacitors by $10^3$ to normalise them. The frequency scaling factor of $1/10^4\pi$ means that both inductors and capacitors must multiplied additionally by $10^4\pi$ to normalise them. In this case the frequency response of the normalised circuit is the same at 1 rad/sec as at 5 kHz in the original circuit. If the circuit voltages are to remain unchanged then all circuit currents including any VCCS must be divided by the impedance scale factor.

The INAM can now be filled with reference to the rules given at the beginning of this section. Recall that a self admittance identified with element $y_{kl}^{ij}$ is, for example, $y_{13}^{13}$ for the 1 F capacitor. The VCCS is this case is $y_{24}^{43}$. The INAM at 2 rad/sec is therefore given by

$$\begin{bmatrix} 1+1/0.794+j2 & -1/0.794 & -j2 & -1 \\ -1/0.794 & 1/0.794+j2.52 & 0 & -j\,2.52 \\ -j2 & -2500 & 10^3+j2 & 2500-10^3 \\ -1 & 2500-j2.52 & -10^3 & 1+10^3-2500+j2.52 \end{bmatrix}$$

$$= \begin{bmatrix} 2.26+j2 & -1.26 & -j2 & -1 \\ -1.26 & 1.26+j2.52 & 0 & -j2.52 \\ -j2 & -2500 & 1000+j2 & 1500 \\ -1 & 2500-j2.52 & -1000 & -1499+j2.52 \end{bmatrix}$$

The objectives of normalising have been achieved since the elemtns of this matrix are generally colose to unity. The one exception is the group of elements derived from the inserted 1 Ω output resistance of the amplifier. This filter incorporating a real operational amplifier will also be solved, using the computer programme in Section 4.4.

## 4.2 ADMITTANCE PARAMETERS

### Multiterminal Networks

Many circuit elements, for example transistors and operational amplifiers, have more terminals than the standard branch which has only two. So far we have assumed that such elements can be internally modelled by equivalent circuits consisting of two terminal branches together with coupling between branches. This approach leads to complicated models with many internal nodes of no interest to the circuit designer which nevertheless substantially increase the order of matrix solutions.

In this section we will consider the direct inclusion of multiterminal elements or networks in NAM analysis by means of admittance parameters. Two main benefits are gained by this approach. As well as the reduction in the order of the NAM by the elimination of insignificant nodes, it is also possible to repeatedly use computed sets of parameters for devices or sub-networks which recur throughout a large circuit.

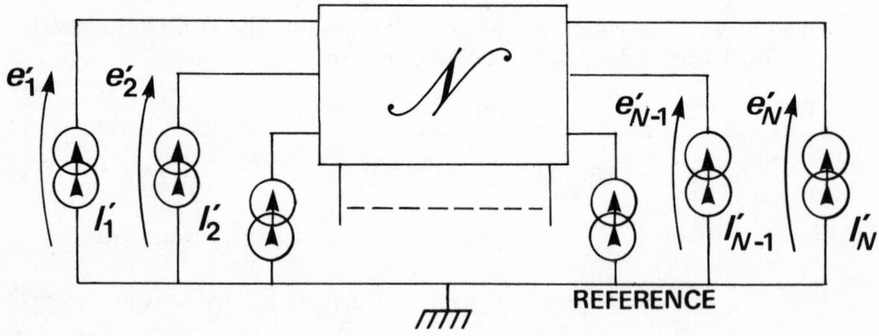


Fig. 4.5 – $N$-terminal network.

Consider the arbitrary network with $N$ terminals in Fig. 4.5. For any circuit consisting of standard branches, we should be able to write down the elements of the INAM matrix by inspection. The matrix equation

$$\mathbf{I}_\mathrm{I}' = \mathbf{Y}_\mathrm{I}' \, \mathbf{e}_\mathrm{I}'$$

relates source currents and the unknown terminal node voltages by the indefinite nodal admittance matrix. Expanding the matrices yields

$$\begin{bmatrix} I'_1 \\ I'_2 \\ \vdots \\ I'_N \end{bmatrix} = \begin{bmatrix} y_{11} & y_{12} & & y_{1N} \\ \cdot & & & \cdot \\ \vdots & & & \vdots \\ y_{N1} & \cdot & \cdot & y_{NN} \end{bmatrix} \begin{bmatrix} e'_1 \\ \cdot \\ \vdots \\ e'_N \end{bmatrix}.$$

The elements of $\mathbf{Y}'_\mathrm{I}$ are known as the indefinite admittance parameters. Each element $y_{ij}$ can be determined from measurements using the definition

$$y_{ij} = \frac{I'_i}{e'_j} \quad \text{for all } e'_k = 0, k \neq j.$$

Measurements are made under small-signal conditions at any required frequency. The real difficulties arise in the provision of a reliable high-frequency short circuit at unexcited nodes while measuring current. Additional circuit elements associated with probes, bias power supplies, mounting capacitances, etc. must all be included within the INAM actually measured. The values of the INAM elements for the $N$-terminal network only must then be extracted by computation.

Two sets of operations must be defined before admittance parameters can be used to combine networks in parallel. Nodes can either be added or removed from a network. Addition is mainly used to provide nodes for connection. The more usual technique is node removal, or *suppression*, in which networks are much reduced in size for insertion into nodal analysis and for storage.

**Node Addition**

There are two ways in which it is important to add nodes to the multiterminal element.

*Reference node*

The conversion from the NAM to the INAM requires the introduction of the row and column associated with the reference node. The NAM for the amplifier in Fig. 4.6 is given by

$$\begin{bmatrix} G_1 & 0 \\ g_\mathrm{m} & G_2 \end{bmatrix}$$

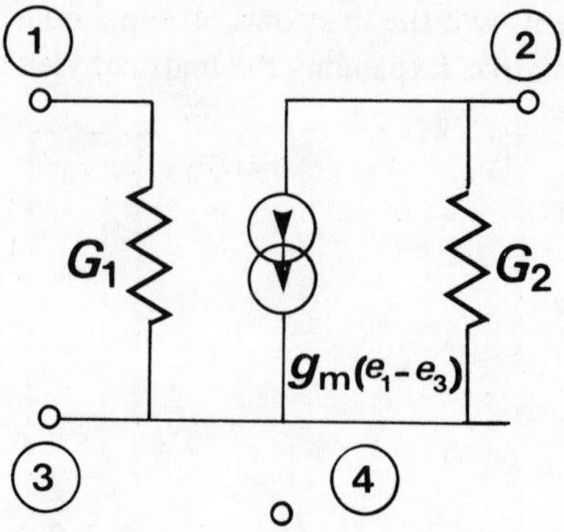


Fig. 4.6 – Node addition

Adding the third row and column such that all rows and columns sum to zero yields the INAM

$$\begin{bmatrix} G_1 & 0 & -G_1 \\ g_m & G_2 & -g_m - G_2 \\ -G_1 - g_m & -G_2 & G_1 + G_2 + g_m \end{bmatrix}$$

*Unconnected nodes*

Addition of an unconnected node is necessary if networks are to be combined or if elements are to be added to the existing network. Since no admittances are connected to the node all INAM elements identified with the node are zero. For example, the addition of a fourth node to the amplifier circuit in Fig. 4.6 results in the INAM.

$$\begin{bmatrix} G_1 & 0 & -G_1 & 0 \\ g_m & G_2 & -g_m - G_2 & 0 \\ -g_m - G_1 & -G_2 & G_1 + G_2 + g_m & 0 \\ 0 & 0 & 0 & 0 \end{bmatrix}.$$

**Node removal**

There are three main methods by which terminal or internal nodes are removed from a multiterminal element.

*Connection to reference node*

If terminal ⓘ is connected to the reference node then $e_i' = 0$. Column $i$ of the INAM may therefore be deleted since all elements in that column are multiplied by zero. Any convenient row of the matrix may be deleted to give a NAM. Usually source currents are all assumed to originate from the reference node, and deletion of row $i$ avoids exciting that node by the negative of the sum of all source currents. However, source current $I_{jk}$ from node ⓚ to node ⓙ may be specified and it is then more convenient to delete row $k$ instead.

*Connection of two nodes*

When two nodes ⓘ and ⓙ are connected together $e_i' = e_j' = e_{ij}'$. The source current $I_{ij}'$ at the combined node equals the sum of the individual source currents $I_i' + I_j'$. Consider, for example, the 4X4 matrix equation (ignoring the partitions which are used later),

$$\left[\begin{array}{cc|cc} y_{11} & y_{12} & y_{13} & y_{14} \\ y_{21} & y_{22} & y_{23} & y_{24} \\ \hline y_{31} & y_{32} & y_{33} & y_{34} \\ y_{41} & y_{42} & y_{43} & y_{44} \end{array}\right] \left[\begin{array}{c} e_1' \\ e_2' \\ \hline e_3' \\ e_4' \end{array}\right] = \left[\begin{array}{c} I_1' \\ I_2' \\ \hline I_3' \\ I_4' \end{array}\right] . \tag{4.9}$$

When nodes ② and ③ are connected $e_{23}' = e_2' = e_3'$, $I_{23}' = I_2' + I_3'$, and

$$\begin{bmatrix} y_{11} & y_{12} + y_{13} & y_{14} \\ y_{21} + y_{31} & y_{22} + y_{23} + y_{32} + y_{33} & y_{24} + y_{34} \\ y_{41} & y_{42} + y_{43} & y_{44} \end{bmatrix} \begin{bmatrix} e_1' \\ e_{23}' \\ e_4' \end{bmatrix} = \begin{bmatrix} I_1' \\ I_{23}' \\ I_4' \end{bmatrix} .$$

*Nullators and norators*

Nullators and norators are two terminal circuit elements defined by the following equations

$$\text{Nullator} \qquad v = 0 \qquad\qquad i = 0 \,,$$

$$\text{Norator} \qquad v = \text{arbitrary} \qquad\qquad i = \text{arbitrary} \,.$$

When used in pairs they provide a more general reduction of the INAM than that provided by connecting two nodes together.

The nullator is a short-circuit that takes no current. Two nodes connected by a nullator are therefore constrained to have the same voltage, but currents remain separated at the individual nodes. Consider again the 4X4 matrix equation (4.9). If a nullator is connected from node ② to node ③ then $e_2' = e_3' = e_{23}'$. Columns 2 and 3 of the matrix $\mathbf{Y}'$ are therefore added and $e_{23}'$ used in place of $e_2'$ and $e_3'$ in e', that is:

$$\begin{bmatrix} y_{11} & y_{12}+y_{13} & y_{14} \\ y_{21} & y_{22}+y_{23} & y_{24} \\ y_{31} & y_{32}+y_{33} & y_{34} \\ y_{41} & y_{42}+y_{43} & y_{44} \end{bmatrix} \begin{bmatrix} e_1' \\ e_{23}' \\ e_4' \end{bmatrix} = \begin{bmatrix} I_1' \\ I_2' \\ I_3' \\ I_4' \end{bmatrix}$$

The norator is an open circuit which allows arbitrary current to flow between the connected nodes. Voltages therefore remain separated. If a norator is connected between nodes ① and ② then the total current at the nodes is available to either and $I_{12}' = I_1' + I_2'$. Votages $e_1'$ and $e_2'$ exist separately. Rows 1 and 2 of the matrix equation (4.9) must therefore be added.

The combined effect of the nullator-norator combination is therefore given by

$$\begin{bmatrix} y_{11}+y_{21} & y_{12}+y_{13}+y_{22}+y_{23} & y_{14}+y_{24} \\ y_{31} & y_{32}+y_{33} & y_{34} \\ y_{41} & y_{42}+y_{43} & y_{44} \end{bmatrix} \begin{bmatrix} e_1' \\ e_{23}' \\ e_4' \end{bmatrix} = \begin{bmatrix} I_{12}' \\ I_3' \\ I_4' \end{bmatrix}$$

Nullators and norators are included in nodal admittance matrix analysis by adding columns or rows respectively of the INAM. When they are connected in parallel between the same pair of nodes the matrix reduction is identical to that in the previous section. This should of course be expected since a parallel combination of a nullator and a norator is equivalent to a short circuit.

*Node suppression*

Gauss matrix reduction and network reduction both have the effect of progressively removing nodes from a circuit. Any unwanted internal node or terminal node ⓚ of a multiterminal element can be suppressed using the general term for pivotal reduction,

$$y_{ij} - y_{ik}y_{kj}/y_{kk} \tag{4.10}$$

This expression is derived from Eq. (4.5) taking the general case for node k instead of node ①.

Since the node to be suppressed is not usually driven by a current source, indeed it is often not accessible at all, the node is assumed to be unexcited. The reduction therefore has no effect on the current source vector other than the deletion of the appropriate row. Suppression is completed after use of expression (4.10) by deleting row and column k of $\mathbf{Y}'$ and row k of $\mathbf{e}'$ and $\mathbf{I}'$.

It is also possible to remove all unwanted nodes by matrix operations on the partitioned nodal admittance equations. For example, if we wish to suppress nodes ① and ② from the four-node network defined by Eq. (4.9) we first rewrite the equation in the partitioned form,

$$\begin{bmatrix} \mathbf{Y}_{11} & \mathbf{Y}_{12} \\ \mathbf{Y}_{21} & \mathbf{Y}_{22} \end{bmatrix} \begin{bmatrix} \mathbf{E}_a \\ \mathbf{E}_b \end{bmatrix} = \begin{bmatrix} \mathbf{I}_a \\ \mathbf{I}_b \end{bmatrix} . \tag{4.11}$$

All matrices can be identified by reference to Eq. (4.9) in which the partitions are shown. As before, suppressed nodes are assumed unexcited and $\mathbf{I}_a = \mathbf{0} = [0 \ \ 0]^t$. From the first row of Eq. (4.11)

$$\mathbf{Y}_{11}\mathbf{E}_a + \mathbf{Y}_{12}\mathbf{E}_b = \mathbf{I}_a = \mathbf{0}$$

therefore, $\mathbf{E}_a = -\mathbf{Y}_{11}^{-1}\mathbf{Y}_{12}\mathbf{E}_b$ .

Substituting into the second row,

$$[\mathbf{Y}_{22} - \mathbf{Y}_{21}\mathbf{Y}_{11}^{-1}\mathbf{Y}_{12}]\,\mathbf{E}_b = \mathbf{I}_b \ , \tag{4.12}$$

or more fully,

$$[\mathbf{Y}_{22} - \mathbf{Y}_{21}\mathbf{Y}_{11}^{-1}\mathbf{Y}_{12}] \begin{bmatrix} e_3' \\ e_4' \end{bmatrix} = \begin{bmatrix} I_3' \\ I_4' \end{bmatrix} . \tag{4.13}$$

This expression is readily identified with that for pivotal reduction in Eq. (4.10) noting that $y_{kk}$ corresponds to a $1\times1$ matrix $\mathbf{Y}_{11}$ containing $y_{11}$.

The first part of Example 5.1 (page 223) is a simple application of node suppression in which the admittance parameters of an operational amplifier with negative feedback are determined.

**Networks in parallel**

Consider two networks A and B with an equal number of nodes $N$ and admittance parameters $\mathbf{Y}^A$ and $\mathbf{Y}^B$ respectively. Connect the networks in parallel as shown in Fig. 4.7.

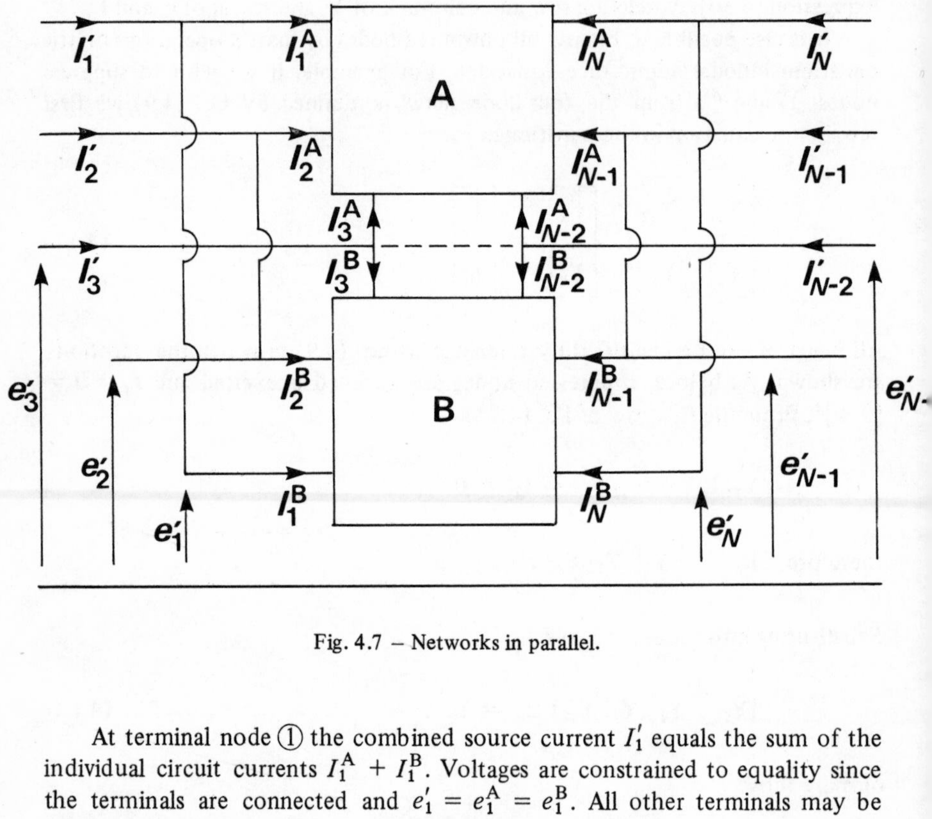


Fig. 4.7 – Networks in parallel.

At terminal node ① the combined source current $I_1'$ equals the sum of the individual circuit currents $I_1^A + I_1^B$. Voltages are constrained to equality since the terminals are connected and $e_1' = e_1^A = e_1^B$. All other terminals may be similarly treated to yield the matrix relationships,

$$\mathbf{I}' = \mathbf{I}^A + \mathbf{I}^B ,$$

$$\mathbf{e}' = \mathbf{e}^A = \mathbf{e}^B .$$

Since $\mathbf{I}^A = \mathbf{Y}^A\mathbf{e}^A$ and $\mathbf{I}^B = \mathbf{Y}^B\mathbf{e}^B$,

$$\mathbf{I}' = (\mathbf{Y}^A + \mathbf{Y}^B)\,\mathbf{e}' = \mathbf{Y}'\,\mathbf{e}' .$$

Therefore, $\mathbf{Y}' = \mathbf{Y}^A + \mathbf{Y}^B$ .

The admittance parameters of the two networks connected in parallel must therefore be added to obtain the parameters of the combined network. For this to be done the number of terminals $N$ must be the same in each circuit. Unconnected nodes are therefore added to each circuit admittance matrix in the form of rows and columns with all elements equal to zero. The matrices $\mathbf{Y}^A$ and $\mathbf{Y}^B$ can then be added. However, the main application of this technique is to insert small sub-networks, usually transistor models, into large circuits, and it is not necessary to add zero matrix elements provided that the relevant nodes are clearly identified.

**$N$-Port networks**

A network port is defined as a pair of terminals for which the current entering one terminal equals the current leaving the other. Multiterminal networks may be treated as $N$-port networks with the terminals arranged in pairs. Port admittance parameters may then be defined and measured in the same way as idefinite admittance parameters. The form of the port equations $\mathbf{Y}_p\mathbf{v} = \mathbf{i}$ is the same as for the NAM but with the advantage that the number of variables involved is one half that of the NAM terminal description. The $N$-port network in Fig. 4.8 has $2N$ terminals grouped in pairs and numbered with port voltages and currents defined according to conventions used here and throughout the next chapter.

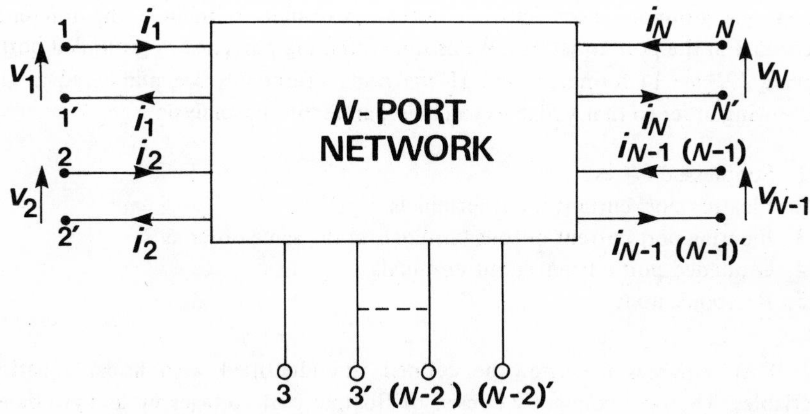


Fig. 4.8 – N-port network.

Conversion from the NAM description of a network to the port description is quite straightforward even though the matrix expressions are involved. The technique is important as a link between the two methods of analysis, adding flexibility in particular to two-port analysis which is treated in the following chapter. Two-port analysis can be more accurate than NAM methods and is particularly efficient for filter networks including those with distributed circuit elements. There are, however, limitations on the topology of circuits that can be analysed, but these can be overcome by combining two-port analysis with NAM circuit reduction technique.

The most common conversion is from a three-terminal NAM description of a network to a two-port description or more generally from $N$+1 terminals to $N$-ports. The $N$-port in Fig. 4.8 can be considered in this way by connecting all the port terminals $1', 2', \ldots N'$ together as the reference node. The port currents $i_1, i_2, \ldots i_N$, and voltages $v_1, v_2, \ldots v_N$ then correspond to the node currents $I'_1, I'_2, \ldots I'_N$, and voltages $e'_1, e'_2 \ldots e'_N$ respectively. The port equations $\mathbf{Y_p v = i}$ and node equations $\mathbf{Y'e' = I'}$ are then identical and $\mathbf{Y_p = Y'}$. The following general conversion procedure includes this special case.

*NAM to N-Port conversion*

Consider the multiterminal network in Fig. 4.9. Suppressed nodes need not be considered here since Eq. (4.12) in the previous section was derived to remove them from the NAM. They will, however, be present in the network, and the following conversion assumes that the NAM has previously been reduced by node suppression. Ports are assumed to be of two types. Floating ports consist of two network terminals not connected to the reference. Grounded ports consist of one network terminal and a connection to the reference terminal. At least one grounded port must be present in order to include the reference terminal in the port equations. We assume $l$ floating ports and $m$ grounded ports, giving $(2l+m+1)$ terminals and $(l+m)$ ports. Terminals are numbered in the following order so that variables can be separated in the analysis.

1. Suppressed nodes.
2. Floating port current input terminals.
3. Floating port current output terminals in the same order as 2.
4. Grounded port current input terminals.
5. Reference node.

Port variables may now be defined and identified with nodal equation variables. The port voltages $\mathbf{v}$ consist of floating port voltages $\mathbf{v_F}$ and grounded port voltages $\mathbf{v_R}$, therefore $\mathbf{v} = [\mathbf{v_F}\ \mathbf{v_R}]^t$. From Fig. 4.9,

$$\mathbf{v_F} = [(e'_1 - e'_{l+1})(e'_2 - e'_{l+2}) \ldots (e'_l - e'_{2l})]^t$$

$$\mathbf{v_R} = [e'_{2l+1} \ldots . e'_{2l+m}]^t \,.$$

The terminal voltage in $\mathbf{v_F}$ may be separated by defining $\mathbf{v_p} = [e'_1 \ldots . e'_l]^t$ and $\mathbf{v_N} = [e'_{l+1} \ldots . e'_{2l}]^t$ so that $\mathbf{v_F} = \mathbf{v_P} - \mathbf{v_N}$.

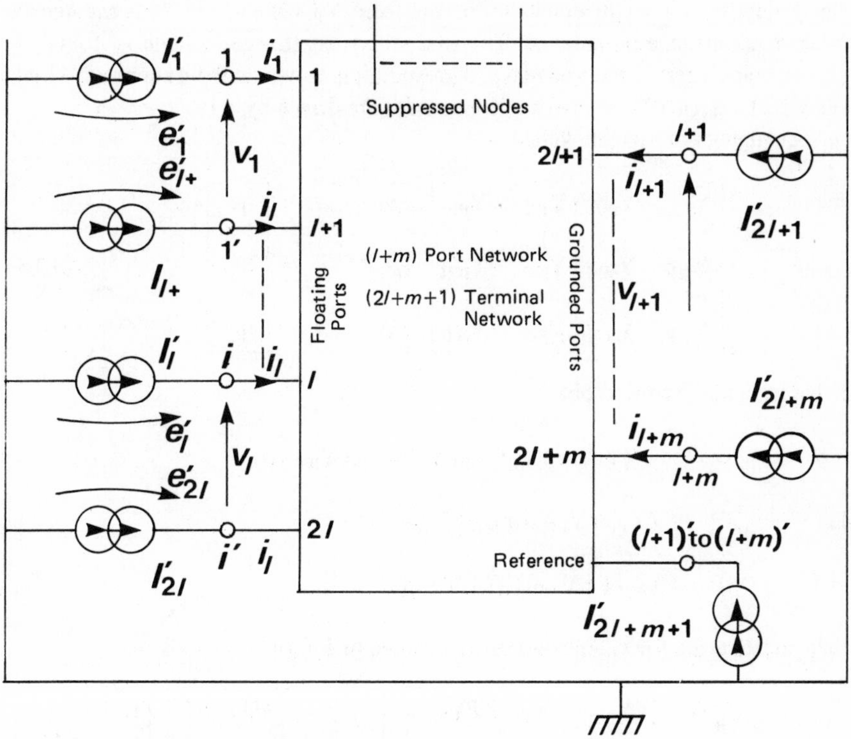


Fig. 4.9 – Terminal to port conversion.

The port currents **i** are equal to the nodal currents, and from Fig. 4.9 we can directly obtain

$$\mathbf{i}_F = [I'_1 \ldots . I'_l]^t$$

$$= -[I'_{l+1} \ldots . I'_{2l}]^t$$

$$\mathbf{i}_R = [I'_{2l+1} \ldots . I'_{2l+m}]^t$$

where $\mathbf{i}_F$ and $\mathbf{i}_R$ are the floating and grounded port currents respectively.

The nodal equations $\mathbf{Y}'\ \mathbf{e}' = \mathbf{I}'$ incorporating the port variables defined above can now be rewritten in partitioned form as

$$\begin{bmatrix} \mathbf{Y}_{PP} & \mathbf{Y}_{PN} & \mathbf{Y}_{PR} \\ \mathbf{Y}_{NP} & \mathbf{Y}_{NN} & \mathbf{Y}_{NR} \\ \mathbf{Y}_{RP} & \mathbf{Y}_{RN} & \mathbf{Y}_{RR} \end{bmatrix} \begin{bmatrix} \mathbf{v}_P \\ \mathbf{v}_N \\ \mathbf{v}_R \end{bmatrix} = \begin{bmatrix} \mathbf{i}_F \\ -\mathbf{i}_F \\ \mathbf{i}_R \end{bmatrix} . \qquad (4.14)$$

The objective is a set of equations of the form $\mathbf{Y_P\, v = i}$ where $\mathbf{Y_P}$ is the matrix of port admittance parameters. The first step is to change variable $\mathbf{v_P}$ to $\mathbf{v_F} = \mathbf{v_P} - \mathbf{v_N}$, and then the set of nodes corresponding to $\mathbf{v_N}$ must be suppressed by a version of Eq. (4.12) noting that these nodes are driven by $-\mathbf{i_F}$.

Changing the variable yeilds

$$\begin{bmatrix} \mathbf{Y_{PP}} & \mathbf{Y_{PP}+Y_{PN}} & \mathbf{Y_{PR}} \\ \mathbf{Y_{NP}} & \mathbf{Y_{NN}+Y_{NP}} & \mathbf{Y_{NR}} \\ \mathbf{Y_{RP}} & \mathbf{Y_{RN}+Y_{RP}} & \mathbf{Y_{RR}} \end{bmatrix} \begin{bmatrix} \mathbf{v_P - v_N} \\ \mathbf{v_N} \\ \mathbf{v_R} \end{bmatrix} = \begin{bmatrix} \mathbf{i_F} \\ -\mathbf{i_F} \\ \mathbf{i_R} \end{bmatrix} . \tag{4.15}$$

Solve the middle row to obtain

$$\mathbf{v_N} = -(\mathbf{Y_{NN}+Y_{NP}})^{-1}(\mathbf{i_F + Y_{NP}v_P + Y_{NR}v_R}) .$$

Let $\mathbf{P} = (\mathbf{Y_{PP}+Y_{PN}})(\mathbf{Y_{NN}+Y_{NP}})^{-1}$ ,

and $\mathbf{R} = (\mathbf{Y_{RN}+Y_{RP}})(\mathbf{Y_{NN}+Y_{NP}})^{-1}$ ,

then substituting for $\mathbf{v_N}$ in rows one and three of Eq. (4.15) yields

$$\begin{bmatrix} \mathbf{Y_{PP} - PY_{NP}} & \mathbf{Y_{PR} - PY_{NR}} \\ \mathbf{Y_{RP} - RY_{NP}} & \mathbf{Y_{RR} - RY_{NR}} \end{bmatrix} \begin{bmatrix} \mathbf{v_F} \\ \mathbf{v_R} \end{bmatrix} = \begin{bmatrix} \mathbf{U+P} & \mathbf{0} \\ \mathbf{R} & \mathbf{U} \end{bmatrix} \begin{bmatrix} \mathbf{i_F} \\ \mathbf{i_R} \end{bmatrix} . \tag{4.16}$$

where $\mathbf{U}$ is a unit matrix of appropriate order and $\mathbf{0}$ is an all zero matrix. Multiplication of the first row of Eq. (4.16) by $\mathbf{W} = (\mathbf{U+P})^{-1}$ and substitution into the second row $\mathbf{R\, i_f}$ yields finally,

$$\begin{bmatrix} \mathbf{W(Y_{PP} - PY_{NP})} & \mathbf{W(Y_{PR} - PY_{NR})} \\ \mathbf{Y_{RP} - R(Y_{NP} + WY_{PP} - WPY_{NP})} & \mathbf{Y_{RR} - R(Y_{NR} + WY_{PR} - WPY_{NR})} \end{bmatrix} \begin{bmatrix} \mathbf{v_F} \\ \mathbf{v_R} \end{bmatrix} = \begin{bmatrix} \mathbf{i_F} \\ \mathbf{i_R} \end{bmatrix} \tag{4.17}$$

This matrix expression gives the set of $(l+m)$ port parameters derived from the $(2l+m)$ terminal NAM. The conversion is general and applies to any combination of terminals provided that $m \geqslant 1$. If there are no floating ports, $l=0$ and $\mathbf{Y_{RR}}$ is the only partition to exist in the NAM, Eq. (4.17) then reduces to $\mathbf{Y_{RR}\, v_R} = \mathbf{i_R}$ and, as we noted in the previous section, $\mathbf{Y_{RR}} = \mathbf{Y}'$. Use of the NAM with the reference node and one terminal as one of the ports ensures that the matrix inversions $(\mathbf{Y_{NN}+Y_{NP}})^{-1}$ and $(\mathbf{U+P})^{-1}$ both exist.

The conversion most used is that from four terminals to two ports. A simplified version of the port matrix in Eq. (4.17) for this particular case is derived in the following example. The complication involved even for the simple resistor network in this example demonstrates once again the necessity for computation in this work.

Practical application of port analysis is usually confined to two-port analysis which is the subject of the next chapter.

*Example* 4.2

Determine the two-port admittance parameters for the resistor network in Fig. 4.10

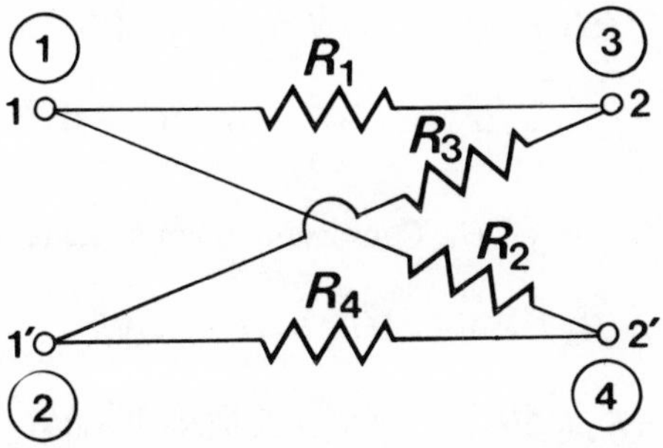


Fig. 4.10 – Resistor network for Example 4.2.

Terminal and port numbering in Fig. 4.10 has been carried out in accordance with Fig. 4.9, and the NAM may be written down by inspection with terminal ④ as reference. Using conductances $G_1$ to $G_4$ equal to $1/R_1$ to $1/R_4$ respectively, the NAM is given by

$$\mathbf{Y}' = \left[\begin{array}{c|c|c} y_{11} & y_{12} & y_{13} \\ \hline y_{21} & y_{22} & y_{23} \\ \hline y_{31} & y_{32} & y_{33} \end{array}\right] = \begin{bmatrix} G_1+G_2 & 0 & -G_1 \\ 0 & G_3+G_4 & -G_3 \\ -G_1 & -G_3 & G_1+G_3 \end{bmatrix} . \quad (4.18)$$

Partitions inserted in $\mathbf{Y}'$ show that the sub-matrices corresponding to those in Eq. (4.14) are all 1×1 with suffixes P,N, and R, equal to 1, 2, and 3 respectively, identifying the elements here. Since this case is of practical importance the two-port admittance matrix will first be derived in terms of the elements of $\mathbf{Y}'$. By substitution therefore,

$$\mathbf{P} = [(y_{11}+y_{12})/(y_{21}+y_{22})] ,$$

$$\mathbf{R} = [(y_{31}+y_{12})/(y_{21}+y_{22})] ,$$

$$\mathbf{W} = [(y_{21}+y_{22})/(y_{11}+y_{12}+y_{21}+y_{22})] ,$$

and from Eq. (4.17) the two-port parameters are

$$\begin{bmatrix} \dfrac{y_{11}y_{22}-y_{12}y_{21}}{y_{11}+y_{12}+y_{21}+y_{22}} & \dfrac{y_{13}(y_{21}+y_{22})-y_{23}(y_{11}+y_{12})}{y_{11}+y_{12}+y_{21}+y_{22}} \\ \dfrac{y_{31}(y_{12}+y_{22})-y_{32}(y_{11}+y_{21})}{y_{11}+y_{12}+y_{21}+y_{22}} & y_{33}-\dfrac{(y_{31}+y_{32})(y_{13}+y_{23})}{y_{11}+y_{12}+y_{21}+y_{22}} \end{bmatrix}$$

Substitution for conductances yields

$$\begin{bmatrix} (G_1+G_2)(G_3+G_4)/G & (G_2G_3-G_1G_4)/G \\ (G_2G_3 \quad -G_1G_4)/G & (G_1+G_3)(G_2+G_4)/G \end{bmatrix},$$

where $G = G_1 + G_2 + G_3 + G_4$. Conversion of conductances to resistances gives

$$\begin{bmatrix} (R_1+R_2)(R_3+R_4)/R & (R_1R_4-R_2R_3)/R \\ (R_1R_4-R_2R_3)/R & (R_1+R_3)(R_2+R_4)/R \end{bmatrix},$$

where $R = R_1R_2R_3 + R_1R_2R_4 + R_1R_3R_4 + R_2R_3R_4$.

This result can be confirmed by deriving each element $y_{jk}$ directly from its definition $(i_j/v_k)$ with all $v_l = 0,\ l \neq k$. Fig. 4.11 shows the circuit for measurement or calculation of $y_{12}$.

Since port 1 is short-circuit,

$$e_3 = \frac{R_2R_4/(R_2+R_4)}{R_1R_3/(R_1+R_3)\ +\ R_2R_4/(R_2+R_4)}\,v_2\ .$$

Summing currents at terminal ①,

$$i_1 = e_3/R_2-(v_2-e_3)/R_1\ .$$

Therefore,

$$y_{12} = \frac{i_1}{v_2} = \frac{R_2R_4/(R_2+R_4)}{R_1R_3/(R_1+R_3)\ +\ R_2R_4/(R_2+R_4)}\left(\frac{1}{R_2}+\frac{1}{R_1}\right)-\frac{1}{R_1}$$

$$= (R_1R_4-R_2R_3)/(R_1R_2R_3+R_1R_2R_4+R_1R_3R_4+R_2R_3R_4)\ .$$

Element $y_{22}$ can be derived from the same circuit as $i_2/v_2$. This is simply the conductance between terminal ② and ②′, that is,

$$1/\{R_1 R_3/(R_1 + R_3) + R_2 R_4/(R_2 + R_4)\} .$$

Elements $y_{11}$ and $y_{21}$ are obtained by application of a voltage $v_1$ at port 1 and a short-circuit at port 2. This is equivalent in Fig. 4.11 to interchanging $R_2$ and $R_3$ and the port numbering. Thus $y_{21}$ is readily observed to be equal to $y_{12}$, as we should expect for a passive circuit, and $y_{11}$ can be obtained from $y_{22}$ by interchanging $R_2$ and $R_3$.

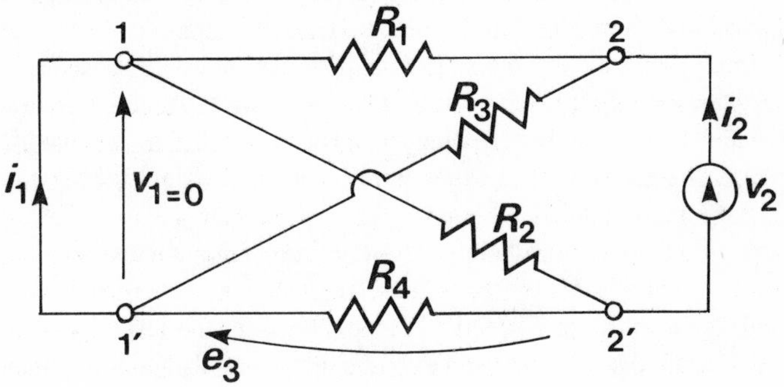


Fig. 4.11 – Circuit for $y_{12}$ and $y_{22}$ measurement.

## 4.3 SMALL-SIGNAL MODELS

Large signal d.c. models for bipolar and MOS transistors were derived in Chapter 3 as three-terminal networks containing diodes, dependent sources, and resistors. These models were then linearised at each step of an iterative solution for node voltages. However, the linearised d.c. model had little significance other than its role as an important intermediate step in the analysis because we were primarily concerned with large signals. If we now choose to analyse a circuit in which voltages differ from the d.c. bias voltages determined from non-linear d.c analysis, the linearised d.c model will adequately describe the behaviour of the original non-linear circuit provided that the voltage increments are small. Assuming, therefore, that the circuit operates within the range of validity of the linearised model, linear small-signal analysis may be used instead of non-linear analysis.

Since the circuit can now be assumed linear at its operating point the principle of superposition may be invoked to separate the response due to the original fixed sources from that due to the small-signal increments. An incremental or small-signal model of the circuit can therefore be generated which is valid only for small-signal analysis. The fixed sources contributing to the d.c. voltages at the operating point in the large-signal model are all set to zero, and the resulting small-signal model is excited by the sources for small-signal analysis.

The main application of this separation is in the determination of the steady-state frequency response by linear a.c. analysis. The small-signal sources required are then sinusoidal and all sources are of the same frequency. The frequency response is obtained by repeated analysis at different frequencies. The linearised d.c. model only contributes the conductances to the circuit for a.c. analysis. Any reactive non-linear components, for example the capacitors in a transistor, must be separately linearised at the circuit operating point and added to the small-signal model. Linear capacitors, inductors, and complex dependent sources must also be added to the linearised model to obtain the linear circuit for analysis.

The nodal admittance matrix methods described in this chapter and advanced developments of them, and two-port analysis in Chapter 5 are particularly suited to small-signal a.c. analysis. Before they can be used, however, it is necessary to provide small-signal a.c. models for all active devices. If these models are to be valid over as wide a frequency range as possible they must include not only the intrinsic and parasitic capacitances within the device but also external parasitic components from the connections and packaging.

Accurate transistor modelling at very high frequencies is very difficult to achieve as it involves numerous high-frequency measurements followed by considerable data reduction. Detailed discussion must be left to specialist texts. Here we will limit discussion to relatively simple models which will nevertheless give valid results but over a more limited frequency range.

**Transistor Capacitance Models**

The Ebers-Moll bipolar transistor model and the MOS transistor model from Section 3.4 may both be modified to include capacitors which are non-linear and either voltage or current dependent.

*Bipolar Transistor*

Four capacitors must be added to the Ebers-Moll model in Fig. 3.14 to account for diffusion and transition capacitors associated with each diode. The dynamic model incorporating these capacitors is shown in Fig. 4.12.

$C_{DE}$ and $C_{DC}$ are the emitter and collector diffusion capacitances. In the active regions these capacitors determine the transistor cut-off frequency $f_T$. In the forward active region the emitter diffusion capacitance $C_{DE}$ is given in terms of the cut-off frequency in the forward region $f_{TF}$ and diode current $j_1$ by,

$$C_{DE} = \theta_E j_1 / 2\pi f_{TF} \ . \tag{4.19}$$

Similarly in the inverse active region,

$$C_{DC} = \theta_C j_2 / 2\pi f_{TR} \ . \tag{4.20}$$

$\theta_E$ and $\theta_C$ were defined in Section 3.4 when introduced in Eqs. (3.22) and

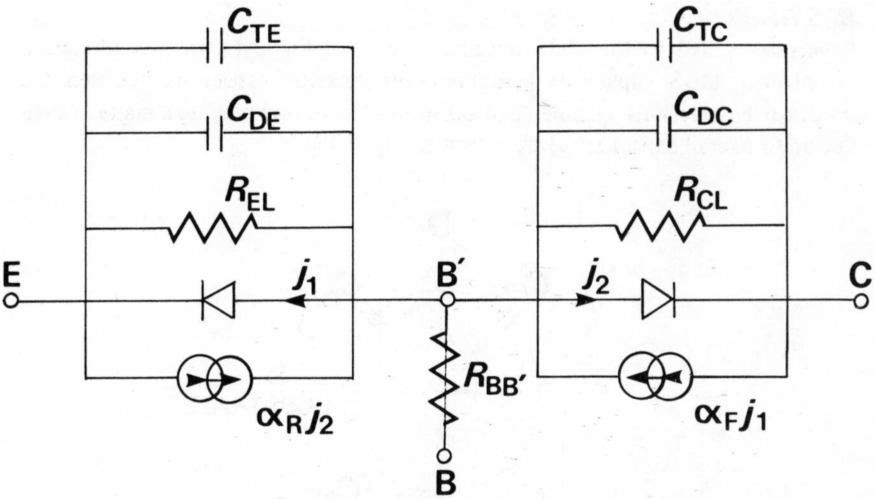


Fig. 4.12 – Dynamic Ebers-Moll model for bipolar transistor.

(3.23). $C_{DE}$ and $C_{DC}$ can therefore be determined by measurement of $f_T$ in the forward and inverse active regions respectively at a specified current and temperature.

$C_{TE}$ and $C_{TC}$ are the emitter and collector transition capacitances. The capacitors are voltage-dependent and are defined for reverse-biased diodes by,

$$C_{TE} = C_{E0}/(1 - v_{B'E}\, V_{ZE})^{n_E} \qquad (4.21)$$

$$C_{TC} = C_{C0}/(1 - v_{B'C}\, V_{ZC})^{n_C} \qquad (4.22)$$

where $C_{E0}$ and $C_{C0}$ are the capacitances at zero bias voltage, $V_{ZE}$ and $V_{ZC}$ are the contact potentials, and $n_E$ and $n_C$ are grading constants at the emitter and collector respectively. Equations (4.21) and (4.22) are valid for $v_{B'E} < V_{ZE}$ and $v_{B'C} < V_{ZC}$ which includes forward bias up to the contact potential (approximately 0.6 V for silicon). However, capacitance levels off to a maximum value in this region and does not go to infinity as predicted by these simplified equations. The grading constants theoretically vary from $\frac{1}{3}$ for an abrupt junction to $\frac{1}{2}$ for a linear junction.

Measurement of capacitance is straightforward for reverse-biased diodes since the parallel resistance is large. Assuming as an engineering approximation that $n_E$ equals $\frac{1}{2}$, the zero bias capacitance $C_{E0}$ and contact potential $V_{ZE}$ can be obtained from a least squares fit to the linear relationship between $C_{TE}^2$ and $v_{B'E}$. The collector diode can be similarly treated.

*MOS Transistor*

Capacitors added to the MOS transistor model of Fig. 3.17 are a combination of intrinsic MOS capacitors and numerous parasitic capacitors between the transistor connections, channel, and substrate. Capacitors making a major contribution to overall capacitance, are shown in Fig. 4.13.

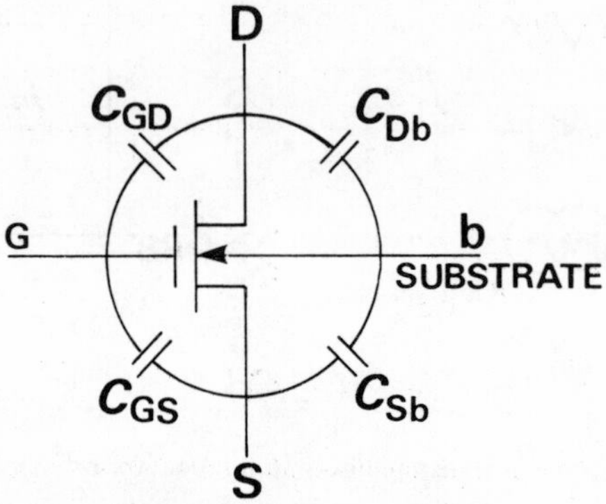


Fig. 4.13 – MOS transistor capacitors.

Gate to drain and gate to source capacitances are predominantly intrinsic MOS capacitors which are voltage-dependent and given by

$$C_{GS} = \tfrac{2}{3}\, C_{ox} \left[ 1 - \frac{(v_{GD} - v_T)^2}{(v_{GS} - v_T + v_{GD} - v_T)^2} \right]$$

$$C_{GD} = \tfrac{2}{3}\, C_{ox} \left[ 1 - \frac{(v_{GS} - v_T)^2}{(v_{GS} - v_T + v_{GD} - v_T)^2} \right]$$

in the non-saturation region. In saturation $C_{GS} = \frac{2}{3} C_{ox}$ and $C_{GD} = 0$. Parasitic capacitors must be added to these intrinsic capacitors. They are generally smaller and not significantly voltage-dependent. Within the MOS transistor the main parasitic contributions are due to the thick-oxide capacitor from the gate-metal to source and drain, and are also due to any overlap of the thin oxide layer.

The source and drain to substrate capacitances $C_{Sb}$ and $C_{Db}$ are due to reverse-biased PN junction diodes set up between source and drain regions and substrate. They are depletion capacitors identical to the bipolar transistor transition capacitors which vary with reverse bias voltage according to Eqs. (4.21) and (4.22). Similar equations may be set up for the diodes to substrate, and since they are reverse-biased measurements may be made as before to define the equation constants.

**Small-Signal Transistor $\pi$ Models**

Small-signal $\pi$ models for the normal active operating region of bipolar and MOS transistors are required for linear a.c. analysis. Satisfactory models for frequencies up to about 30 MHz are provided by a combination of the linearised d.c. models from Chapter 3 and the capacitance models from the previous section. At higher frequencies parasitic capacitances and inductances become increasingly important, and models incorporating them can be effective to 1 GHz.

The linearised small-signal d.c. model of the bipolar transistor in Fig. 4.14 is derived from Fig. 3.13(b) in Chapter 3 by omitting the fixed sources.

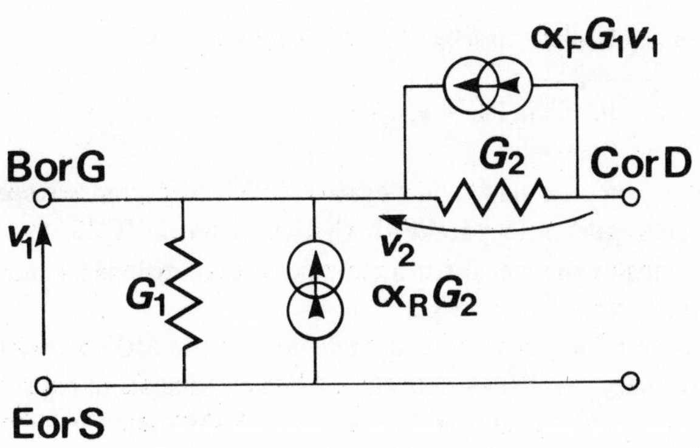


Fig. 4.14 – Linearised small-signal d.c. model.
B C E refers to a bipolar transistor.
G D S refers to a MOS transistor for which $\alpha_F = \alpha_R = 1$.

Fig. 4.14 also applies to the MOS transistor since it may be derived from Fig. 3.18(b) while noting that $\alpha_F = \alpha_R = 1$. Three additional sets of components are required to complete the small-signal model.

Leakage resistors $R_{EL}$ and $R_{CL}$, and the bulk base resistor $R_{BB'}$ from Fig. 3.14 must be added to the bipolar transitors model. Leakage resistors also exist for the MOS transistor, but they are usually assumed to be sufficiently large to be negligible. For the present $R_{BB'}$ will be omitted so that bipolar and MOS transistors can be treated together. It will reappear in series with the base terminal before parasitic components are added to the model. $R_{EL}$ and $R_{CL}$ will be incorporated as the conductances $G_{1L}$ and $G_{2L}$ in parallel with $G_1$ and $G_2$ respectively.

The Early effect due to base width spreading in bipolar transistors and changes of channel length in MOS transistors both cause an increase in the device output current at the collector or drain. In non-linear d.c. analysis this effect was incorporated in the model by direct modification to the current equations.

For the npn bipolar transistor the Early effect was modelled by an increase in common emitter current gain given by the equation,

$$\beta_N = \beta_{NO} + v_{CB}/v_N ,$$

from which Eq. (3.36) was derived. In the normal active operating region the collector current at a collector-base voltage $v_{CB}$ equals $i_B\beta_N$ and at $v_{CB} = 0$, it equals $i_B\beta_{NO}$. The Early effect is equivalent to a conductance $G_E$ equal to the slope of the output characteristic plotted as a graph of $|i_c|$ against $v_{CB}$.

Therefore,

$$G_E = (i_B\beta_N - i_B\beta_{NO})/v_{CB} = i_B/V_N$$

$$= |i_C/(\beta_{NO}V_N + v_{CB})| , \tag{4.23}$$

where $i_B$ and $i_C$ are the base and negative collector currents respectively. The conductance $G_E$ appears in parallel with the forward VCCS $\propto_F G_1$ and $G_2$. It is defined in terms of the transistor d.c. operating conditions $i_C$ and $v_{CB}$, and the d.c. model parameters $\beta_{NO}$ and $V_N$.

The corresponding active operating region of the MOS transistor is known as the saturation region. (Note that the MOS non-saturation region corresponds to the bipolar transistor saturated region.) The MOS drain current in saturation from Eq. (3.43) is given by

$$i_D = -K(1 + \gamma v_{DS})(v_{GS} - v_T)^2 .$$

Partial differentiation with respect to $v_{DS}$ with $v_{GS}$ constant yields the MOS drain output conductance. $G_D$, that is,

$$G_D = \frac{\partial i_D}{\partial v_{DS}} = -\gamma K(v_{GS} - v_T)^2$$

$$= |i_D/(1/\gamma + v_{DS})| . \tag{4.24}$$

The drain conductance is therefore identical in form to the Early conductance but appears between drain and source, thus maintaining zero gate current. The conductance is defined as before in terms of the d.c. operating conditions and the d.c. model parameter $\gamma$.

Transistor capacitances were discussed in the previous section. If we exclude the MOS parasitic capacitance to substrate for the present, capacitors $C_1$ and $C_2$ in parallel with conductances $G_1$ and $G_2$ respectively, will model either bipolar or MOS transistor capacitance.

The complete small-signal model incorporating all capacitors and conductances is shown in Fig. 4.15

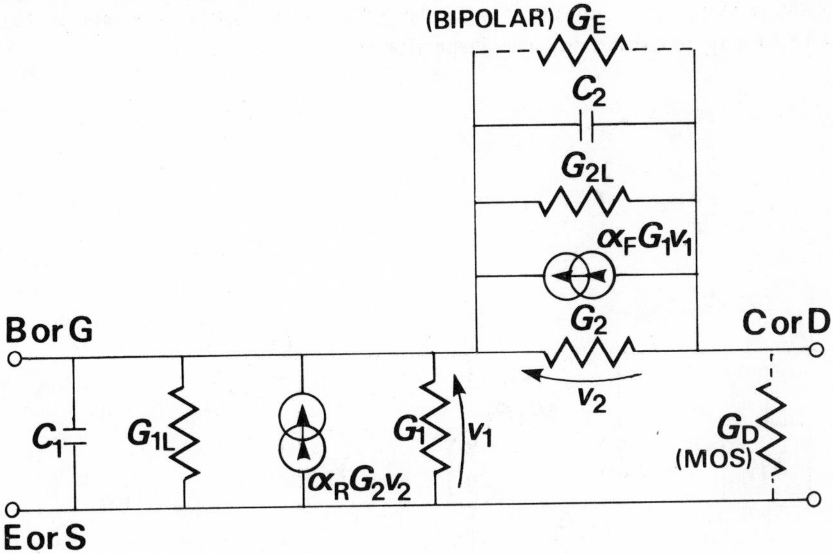


Fig. 4.15 – Complete small-signal model.
B C E bipolar transistor ($G_D$ onitted).
G D S MOS transistor $\alpha_F = \alpha_R = 1$ ($G_E$, $G_{1L}$, and $G_{2L}$ omitted).

The NAM for the small-signal model in Fig. 4.15 can be written down by inspection, with emitter/source references as

$$\begin{bmatrix} G_1(1-\alpha_F)+G_2(1-\alpha_R)+G_{1L}+G_{2L}+G_E+sC_1+sC_2 & G_2(\alpha_R-1)-G_{2L}-G_E-sC_2 \\ G_1\alpha_F-G_2-G_{2L}-G_E-sC_2 & G_2+G_{2L}+G_E+G_D+sC_2 \end{bmatrix} . \tag{4.25}$$

This NAM can be regarded as the admittance parameters of the transistor in common emitter/source. Also since the model is a three-terminal network which can be connected as a two-port, the admittance parameters are equal to the two-port admittance parameters.

In order that this model can be put into the more familiar $\pi$ form the admittance parameters must be reinterpreted to generate an equivalent $\pi$ circuit. We will first look at the problem by finding the equivalent circuit of any set of admittance parameters given by the general matrix equation for three terminals,

$$\begin{bmatrix} y_{11} & y_{12} \\ y_{21} & y_{22} \end{bmatrix} \begin{bmatrix} e'_1 \\ e'_2 \end{bmatrix} = \begin{bmatrix} I'_1 \\ I'_2 \end{bmatrix} . \tag{4.26}$$

It can easily be checked that the equivalent circuit model in Fig, 4.16(a) has the same NAM as Eq. (4.26). If the circuit contains only passive components the NAM is symmetric and $y_{12} = y_{21}$. The NAM of the passive $\pi$ model in Fig. 4.16(b) may be written down by inspection as

$$\begin{bmatrix} y_{11} & y_{12} \\ y_{12} & y_{22} \end{bmatrix}.$$

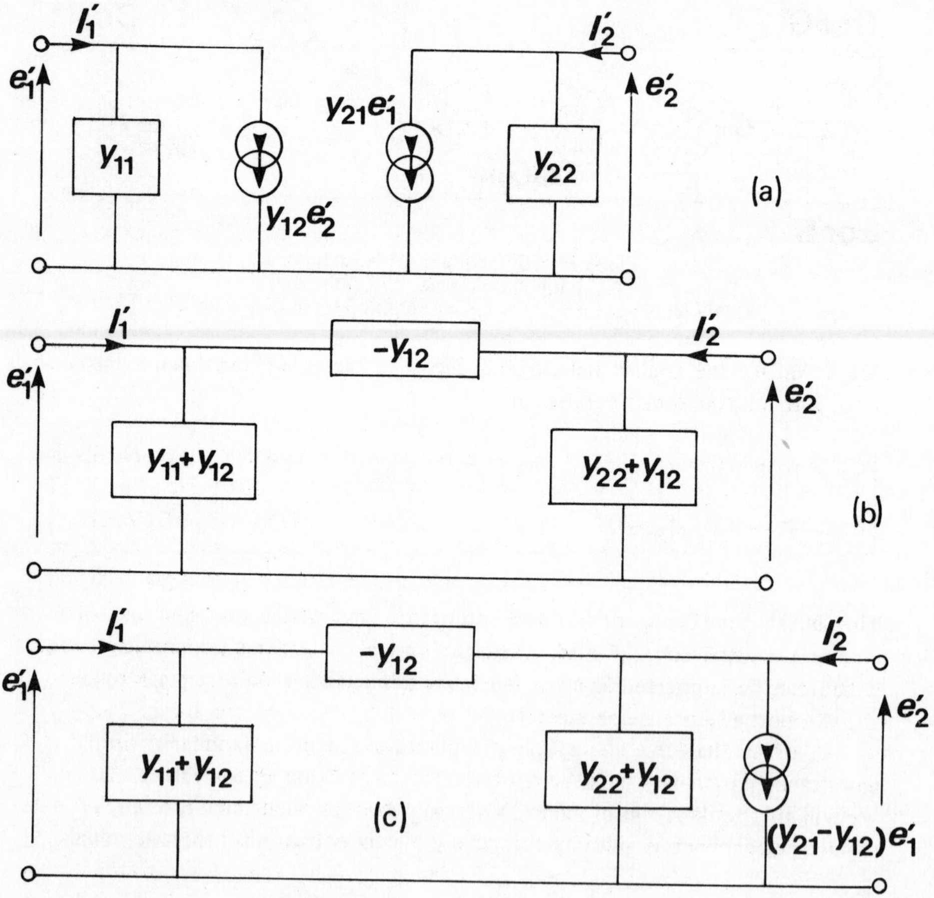


Fig. 4.16 – Admittance parameter models. (a) Two-generator model (b) Passive model for $y_{12} = y_{21}$; (c) One-generator model.

A $\pi$ model for a transistor with only one VCCS may be derived by separating the active and passive parts of the NAM, that is,

$$\mathbf{Y}' = \begin{bmatrix} y_{11} & y_{12} \\ y_{21} & y_{22} \end{bmatrix} = \begin{bmatrix} y_{11} & y_{12} \\ y_{12} & y_{22} \end{bmatrix} + \begin{bmatrix} 0 & 0 \\ (y_{21}-y_{12}) & 0 \end{bmatrix} .$$

The separated parts of $\mathbf{Y}'$ can be recognised as the passive model of Fig. 4.16(b) connected in parallel with a VCCS with transfer admittance $(y_{21} - y_{12})$. The one-generator equivalent circuit model in Fig. 4.16(c) can therefore be derived and the NAM confirmed.

The elements of the one-generator model in Fig. 4.16(c) can now be identified for the transistor NAM of matrix (4.25). Recall that $G_D = 0$ and $R_{BB'}$ has been omitted for a bipolar transistor, and that $G_E = 0$, $G_{1L} = G_{2L} = 0$, and $\alpha_R = \alpha_F = 1$ for a MOS transistor. Admittance parameter $\pi$ models for the two transistor types are shown in Fig. 4.17(a) and 4.17(b). The more familiar hybrid $\pi$ model for the bipolar transistor is an approximation based on the two-generator equivalent circuit of Fig. 4.16(a). The real part of $y_{12}$ due to resistive conductances and the reverse current gain $\alpha_R$ are assumed negligible. The capacitor $C_2$, however, still appears in the model connected between base and collector instead of being modelled as part of the reverse transfer admittance $y_{12}$. The hybrid $\pi$ bipolar transistor model shown in Fig. 4.17(c) is then obtained.

All parameters of the transistor $\pi$ models discussed here are available from d.c. linearisation and capacitance models evaluated at a specified set of d.c. operating conditions for the transistor. All small-signal model elements change as these conditions are varied, and small-signal analysis must be preceded by non-linear d.c. analysis, or at least design, so that the transistor model can be evaluated. Alternatively, small-signal models may be determined by direct measurement or derived as poor approximation from manufacturers' data.

Two main faults of the $\pi$ models must be corrected before they can be used at frequencies greater than about 30 MHz. At frequencies beyond the transistor $f_T$, the current gain $\alpha_F$ for bipolar transistors introduces a phase shift, and at higher frequencies the magnitude of $\alpha_F$ also reduces. Excess phase shift at constant amplitude is corrected for a phase delay of $\tau$ secs. by use of $g_m e^{-j\omega\tau}$ in place of $g_m$. Not all computer programmes allow complex transfer admittance, and it is possible to modify the passive component values in the model while retaining a real value for $g_m$ by the technique of optimisation. Adjustment to the model parameters is then made by a computer to obtain the best fit to measured data. The objective is to modify the available model parameters so that excess phase shift is introduced.

The technique of optimisation can also be used to correct the basic $\pi$ models for use at frequencies up to and beyond 1 GHz. At very high frequencies

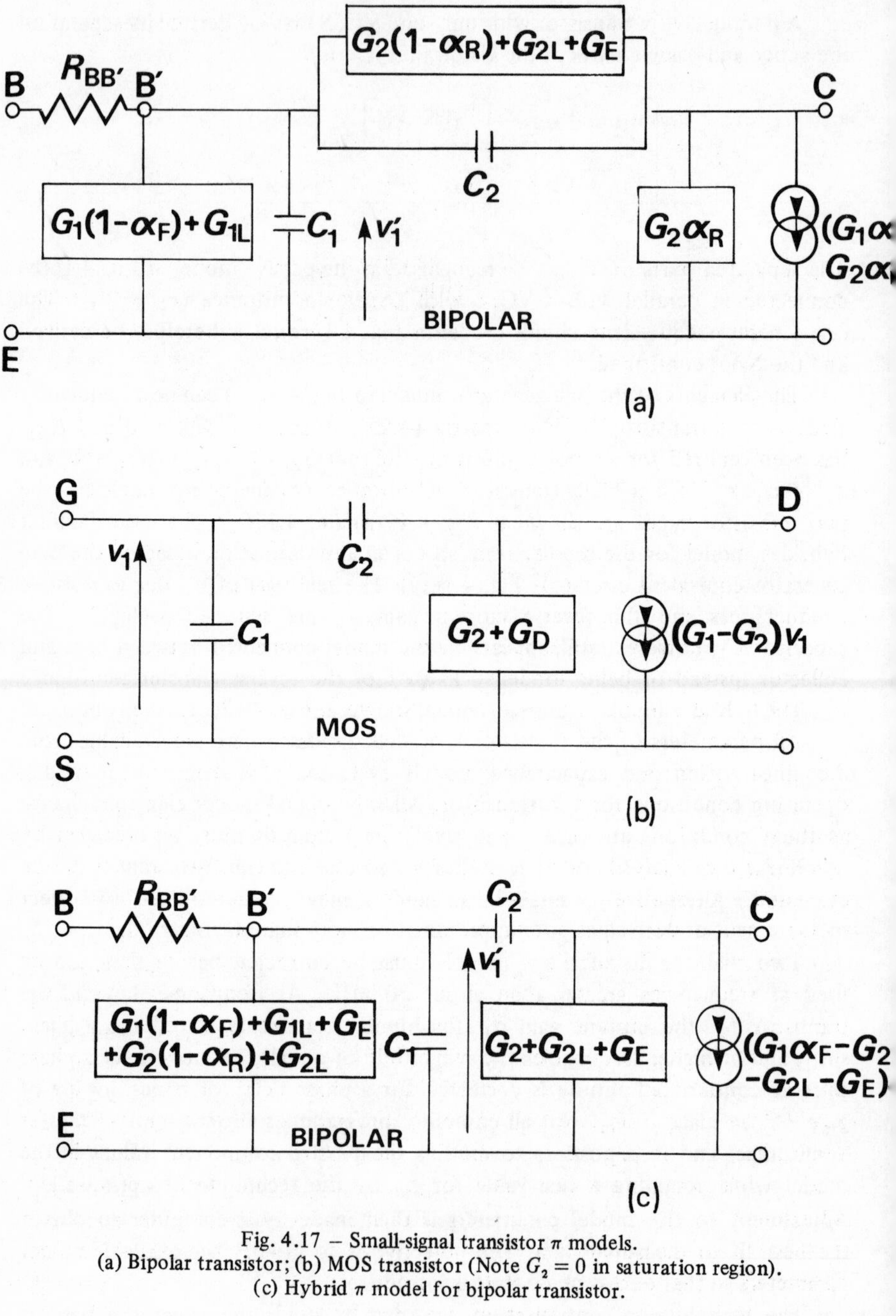


Fig. 4.17 – Small-signal transistor $\pi$ models.
(a) Bipolar transistor; (b) MOS transistor (Note $G_2 = 0$ in saturation region).
(c) Hybrid $\pi$ model for bipolar transistor.

the parasitic capacitances and inductances of the transistor connections have a major effect on response, and it is essential to include them in an analysis. The simple low-frequency $\pi$ model is modified by the addition of one or more rings of parasitic components as shown in Fig. 4.18.

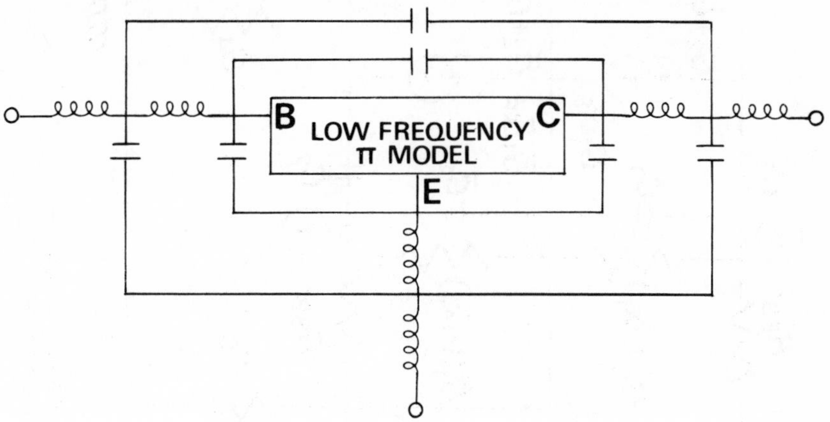


Fig. 4.18 – Parasitic components added to the transitor $\pi$ model.

Parasitic components are identified by fitting measured parameter data, for example two-port admittance or scattering parameters, by optimisation of the model components starting from the low-frequency model and estimated parasitics. This subject cannot be pursued here, but typical component values involved are less than 1 pF or 0.0.1 $\mu$H.

**Operational Amplifier Models**

Operational amplifiers are direct-coupled high-again differential amplifiers with internally or externally compensated frequency response characteristics which ensure stability in most negative feedback applications. Small-signal models may be developed directly from the internal transistor configuration. However, general purpose models must inevitably be a compromise between the complexity of that approach and the convenience of a simple model. The emphasis here is on the latter as most operational amplifier circuits operate at relatively low frequencies.

The most common input circuit to an operational amplifier consists of a long-tailed pair differential d.c. amplifier together with one or more constant-current generators. A simplified bipolar transistor input circuit is shown in Fig. 4.19(a). The small-signal model in Fig. 4.19(b) incorporates hybrid $\pi$ transistor models from Fig. 4.17(c) for the two input transistors and a very high resistance equal to the current generator collector output resistance.

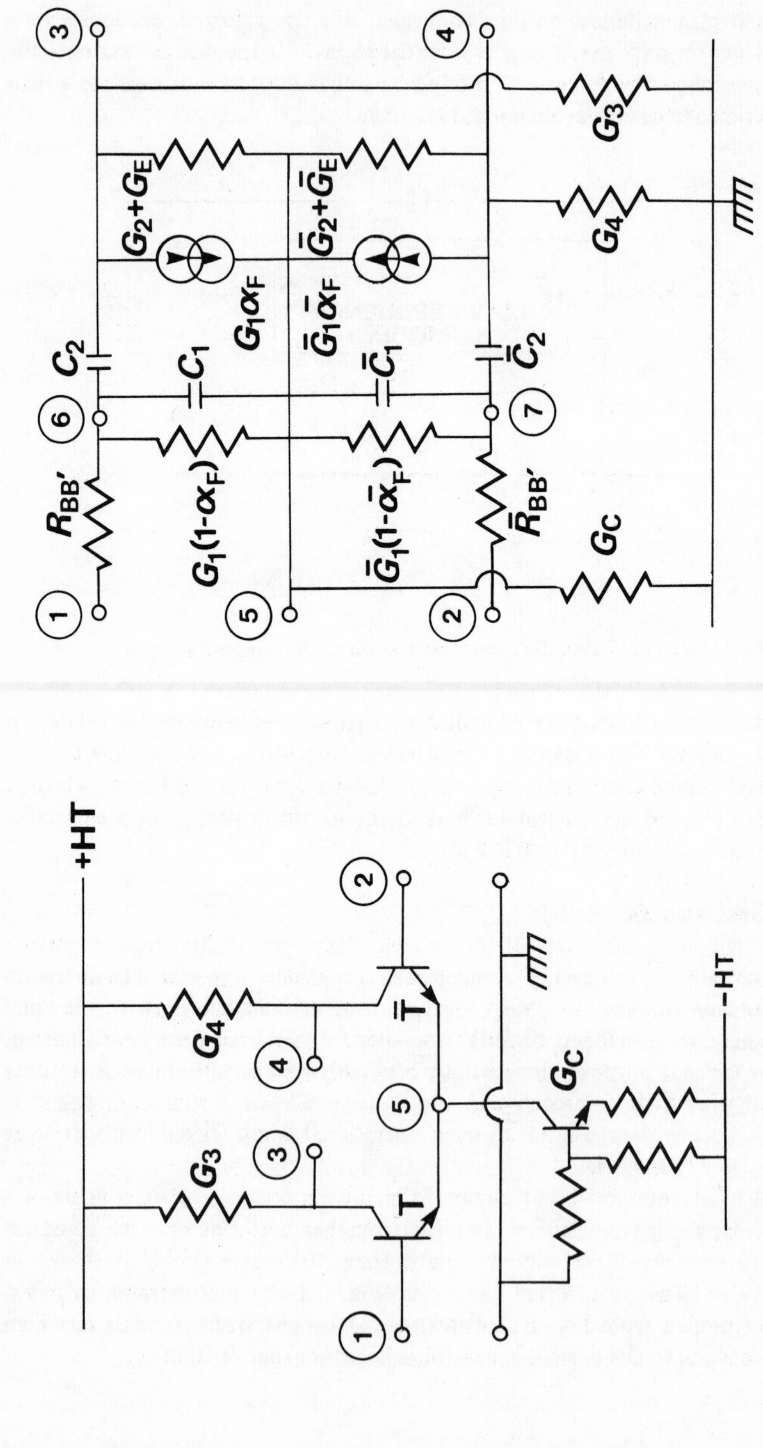


Fig. 4.19 – Simplified operational amplifier input stage.
(a) Differential amplifier; (b) Small-signal model.

Exact analysis of the long-tailed pair differential amplifier can be found in most texts on electronic or microelectronic circuits. Here we are examining the small-signal model of the amplifier for guidance towards a suitable general input circuit. If we neglect the effect of feedback to the input nodes ① and ②, and consider that we have only those two nodes and ground available for measurement, we must base the model on the NAM for nodes ① and ②. Assuming the input circuit to be passive the NAM is given by

$$\begin{bmatrix} y_{11} & y_{12} \\ y_{12} & y_{22} \end{bmatrix} \begin{bmatrix} e'_1 \\ e'_2 \end{bmatrix} = \begin{bmatrix} I'_1 \\ I'_2 \end{bmatrix} .$$

The admittances from each input terminal to ground and from terminal to terminal are easily identified as those shown in Fig. 4.20.

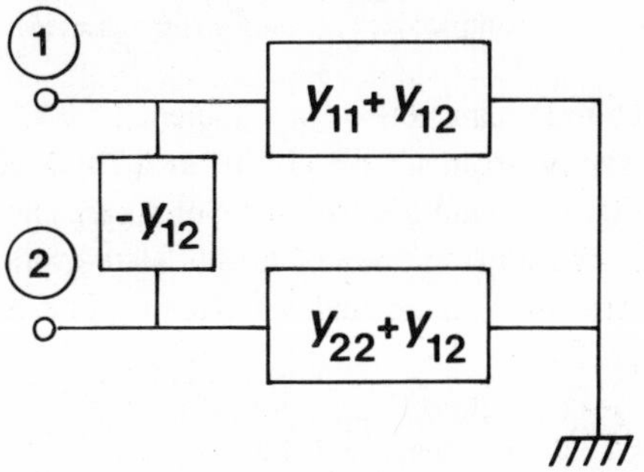


Fig. 4.20 – Input admittance model.

In the small-signal model of Fig. 4.19(b) seven nodes are actually present, and the NAM for the circuit can be written down. At any specified frequency nodes ③ to ⑦ can be suppressed, using a version of Eq. (4.13). Over a restricted range of frequency it should be possible to identify the admittances of the input model as parallel combinations of a resistor and capacitor. Also in many low-frequency applications the capacitors are omitted altogether.

Output stages of operational amplifiers operate in push-pull under large-signal conditions with one transistor delivering current to the load while the other transistor is off. The small-signal model, therefore, can only be an approximation, and it usually contains a single resistor to model the mean output resistance.

The frequency response of an uncompensated amplifier consists of three or more low-pass RC filters with various cut-off frequencies giving a response of the form shown in Fig. 4.21. The compensated amplifier is arranged to have a gain fall off of 6 dB per octave to ensure stability. For many purposes modelling of that slope by a single RC low pass filter is sufficient.

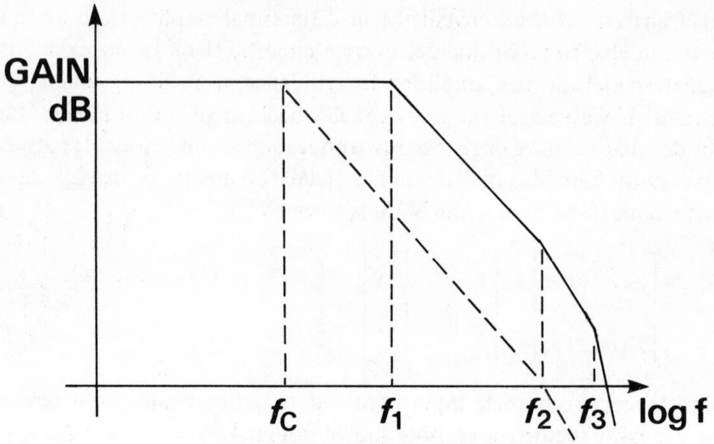


Fig. 4.21 – Operational amplifier frequency response.
——— uncompensated; – – – – – compensated.

A general purpose model can now be assembled. In Fig. 4.22 a 6dB/octave fall-off in frequency response from a 3 dB cut-off at $f_c$ has been inserted between the input stage from Fig. 4.20 and a simple output stage. The CR time constant required is given by $CR = 1/2\pi f_c$. Values of C and R are quite arbitrary, and the choice of one ohm is intended for normalised circuits. The NAM for the model is given by

$$\begin{bmatrix} y_{11} & y_{12} & 0 & 0 \\ y_{12} & y_{22} & 0 & 0 \\ -A & A & 1+s\omega_c & 0 \\ 0 & 0 & -y_{44} & y_{44} \end{bmatrix}.$$

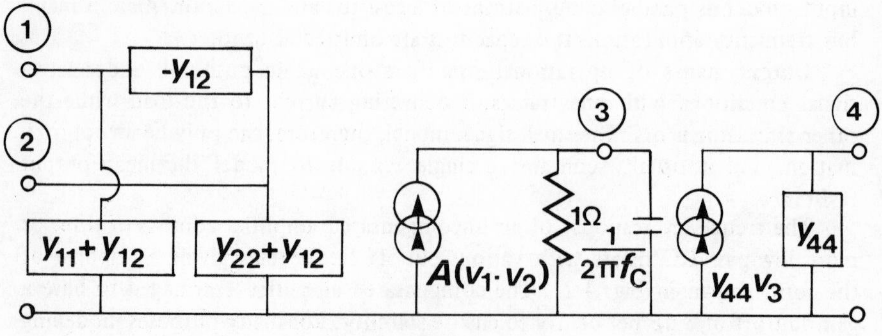


Fig. 4.22 – Operational amplifier model. Gain $A$, Cut-off frequency $f_c$.

## 4.4 COMPUTER PROGRAMMING

We are now in a position to write a reasonably efficient a.c. analysis computer program using the nodal admittance matrix methods of this chapter. The restriction of most versions of BASIC to real matrix operations is, surprisingly, not a source of much difficulty, and within the 15K bytes of available memory it has been possible to provide the convenience of on-line circuit input, modification, and output as well as providing full matrix inversion of the complex NAM. Subroutines for sensitivity analysis which are based on use of the elements of this inverse and will be described in Chapter 8, can therefore also be included.

### A.C. Circuit Analysis using ACNLU

ACNLU is an a.c. circuit analysis program written in BASIC for on-line computation from time-shared terminals. A five-node, twenty-branch circuit will run in approximately 12K bytes of memory using Data General Basic 3.8 on the Nova 1220 computer. The program listing is given in the Appendix to this chapter with the exception of sensitivity subroutines 6000 and 8000 which are described and listed in Chapter 8.

The circuit description in initially entered into the program as a list of component types, and node connections. These are stored for subsequent modification by the user during an analysis run. The NAM and source matrices are filled, using Rules 1 and 2 from Section 4.1. Separate NAM matrices are filled for resistive, inductive, and capacitive admittance terms so that multiplication by frequency $\omega$ or $1/\omega$ does not involve refilling each matrix during a frequency response computation. Matrix **R** contains the conductance and real transfer admittances, **C** contains capacitive admittances, and **L** contains inductive admittances. The NAM is then assembled as the matrices $\mathbf{R} + j\mathbf{B}$ where $\mathbf{B} = \omega\mathbf{C} - \mathbf{L}/\omega$.

The inverse of the NAM stored as the two real matrices **R** and **B**, proceeds through the LU factorisation into the same matrix storage, and is directly programmed from Section 4.1. The full matrix inversion into $\mathbf{D} + j\mathbf{F}$ is then obtained by $N$ (the number of nodes) separate solutions involving forward and backward substitution. Various outputs are available including node voltages, gain and phase, input and output impedances, sensitivities and the inverse NAM. The solution may then be repeated with the original circuit, frequencies, sensitivities, inputs, or outputs modified.

### Program Inputs

*Components – Statement Numbers 180 to 440*

Components are initially entered into the program in the form of a list of component types, values, and node connections. Resistors must be numbered from 1 to R9, inductors from 1 to L9, capacitors from 1 to C9, and transfer admittances from 1 to G9. No other components are allowed at present. Resistor values and node connections are entered into matrix S(R9+1,2). Inductors are similarly entered into matrix M(L9+1,2), and capacitors are entered into matrix A(C9+1,2). VCCS dependent sources have four node connections, value, and a type number which allows entry of transfer conductance, capacitive transfer admittance, or inductive transfer admittance. VCCS are stored in matrix G(G9+1,5).

*Frequencies – Statement Numbers 450 to 470*
Frequencies for analysis are entered as the four inputs F0, F1, F2, F3. Analysis is carried out at F3 frequencies beginning at F0. Linear or logarithmic scales are generated from the expression $F_{n+1} = (F_n + F_1)*F_2$. and converted to radians/sec, using P2 = $2\pi$.

*Output Selection – Statement Numbers 480 to 580*
The following outputs contained in N$(12) may be selected.

1. NV Node voltages.
2. SE Node voltages and node voltage sensitivity to listed components.
3. YM Nodal admittance matrix in form **R** + j**B**.
4. ZM Inverse NAM in form **D** + j**F**.
5. DB Gain, phase, input, and output impedance.
6. SN Gain, phase, and sensitivities to listed components.

These commands are decoded as integers 1 to 6 in I7 and govern the organisation of results after NAM inversion.

*Sources – Statement Numbers 590 to 720*
A circuit with $N$ nodes (at present $N \leqslant 10$) has a maximum of $N$ current sources connected one between each node and reference. For NV and SE outputs only, I5 sources are stored as real and imaginary parts in the matrices I(N,1) and J(N,1) respectively. The sources are initially entered as magnitude and phase.

*Input and Output Nodes – Statement Numbers 730 to 760*
N1 is the input node and N2 is the output node.

*Sensitivity Component List – Statement Numbers 770 to 1050*
T9 (at present T9 $\leqslant$ 9) components R,L,C, or GM may be listed for sensitivity computation. The component list may subsequently be modified and the analysis repeated when more than 9 component sensitivities are required. Components are simply listed in the form R,7 L,2 GM,1 for example, each on a separate input line. These are initially stored in matrix U(T9,6) with component type numbers 1 to 4 for R,L,C, and GM respectively in column 0, and component number in column 1. The matrix is later filled with component value and node connections derived from the component list.

**Analysis and Output**

*Filling the NAM – Statement Numbers 1060 to 1550*

The NAM is stored as three matrices R(N,N), L(N,N), and C(N,N) containing the resistive, inductive, and capacitive admittances respectively. Each passive component and each VCCS from the component matrices is entered into the appropriate NAM matrix in four positions according to Rule 1 from Section 4.1. This process is repeated at the beginning of each solution so that the latest modifications to the component list are incorporated.

*Filling the Sensitivity List – Statement Numbers 1560 to 1860*

Modification of the components or of the component sensitivity list affects the matrix U(T9,6). This is therefore refilled for each analysis run using columns 0 and 1 which hold a simplified listing. Full details of the sensitivity part of this program are given in Chapter 8, including the information required to fill columns 2 to 6 of this matrix.

*NAM Inversion – Subroutine 4000, Statment Numbers 4000 to 4880*

The NAM is assembled for inversion at a frequency W rad/sec. as matrices R(N,N) + j B(N,N). If W=0, that is for d.c analysis, the subroutine returns matrix D(N,N) equal to the inverse of R(N,N) and sets F(N,N) equal to zero. RC networks are thus correctly treated but inductors are ignored. In an a.c. analysis with W>0, NAM inversion proceeds by use of the LU factorisation.

After storage of the elements of **R** in matrix **X**, the NAM inversion performs the LU factorisation of **R** + j**B** into the same storage locations. (Statements 4250 and 4480). Forward substitution (4490 to 4640) and backward substitution (4650 to 4730) is repeated $N$ times to obtain the full inverse as **D** + j**F**. Each column $i$ of **D** + j**F** is generated by solution of $(\mathbf{R} + \mathrm{j}\mathbf{B})\mathbf{e} = \mathbf{I}$ with $I_k = 0, k \neq i$, and $I_i = 1$. After inversion **R** is refilled.

*Outputs – Statement Numbers 1940 to 2700*

Each of the six outputs previously selected are obtained by organising the inverse NAM. This is carried out at each frequency with suitable headings generated on the first analysis only. Subsequent results are then tabulated and unrestricted in number since they are not stored.

The real and imaginary parts of the node voltages are obtained as E(N,1) +$j$V(N,1). Subroutines 6000 and 8000 are called to generate node voltage, and gain and phase sensitivities respectively. These subroutines are described in Chapter 8.

**Operation Control**

*Commands – Statements 2710 to 3050*

The following commands contained in L$(20) govern the action of the programme after each analysis.

1. ER End the computer run.

2. NC Modify the component list
3. NF Modify the frequencies for analysis
4. NO Modify the outputs required
5. NI Modify the current sources
6. YM Print the last computed matrices **R**, $\omega$**C**, **L**/$\omega$
7. ZM Print the last computed inverse NAM **D** + j**F**
8. NN Modify the input and output nodes
9. NS Modify the sensitivity list
10. GO Run the analysis.

These commands are decoded as integers 1 to 10 into I6 which controls the programme operation.

*Other Programme Variables*

| | |
|---|---|
| Control variables | I8, I9 |
| Calculation space | C8, G8, I5, L5, L6, L7, L8, R5, R6, R7, R8, S6, S7, S8, S9, N$(2). B(10,10), P(1,10), Q(1,10), T(10,1) |

*Example* 4.3

Analyse the filter circuit in Fig. 4.4 from Example 4.1 using the computer program ACNLU. Determine frequency response, input and output impedance from 1 kHz to 10 kHz when a type 741 operational amplifier with feedback is used to provide the voltage gain required.

The filter is redrawn for analysis in Fig. 4.23 using 1 kΩ and 1.5 kΩ feedback resistors to the negative input of the operational amplifier in order to define voltage gain. The amplifier model has been derived from published data assuming that impedances from the input terminals to reference ground have negligible effect. The cut-off frequency of 10 Hz has been achieved by using a 1 kΩ resistor in parallel with a 15.9 μF capacitor thus maintaining impedances at roughly similar values throughout the circuit. The voltage gain of $10^5$ converts to a VCCS of 100S using the 1 kΩ resistor. The second voltage gain of unity converts to a VCCS of 0.0133S using the 75 Ω amplifier output impedance.

The ACNLU computer run for this example is reproduced in Fig. 4.24. The NAM and inverse NAM at the last computed frequency are also given. Response is fairly close to the theoretical values for an ideal amplifier which may be directly computed from the transfer function $F(s) = A/\{1 + [R_1C_1(1-A) + (R_1+R_2)C_2]\,s + R_1R_2C_1C_2s^2)\}$. The computer run continues by changing the output to gain sensitivity, and will be completed in Example 8.7, Chapter 8.

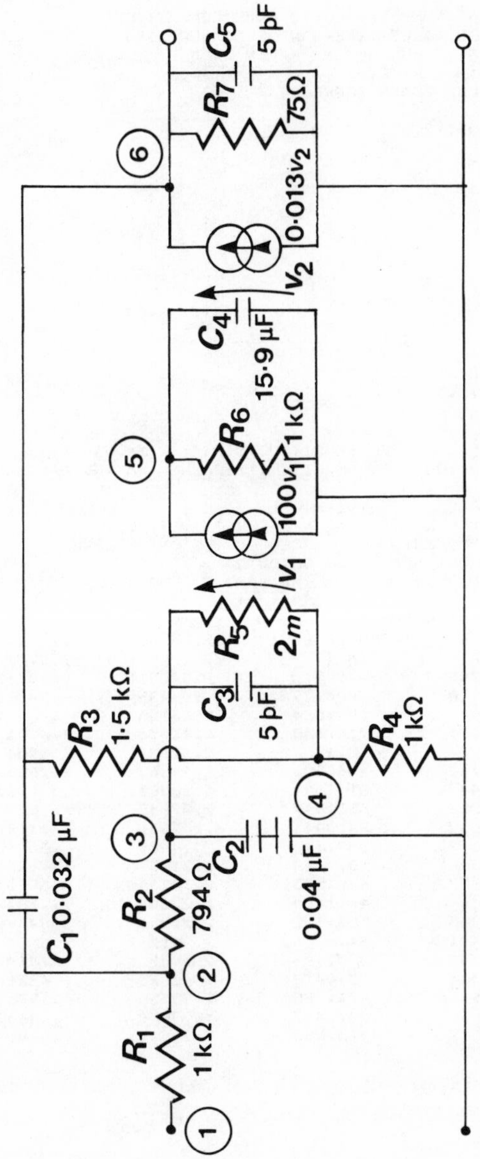


Fig. 4.23 – Filter circuit from Fig. 4.4 prepared for analysis.

```
AC NAM ANALYSIS
RLC INPUTS: COMPONENT NUMBER,VALUE,2 NODE CONNECTIONS
GM  INPUTS: GM NUMBER,VALUE,TYPE,4 NODE CONNECTIONS
     TYPE: 1 FOR GM, 2 FOR GM/JW, 3 FOR JWGM
    NODES: SEND +, - , RECEIVE +, -
I   INPUTS: NODE,VALUE,PHASE DEGREES
NO.NODES        ? 6
NO. R,L,C,GM    ? 7,0,5,2
R ? 1,1E3,1,2
R ? 2,794,2,3
R ? 3,1.5E3,6,4
R ? 4,1E3,4,0
R ? 5,2E6,3,4
R ? 6,1E3,5,0
R ? 7,75,6,0
C ? 1,.032E-6,2,6
C ? 2,.04E-6,3,0
C ? 3,5E-12,3,4
C ? 4,15.9E-6,5,0
C ? 5,5E-12,6,0
GM ? 1,100,1,3,4,5,0
GM ? 2,.01333,1,5,0,6,0
F-LOW,F+INC,F*MULT,NO.F
 ? 1E3,1E3,1,10
OUTPUTS: NV NODE VOLTS       SE SENSITIVITY+NV       YM NAM
         DB GAIN ZIN ZOUT    SN SENSITIVITY+GAIN     ZM INV NAM
OUTPUT ? DB
INPUT NODE,OUTPUT NODE          ? 1,6
NEXT COMMAND: ER END RUN         NS NEW SENSITIVITY LIST
MODIFICATION: NC COMPONENTS      NF FREQUENCIES  NI SOURCES
            : NN IN/OUT NODES    NO OUTPUTS
      OUTPUT: YM 1/R WC 1/WL     ZM INV(NAM)      GO ANALYSE
NEXT ? GO
INPUT NODE  1    OUTPUT NODE  6
HZ.            GAIN DB        PHASE DEG      RP             IP
               ZIN RP         ZIN IP         ZOUT RP        ZOUT IP
 1000           8.22209       -9.00016        2.54521       -.403128
               -11129.8        9900.79         .904435      -.943896
 2000           9.00947       -19.9854        2.65155       -.964318
               -4478.91        916.648        1.65036       -.660372
 3000           10.2244       -35.97          2.62627       -1.906
               -1862.61       -446.521        1.9456        -.266309
 4000           11.2442       -61.1504        1.76083       -3.19639
               -715.526       -682.449        2.07732        6.82803E-02
 5000           10.5032        86.4441       -.207831       -3.34444
               -132.117       -687.175        2.14689        .354462
 6000           7.7837         60.2018       -1.21757       -2.12615
               200.357        -639.533        2.18933        .609346
 7000           4.6339         44.1604       -1.22307       -1.18774
               406.443        -583.552        2.21856        .843908
 8000           1.77491        34.4206       -1.01193       -.693419
               542.541        -530.906        2.2409         1.06478
 9000          -.711004        28.0417       -.813235       -.433164
               636.94         -484.298        2.25952        1.27607
 10000         -2.87978        23.5402       -.658076       -.286689
               705.017        -443.783        2.27614        1.48042
NEXT ? ZM
```

```
INV(NAM) REAL PARTS

 705.017      -294.983      -377.883      -1319.03      .022002
 1.16015
-294.983      -294.983      -377.883      -1319.03      .022002
 1.16015
-236.958      -236.958      -283.593       106.854     -1.77355E-03
 .892943
-236.472      -236.472      -292.946       125.811      7.9105E-03
 .910589
-689.568      -689.568      -763.418      -1386.11      2.26293E-02
-72.6485
-591.182      -591.182      -732.375      -1185.47      1.97836E-02
 2.27614

INV(NAM) IMAG PARTS

-443.783      -443.783      -916.557       927.752     -1.54578E-02
 1.6478
-443.783      -443.783      -916.557       927.752     -1.54578E-02
 1.6478
 29.0783       29.0783      -350.635       714.531     -1.19163E-02
-.133814
 35.9669       35.9669      -342.998       728.36      -1.21421E-02
 .591947
 47.95         47.95        -936.922       1896.11     -.031603
 1.69362
 89.9226       89.9226      -857.494       1820.92     -3.03507E-02
 1.48042

NEXT ? YM
1/R MATRIX

 .001         -.001          0             0             0             0
-.001          2.25945E-03  -1.25945E-03   0             0             0
 0            -1.25945E-03   1.25995E-03  -5E-07         0             0
 0             0            -5E-07         1.66717E-03   0
-6.66667E-04
 0             0            -100           100           .001
 0
 0             0             0            -6.66667E-04  -.01333
 .014

WC MATRIX

 0             0             0             0             0             0
 0             2.01062E-03   0             0             0
-2.01062E-03
 0             0             2.51359E-03  -3.14159E-07   0             0
 0             0            -3.14159E-07   3.14159E-07   0             0
 0             0             0             0             .999027
 0
 0            -2.01062E-03   0             0             0
 2.01093E-03

1/WL MATRIX

 0             0             0             0             0             0
 0             0             0             0             0             0
 0             0             0             0             0             0
 0             0             0             0             0             0
 0             0             0             0             0             0
 0             0             0             0             0             0

NEXT ?
```

Fig. 4.24 – Computer run for Example 4.3.

Extensive examples demonstrating the facilities available in the program ACNLU cannot be given here, although a second example mainly concerned with node voltage sensitivity will be found in Chapter 8, Example 8.6. A more complete description with further examples will be available in the user's manual. However, the program inputs are well defined by conversational input instructions, and the program should be used to analyse circuits familiar to the user. Accuracy can then be assessed and the available outputs can be explored with little if any need for a manual.

## 4.5 CONCLUSION

The nodal admittance matrix methods of circuit analysis in this chapter form the basis of most computer programmes primarily intended for frequency response evaluation. Modifications and extensions of basic methods using inversion algorithms which take advantage of the characteristics of circuit matrices, or are adapted to very large networks, can be considerably more efficient. Nevertheless recent research has moved away from numerical NAM inversion towards the transfer function methods of Chapter 6 and the state space methods of Chapter 7.

## FURTHER READING

Calahan, D. A., (1972), *Computer Aided Network Design,* revised edition, Chapters 2 and 10, McGraw-Hill, New York.

Desoer, C. A. and Kuh, E. S., (1969), *Basic Circuit Theory,* Chapter 10, McGraw-Hill, New York.

## PROBLEMS

4.1 Fill the nodal admittance and current source matrices for the networks in Figs. 1.23, 1.20, 4.25, and 1.21. (In Fig. 4.25 use simplified transistor models and scale the type 741 amplifier components by the factors $10^4$ for impedance and 2500 for frequency. In Fig. 1.21 use $M = 0.75$).

4.2 The solution for node voltages in a network is given by $\mathbf{e}' = [\mathbf{Y}']^{-1}\mathbf{I}'$, where $\mathbf{I}'$ is the column vector of node to ground fixed current sources. In a system with input node ⓘ and output node ⓙ and open-circuit output derive the following expressions.

$$\text{Voltage gain} = \frac{e_j'}{e_i'} = \frac{\Delta_{ij}}{\Delta_{ii}} \qquad \text{Transfer impedance} = \frac{e_j'}{I_i'} = \frac{\Delta_{ij}}{\Delta}$$

$$\text{Input impedance} = \frac{e_i'}{I_i'} = \frac{\Delta_{ii}}{\Delta}$$

$\Delta_{ij}$ is cofactor $i,j$ of $\mathbf{Y}'$ and $\Delta$ is $|\mathbf{Y}'|$.

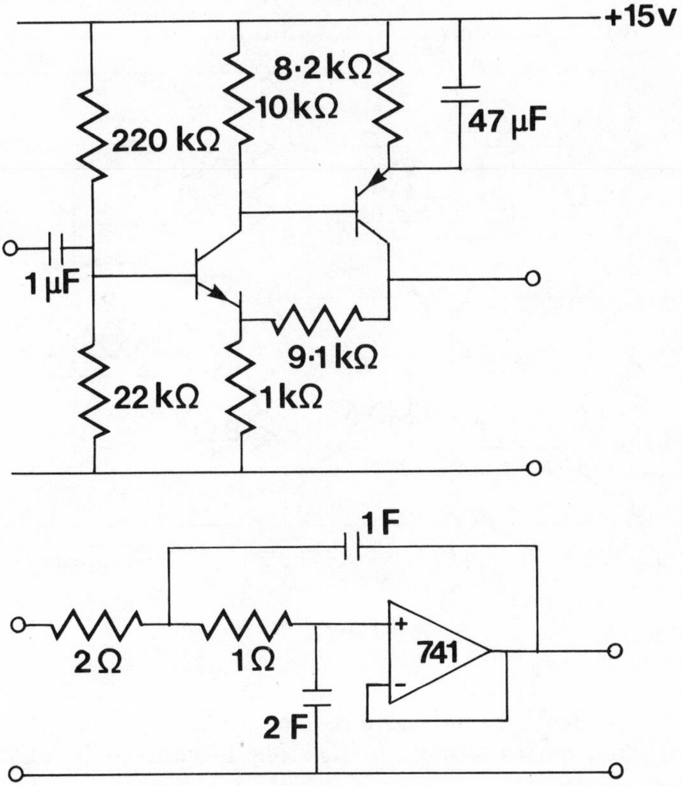


Fig. 4.25 – Problem 4.1.

4.3 Write down the NAM for the network in Fig. 4.26.
Derive the NAM and INAM
(a) with node ③ connected to reference node,
(b) also with node ⑤ connected to node ②.
(c) also with node ① suppressed,
(d) and finally with node ④ as the reference node.

4.4 Use the original NAM from Problem 4.3 and excite the network in Fig. 4.26 by fixed 1 A sources connected from the reference node to nodes ②, ③, ④, and ⑤. Carry out Gauss matrix reduction and LU factorisation of the NAM and solve for node voltages.

4.5 Determine the 3-port matrix of the network in Fig. 4.26 where the ports 1 to 3 respectively are node ② to reference, node ⑤ to reference, and node ③ to node ④.

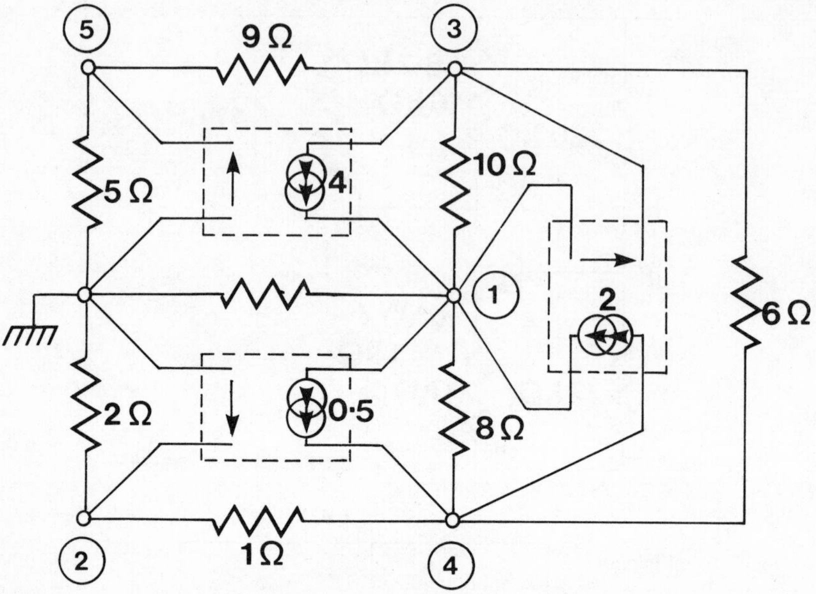


Fig. 4.26 – Problems 4.3 to 4.5.

4.6 Fill the NAM for the bandpass filter in Fig. 8.9. The centre frequency of the passband is 2805 Hz. Normalise frequency to 1 rad/sec and the resistors to 1 Ω.

4.7 Show that the nullator-norator network in Fig. 4.27 represents an inverting operational amplifier with gain set by feedback to $R_4/R_3$. Develop a similar model for a differential amplifier with input and output impedances. ($\omega_t$ is the unity gain frequency).

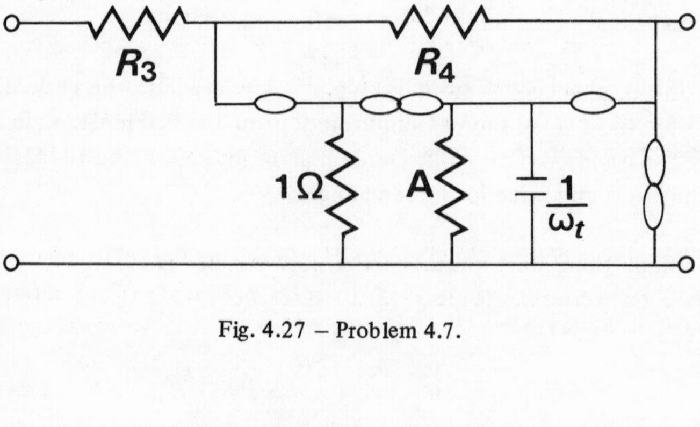


Fig. 4.27 – Problem 4.7.

4.8 Determine nullator-norator models for each of the ideal dependent sources from Table 1.1 and develop models for the four impedance converters and inverters, and also the mutual inductor.

4.9 The transistor amplifier in Fig. 3.26(c) operates at a collector-base voltage of about 5 V and a base-emitter voltage of about 0.7 V. The linearised resistive components in the transistor model were derived in Problem 3.3. Determine the capacitive elements of the model assuming $f_{TF}$ = 100 MHz, $f_{TR}$ = 5 MHz, $C_{E0}$ = 50pF, and $C_{C0}$ = 10pF. If the amplifier emitter resistor is decoupled by a 100 $\mu$F capacitor, evaluate the frequency response of the amplifier.

4.10 Determine the component values of the hybrid-$\pi$ model for the transistor in Problem 4.9 and write down its NAM (omitting $R_{BB'}$).

4.11 Determine the voltage gain transfer function of the gyrator circuit in Fig. 4.28. The amplifiers are differential VCCS, and the given transconductance applies to each output with the appropriate sign. Evaluate the frequency response from 1 MHz to 10 MHz.

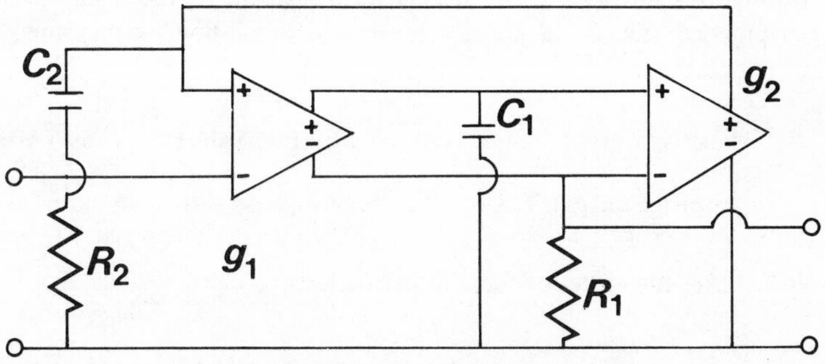


Fig. 4.28 – Problem 4.11.
$R_1 = 700\ \Omega, R_2 = 64\ \Omega, C_1 = C_2 = 21$ pF., $g_1 = g_2 = 650\ \mu$S.

*Computing*

4.12 Determine the frequency response, input, and output impedances of a number of familiar circuits, for example those in Figs. 1.20, 1.21, 4.25, 4.28, 8.9. Explore all the available outputs of program ACNLU and compare results with those obtained by other programs.

4.13 The circuit in Fig 4.29 is a 6 MHz television sound bandpass filter in which the integrated circuit amplifiers have been modelled as ideal VCCS. Each half of the circuit behaves as a tuned circuit using two differential

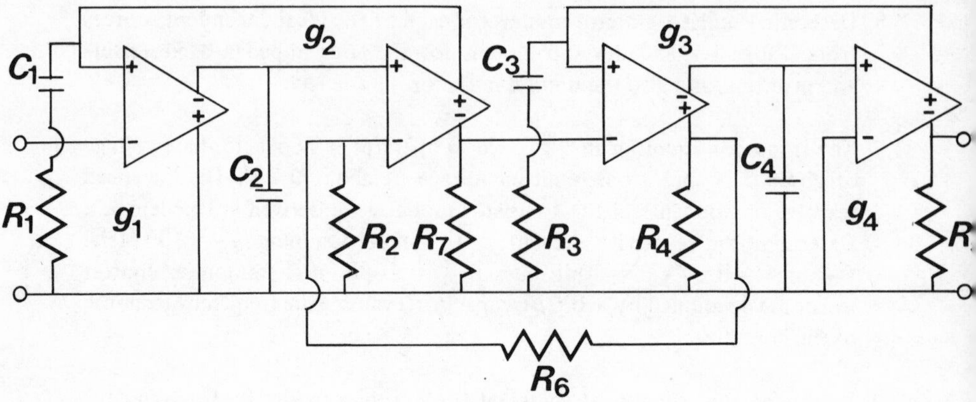


Fig. 4.29 – Problem 4.13.
$R_1 = R_3 = 184\ \Omega, R_2 = 327\ \Omega, R_4 = 161\ \Omega, R_5 = 1\ \text{k}\Omega, R_6 = 980\ \Omega,$
$R_7 = 144\ \Omega, C_1 = C_2 = C_3 = C_4 = 15\ \text{pF}.$

VCCS amplifiers as a gyrator. Coupling between the two tuned circuits is through $R_7$ and is varied by feedback through the $R_6$ to $R_2$ voltage divider. In practice the VCCS are not ideal, and the following measured values were obtained.

Positive output VCCS 646 $\mu$S with phase shift 0.64° at 6 MHz.

Negative output VCCS 652 $\mu$F with phase shift 2.34° at 6 MHz.

Investigate the effect of these imperfections.

4.14 Add graph plotting subroutines to program ACNLU suited to the computer facilities available. One necessary modification to ACNLU, in addition to storage of the plotted data, is a method of avoiding 360° steps in the phase output occurring as a result of use of the standard function ATN(X).

4.15 Add capacitors to the bipolar transistor modelling program from Problem 3.12.

4.16 Add nullators and norators to the list of components acceptable to program ACNLU. A more general set of active components allowed as input to ANCLU could be provided through use of the models from Problem 4.8 (including voltage sources by gyration of current sources).

# APPENDIX

```
0010 PRINT "AC NAM ANALYSIS"
0020 PRINT "RLC INPUTS: COMPONENT NUMBER,VALUE,2 NODE CONNECTIONS"
0030 PRINT "GM  INPUTS: GM NUMBER,VALUE,TYPE,4 NODE CONNECTIONS"
0040 PRINT "      TYPE: 1 FOR GM, 2 FOR GM/JW, 3 FOR JWGM"
0050 PRINT "     NODES: SEND +, - , RECEIVE +, - "
0060 PRINT "I  INPUTS: NODE,VALUE,PHASE DEGREES"
0070 DIM U[9,6],N$[2],M$[12],L$[20]
0080 DIM P[1,10],Q[1,10]
0090 LET P2=8*ATN(1)
0100 LET I6=0
0110 LET T9=0
0120 LET M$="NVSEYMZMDBSN"
0130 LET L$="ERNCNFNONIYMZMNNNSGO"
0140 PRINT "NO.NODES",        N = number of nodes
0150 INPUT N
0160 DIM I[N,1],J[N,1],E[N,1],P[1,N],Q[1,N],T[N,1],V[N,1]
0170 DIM B[N,N],C[N,N],D[N,N],F[N,N],L[N,N],R[N,N]
0180 PRINT "NO. R,L,C,GM",
0190 INPUT R9,L9,C9,G9
0200 DIM S[R9+1,2],M[L9+1,2],A[C9+1,2],G[G9+1,5]
0210 LET R8=R9      R8 = no of resistors
0220 LET L8=L9      L8 ... 8 ind
0230 LET C8=C9      C8 ... cap
0240 LET G8=G9      g8 ... sources
0250 GOTO 0280      g8 ...
0260 PRINT "NO. MODIFIED R,L,C,GM";
0270 INPUT R8,L8,C8,G8
0280 FOR S9=1 TO R8
0290   PRINT "R";
0300   INPUT S8,S[S8,0],S[S8,1],S[S8,2]        store R in S    S8 =
0310 NEXT S9                    p    q                         COMP NUMBER
0320 FOR S9=1 TO L8
0330   PRINT "L";
0340   INPUT S8,M[S8,0],M[S8,1],M[S8,2]        store L in m
0350 NEXT S9
0360 FOR S9=1 TO C8
0370   PRINT "C";
0380   INPUT S8,A[S8,0],A[S8,1],A[S8,2]        store C in A
0390 NEXT S9
0400 FOR S9=1 TO G8
0410   PRINT "GM";
0420   INPUT S8,G[S8,0],G[S8,1],G[S8,2],G[S8,3],G[S8,4],G[S8,5]   store gfs in G
0430 NEXT S9
0440 IF I6>1 THEN GOTO 2710     1st RUN I6=0
0450 PRINT "F-LOW,F+INC,F*MULT,NO.F"
0460 INPUT F0,F1,F2,F3     F3 N° OF FREQ CALCS.
0470 IF I6>1 THEN GOTO 2710
0480 IF I6>0 THEN GOTO 0510
0490 PRINT "OUTPUTS: NV NODE VOLTS     SE SENSITIVITY+NV     YM NAM"
0500 PRINT "         DB GAIN ZIN ZOUT  SN SENSITIVITY+GAIN   ZM INV NAM"
0510 PRINT "OUTPUT";
0520 INPUT N$      GIVE OUTPUT REQUIRED
0530 FOR I9=1 TO 6
0535   LET I7=I9
0540   IF N$=M$[2*I7-1,2*I7] THEN GOTO 0580     SCANS OUTPUT CODES
0550 NEXT I9
0560 PRINT "UNRECOGNISED OUTPUT RETYPE";   NO MATCH
0570 GOTO 0520
0580 IF I6>1 THEN GOTO 0770
0590 IF I7>2 THEN GOTO 0730
0600 MAT I=ZER
0610 MAT J=ZER
0620 PRINT "NO. I SOURCES",
```

```
0630 GOTO 0650
0640 PRINT "NO. MODIFIED I SOURCES",
0650 INPUT I5
0660 FOR S9=1 TO I5
0670   PRINT "I";
0680   INPUT S8,S7,S6
0690   LET I[S8,1]=S7*COS(S6*P2/360)
0700   LET J[S8,1]=S7*SIN(S6*P2/360)
0710 NEXT S9
0720 IF I6>1 THEN GOTO 2710
0730 IF I7<5 THEN GOTO 0770
0740 PRINT "INPUT NODE,OUTPUT NODE",
0750 INPUT N1,N2
0760 IF I6>1 THEN GOTO 2710
0770 IF I7=2 THEN GOTO 0790
0780 IF I7<>6 THEN GOTO 2710
0790 IF T9>0 THEN GOTO 2710
0810 PRINT "SENSITIVITY COMPONENT LIST"
0820 PRINT "INPUT R L C OR GM, COMPONENT NUMBER"
0830 PRINT "NO.SENSITIVITIES ";
0840 INPUT T9
0850 IF T9<10 THEN GOTO 0880
0860 PRINT "MAX 9 RE-INPUT",
0870 GOTO 0840
0880 MAT U=ZER[T9,6]
0890 PRINT "CPT.TYPE, NO."
0900 FOR S9=1 TO T9
0910   INPUT N$,S8
0920   LET S7=1
0930   IF N$="R" THEN GOTO 1020
0940   LET S7=2
0950   IF N$="L" THEN GOTO 1020
0960   LET S7=3
0970   IF N$="C" THEN GOTO 1020
0980   LET S7=4
0990   IF N$="GM" THEN GOTO 1020
1000   PRINT "UNRECOGNISED CPT. RE-INPUT"
1010   GOTO 0910
1020   LET U[S9,0]=S7
1030   LET U[S9,1]=S8
1040 NEXT S9
1050 GOTO 2710
1060 REM FILL RLC MATRICES
1070 LET I5=0
1080 DIM R[N,N],L[N,N],C[N,N]
1090 MAT R=ZER
1100 MAT L=ZER
1110 MAT C=ZER
1120 FOR S9=1 TO R9
1130   LET R8=S[S9,1]
1140   LET L8=S[S9,2]
1150   LET C8=R8
1160   LET G8=L8
1170   LET S8=1/S[S9,0]
1180   LET R[R8,C8]=R[R8,C8]+S8
1190   LET R[L8,G8]=R[L8,G8]+S8
1200   LET R[L8,C8]=R[L8,C8]-S8
1210   LET R[R8,G8]=R[R8,G8]-S8
1220   IF I5=1 THEN GOTO 1550
1230 NEXT S9

*
```

```
1240 FOR S9=1 TO L9
1250   LET R8=M[S9,1]
1260   LET L8=M[S9,2]
1270   LET C8=R8
1280   LET G8=L8
1290   LET S8=1/M[S9,0]
1300   LET L[R8,C8]=L[R8,C8]+S8
1310   LET L[L8,G8]=L[L8,G8]+S8
1320   LET L[L8,C8]=L[L8,C8]-S8
1330   LET L[R8,G8]=L[R8,G8]-S8
1340   IF I5=1 THEN GOTO 1550
1350 NEXT S9
1360 FOR S9=1 TO C9
1370   LET R8=A[S9,1]
1380   LET L8=A[S9,2]
1390   LET C8=R8
1400   LET G8=L8
1410   LET S8=A[S9,0]
1420   LET C[R8,C8]=C[R8,C8]+S8
1430   LET C[L8,G8]=C[L8,G8]+S8
1440   LET C[L8,C8]=C[L8,C8]-S8
1450   LET C[R8,G8]=C[R8,G8]-S8
1460   IF I5=1 THEN GOTO 1550
1470 NEXT S9
1475 LET I5=1
1480 FOR S9=1 TO G9
1490   LET R8=G[S9,5]
1500   LET L8=G[S9,4]
1510   LET C8=G[S9,2]
1520   LET G8=G[S9,3]
1530   LET S8=G[S9,0]
1540   ON G[S9,1] THEN GOTO 1180, 1300, 1420
1550 NEXT S9
1560 REM FILL SENSITIVITY LIST
1570 IF T9=0 THEN GOTO 1870
1580 FOR S9=1 TO T9
1590   LET S8=U[S9,1]
1600   IF U[S9,0]<1 THEN GOTO 1800
1610   ON U[S9,0] THEN GOTO 1620, 1680, 1740, 1800
1620   LET U[S9,2]=S[S8,1]
1630   LET U[S9,3]=S[S8,2]
1640   LET U[S9,4]=S[S8,1]
1650   LET U[S9,5]=S[S8,2]
1660   LET U[S9,6]=S[S8,0]
1670   GOTO 1860
1680   LET U[S9,2]=M[S8,1]
1690   LET U[S9,3]=M[S8,2]
1700   LET U[S9,4]=M[S8,1]
1710   LET U[S9,5]=M[S8,2]
1720   LET U[S9,6]=M[S8,0]
1730   GOTO 1860
1740   LET U[S9,2]=A[S8,1]
1750   LET U[S9,3]=A[S8,2]
1760   LET U[S9,4]=A[S8,1]
1770   LET U[S9,5]=A[S8,2]
1780   LET U[S9,6]=A[S8,0]
1790   GOTO 1860
1800   LET U[S9,2]=G[S8,5]
1810   LET U[S9,3]=G[S8,4]
1820   LET U[S9,4]=G[S8,2]
1830   LET U[S9,5]=G[S8,3]
1840   LET U[S9,6]=G[S8,0]
1850   LET U[S9,0]=-G[S8,1]
1860 NEXT S9

*
```

```
1870 REM SOLUTION
1880 LET W=F0*P2
1890 FOR I9=1 TO F3
1900   IF I9=1 THEN GOTO 1920
1910   LET W=(W+F1*P2)*F2
1920   GOSUB 4000
1940   REM OUTPUTS
1950   ON I7 THEN GOTO 1960, 1960, 2330, 2330, 2500, 2500
1960   IF I9>1 THEN GOTO 2180
1970   ON I7 THEN GOTO 1980, 2030
1980   PRINT 'NODE VOLTS 1 TO ';N;' GIVEN AS'
1990   PRINT 'REAL PARTS NV1 NV2 --- NV5'
2000   PRINT '                NV6 NV7 -- NV10'
2010   PRINT 'IMAG.PARTS NV1 NV2 --- ETC'
2020   GOTO 2180
2030   PRINT 'NODE VOLTAGES AND SENSITIVITIES GIVEN AS'
2040   PRINT 'REAL PARTS NV',
2050   FOR S9=1 TO T9
2060     IF U[S9,0]<1 THEN GOTO 2140
2070     ON U[S9,0] THEN GOTO 2080, 2100, 2120
2080     PRINT 'R';U[S9,1],
2090     GOTO 2150
2100     PRINT 'L';U[S9,1],
2110     GOTO 2150
2120     PRINT 'C';U[S9,1],
2130     GOTO 2150
2140     PRINT 'GM';U[S9,1],
2150   NEXT S9
2160   PRINT
2170   PRINT 'THEN IMAG.PARTS SIMILARLY'
2180   PRINT W/P2;' HZ.'
2190   MAT E=D*I
2200   MAT V=F*J
2210   MAT E=E-V
2220   MAT T=F*I
2230   MAT V=D*J
2240   MAT V=V+T
2250   IF I7=2 THEN GOTO 2310
2260   MAT P=TRN(E)
2270   MAT PRINT P
2280   MAT P=TRN(V)
2290   MAT PRINT P
2300   GOTO 2700
2310   GOSUB 6000
2320   GOTO 2700
2330   IF I9>1 THEN GOTO 2380
2340   ON I7-2 THEN GOTO 2350, 2370
2350   PRINT 'NODAL ADMITTANCE MATRICES'
2360   GOTO 2380
2370   PRINT 'INVERSE NODAL ADMITTANCE MATRICES'
2380   PRINT W/P2;' HZ.'
2390   ON I7-2 THEN GOTO 2400, 2450
2400   PRINT 'REAL PARTS'
2410   MAT PRINT R
2420   PRINT 'IMAG.PARTS'
2430   MAT PRINT B
2440   GOTO 2700
2450   PRINT 'REAL PARTS'
2460   MAT PRINT D
2470   PRINT 'IMAG.PARTS'
2480   MAT PRINT F
2490   GOTO 2700
2500   IF I9>1 THEN GOTO 2570
2510   PRINT 'INPUT NODE ';N1;'  OUTPUT NODE ';N2
2520   PRINT 'HZ.','GAIN DB','PHASE DEG','RP','IP'
```

*

```
2530     ON I7-4 THEN GOTO 2540, 2560
2540     PRINT " ","ZIN RP","ZIN IP","ZOUT RP","ZOUT IP"
2550     GOTO 2570
2560     PRINT "CPT.NO.","DB SEN","DEG SEN","RP SEN","IP SEN"
2570     LET S6=D[N2,N1]
2580     LET S7=F[N2,N1]
2590     LET S8=D[N1,N1]
2600     LET S9=F[N1,N1]
2610     LET C8=(S6*S8+S7*S9)/(S8*S8+S9*S9)
2620     LET G8=(S7*S8-S6*S9)/(S8*S8+S9*S9)
2630     LET L8=57.2958*ATN(G8/C8)
2640     LET R8=4.34295*LOG(C8*C8+G8*G8)
2650     PRINT W/P2,R8,L8,C8,G8
2660     IF I7=6 THEN GOTO 2690
2670     PRINT " ",S8,S9,D[N2,N2],F[N2,N2]
2680     GOTO 2700
2690     GOSUB 8000
2700 NEXT I9
2710 IF I6>0 THEN GOTO 2760
2720 PRINT "NEXT COMMAND: ER END RUN        NS NEW SENSITIVITY LIST"
2730 PRINT "MODIFICATION: NC COMPONENTS     NF FREQUENCIES  NI SOURCES"
2740 PRINT "            : NN IN/OUT NODES   NO OUTPUTS"
2750 PRINT "      OUTPUT: YM 1/R WC 1/WL    ZM INV(NAM)      GO ANALYSE"
2760 PRINT "NEXT";
2770 INPUT N$
2780 FOR I9=1 TO 10
2785   LET I6=I9
2790   IF N$=L$[2*I6-1,2*I6] THEN GOTO 2830
2800 NEXT I9
2810 PRINT "UNRECOGNISED COMMAND RETYPE";
2820 GOTO 2770
2830 IF I6>5 THEN GOTO 2850
2840 ON I6 THEN GOTO 3050, 0260, 0450, 0510, 0640
2850 ON I6-5 THEN GOTO 2860, 2980, 0740, 0830, 1060
2860 PRINT "1/R MATRIX"
2870 MAT PRINT R
2880 PRINT
2890 PRINT "WC MATRIX"
2900 MAT B=(W)*C
2910 MAT PRINT B
2920 PRINT
2930 PRINT "1/WL MATRIX"
2940 MAT B=(1/W)*L
2950 MAT PRINT B
2960 PRINT
2970 GOTO 2760
2980 PRINT "INV(NAM) REAL PARTS"
2990 MAT PRINT D
3000 PRINT
3010 PRINT "INV(NAM) IMAG PARTS"
3020 MAT PRINT F
3030 PRINT
3040 GOTO 2760
3050 STOP

*
```

```
4000 REM SUBROUTINE - NAM INVERT R+J(WC-1/WL) INTO D+JF
4010 DIM B[N,N],D[N,N],F[N,N],X[N,N]
4020 IF W=0 THEN GOTO 4070
4030 MAT D=(W)*C
4040 MAT F=(1/W)*L
4050 MAT B=D-F
4060 GOTO 4100
4070 MAT F=ZER
4080 MAT D=INV(R)
4090 GOTO 4880
4100 IF I7=3 THEN GOTO 4880
4110 IF I9>1 THEN GOTO 4260
4120 MAT X=R
4250 REM LU FACTORIZATION
4260 FOR R5=2 TO N
4270   LET L5=R[1,1]
4280   LET L6=B[1,1]
4290   LET L7=R[1,R5]
4300   LET R[1,R5]=(L7*L5+B[1,R5]*L6)/(L5^2+L6^2)
4310   LET B[1,R5]=(B[1,R5]*L5-L7*L6)/(L5^2+L6^2)
4320 NEXT R5
4330 FOR R6=2 TO N
4340   FOR R5=2 TO N
4350     LET R7=R5-1
4360     IF R6<R5 THEN LET R7=R6-1
4370     FOR R8=1 TO R7
4380       LET R[R5,R6]=R[R5,R6]-R[R5,R8]*R[R8,R6]+B[R5,R8]*B[R8,R6]
4390       LET B[R5,R6]=B[R5,R6]-R[R5,R8]*B[R8,R6]-B[R5,R8]*R[R8,R6]
4400     NEXT R8
4410     IF R6<=R5 THEN GOTO 4470
4420     LET L5=R[R5,R5]
4430     LET L6=B[R5,R5]
4440     LET L7=R[R5,R6]
4450     LET R[R5,R6]=(L7*L5+B[R5,R6]*L6)/(L5^2+L6^2)
4460     LET B[R5,R6]=(B[R5,R6]*L5-L7*L6)/(L5^2+L6^2)
4470   NEXT R5
4480 NEXT R6
4490 REM FORWARD SUBSTITUTION
4500 MAT D=IDN
4510 MAT F=ZER
4520 FOR R6=1 TO N
4530   FOR R5=1 TO N
4540     FOR R7=1 TO R5-1
4550       LET D[R5,R6]=D[R5,R6]-R[R5,R7]*D[R7,R6]+B[R5,R7]*F[R7,R6]
4560       LET F[R5,R6]=F[R5,R6]-B[R5,R7]*D[R7,R6]-R[R5,R7]*F[R7,R6]
4570     NEXT R7
4580     LET L5=R[R5,R5]
4590     LET L6=B[R5,R5]
4600     LET L7=D[R5,R6]
4610     LET D[R5,R6]=(L7*L5+F[R5,R6]*L6)/(L5^2+L6^2)
4620     LET F[R5,R6]=(F[R5,R6]*L5-L7*L6)/(L5^2+L6^2)
4630   NEXT R5
4640 NEXT R6
4650 REM BACKWARD SUBSTITUTION
4660 FOR R6=1 TO N
4670   FOR R5=N-1 TO 1 STEP -1
4680     FOR R7=N TO R5+1 STEP -1
4690       LET D[R5,R6]=D[R5,R6]-R[R5,R7]*D[R7,R6]+B[R5,R7]*F[R7,R6]
4700       LET F[R5,R6]=F[R5,R6]-B[R5,R7]*D[R7,R6]-R[R5,R7]*F[R7,R6]
4710     NEXT R7
4720   NEXT R5
4730 NEXT R6
4740 REM REFILL R
4750 MAT R=X
4880 RETURN

*
```

Fig. 4.30 – Program ACNLU.

CHAPTER 5

# Two-Port Analysis

Circuit analysis methods in the previous three chapters give a solution for all voltages and currents in a network without regard to the information required by the designer. In many circuits, particularly those of filter networks and amplifiers, the analysis requirement is satisfied by evaluation of the frequency reponse of the circuit gain together with input and output impedances. Two-port analysis is a convenient alternative technique which does not give solutions for intermediate circuit nodes or branches. Analysis therefore proceeds more rapidly, and in many cases more accurately than analysis involving the inversion of large matrices. Two-port analysis is also important since it can be extended to include circuits involving transmission lines and other distributed circuit elements.

## 5.1 TWO-PORTS

### *N*-terminal Networks

The nodal admittance matrix in Chapter 4 refers to a circuit with $n_N$ nodes which may be excited by a set of current sources each connected between a node and the reference node. The nodal admittance matrix $\mathbf{Y}'$ of a network is easily obtained by inspection of the network assuming that all circuit nodes are externally available for excitation. The matrix equation $\mathbf{I}' = \mathbf{Y}'\mathbf{e}'$ must then be solved by inversion of $\mathbf{Y}'$ to find the node voltages $\mathbf{e}'$, and from them all branch voltages and currents.

An $N$-terminal network may also be regarded more generally as a multiterminal element where the $N$ available connections are the terminals of the element. These do not necessarily include all the available nodes. The transistor, for example, usually has three terminals (emitter, base, and collector) with no internal discrete components. An operational amplifier, however, has five or more terminals with many internal nodes not available for connection. In Chapter 4 the Y-parameters of an $N$-terminal network were defined and the technique and node suppression was developed to reduce the order of the NAM for such networks. Reduced networks were then included in an analysis with only the nodes required for solution appearing in $\mathbf{Y}'$.

The general conversion, given in Chapter 4, from the NAM to an $N$-port admittance matrix is another technique for reducing the order of analysis matrices. In two-port analysis the only nodes of interest are at the input and output of a network. All other nodes are therefore suppressed and the two-port admittance parameters obtained as a 2X2 matrix $\mathbf{Y}_p$. The network in Fig. 5.1 has three terminals and results from reduction of the NAM. More generally the four-terminal network in Fig. 5.2 is obtained by NAM to $N$-port conversion applied when there is not a common node between input and output. Both situations are of course similar if the three-terminal network is drawn with four terminals as shown by the extension to the network in Fig. 5.1. The two-port Y-parameters for any network can be found by the techniques of Chapter 4.

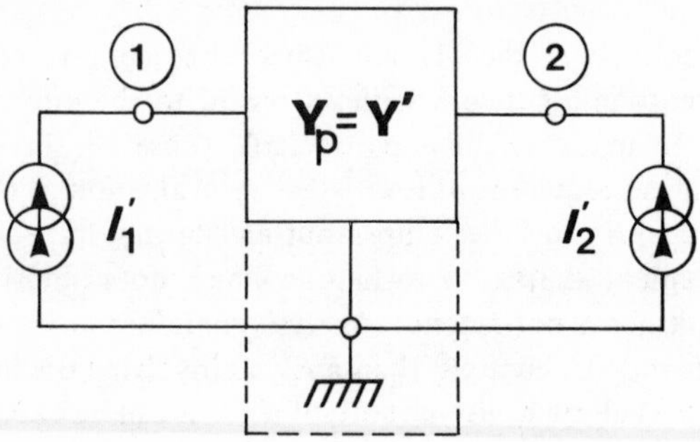


Fig. 5.1 – A three-terminal network.

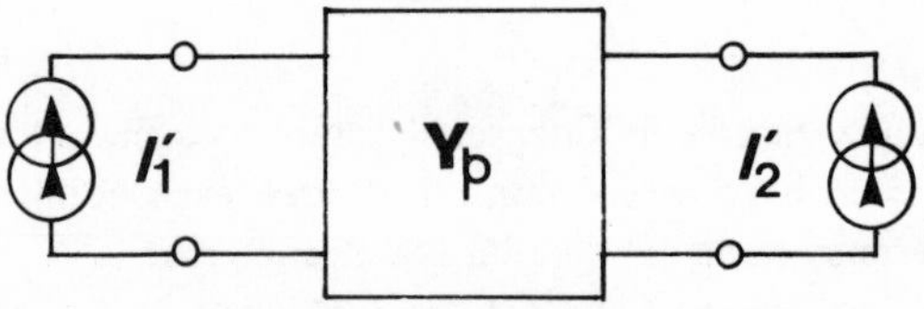


Fig. 5.2 – A four-terminal network.

A *port* was defined in Chapter 4 as a pair of network terminals in which the current entering one terminal is equal to the current leaving the other terminal at all times. The four-terminal networks in Figs. 5.1 and 5.2 clearly operate within this definition. The two terminals connected to $I_1$ are labelled 1 and 1$'$ and called port 1. Similarly the two terminals 2 and 2$'$ are connected to $I_2'$ and called port 2. The two-port network in Fig. 5.3 applies to both three-terminal and four-terminal networks, and the voltage and current directions shown are those adopted in this book. By convention port 1 is assumed to be the input port and port 2 is assumed to be the output port. Two-port analysis is based on the assumption that two-ports may be interconnected while maintaining the equality condition of terminal currents at each port.

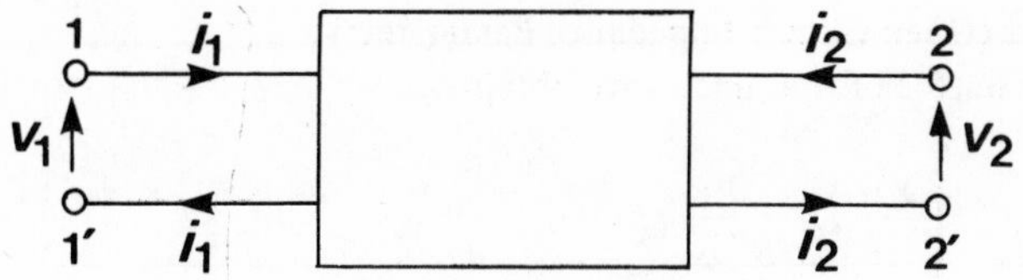


Fig. 5.3 – A two-port.

**Y-parameters (Short Circuit Admittance Parameters)**

The Y-parameters of a network were defined in Chapter 4 in terms of terminal admittances under short circuit conditions. When the network is reduced to a two-port the 2X2 admittance matrix is given by

$$\mathbf{Y} = \begin{bmatrix} y_{11} & y_{12} \\ y_{21} & y_{22} \end{bmatrix} .$$

The relationship between the port voltage and currents is expressed in full by the simultaneous equations

$$i_1 = y_{11}v_1 + y_{12}v_2$$

$$i_2 = y_{21}v_1 + y_{22}v_2 . \tag{5.1}$$

Since the circuit is assumed linear, Eqs. (5.1) are linear and the admittances $y_{11}, y_{12}, y_{21}$, and $y_{22}$ are functions of the Laplace variable $s$. They may be evaluated at a given frequency by the substitution $s = j\omega$ giving the admittances as complex numbers.

The Y-parameters are defined and measured under the conditions of short circuit at either the input or output port. When the output is short circuit $v_2 = 0$ and

$$y_{11} = (i_1/v_1)_{v_2=0}, \quad y_{21} = (i_2/v_1)_{v_2=0} .$$

$y_{11}$ is therefore defined as the input admittance and $y_{21}$ is defined as the forward transfer admittance with output short circuit. The parameters are referred to as the short circuit input admittance and short circuit forward transfer respectively. When $v_1 = 0$, the input is short circuit and

$$y_{12} = (i_1/v_2)_{v_1=0}, \quad y_{22} = (i_2/v_2)_{v_1=0} .$$

$y_{12}$ is the short circuit reverse transfer admittance. $y_{22}$ is the short circuit output admittance.

**Z-parameters (Open Circuit Impedance Parameters)**
If we solve Eq. (5.1) for $v_1$ and $v_2$ we obtain

$$v_1 = \frac{y_{22}}{\Delta} i_1 - \frac{y_{12}}{\Delta} i_2$$

$$v_2 = \frac{y_{21}}{\Delta} i_1 + \frac{y_{11}}{\Delta} i_2 \tag{5.2}$$

where $\Delta = y_{11}y_{22} - y_{12}y_{21}$. Eq. (5.2) may be rewritten

$$v_1 = z_{11}i_1 + z_{12}i_2$$

$$v_2 = z_{21}i_1 + z_{22}i_2 \tag{5.3}$$

and the Z-parameters are assembled in the 2X2 matrix

$$\mathbf{Z} = \begin{bmatrix} z_{11} & z_{12} \\ z_{21} & z_{22} \end{bmatrix}$$

When $i_1$ or $i_2$ are constrained to zero either the input or the output is open circuit. The Z-parameters are then defined in the same way as Y-parameters by the following expressions:

$$z_{11} = (v_1/i_1)_{i_2=0}, \quad z_{12} = (v_1/i_2)_{i_1=0} ,$$

$$z_{21} = (v_2/i_1)_{i_2=0}, \quad z_{22} = (v_2/i_2)_{i_1=0} .$$

Therefore,

$z_{11}$ is the open circuit input impedance,

$z_{12}$ is the open circuit reverse transfer impedance,

$z_{21}$ is the open circuit forward transfer impedance, and

$z_{22}$ is the open circuit output impedance.

**A-parameters (Transmission or Chain Parameters)**
The simultaneous Eqs. (5.1) and (5.2) utilise either the two-port voltages or the currents as independent variables. The remaining currents in Y-parameters or

voltages in Z-parameters are dependent variables. In A-parameters the two output variables, $v_2$ and $i_2$ are selected as independent. The relationship between input and output is then expressed by the simultaneous equations.

$$v_1 = a_{11}v_2 - a_{12}i_2$$

$$i_1 = a_{21}v_2 - a_{22}i_2 \,, \tag{5.4}$$

or in matrix form

$$\begin{bmatrix} v_1 \\ i_1 \end{bmatrix} = \begin{bmatrix} a_{11} & a_{12} \\ a_{21} & a_{22} \end{bmatrix} \begin{bmatrix} v_2 \\ -i_2 \end{bmatrix} . \tag{5.5}$$

The A-parameters are assembled in the 2X2 matrix

$$\mathbf{A} = \begin{bmatrix} a_{11} & a_{12} \\ a_{21} & a_{22} \end{bmatrix} .$$

The use of $-i_2$ in Eq. (5.5) in place of $i_2$ is accepted practice, but this is not universally adopted, particularly in older texts. As we shall see in Section 5.3 the reason for using $-i_2$ is one of mathematical convenience when two-ports are connected in cascade

The A-parameters are defined by the following expressions

$$a_{11} = (v_1/v_2)_{i_2=0}, \quad a_{12} = (v_1/-i_2)_{v_2=0} \,,$$

$$a_{21} = (i_1/v_2)_{i_2=0}, \quad a_{22} = (i_1/-i_2)_{v_2=0} \,.$$

The parameter dimensions are therefore mixed with $a_{11}$ and $a_{22}$ dimensionless $a_{12}$ is a transfer impedance and $a_{21}$ is a transfer admittance. Measurements or definitions require the output terminal to be either open circuit ($i_2 = 0$) or short circuit ($v_2 = 0$), and excitation to be applied at the input. Therefore,

$a_{11}$ is the inverse of the open circuit forward voltage gain,

$a_{12}$ is the inverse of the short circuit forward transfer admittance,

$a_{21}$ is the inverse of the open circuit forward transfer impedance, and

$a_{22}$ is the inverse of the short circuit forward current gain.

**B-parameters (Inverse Transmission Parameters)**
If we solve Eq. (5.4) for $v_2$ and $i_2$ we obtain

$$v_2 = \frac{a_{22}}{\Delta} v_1 - \frac{a_{12}}{\Delta} i_1$$

$$i_2 = \frac{a_{21}}{\Delta} v_1 - \frac{a_{11}}{\Delta} i_1 \, , \tag{5.6}$$

where $\Delta = a_{11}a_{22} - a_{12}a_{21}$. Eq. (5.6) may be rewritten

$$v_2 = b_{11}v_1 + b_{12}(-i_1)$$

$$i_2 = b_{21}v_1 + b_{22}(-i_1) \, , \tag{5.7}$$

and the B-parameters are assembled in the 2X2 matrix

$$\mathbf{B} = \begin{bmatrix} b_{11} & b_{12} \\ b_{21} & b_{22} \end{bmatrix} .$$

The B-parameters are defined by the expressions,

$$b_{11} = (v_2/v_1)_{i_1=0}, \quad b_{12} = (v_2/-i_1)_{v_1=0} \, ,$$

$$b_{21} = (i_2/v_1)_{i_1=0}, \quad b_{22} = (i_2/-i_1)_{v_2=0} \, .$$

These are similar to those defined for A-parameters, but the network is excited at the output with the input open or short circuit, therefore

$b_{11}$ is the inverse of the open-circuit reverse voltage gain,

$b_{12}$ is the inverse of the short circuit reverse transfer admittance,

$b_{21}$ is the inverse of the open circuit reverse transfer impedance, and

$b_{22}$ is the inverse of the short circuit reverse current gain.

**H-parameters (Hybrid parameters)**
Two-ports have four associated variables, $i_1$, $v_1$, $i_2$ and $v_2$. These have been selected in pairs as independent variables, and so far four pairs have been used in defining parameters. The variable pairs used are $(v_1, v_2)$, $(i_1, i_2)$, $(v_2, i_2)$ and

$(v_1, i_1)$ for the Y,Z, A, and B-parameters respectively. The total number of combinations available is ${}_4C_2 = 6$. The remaining pairs $(i_1, v_2)$ and $(v_1, i_2)$ are used as independent variables in the H and G-parameters respectively,

Selecting $(i_1, v_2)$ we can write

$$v_1 = h_{11}i_1 + h_{12}v_2$$

$$i_2 = h_{21}i_1 + h_{22}v_2 \ , \tag{5.8}$$

and the H-parameters are assembled in the 2X2 matrix

$$\mathbf{H} = \begin{bmatrix} h_{11} & h_{12} \\ h_{21} & h_{22} \end{bmatrix} .$$

The H-parameters are defined by the expressions,

$$h_{11} = (v_1/i_1)_{v_2=0}, \quad h_{12} = (v_1/v_2)_{i_1=0} \ ,$$

$$h_{21} = (i_2/i_1)_{v_2=0}, \quad h_{22} = (i_2/v_2)_{i_1=0} \ .$$

Therefore,

$h_{11}$ is the short circuit input impedance,

$h_{12}$ is the open circuit reverse voltage gain,

$h_{21}$ is the short circuit forward current gain, and

$h_{22}$ is the open circuit output admittance.

**G-parameters (Inverse Hybrid Parameters)**

Solving Eq. (5.8),

$$v_2 = -\frac{h_{21}}{\Delta} v_1 + \frac{h_{22}}{\Delta} i_2$$

$$i_1 = \frac{h_{22}}{\Delta} v_1 - \frac{h_{12}}{\Delta} i_2 \ . \tag{5.9}$$

where $\Delta = h_{11}h_{22} - h_{12}h_{21}$. Eq. (5.9) may be rewritten

$$\begin{aligned} i_1 &= g_{11}v_1 + g_{12}i_2 \\ v_2 &= g_{21}v_1 + g_{22}i_2 \ , \end{aligned} \qquad (5.10)$$

and the inverse hybrid matrix is given by

$$\mathbf{G} = \begin{bmatrix} g_{11} & g_{12} \\ g_{21} & g_{22} \end{bmatrix} .$$

G-parameters are defined by the expressions,

$$g_{11} = (i_1/v_1)_{i_2=0}, \quad g_{12} = (i_1/i_2)_{v_1=0} \ ,$$

$$g_{21} = (v_2/v_1)_{i_2=0}, \quad g_{22} = (v_2/i_2)_{v_1=0} \ .$$

Therefore,

$g_{11}$ is the open circuit input admittance,

$g_{12}$ is the short circuit reverse current gain,

$g_{21}$ is the open circuit forward voltage gain, and

$g_{22}$ is the short circuit output impedance.

**Parameter Conversion**

We now have six sets of parameters which describe the behaviour of the same network. Clearly the parameters must be related, and conversion between any two sets of parameters, which is frequently necessary in circuit analysis, must be tabulated.

Several conversions have already been derived. These were in fact the matrix inversions which relate Y and Z, A and B, and H and G-parameters. The remaining conversions are easily derived by manipulating the simultaneous equations. The Y-parameters, for example, from Eq. (5.1) are

$$\begin{bmatrix} i_1 \\ i_2 \end{bmatrix} = \begin{bmatrix} y_{11} & y_{12} \\ y_{21} & y_{22} \end{bmatrix} \begin{bmatrix} v_1 \\ v_2 \end{bmatrix} . \qquad (5.11)$$

The conversions to H-parameters may be derived as follows.

Rearrange Eqs. (5.11) to obtain $(i_1, v_2)$ as independent variables.

$$\begin{bmatrix} v_1 \\ i_2 - y_{21}v_1 \end{bmatrix} = \begin{bmatrix} 1/y_{11} & -y_{12}/y_{11} \\ 0 & y_{22} \end{bmatrix} \begin{bmatrix} i_1 \\ v_2 \end{bmatrix}. \tag{5.12}$$

Substitute from the upper equation to eliminate $v_1$ in the lower equation in Eqs. (5.12) to yield

$$\begin{bmatrix} v_1 \\ i_2 \end{bmatrix} = \begin{bmatrix} 1/y_{11} & -y_{12}/y_{11} \\ y_{21}/y_{11} & \Delta/y_{11} \end{bmatrix} \begin{bmatrix} i_1 \\ v_2 \end{bmatrix},$$

where $\Delta = y_{11}y_{22} - y_{12}y_{21}$: Therefore,

$$\mathbf{H} = \begin{bmatrix} h_{11} & h_{12} \\ h_{21} & h_{22} \end{bmatrix} = \begin{bmatrix} 1/y_{11} & -y_{12}/y_{11} \\ y_{21}/y_{11} & \Delta/y_{11} \end{bmatrix}. \tag{5.13}$$

All other parameter conversions can be similarly derived, and a few at least should be worked as problems.

A conversion table of thirty entries giving the full set of parameter conversions is normally given at this point. However, this disguises the numerical relationships within the table which lead to efficient computer programming. These relationships are also interesting since the grouping of numerically equal conversions illustrates the principle of duality, and the cyclic nature of many of the conversions can be demonstrated.

Table 5.1 giving the two-port parameter conversions is therefore arranged with the 19 necessary entries in five rows. The eight A-parameter conversions form the first section. The three basic conversions required for all inter-G,H,Y and Z conversions form the centre section. The little used B-parameter conversions are given in the final section.

**Terminated Two-Ports**

The two-port network in Fig. 5.3, analysed in terms of the port admittance matrix, was assumed to be excited at its input and output ports by means of current sources. The matrix equation $\mathbf{i} = \mathbf{Y}_p\mathbf{v}$ was then solved for port voltages. However, the two-port matrix of Y and other parameters relates input and output voltages and currents and leaves the sources undefined. The sources are not therefore limited to ideal current sources, and in general either voltage or current

**Table 5.1** – Two-Port Parameter Conversions

A – Parameter Conversions

AY
$$\begin{bmatrix} p_{22}/p_{12} & -\Delta/p_{12} \\ -1/p_{12} & p_{11}/p_{12} \end{bmatrix}$$

AG
$$\begin{bmatrix} p_{21}/p_{11} & -\Delta/p_{11} \\ -1/p_{11} & p_{12}/p_{11} \end{bmatrix}$$

AH
$$\begin{bmatrix} p_{12}/p_{22} & \Delta/p_{22} \\ -1/p_{22} & p_{21}/p_{22} \end{bmatrix}$$

AZ ZA
$$\begin{bmatrix} p_{11}/p_{21} & \Delta/p_{21} \\ 1/p_{21} & p_{22}/p_{21} \end{bmatrix}$$

YA
$$\begin{bmatrix} -p_{22}/p_{21} & -1/p_{21} \\ -\Delta/p_{21} & -p_{11}/p_{21} \end{bmatrix}$$

GA
$$\begin{bmatrix} 1/p_{21} & p_{22}/p_{21} \\ p_{11}/p_{21} & \Delta/p_{21} \end{bmatrix}$$

HA
$$\begin{bmatrix} -\Delta/p_{21} & -p_{11}/p_{21} \\ -p_{22}/p_{21} & -1/p_{21} \end{bmatrix}$$

AB BA
$$\begin{bmatrix} p_{22}/\Delta & p_{12}/\Delta \\ p_{21}/\Delta & p_{11}/\Delta \end{bmatrix}$$

G,H,Y,Z – Parameter Conversions

GH HG YZ ZY
$$\begin{bmatrix} p_{22}/\Delta & -p_{12}/\Delta \\ -p_{21}/\Delta & p_{11}/\Delta \end{bmatrix}$$

GZ ZG HY YH
$$\begin{bmatrix} 1/p_{11} & -p_{12}/p_{11} \\ p_{21}/p_{11} & \Delta/p_{11} \end{bmatrix}$$

GY YG HZ ZH
$$\begin{bmatrix} \Delta/p_{22} & p_{12}/p_{22} \\ -p_{21}/p_{22} & 1/p_{22} \end{bmatrix}$$

B – Parameter Conversions

| | | | |
|---|---|---|---|
| BY $\begin{bmatrix} p_{11}/p_{12} & -1/p_{12} \\ -\Delta/p_{12} & p_{22}/p_{12} \end{bmatrix}$ | BH $\begin{bmatrix} p_{12}/p_{11} & 1/p_{11} \\ -\Delta/p_{11} & p_{21}/p_{11} \end{bmatrix}$ | BG $\begin{bmatrix} p_{21}/p_{22} & -1/p_{22} \\ \Delta/p_{22} & p_{12}/p_{22} \end{bmatrix}$ | BZ $\begin{bmatrix} p_{22}/p_{21} & 1/p_{21} \\ \Delta/p_{21} & p_{11}/p_{21} \end{bmatrix}$ |
| YB $\begin{bmatrix} -p_{11}/p_{12} & -1/p_{12} \\ -\Delta/p_{12} & -p_{22}/p_{12} \end{bmatrix}$ | HB $\begin{bmatrix} 1/p_{12} & p_{11}/p_{12} \\ p_{22}/p_{12} & \Delta/p_{12} \end{bmatrix}$ | GB $\begin{bmatrix} -\Delta/p_{12} & -p_{22}/p_{12} \\ -p_{11}/p_{12} & -1/p_{12} \end{bmatrix}$ | ZB $\begin{bmatrix} p_{22}/p_{12} & \Delta/p_{12} \\ 1/p_{12} & p_{11}/p_{12} \end{bmatrix}$ |

Notes i Conversion YA means FROM Y TO A.

ii $\begin{bmatrix} p_{11} & p_{12} \\ p_{21} & p_{22} \end{bmatrix}$ and determinant $\Delta$ refer to the FROM matrix.

sources with finite impedance may be specified. It is normally assumed that the input of a two-port is excited by a fixed voltage source of finite impedance $Z_s$ while the output is loaded by a finite passive load impedance $Z_L$. The circuit is given in Fig. 5.4.

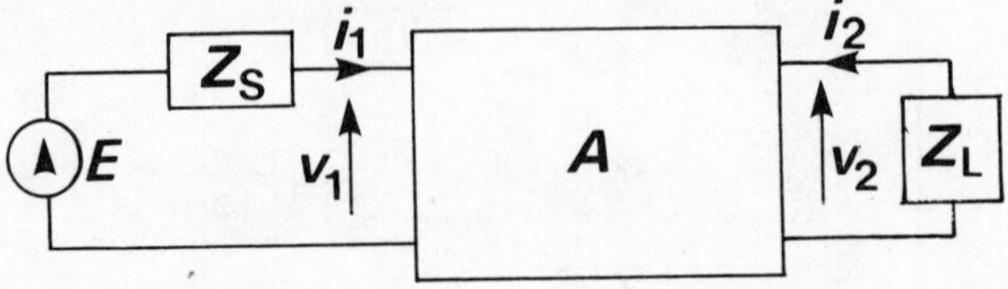


Fig. 5.4 – Excited and Loaded Two-Port.

The following properties of the two-port operating as an amplifier may be defined.

Voltage gain $= v_2/v_1$

Current gain $= i_2/i_1$

Input impedance $= v_1/i_1$

Output impedance $= v_2(Z_L{=}\infty)/\{-i_2\}\,(Z_L{=}0)$ .

From the A-parameters of the two-port,

$$v_1 = a_{11}v_2 - a_{12}i_2 \ , \tag{5.14}$$

$$i_1 = a_{21}v_2 - a_{22}i_2 \ . \tag{5.15}$$

Also $$E - v_1 = Z_s i_1 \ , \tag{5.16}$$

$$v_2 = -Z_L i_2 \ . \tag{5.17}$$

Substituting for $i_2$ in Eq. (5.14) from (5.17) yields

$$\text{Voltage gain} = \frac{v_2}{v_1} = \frac{Z_L}{a_{11}Z_L + a_{12}} \ . \tag{5.18}$$

Substituting similarly into Eq. (5.15) yields

$$\text{Current gain} = \frac{i_2}{i_1} = \frac{-1}{a_{21}Z_L + a_{22}} \ . \tag{5.19}$$

Dividing Eqs. (5.14) and (5.15) while substituting for $v_2$ from Eq. (5.17) yields

$$\text{Input impedance} = \frac{v_1}{i_1} = \frac{a_{11}Z_L + a_{12}}{a_{21}Z_L + a_{22}} \ . \tag{5.20}$$

When $Z_L = \infty$, $i_2 = 0$, hence

$$v_1 = a_{11}v_2$$

$$i_1 = a_{21}v_2 \ .$$

Substituting into Eqn (5.16)

$$v_2(Z_L = \infty) = \frac{E}{a_{11} + a_{21}Z_s} \ .$$

When $Z_L = 0, v_2 = 0$, hence

$$v_1 = -a_{12}i_2$$

$$i_1 = -a_{22}i_2 \ .$$

Substituting into Eq. (5.16)

$$-i_2(Z_L{=}0) = \frac{E}{a_{12} + a_{22}Z_s} \ .$$

Therefore,

$$\text{Output impedance} = \frac{v_2(Z_L{=}\infty)}{-i_2(Z_L{=}0)} = \frac{a_{12} + a_{22}Z_s}{a_{11} + a_{21}Z_s} \ . \tag{5.21}$$

Performance properties of two-ports in terms of A-parameters have been derived and may be computed from Eqs. (5.18) to (5.21). These expressions may be converted to all other parameter sets, using Table 5.1. Results are summarised in Table 5.2.

If a source impedance $Z_s'$ and a load impedance $Z_L'$ are included within the circuit for which the A-parameters have been obtained, then the expressions (5.18) to (5.21) may be simplified by setting $Z_s = 0$ and $Z_L = \infty$. Therefore,

$$\text{Voltage gain} = 1/a_{11} \ . \tag{5.22}$$

**Table 5.2** – Performance of Terminated Two-Ports
$Z_L$ = load impedance, $Z_s$ = source impedance

| | Voltage Gain | Current Gain | Input Impedance | Output Impedance |
|---|---|---|---|---|
| A | $\dfrac{Z_L}{a_{12}+a_{11}Z_L}$ | $\dfrac{-1}{a_{22}+a_{21}Z_L}$ | $\dfrac{a_{12}+a_{11}Z_L}{a_{22}+a_{21}Z_L}$ | $\dfrac{a_{12}+a_{22}Z_s}{a_{11}+a_{21}Z_s}$ |
| B | $\dfrac{\Delta Z_L}{b_{12}+b_{22}Z_L}$ | $\dfrac{-\Delta}{b_{11}+b_{21}Z_L}$ | $\dfrac{b_{12}+b_{22}Z_L}{b_{11}+b_{21}Z_L}$ | $\dfrac{b_{12}+b_{11}Z_s}{b_{22}+b_{21}Z_s}$ |
| H | $\dfrac{-h_{21}Z_L}{h_{11}+\Delta Z_L}$ | $\dfrac{h_{21}}{1+h_{22}Z_L}$ | $\dfrac{h_{11}+\Delta Z_L}{1+h_{22}Z_L}$ | $\dfrac{h_{11}+Z_s}{\Delta+h_{22}Z_s}$ |
| G | $\dfrac{g_{21}Z_L}{g_{22}+Z_L}$ | $\dfrac{-g_{21}}{\Delta+g_{11}Z_L}$ | $\dfrac{g_{22}+Z_L}{\Delta+g_{11}Z_L}$ | $\dfrac{g_{22}+\Delta Z_s}{1+g_{11}Z_s}$ |
| Y | $\dfrac{-y_{21}Z_L}{1+y_{22}Z_L}$ | $\dfrac{y_{21}}{y_{11}+\Delta Z_L}$ | $\dfrac{1+y_{22}Z_L}{y_{11}+\Delta Z_L}$ | $\dfrac{1+y_{11}Z_s}{y_{22}+\Delta Z_s}$ |
| Z | $\dfrac{z_{21}Z_L}{\Delta+z_{11}Z_L}$ | $\dfrac{-z_{21}}{z_{22}+Z_L}$ | $\dfrac{\Delta+z_{11}Z_L}{z_{22}+Z_L}$ | $\dfrac{\Delta+z_{22}Z_s}{z_{11}+Z_s}$ |

Since the output current through the actual load $Z'_L$ is $-v_2/Z'_L$,

$$\text{Current gain} = -1/a_{21}Z'_L\ . \tag{5.23}$$

Input impedance includes $Z'_s$ in series with it, that is

$$(\text{Input impedance} + Z'_s) = a_{11}/a_{21}\ . \tag{5.24}$$

Similarly output impedance includes $Z'_L$ in parallel, that is,

$$(\text{Output impedance} \,//\, Z'_L) = a_{12}/a_{11}\ . \tag{5.25}$$

In practice the simplified expressions (5.22) to (5.25) are more convenient than the full expressions in Table 5.2 since source and load impedances when present are usually resistive and easily accounted. As before these expressions may be converted to all other parameters using Table 5.1. Results are summarised in Table 5.3.

**Table 5.3** – Performance of Two-Ports simplified for $Z_L = \infty, Z_s = 0$

| | Voltage Gain | Current Gain | Input Impedance ($+ Z_s'$) | Output Impedance ($//Z_L'$) |
|---|---|---|---|---|
| A | $1/a_{11}$ | $1/a_{21}Z_L'$ | $a_{11}/a_{21}$ | $a_{12}/a_{11}$ |
| B | $\Delta/b_{22}$ | $-\Delta/b_{21}Z_L'$ | $b_{22}/b_{21}$ | $b_{12}/b_{21}$ |
| H | $-h_{21}/\Delta$ | $h_{21}/h_{22}Z_L$ | $\Delta/h_{22}$ | $h_{11}/\Delta$ |
| G | $g_{21}$ | $-g_{21}/g_{11}Z_L'$ | $1/g_{11}$ | $g_{22}$ |
| Y | $-y_{21}/y_{22}$ | $y_{21}/\Delta Z_L'$ | $y_{22}/\Delta$ | $1/y_{22}$ |
| Z | $z_{21}/z_{11}$ | $-z_{21}/Z_L'$ | $z_{11}$ | $\Delta/z_{11}$ |

*Example* 5.1

A differential amplifier has the following characteristics. Input impedance = 100 kΩ, output impedance = 100 Ω, Voltage gain = 50. Determine the two-port Y-parameters for the amplifier when connected as shown in Fig. 5.5 and find the input impedance, output impedance, and voltage gain for the unloaded two-port.

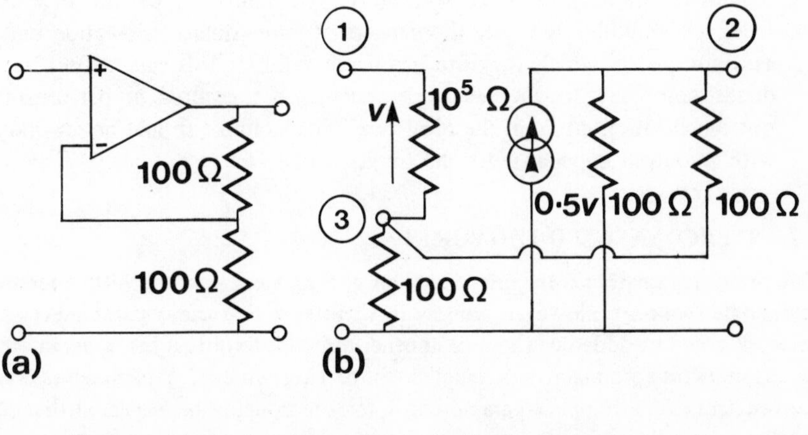


Fig. 5.5 – (a) Circuit for Example 5.1; (b) Circuit redrawn for analysis.

Filling the 3X3 nodal admittance matrix $\mathbf{Y}'$ according to the rules given in Chapter 4 yields

$$\mathbf{Y}' = \begin{bmatrix} 10^{-5} & 0 & -10^{5} \\ -0.5 & 2 \times 10^{-2} & 0.49 \\ -10^{-5} & -10^{-2} & 2.001 \times 10^{-2} \end{bmatrix} .$$

Supress node ③ to give the NAM for the two-port.

$$\mathbf{Y}' = \begin{bmatrix} 10^{-5}(1-1/2001) & -10^{-2}/2001 \\ -0.5 + 10^{-5} \times \dfrac{49}{2.001} & 2 \times 10^{-2} + 10^{-2} \times \dfrac{49}{2.001} \end{bmatrix}$$

$$= \begin{bmatrix} 0.9995 \times 10^{-5} & -0.49975 \times 10^{-5} \\ -0.499755 & 0.264878 \end{bmatrix} .$$

Using Table 5.3,

$$\text{Voltage gain} = -y_{21}/y_{22} = 1.887$$

$$\text{Input impedance} = y_{22}/\Delta = 17.67 \times 10^{5}\ \Omega$$

$$\text{Output impedance} = 1/y_{22} = 3.775\ \Omega .$$

The solution was worked using a calculator so that the value of the determinant $\Delta$ could be obtained accurately. The values for performance as a feedback amplifier illustrate the dangers of approximate calculation unless assumptions on which they are based are valid. In this case output impedance compared to the feedback resistors has resulted in performance noticeably different from the ideal case. The example should be reworked, with an output impedance for the amplifier of 1 Ω.

## 5.2 INTERCONNECTED TWO-PORTS

Two -port paramaters were introduced in the previous section with reference to a single two-port. However, we saw in Chapter 4 that the Y-parameters of a network could be added to those of another network to obtain the Y-parameters of the networks connected in parallel. Since the two-port Y-parameters were derived from *N*-terminal Y-parameters it may reasonably be expected that the parallel connection of ports implies the addition of parameters as before.

We will similarly find that the other parameter sets each have a specific function when two-ports are connected together in different ways.

**Parallel-Parallel Connection**

When two-port connections are connected in parallel at both input and output as shown in Fig. 5.6,

$$\begin{bmatrix} v_1 \\ v_2 \end{bmatrix} = \begin{bmatrix} v'_1 \\ v'_2 \end{bmatrix} = \begin{bmatrix} v''_1 \\ v''_2 \end{bmatrix}, \text{ and } \begin{bmatrix} i_1 \\ i_2 \end{bmatrix} = \begin{bmatrix} i'_1 + i''_1 \\ i'_2 + i''_2 \end{bmatrix}. \tag{5.26}$$

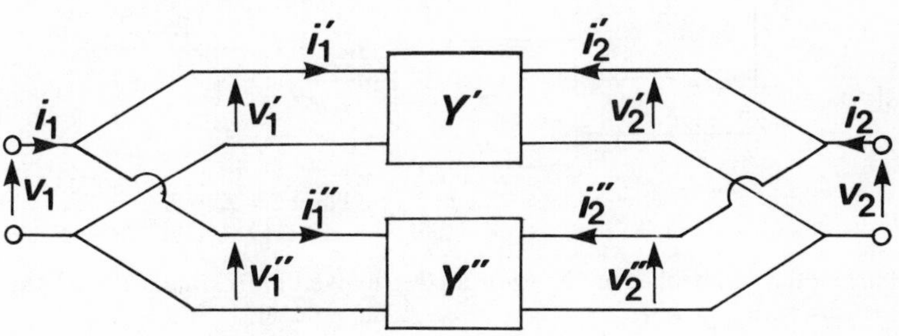


Fig. 5.6 – Parallel-Parallel Connected Two-Ports **Y** = **Y**′ + **Y**″.

The Y-parameter equations for the two-ports are given individually by

$$\begin{bmatrix} i'_1 \\ i'_2 \end{bmatrix} = \begin{bmatrix} y'_{11} & y'_{12} \\ y'_{21} & y'_{22} \end{bmatrix} \begin{bmatrix} v'_1 \\ v'_2 \end{bmatrix}, \text{ and } \begin{bmatrix} i''_1 \\ i''_2 \end{bmatrix} = \begin{bmatrix} y''_{11} & y''_{12} \\ y''_{21} & y''_{22} \end{bmatrix} \begin{bmatrix} v''_1 \\ v''_2 \end{bmatrix}.$$

Substituting into expressions (5.26),

$$\begin{bmatrix} i_1 \\ i_2 \end{bmatrix} = \begin{bmatrix} y'_{11} + y''_{11} & y'_{12} + y''_{12} \\ y'_{21} + y''_{21} & y'_{22} + y''_{22} \end{bmatrix} \begin{bmatrix} v_1 \\ v_2 \end{bmatrix}.$$

The Y-parameters of the combined network in parallel-parallel connection are therefore given by the sum of the Y-parameters of the individual networks, that is,

$$\mathbf{Y} = \mathbf{Y}' + \mathbf{Y}''. \tag{5.27}$$

**Series-Series Connection**

When two-port networks are connected in series at both input and output as shown in Fig. 5.7,

$$\begin{bmatrix} v_1 \\ v_2 \end{bmatrix} = \begin{bmatrix} v_1' + v_1'' \\ v_2' + v_2'' \end{bmatrix}, \text{ and } \begin{bmatrix} i_1 \\ i_2 \end{bmatrix} = \begin{bmatrix} i_1' \\ i_2' \end{bmatrix} = \begin{bmatrix} i_1'' \\ i_2'' \end{bmatrix}. \tag{5.28}$$

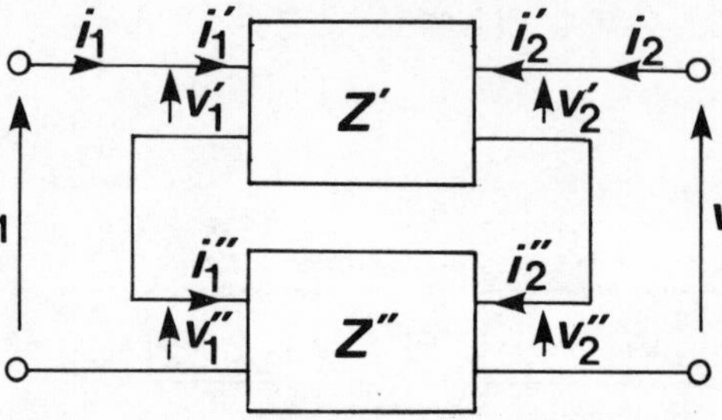


Fig. 5.7 – Series-Series Connected Two-Ports $\mathbf{Z} = \mathbf{Z}' + \mathbf{Z}''$.

The Z-parameter equations for the two-ports are given individually by

$$\begin{bmatrix} v_1' \\ v_2' \end{bmatrix} = \begin{bmatrix} z_{11}' & z_{12}' \\ z_{22}' & z_{22}' \end{bmatrix} \begin{bmatrix} i_1' \\ i_2' \end{bmatrix}, \text{ and } \begin{bmatrix} v_1'' \\ v_2'' \end{bmatrix} = \begin{bmatrix} z_{11}'' & z_{12}'' \\ z_{21}'' & z_{22}'' \end{bmatrix} \begin{bmatrix} i_1'' \\ i_2'' \end{bmatrix}.$$

Substituting into expressions (5.28)

$$\begin{bmatrix} v_1 \\ v_2 \end{bmatrix} = \begin{bmatrix} z_{11}' + z_{11}'' & z_{12}' + z_{12}'' \\ z_{21}' + z_{21}'' & z_{22}' + z_{22}'' \end{bmatrix} \begin{bmatrix} i_1 \\ i_2 \end{bmatrix}.$$

The Z-parameters of the combined network in series-series connection are therefore given by the sum of the Z-parameters of the individual networks, that is,

$$\mathbf{Z} = \mathbf{Z}' + \mathbf{Z}''. \tag{5.29}$$

**Series-Parallel Connection**

When two-port networks are connected with inputs in series and outputs in parallel as shown in Fig. 5.8,

$$\begin{bmatrix} v_1 \\ i_2 \end{bmatrix} = \begin{bmatrix} v_1' + v_1'' \\ i_2' + i_2'' \end{bmatrix}, \text{ and } \begin{bmatrix} i_1 \\ v_2 \end{bmatrix} = \begin{bmatrix} i_1' \\ v_2' \end{bmatrix} = \begin{bmatrix} i_1'' \\ v_2'' \end{bmatrix}. \tag{5.30}$$

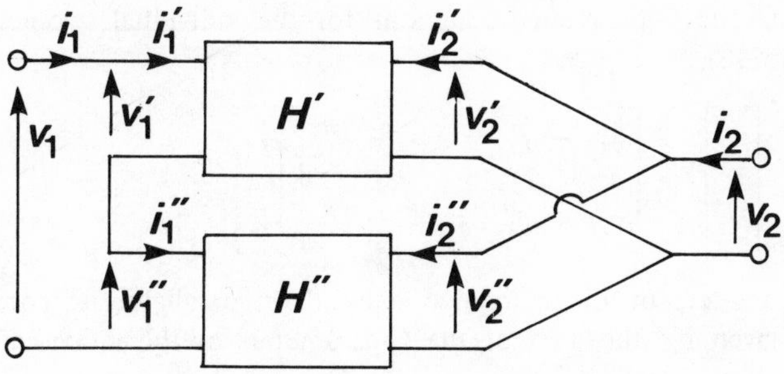


Fig. 5.8 – Series-Parallel Connected Two-Ports $\mathbf{H} = \mathbf{H}' + \mathbf{H}''$.

Substituting the H-parameter equation for the individual two-ports into expressions (5.30),

$$\begin{bmatrix} v_1 \\ i_2 \end{bmatrix} = \begin{bmatrix} h'_{11} + h''_{11} & h'_{12} + h''_{12} \\ h'_{21} + h''_{21} & h'_{22} + h''_{22} \end{bmatrix} \begin{bmatrix} i_1 \\ v_2 \end{bmatrix} .$$

The H-parameters of the combined network in series-parallel connection are therefore given by the sum of the H-parameter of the individual networks, that is,

$$\mathbf{H} = \mathbf{H}' + \mathbf{H}'' . \tag{5.31}$$

**Parallel-Series Connection**

When two-port networks are connected with inputs in parallel and outputs in series as shown in Fig. 5.9,

$$\begin{bmatrix} i_1 \\ v_2 \end{bmatrix} = \begin{bmatrix} i'_1 + i''_1 \\ v'_2 + v''_2 \end{bmatrix}, \text{and} \begin{bmatrix} v_1 \\ i_2 \end{bmatrix} = \begin{bmatrix} v'_1 \\ i'_2 \end{bmatrix} = \begin{bmatrix} v''_1 \\ i''_2 \end{bmatrix} . \tag{5.32}$$

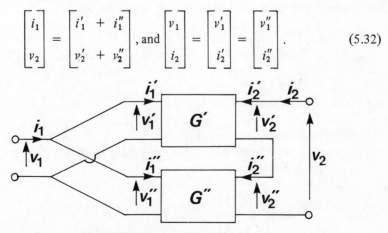

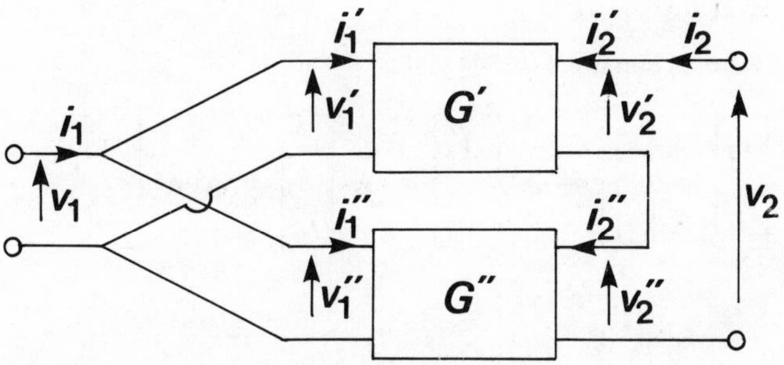


Fig. 5.9 – Parallel-Series Connected Two-Port $\mathbf{G} = \mathbf{G}' + \mathbf{G}''$.

Substituting the G-parameter equations for the individual two-ports into expressions (5.32),

$$\begin{bmatrix} i_1 \\ v_2 \end{bmatrix} = \begin{bmatrix} g'_{11} + g''_{11} & g'_{12} + g''_{12} \\ g'_{21} + g''_{21} & g'_{22} + g''_{22} \end{bmatrix} \begin{bmatrix} v_1 \\ i_2 \end{bmatrix} .$$

The G-parameters of the combined network in parallel-series connection are therefore given by the sum of the G-parameters of the individual networks, that is,

$$\mathbf{G} = \mathbf{G}' + \mathbf{G}'' . \tag{5.33}$$

**Cascade Connection**

When networks are connected in cascade as shown in Fig. 5.10,

$$\begin{bmatrix} v_1 \\ i_1 \end{bmatrix} = \begin{bmatrix} v'_1 \\ i'_1 \end{bmatrix} , \begin{bmatrix} v_2 \\ -i_2 \end{bmatrix} = \begin{bmatrix} v''_2 \\ -i''_2 \end{bmatrix} , \begin{bmatrix} v''_1 \\ i''_1 \end{bmatrix} = \begin{bmatrix} v'_2 \\ -i'_2 \end{bmatrix} . \tag{5.34}$$

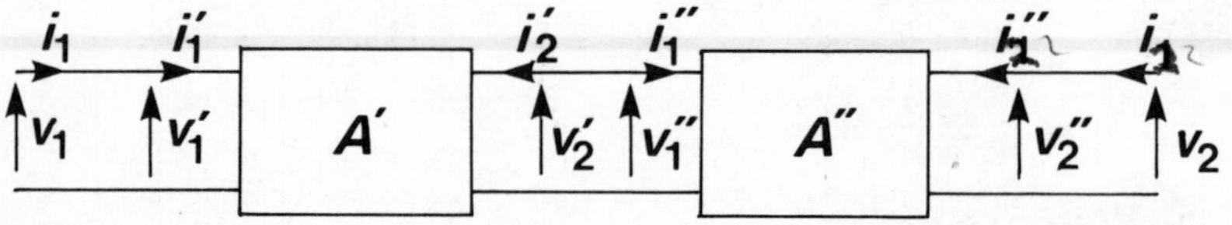


Fig. 5.10 – Cascade Connected Two-Ports $\mathbf{A} = \mathbf{A}'\mathbf{A}''$ $\mathbf{B} = \mathbf{B}''\mathbf{B}'$.

The A-parameters for the individual teo-ports are given by

$$\begin{bmatrix} v'_1 \\ i'_1 \end{bmatrix} = \mathbf{A}' \begin{bmatrix} v'_2 \\ -i'_2 \end{bmatrix} , \text{ and } \begin{bmatrix} v''_1 \\ i''_1 \end{bmatrix} = \mathbf{A}'' \begin{bmatrix} v''_2 \\ -i''_2 \end{bmatrix} .$$

Substitution into expressions (5.34) yields,

$$\begin{bmatrix} v_1 \\ i_1 \end{bmatrix} = \begin{bmatrix} v'_1 \\ i'_1 \end{bmatrix} = \mathbf{A}' \begin{bmatrix} v'_2 \\ -i'_2 \end{bmatrix} = \mathbf{A}' \begin{bmatrix} v''_1 \\ i''_1 \end{bmatrix} = \mathbf{A}'\,\mathbf{A}'' \begin{bmatrix} v''_2 \\ -i''_2 \end{bmatrix}$$

$$= \mathbf{A}'\,\mathbf{A}'' \begin{bmatrix} v_2 \\ -i_2 \end{bmatrix} .$$

The A-parameters of the combined network is cascade connections are given by the product of the A-parameters of the individual networks, that is,

$$\mathbf{A} = \mathbf{A}' \ \mathbf{A}'' \ . \tag{5.35}$$

**Reverse Cascade Connection**

The B-parameters of the individual networks in Fig. 5.10 are given by

$$\begin{bmatrix} v_2' \\ i_2' \end{bmatrix} = \mathbf{B}' \begin{bmatrix} v_1' \\ -i_1' \end{bmatrix} \text{ and, } \begin{bmatrix} v_2'' \\ i_2'' \end{bmatrix} = \mathbf{B}'' \begin{bmatrix} v_1'' \\ -i_1'' \end{bmatrix} .$$

Substitution into expressions (5.34) with some attention to signs yields,

$$\begin{bmatrix} v_2 \\ i_2 \end{bmatrix} = \mathbf{B}'' \ \mathbf{B}' \begin{bmatrix} v_1 \\ -i_1 \end{bmatrix} .$$

The B-parameters of the combined network in cascade connection are given by

$$\mathbf{B} = \mathbf{B}'' \ \mathbf{B}' \ . \tag{5.36}$$

The order of the individual B-parameter terms in Eq. (5.36) are reversed compared to A-parameters in Eq. (5.35), and the circuit treated from output to input. As we shall see in the following section, some active components do not possess finite B-parameters since the A-parameter determinant equals zero. Circuits are therefore normally cascaded by means of A-parameters in the forward direction of the amplifiers. Use of B-parameters is minimal and in practice they are applied only in special cases.

**Validity**

The definition of a port specifically states that the current entering one terminal of a port equals the current leaving the other terminal. When two-ports are interconnected this equality must be maintained if the expressions derived for the parameters of the combined two-port are to be valid. In the cascade connection validity is automatically satisfied, but in series and parallel connections this is not necessarily true.

Two-port interconnection using Z, Y, H, and G-parameters applies to circuits with inputs and outputs in either series or parallel connection. Both input and output must be checked for validity. Four tests are illustrated in Fig. 5.11. In each case the remaining ports are connected in series or parallel as required, and excited. The condition for validity in each case is for the voltage $v=0$. Each two-port connection requires two tests so that validity may be chacked at

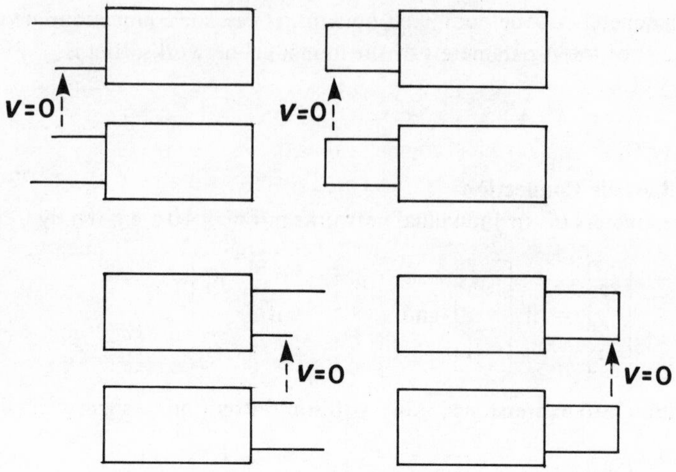


Series Connected Ports Parallel Connected Ports

Fig. 5.11 – Validity Tests.

both input and output ports. It is necessary for both tests to give $v$=0 for the connection to be valid. Two simple circuit connections which violate one or both validity tests are shown in Fig. 5.12.

These validity tests, known as Brune's criterion, are necessary and sufficient only for reciprocal networks. However they may be satisfactorily applied to most active networks including all three-terminal configurations.

**Terminal Transformations**

All two-port interconnections so far discussed in this section operate between two or more circuits. A number of transformations which result from changing

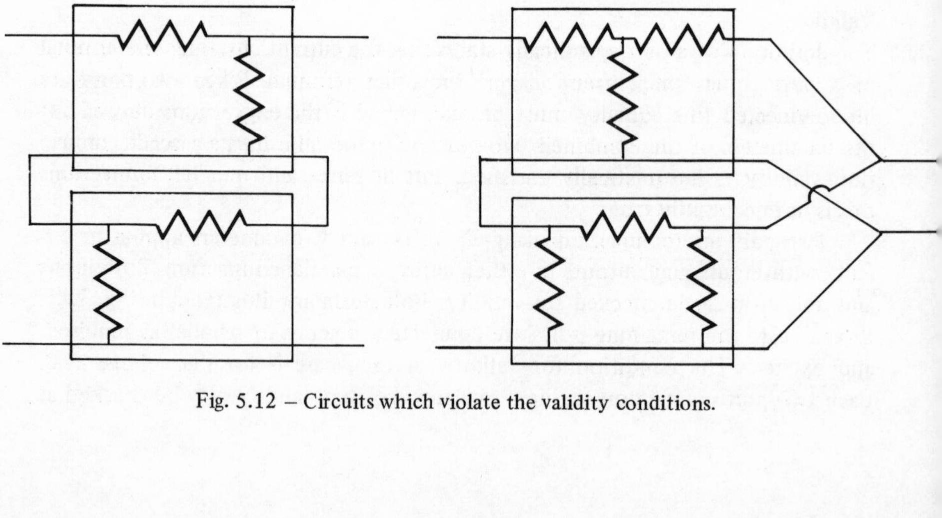

Fig. 5.12 – Circuits which violate the validity conditions.

the terminal connections of a single two-port also have useful applications in the analysis of electronic circuits. Each transformation is easily derived by redefinition of the circuit variables.

*Reversal of Input and Output Connections*

Input and output connection reversal is illustrated in Fig. 5.13. The two-port shown is a transistor to emphasize that the most useful application of these transformations is the conversion from common-emitter to common-base or common-collector parameters.

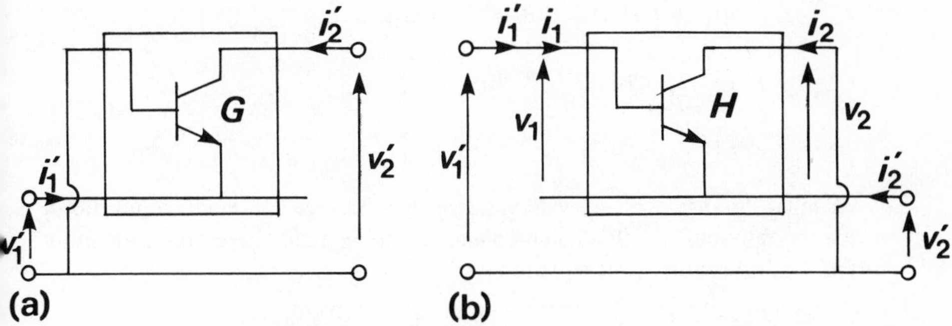


Fig. 5.13 – Terminal Transformations.
(a) Reversal of Input Connections; (b) Reversal of Output connections.

The common-emitter to common-collector transformation is shown in fig. 5.13(b) with voltages and currents added for both connections. The relation between the new variables $v'_1$, $i'_1$, $v'_2$, and $i'_2$ for common-collector and the variables $v_1$, $i_1$, $v_2$, and $i_2$ for common-emitter can be written down from inspection of the figure, yielding the equations,

$$v'_1 = v_1 - v_2$$

$$v'_2 = -v_2$$

$$i'_1 = i_1$$

$$i'_2 = -i_1 - i_2 \ .$$

Beginning with H-parameters of the two-port before transformation,

$$\begin{bmatrix} v_1 \\ i_2 \end{bmatrix} = \begin{bmatrix} h_{11} & h_{12} \\ h_{21} & h_{22} \end{bmatrix} \begin{bmatrix} i_1 \\ v_2 \end{bmatrix},$$

and changing the dependent variables yields,

$$\begin{bmatrix} v_1 - v_2 \\ i_1 + i_2 \end{bmatrix} = \begin{bmatrix} h_{11} & h_{12}-1 \\ h_{21}+1 & h_{22} \end{bmatrix} \begin{bmatrix} i_1 \\ v_2 \end{bmatrix}.$$

Correction of the signs requires $-v_2$ and $-(i_1+i_2)$, therefore,

$$\begin{bmatrix} v_1 - v_2 \\ -(i_1 + i_2) \end{bmatrix} = \begin{bmatrix} h_{11} & -(h_{12}-1) \\ -(h_{21}+1) & h_{22} \end{bmatrix} \begin{bmatrix} i_1 \\ -v_2 \end{bmatrix},$$

$$\begin{bmatrix} v_1' \\ i_2' \end{bmatrix} = \begin{bmatrix} h_{11} & 1-h_{12} \\ -h_{21}-1 & h_{22} \end{bmatrix} \begin{bmatrix} i_1' \\ v_2' \end{bmatrix}. \tag{5.37}$$

Similar arrangement of variables for the reversal of input connections which corresponds to the common-emitter to common-base transformation of Fig. 5.13(a) results in the equation

$$\begin{bmatrix} i_1' \\ v_2' \end{bmatrix} = \begin{bmatrix} g_{11} & -g_{12}-1 \\ 1-g_{21} & g_{22} \end{bmatrix} \begin{bmatrix} v_1' \\ i_2' \end{bmatrix}. \tag{5.38}$$

Both transformations can be converted into A-parameters for application in cascade circuits or to other parameters, using Table 5.1.

*Reversal of Input or Output Terminals*

Reversal of input or output terminals is shown in Fig. 5.14. When input terminals are reversed it is clear from the figure that the new variables $v_1'$, $i_1'$, $v_2'$, $i_2'$ are related to those of the original two-port by

$$\begin{aligned} v_1' &= -v_1 & \qquad v_2' &= v_2 \\ i_1' &= -i_1 & \qquad i_2' &= i_2' \end{aligned}.$$

The A-parameters of the original network are

$$\begin{bmatrix} v_1 \\ i_1 \end{bmatrix} = \begin{bmatrix} a_{11} & a_{12} \\ a_{21} & a_{22} \end{bmatrix} \begin{bmatrix} v_2 \\ -i_2 \end{bmatrix}.$$

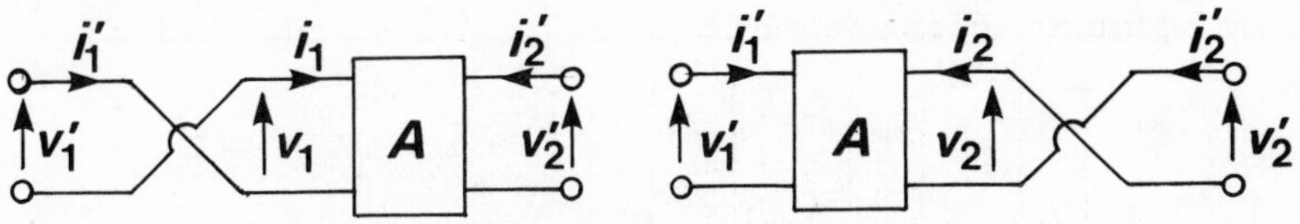


Fig. 5.14 – Reversal of Input or Output Terminals.

Therefore,
$$\begin{bmatrix} -v_1 \\ -i_1 \end{bmatrix} = \begin{bmatrix} -a_{11} & -a_{12} \\ -a_{21} & -a_{22} \end{bmatrix} \begin{bmatrix} v_2 \\ -i_2 \end{bmatrix},$$

$$\begin{bmatrix} v_1' \\ i_1' \end{bmatrix} = \begin{bmatrix} -a_{11} & -a_{12} \\ -a_{21} & -a_{22} \end{bmatrix} \begin{bmatrix} v_2' \\ -i_2' \end{bmatrix},$$

$$\mathbf{A}' = -\mathbf{A} . \tag{5.39}$$

The same equation is obtained for the reversal of output terminals.

Multiplication of all terms in the $A$ matrix by $-1$ could also be achieved with matrix multiplication by

$$\begin{bmatrix} -1 & 0 \\ 0 & -1 \end{bmatrix}.$$

This corresponds to the cascade of the original network with an ideal transformer of turns ratio $n = -1$.

*Reversal of Two-Port*

Interchange of the input and output connections of a two-port is shown in Fig. 5.15. From the figure,

$$v_1' = v_2 \qquad v_2' = v_1$$

$$i_1' = i_2 \qquad i_2' = i_1$$

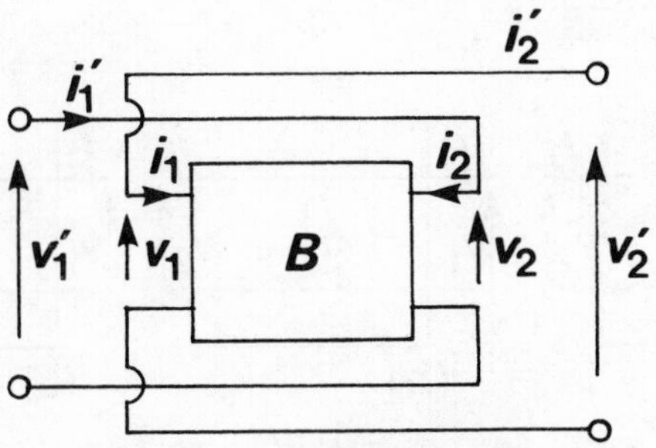


Fig. 5.15 – Reversal of Two-Port.

The B-parameters of a network are

$$\begin{bmatrix} v_2 \\ i_2 \end{bmatrix} = \begin{bmatrix} b_{11} & b_{12} \\ b_{21} & b_{22} \end{bmatrix} \begin{bmatrix} v_1 \\ -i_1 \end{bmatrix},$$

therefore, $$\begin{bmatrix} v_1' \\ i_1' \end{bmatrix} = \begin{bmatrix} b_{11} & b_{12} \\ b_{21} & b_{22} \end{bmatrix} \begin{bmatrix} v_2' \\ -i_2' \end{bmatrix},$$

and the A-parameters of the reversed two-port clearly equal the B-parameters of the original network, that is,

$$\mathbf{A}' = \mathbf{B} = \mathbf{A}^{-1}\ . \tag{5.40}$$

## 5.3 TRANSMISSION PARAMETERS OF CIRCUIT ELEMENTS

In electronic circuits components are connected primarily in cascade with forward connected amplifying elements between input and output. Analysis by means of A-parameters is therefore used, and all components are initially specified in terms of the transmission parameters. This has the added advantages that the cascade connection is always valid and that all basic components possess A-parameters. The use of other circuits implies the use of Z, Y, H, and G-parameters, and conversion using Table 5.1. is necessary. It is also necessary in most cases to convert back to A-parameter after small sections of a circuit have been analysed. This section is therefore written entirely in terms of A-parameters.

### Impedance elements

Resistors, capacitors, and inductors have impedance $Z$ equal to $R$, $1/j\omega C$, and $j\omega L$ respectively, and each have two terminals. They are connected as a two-port in one of the two standard configurations shown in Figs. 5.16(a) and 5.16(b). The equations for each connection are easily derived by inspection, usin Ohm's law and Kirchoff's laws.

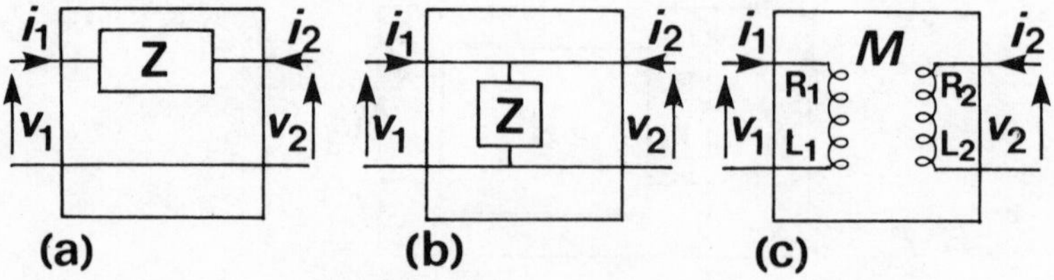


Fig. 5.16 – Lumped Impedance Elements.
(a) Series connected impedance; (b) Parallel connected impedance; (c) Mutual Inductance.

*Series Connected Impedance*
From Fig. 5.16(a),

$$i_1 = -i_2$$

$$v_1 = v_2 - i_2 Z \ ,$$

that is,
$$\begin{bmatrix} v_1 \\ i_1 \end{bmatrix} = \begin{bmatrix} 1 & Z \\ 0 & 1 \end{bmatrix} \begin{bmatrix} v_2 \\ -i_2 \end{bmatrix} . \tag{5.41}$$

The A-parameters of a series connected impedance are therefore

$$\begin{bmatrix} 1 & Z \\ 0 & 1 \end{bmatrix} .$$

*Parallel Connected Impedance*
From Fig. 5.16(b),

$$v_1 = v_2 \ ,$$

$$i_1 = -i_2 + v_2/Z \ ,$$

that is,
$$\begin{bmatrix} v_1 \\ i_1 \end{bmatrix} = \begin{bmatrix} 1 & 0 \\ 1/Z & 1 \end{bmatrix} \begin{bmatrix} v_2 \\ -i_2 \end{bmatrix} . \tag{5.42}$$

The A-parameters of a parallel connected impedance are therefore

$$\begin{bmatrix} 1 & 0 \\ 1/Z & 1 \end{bmatrix} .$$

*Mutual Inductance*
The equations for a mutual inductance connected as in Fig. 5.16(c) were given in Table 1.1 as

$$\begin{aligned} v_2 &= \mathrm{j}\omega M i_1 + (R_2 + \mathrm{j}\omega L_2)\, i_2 \\ v_1 &= (R_1 + \mathrm{j}\omega L_1) i_1 + \mathrm{j}\omega M i_2 \end{aligned} . \tag{5.43}$$

The circuit is already in two-port form and the coefficients in Eqs. (5.43) may be interpreted directly as Z-parameters, that is,

$$\mathbf{Z} = \begin{bmatrix} R_1+j\omega L_1 & j\omega M \\ j\omega M & R_2+j\omega L_2 \end{bmatrix}.$$

Using Table 5.1,

$$\mathbf{A} = \begin{bmatrix} (R_1+j\omega L_1)/j\omega M & (\omega^2 M^2 + \{R_1+j\omega L_1\}\{R_2+j\omega L_2\})/j\omega M \\ 1/j\omega M & (R_2+j\omega L_2)/j\omega M \end{bmatrix}. \tag{5.44}$$

If the analysis is repeated with reference to Fig. 5.17 where the mutual inductance is separated into three parts representing the primary self impedance, the inductive mutual coupling, and the secondary self impedance, the circuit is in cascade and the A-parameters are

$$\mathbf{A} = \begin{bmatrix} 1 & R_1+j\omega L_1 \\ 0 & 1 \end{bmatrix} \begin{bmatrix} 0 & -j\omega M \\ 1/j\omega M & 0 \end{bmatrix} \begin{bmatrix} 1 & R_2+j\omega L_2 \\ 0 & 1 \end{bmatrix}.$$

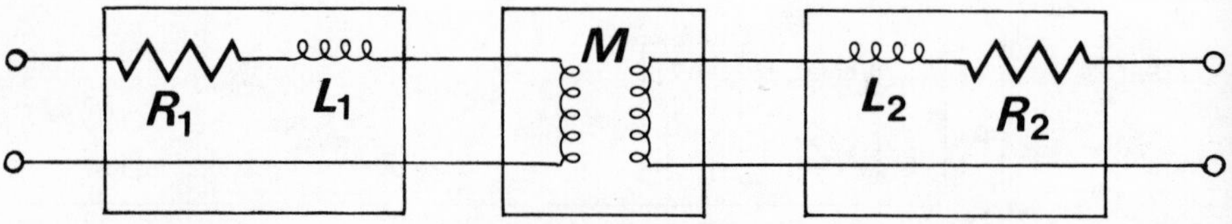


Fig. 5.17 – Mutual Inductance.

The multiplications can be worked to show that this expression is the same as that in Eq. (5.44). The term

$$\begin{bmatrix} 0 & -j\omega M \\ 1/j\omega M & 0 \end{bmatrix}$$

represents the mutual inductance energy coupling, and for convenience in analysis it can be extracted in this form for computation. However, in a physically realizable mutual inductor it cannot be separated from the self inductances since $M^2 \leqslant L_1 L_2$. In an electronic simulation using RC-active networks the isolated term can be closely realized with series input and output impedances close to zero.

The unity-coupled, zero resistance, mutual inductor, which had an undefined admittance matrix in Chapter 2, can be included in two-port analysis, but Y-parameters remain undefined since $a_{12} = 0$. Parallel-parallel connected circuits cannot be analysed in this case without additional elements which ensure $a_{12} \neq 0$.

It is interesting to note that the equivalent circuit for inductive mutual coupling, shown in Fig. 5.18, illustrates how the $(L-M)$ terms arise in equivalent circuits for transformers.

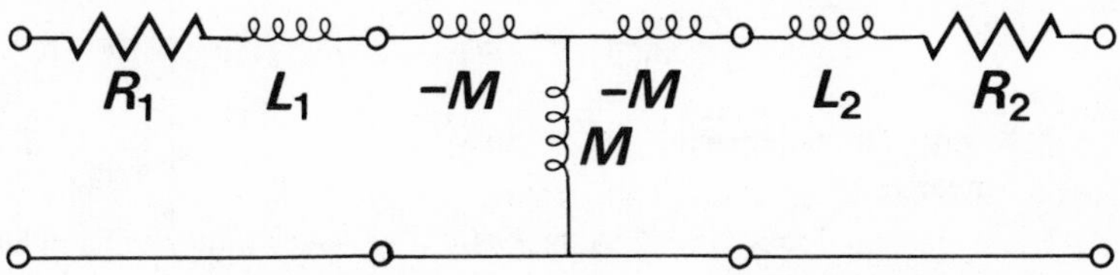


Fig. 5.18 – Equivalent circuit for Mutual Inductance.

**Dependent Sources**

The dependent sources in Table 1.1 are each characterised by a single equation, and the A-parameters may be directly derived. The VCVS, for example, is defined by the equation

$$v_2 = \mu v_1 \ .$$

Also $i_1 = 0$ since the input current of voltage controlled sources is assumed to be zero. The A-parameters are therefore given by

$$\mathbf{A}_{\text{vcvs}} = \begin{bmatrix} 1/\mu & 0 \\ 0 & 0 \end{bmatrix} .$$

The VCCS may be treated similarly to yield

$$\mathbf{A}_{\text{vccs}} = \begin{bmatrix} 0 & -1/g \\ 0 & 0 \end{bmatrix} .$$

The CCVS and CCCS have short circuit input, therefore $v_1 = 0$ and the following A-parameters are easily derived.

$$\mathbf{A}_{\text{ccvs}} = \begin{bmatrix} 0 & 0 \\ 1/r & 0 \end{bmatrix} \qquad \mathbf{A}_{\text{cccs}} = \begin{bmatrix} 0 & 0 \\ 0 & -1/\alpha \end{bmatrix} .$$

**Impedance Converters and Inverters**

Impedance converters and inverters were introduced in Section 1.8 and their characteristic equations were given in Table 1.1. They were treated in more detail in Section 2.4, and it remains here to identify A-parameters from equations already given. This is easily done by inspection, and the resulting parameters are tabulated below

| | Equations from Table 1.1 | A-parameters |
|---|---|---|
| Ideal transformer | $v_1 = \pm n v_2$<br>$-i_2 = \pm n i_1$ | $\begin{bmatrix} \pm n & 0 \\ 0 & \pm 1/n \end{bmatrix}$ |
| Negative impedance converter | $v_1 = \pm n v_2$<br>$-i_2 = \mp n i_1$ | $\begin{bmatrix} \pm n & 0 \\ 0 & \mp 1/n \end{bmatrix}$ |
| Gyrator | $i_1 = \pm g v_2$<br>$-i_2 = \pm g v_1$ | $\begin{bmatrix} 0 & \pm 1/g \\ \pm g & 0 \end{bmatrix}$ |
| Negative impedance inverter | $i_1 = \pm g v_2$<br>$-i_2 = \mp g v_1$ | $\begin{bmatrix} 0 & \mp 1/g \\ \pm g & 0 \end{bmatrix}$ |

Equivalent circuits of these network elements were given in Table 2.1. It is interesting to note, however, that the negative impedance inverter can be modelled by means of the T network shown in Fig. 5.19 which is identical in form to the model of inductive mutual coupling in Fig. 5.18. A similar network of capacitors simulating capacitive mutual coupling completes the set and forms part of the generalized theory of networks.

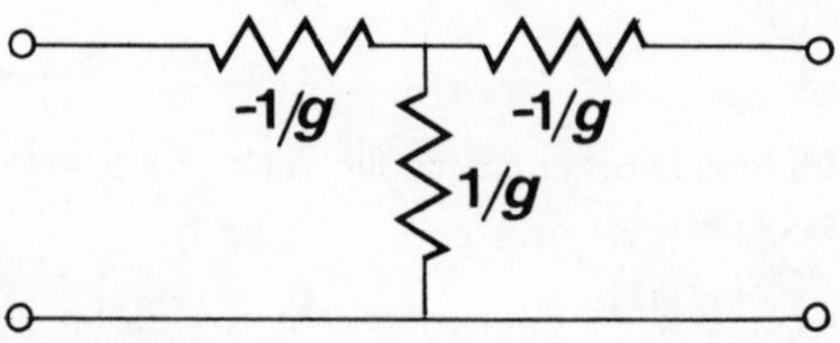


Fig. 5.19 – Equivalent circuit of negative impedance inverter.

**Series and Shunt Connected Two-ports**

Two circuit configurations which are useful for computation, particularly when concerned with distributed networks by two-port analysis, are shown in Fig. 5.20.

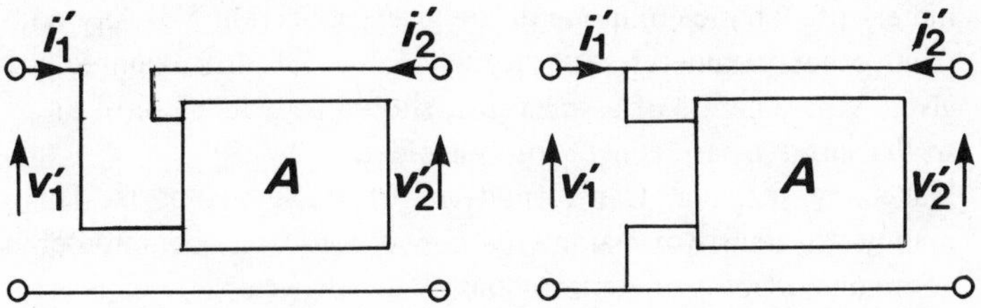


Fig. 5.20 – Series and shunt connected two-ports.

The input impedance of a two-port for infinite load is given in Table 5.3 as $a_{11}/a_{21}$. Using Eq. (5.41), the A-parameters of series connected two-ports are given by

$$A' = \begin{bmatrix} 1 & a_{11}/a_{21} \\ 0 & 1 \end{bmatrix}.$$

Similarly from Eq. (5.42) for the shunt connected two-port,

$$A' = \begin{bmatrix} 1 & 0 \\ a_{21}/a_{11} & 1 \end{bmatrix}.$$

## 5.4 CIRCUIT ANALYSIS

### General Procedure

The previous three sections have covered all the necessary theory for the analysis of any lumped linear circuit which can be drawn as a combination of two-port networks. The analysis follows the same five-step general procedure in each case.

STEP 1 Redraw the circuit in two-port form. The available set of standard connections between two-ports is given in Section 5.2. Terminal transformations of a single two-port add flexibility in the treatment of active devices. At this stage it is necessary to incorporate models of active devices utilizing lumped components where parameters are not available. Preliminary analysis of the models of active devices is often more efficient since it is normal for only a limited number of different transistors, for example, to be used in one circuit. At this point it is opportune to note once again that not all circuits can be drawn in the form of interconnected two-ports.

STEP 2 Determine the two-port parameters of all components, The A-parameters of all basic components are given in Section 5.3. Any conversion from other parameter sets through Table 5.1, for example, from the given Y-parameters of a transistor, should be delayed until Steps 3 or 4 so that unnecessary conversion is avoided.

STEP 3 Make any terminal transformations of single two-ports. This applies mainly to transistor parameter conversion from common-emitter to common-collector or common-base.

STEP 4 Combine groups of circuit components in standard connections. Since circuits consist primarily of cascade components, sections of the circuit in non-cascade arrangements must be cleared to yield a single set of A-parameters.

Predominant connections in electronic circuits other than cascade are parallel-parallel and series-series, involving addition of Y or Z-parameters.

STEP 5 Continue circuit reduction until a single set of two-port paramaters is obtained, and evaluate performance. Repeated application of Steps 3 and 4 will eventually result in a single set of two-port parameters. These are usually A-parameters but may be any one of the six sets of parameters. Performance is evaluated using Table 5.2 or, preferably, Table 5.3.

The general features of two-port analysis are followed when analysis is carried out by hand or by computer. However, the detail of analysis differs in the two cases because the computer follows a predetermined and unique set of analysis steps whereas by hand variation in the order of analysis based on preference or experience can significantly reduce the labour involved. This emphasizes that Step 1 does not in general result in a unique two-port version of a circuit and that the order of reduction in Steps 4 and 5 is also not unique. Application of the analysis method is best illustrated by a few simple examples.

*Example* 5.2

Determine the open-circuit voltage gain of the bridged-T network in Fig. 5.21(a).

The circuit is redrawn in two-port form in Fig. 5.21(b) as required by Step 1 of the analysis. The two-port A-parameters of series and shunt connected impedances are given in Section 5.3 as

$$\begin{bmatrix} 1 & Z \\ 0 & 1 \end{bmatrix} \text{ and } \begin{bmatrix} 1 & 0 \\ 1/Z & 1 \end{bmatrix}$$

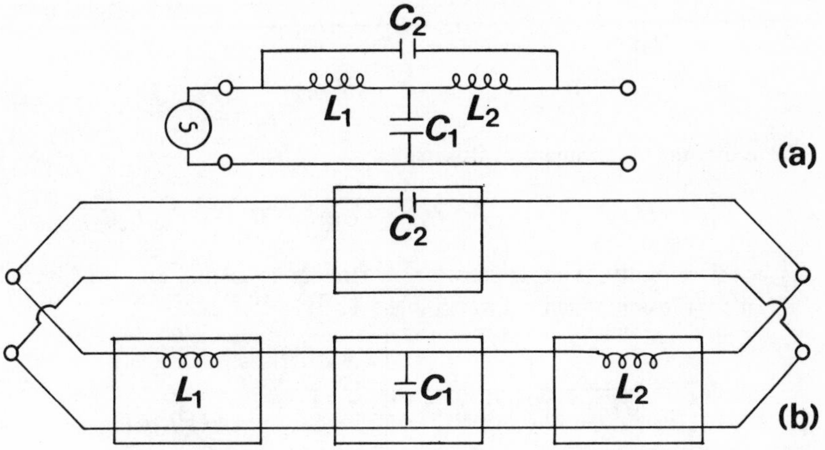


Fig. 5.21 – (a) Circuit for Example 5.2; (b) Redrawn Two-port version.

respectively, and parameters of all components in this example can be written down by inspection. Step 3 does not apply since there are no terminal transformations.

Step 4 requires the formulation of an order for component combination with the objective of producing a set of two-port parameters for the complete network. The bridged-T circuit is very simple and little choice is available. Components $L_1$, $C_1$ and $L_2$ are in cascade and they could be combined using either A or B-parameters, or even a mixture of both. Following convention, the A-parameters are multiplied to obtain the A-parameters of the T network consisting of $L_1$, $C_1$, and $L_2$.

The T network is connected in parallel at both input and output to $C_2$, a series connected impedance. The A-parameters of $C_2$ and of the T network must therefore be converted to Y-parameters so that the two sets of parameters can be added to obtain the Y-parameters of the parallel combination. It is not necessary to re-convert back to A-parameters for evaluation of performance in Step 5 since Table 5.3 gives values for voltage gain, etc. in all sets of parameters.

The A-parameters of the T network are given by

$$\mathbf{A}_\mathrm{T} = \begin{bmatrix} 1 & sL_1 \\ 0 & 1 \end{bmatrix} \begin{bmatrix} 1 & 0 \\ sC_1 & 1 \end{bmatrix} \begin{bmatrix} 1 & sL_2 \\ 0 & 1 \end{bmatrix}$$

$$= \begin{bmatrix} 1 + s^2L_1C_1 & sL_1 \\ sC_1 & 1 \end{bmatrix} \begin{bmatrix} 1 & sL_2 \\ 0 & 1 \end{bmatrix}$$

$$= \begin{bmatrix} 1 + s^2L_1C_1 & s(L_1+L_2) + s^3L_1L_2C_1 \\ sC_1 & 1 + s^2C_1L_2 \end{bmatrix}$$

Note that the determinant of this matrix

$$\Delta = (1+s^2L_1C_1)(1+s^2C_1L_2) - sC_1(sL_2 + sL_1 + s^3L_1L_2C_1)$$

is equal to unity. This is true for all circuits consisting only of lumped linear passive components. From Table 5.1,

$$\mathbf{Y}_T = \frac{1}{s(L_1+L_2) + s^3L_1L_2C_1} \begin{bmatrix} 1 + s^2C_1L_2 & -1 \\ -1 & 1+s^2L_1C_1 \end{bmatrix}$$

The A-parameters of $C_2$ are

$$\mathbf{A}_{C_2} = \begin{bmatrix} 1 & 1/sC_2 \\ 0 & 1 \end{bmatrix},$$

and from Table 5.1,

$$\mathbf{Y}_{C_2} = \begin{bmatrix} sC_2 & -sC_2 \\ -sC_2 & sC_2 \end{bmatrix}.$$

Adding Y-parameters $\mathbf{Y}_T + \mathbf{Y}_{C_2}$ and combining the fractions, $\mathbf{Y}$ equals

$$\begin{bmatrix} \dfrac{1+s^2(C_1L_2+C_2L_1+C_2L_2) + s^4C_1C_2L_1L_2}{s(L_1+L_2) + s^3L_1L_2C_1} & -\dfrac{1+s^2C_2(L_1+L_2) + s^4C_1C_2L_1L_2}{s(L_1+L_2) + s^3L_1L_2C_1} \\ -\dfrac{1+s^2C_2(L_1+L_2) + s^4C_1C_2L_1L_2}{s(L_1+L_2) + s^3L_1L_2C_1} & \dfrac{1+s^2(L_1C_1+C_2L_1+C_2L_2) + s^4C_1C_2L_1L_2}{s(L_1+L_2) + s^3L_1L_2C_1} \end{bmatrix}$$

From Table 5.3, the voltage gain for zero source and infinite load impedance equals $-y_{21}/y_{22}$, that is,

$$\text{Voltage gain} = \frac{1 + s^2C_2(L_1+L_2) + s^4C_1C_2L_1L_2}{1 + s^2(L_1C_1+L_1C_2+L_2C_2) + s^4C_1C_2L_1L_2} \,.$$

Other performance figures are also available from Table 5.3. It is instructive to determine the same transfer function by any other method in order to compare the labour involved.

*Example* 5.3

Determine the voltage gain, input impedance, and output impedance of the transistor amplifier circuit with emitter feedback in Fig. 5.22(a). The transistor hybrid parameters are $h_{ie} = 1$ kΩ, $h_{re} = 4 \times 10^{-4}$, $h_{fe} = 50$, $h_{oe} = 50 \times 10^{-6}$ S.

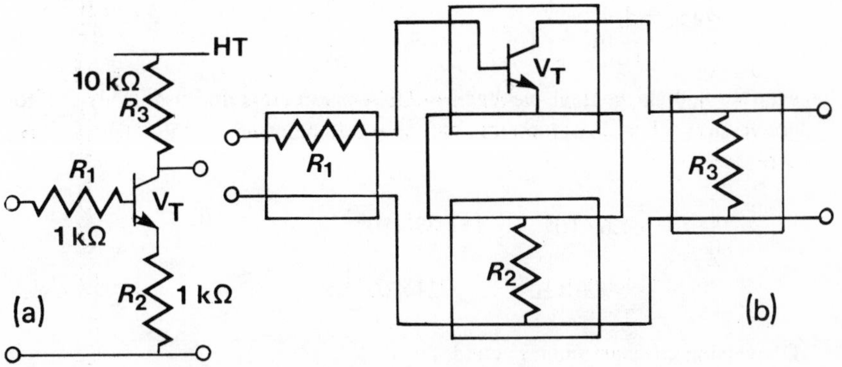


Fig. 5.22 – (a) Transistor amplifier for Example 5.3.
(b) Redrawn circuit for two-port analysis.

The amplifier is redrawn in two-port form in Fig. 5.22(b). The A-parameters of the resistive impedances can be written down as in the previous example. The quoted transistor parameters are the hybrid or H-parameters for the common emitter connection, and it is necessary only to identify $h_{11} = h_{ie}$, $h_{12} = h_{re}$, $h_{21} = h_{fe}$, and $h_{22} = h_{oe}$. Analysis proceeds broadly in the direction of the input to output cascade while clearing in advance sections of the circuit which are connected differently. In this case transistor $V_T$ and $R_2$ have their terminals connected in series at both input and output. They are therefore in SERIES-SERIES connection and it will be necessary to convert the component parameters into Z-parameters for addition. The Z-parameters of the combination of transistor and emitter feedback resistor must then be transformed to A-parameters so that the input to output cascade may be calculated.

The quoted H-parameters of $\mathbf{V_T}$ are

$$\mathbf{H_{VT}} = \begin{bmatrix} 1000 & 4\times10^{-4} \\ 50 & 50\times10^{-6} \end{bmatrix},$$

and conversion to Z-parameters yields

$$\mathbf{Z_{VT}} = \begin{bmatrix} 600 & 8 \\ -10^{6} & 2\times10^{4} \end{bmatrix}.$$

The Z-parameters of the parallel connected 1 kΩ resistor $R_2$ are easily written down directly as

$$\mathbf{Z_{R2}} = \begin{bmatrix} 10^{3} & 10^{3} \\ 10^{3} & 10^{3} \end{bmatrix},$$

even though in general we begin with A-parameters and use Table 5.1 for conversion. The Z-parameters of the combination of $\mathbf{V_T}$ and $R_2$ are therefore,

$$\mathbf{Z_T} = \begin{bmatrix} 1.6\times10^{3} & 1.008\times10^{3} \\ -999\times10^{3} & 21\times10^{3} \end{bmatrix}.$$

Conversion to A-parameters yields,

$$\mathbf{A_T} = \begin{bmatrix} -1.6\times10^{-3} & -1041 \\ -10^{-6} & -21\times10^{-3} \end{bmatrix}.$$

The A-parameters of the complete amplifier are obtained from the cascade of $R_1$, the transistor with emitter feedback resistor, and $R_3$. Therefore

$$\mathbf{A} = \begin{bmatrix} 1 & 10^{3} \\ 0 & 1 \end{bmatrix} \begin{bmatrix} -1.6\times10^{-3} & -1041 \\ -10^{-6} & -21\times10^{-3} \end{bmatrix} \begin{bmatrix} 1 & 0 \\ 10^{-4} & 1 \end{bmatrix}$$

$$= \begin{bmatrix} -0.109 & -1062 \\ -3.1\times10^{-6} & -0.021 \end{bmatrix}.$$

Performance of the amplifier with zero source impedance and open-circuit load impedance obtained from Table 5.3 is summarized as follows.

| | | | |
|---|---|---|---|
| Voltage gain | $= 1/a_{11}$ | = | −9:17 |
| Input impedance | $= a_{11}/a_{21}$ | = | 35 kΩ |
| Output impedance | $= a_{12}/a_{11}$ | = | 9.74 kΩ |

If these results are compared to similar results computed for $R_2 = 0$, the effect of emitter feedback will be observed to have reduced gain, increaeed output impedance, and increased input impedance.

**Wiring-operator circuit descriptions**

Examples 5.2 and 5.3 both concern very simple circuits yet they already illustrate the complexity of the matrix operations necessary to reach a single set of two-port parameters which describe the circuit. Clearly, most practical circuits will require analysis by computer, particularly if a frequency response is involved. Before this can be achieved the computer must either receive a description of the network in two-port form or it must work out the description via a suitable algorithm from a list of component connections similar to that used for nodal analysis. Once a satisfactory two-port description is obtained the matrix manipulations required are relatively straightforward but numerically cumbersome.

The first alternative implies that the designer carries out Step 1 of the general two-port analysis procedure while the computer carries out Steps 2 to 5. In the second alternative the computer carries out all analysis steps, the most difficult being Step 1. In both cases the two-port description must be chosen for easy numerical evaluation by a computer program resulting in a frequency response or possibly symbolic or numerical functions of complex frequency as in Example 5.2. Two-port analysis programs at present generally evaluate the frequency response numerically from a two-port description provided by the designer. Such programs are efficient and convenient to use if they begin from circuit descriptions in terms of *wiring operators.* †

Consider the circuit shown in Fig. 5.23 which consists of an inductor $L_1$ and capacitor $C_1$ in series, connected in parallel with a resistor $R_1$. This can be described by the expression

(L1 S C1)P R1 .

†These were first extensively used in a circuit analysis computer program called MARTHA written in the APL programming language and operated from time-shared computer terminals

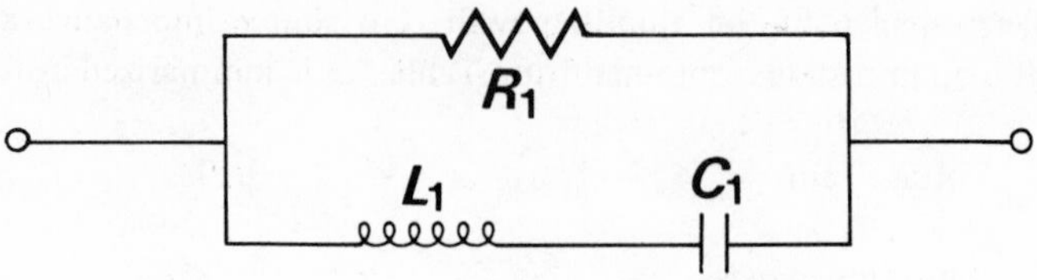


Fig. 5.23 – One-port LCR circuit.

The operators S and P describe how the components are connected either in series or parallel and are called wiring operators. The parenthesis in the expression introduces a hierarchy to enforce the grouping of components.

Wiring operators have the same kind of interpretation as arithmetic operators in an algebraic expression. $A+B$, for example, involves the operator +, and when the expression is evaluated it signifies that the value of variable $A$ is added to the value of variable $B$. When L1 S C1 is evaluated the operator S requires the specific calculation, for example, that the impedance of the series circuit is evaluated from the algebraic expression $j\omega L1 + 1/j\omega C1$.

The operators S and P are *binary* since they describe the network connection between two components. In general they operate between one-ports including sub-networks such as the series L1, C1 combination above. This Chapter is concerned with two-port analysis, and operators which define the connections and terminal transformations from Section 5.2 are summarised in Table 5.4. Operators describing terminal transformations are *unary* since they operate on one two-port only.

It may seem superfluous to introduce a new set of operator names here. However, the computer program MARTHA is written in the programming language APL and contains many more operators and facilities than can be implemented in the BASIC language. For reasons of economy in programming and interpretation, a set of operator names of standard length is a necessity in BASIC. Furthermore the use of parenthesis implies an excessive use of memory which is impossible within the size of program which could be run on most standard minicomputer systems with facilities for time-shared BASIC. Hierarchy without parenthesis is provided partly by using A-parameters with both forward and backward cascade and partly by reference to A-parameters stored in memory.

The wiring operator description of the bridged-T circuit in Fig. 5.21 ia given by

| | |
|---|---|
| ((WS L1)WC C1 WC(WS L2))WPP WS C2 | MARTHA |
| LS1.CC.CP1.LS2.PP.CS2 | 2PORT |

The transistor amplifier in Fig. 5.22 may be similarly described by

| | |
|---|---|
| (WS R1)WC(VT1 WSS R2)WC R3 | MARTHA |
| VT1.SS.RP2.BC.RS1.CC.RP3 | 2PORT |

Wiring operator expressions are evaluated with the following hierarchy.

I Parenthesis; which are cleared commencing from the innermost pair.
II Unary Operators; which apply only to the next component or two-port to the right.
III Binary Operators; which are evaluated right to left (MARTHA) or left to right (2PORT).

Application of the evaluation rules to wiring operator expressions for the bridged-T and transistor amplifier circuits will result in the same procedure as that used in the worked examples except for the evaluation of cascades in reverse order when using MARTHA. More complicated circuits will be defined by their wiring operator expressions in Examples 5.4 and 5.5.

## 5.5 COMPUTER PROGRAMMING

### Linear Circuit Analysis using 2PORT

2PORT is an a.c. circuit analysis program written in BASIC for on-line computation from time-shared terminals. The program is dimensioned for a maximum of 40 analysis frequencies and runs in approximately 12K bytes of memory using Data General BASIC 3.8 on the NOVA 1220 computer. The program listing is given in the Appendix to this Chapter.

2PORT is based on circuit description using wiring operators with circuit values entered separately during the analysis. Each line of circuit description is linked to successive lines by reference to A-parameter stores, and any two-port circuit may be analysed without size restriction. However, there are limitations imposed by the small number of stores at present provided. Efficiency in the circuit description is improved by omitting repeated wiring operators and the lack of parenthesis is largely overcome by the use of a backward cascade operator.

*Program Inputs and Outputs*

Circuit elements available in 2PORT are given in Table 5.5 together with their wiring operator descriptions. This table also summarizes active device models and other inputs provided by 2PORT. Reference to Table 5.5 together with Table 5.4. which summarizes the connections, will enable a user to write two-port circuit descriptions. The program immediately requests component values and

| CIRCUIT CONNECTION | WIRING OPERATOR | |
|---|---|---|
| | MARTHA | BASIC 2 PORT |
| 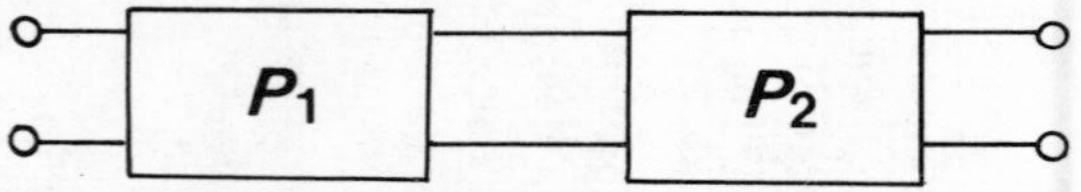  | P1 WC P2 | P1.CC.P2<br>P2.BC.P1 |
| 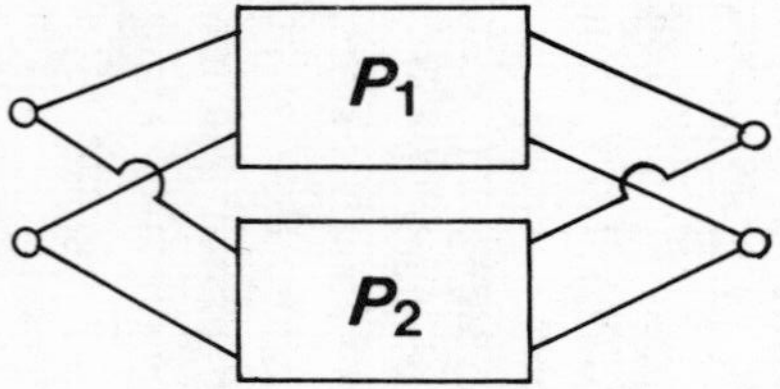  | P1 WPP P2 | P1.PP.P2 |
| 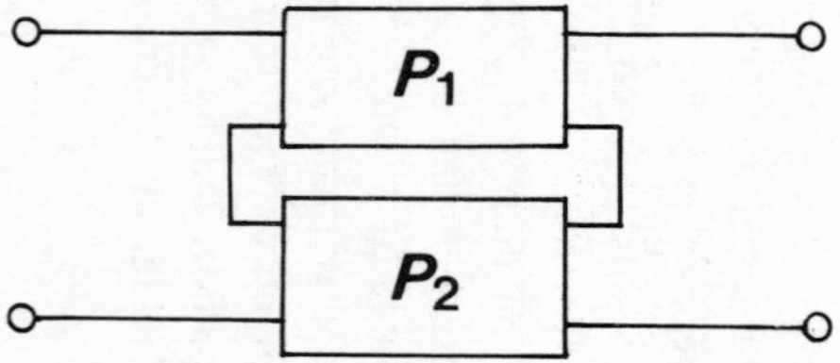  | P1 WSS P2 | P1.SS.P2 |
| 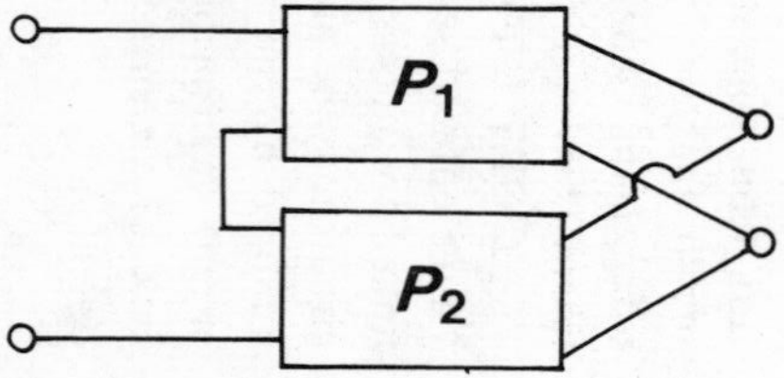  | P1 WSP P2 | P1.SP.P2 |

| | | |
|---|---|---|
| 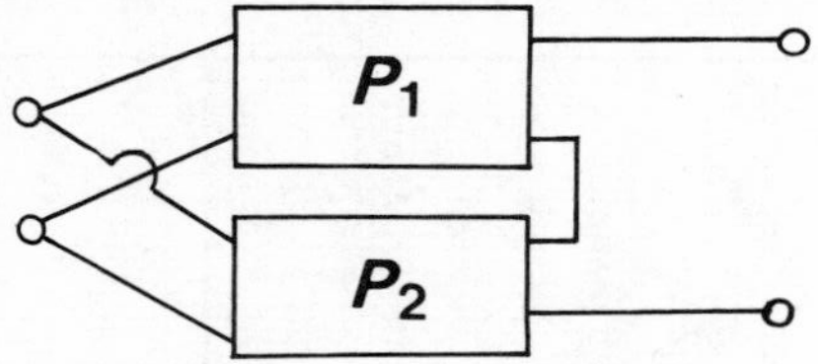 | P1 WPS P2 | P1.PS.P2 |
| 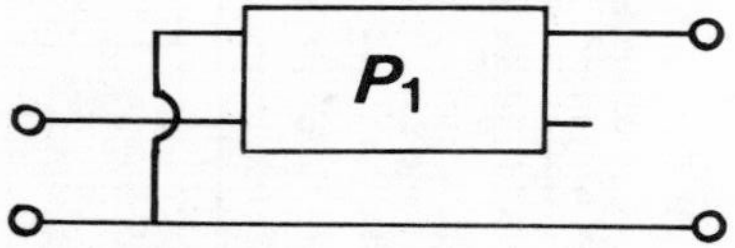 | WROT WN P1 | RI.P1 |
| 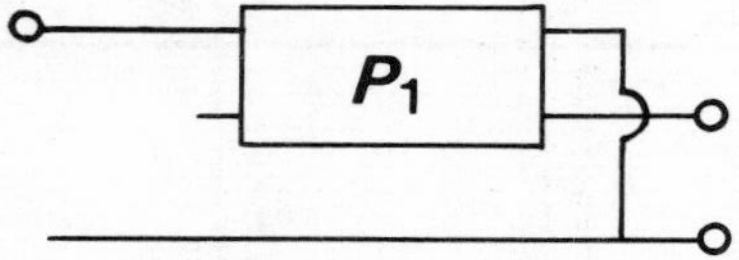 | WN WROT P1 | RO.P1 |
| 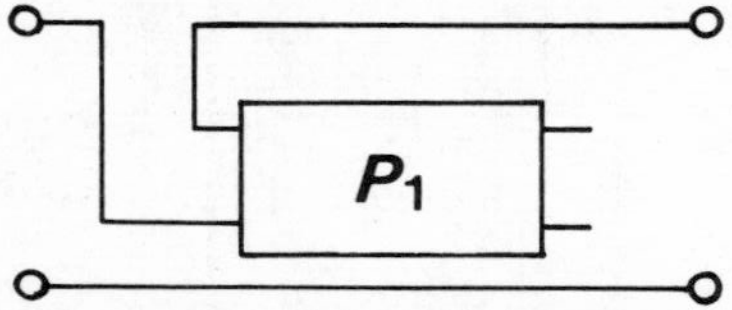 | WS P1 | ZS.P1 |
| 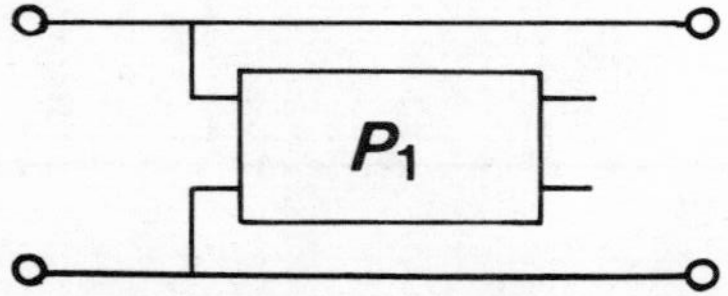 | WP P1 | ZP.P1 |

**Table 5.4** – Two-Port Wiring Operators

| COMPONENT INPUTS (Maximum 15 per circuit line) | | | | | | | | | |
|---|---|---|---|---|---|---|---|---|---|
| RPmn | | 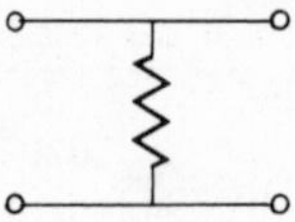 | Resistor | 1 | 2 | Ohms | 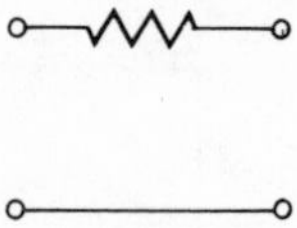 | | RSmn |
| CPmn | | 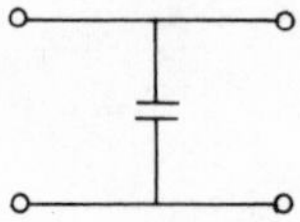 | Capacitor | 3 | 4 | Farads | 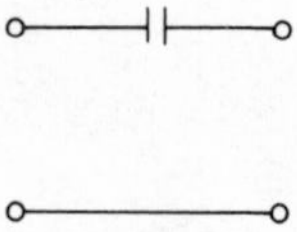 | | CSmn |
| LPmn | | 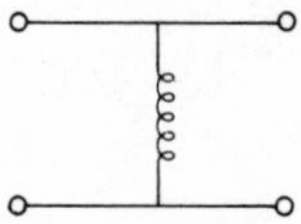 | Inductor | 5 | 6 | Henrys | 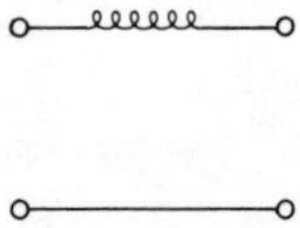 | | LSmn |
| VVmn | | 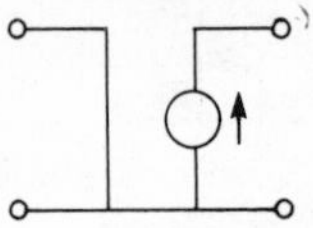 | Voltage Amplifier | 7 | 8 | Transconductance | 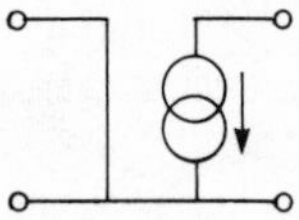 | Amps/Volt | VImn |
| IVmn | Ohms | 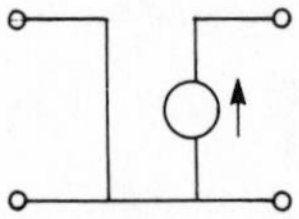 | Transresitance | 9 | 10 | Current Amplifier | 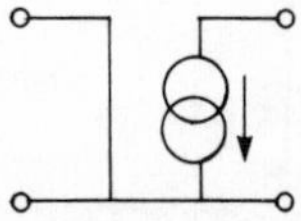 | | IImn |
| TDmn | Secs.. | 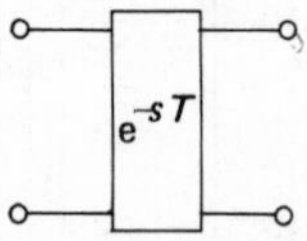  | Time Delay | 11 | 12 | Mutual Inductance | 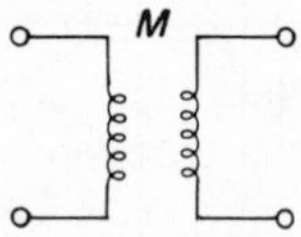  | Henrys | MLmn |
| TRmn | | 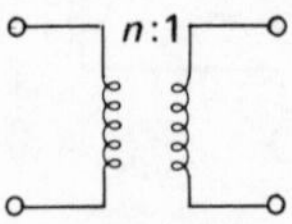  | Perfect Transformer | 29 | 30 | Negative Impedance Converter | 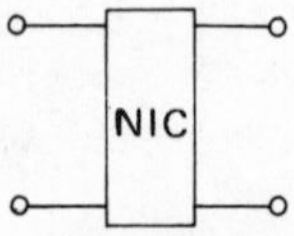  | | NCmn |
| GYmn | Amps/Volt | 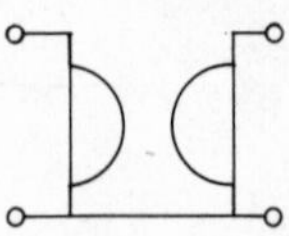 | Gyrator | 31 | 32 | Negative Impedance Inverter | 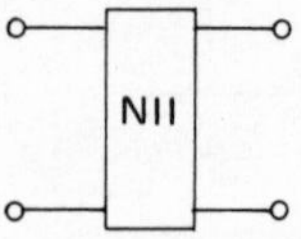  | | NImn |

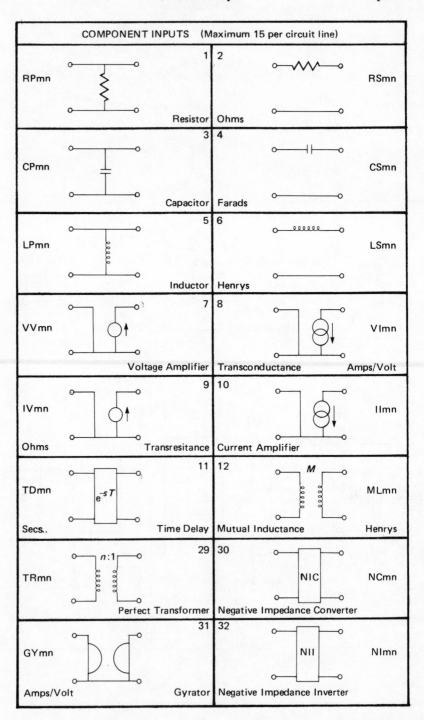

**Table 5.5** – Inputs to program 2PO

**DEVICE INPUTS (Maximum 4 per circuit line)**

13

VTmn

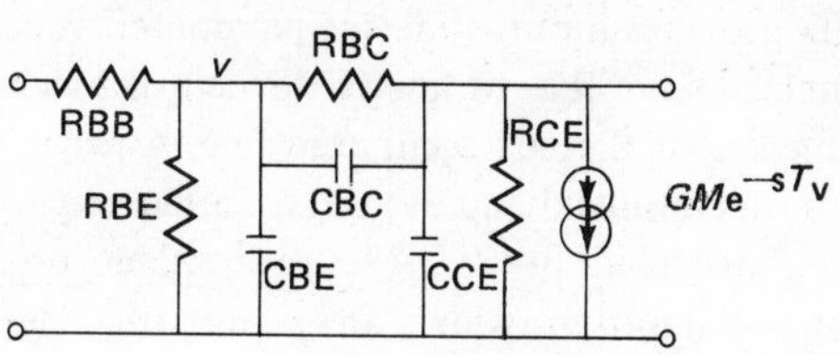


| TRANSISTOR, FET, VALVE MODEL | RBB, CBC, CBE, CCE, T | may be zero |
|---|---|---|
| | RBE, RBC, RCE | may be large |

15

VAmn

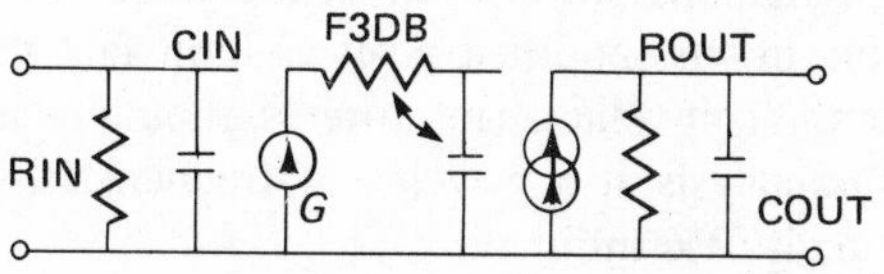


| AMPLIFIER MODEL | F3DB, RIN | may be large |
|---|---|---|
| | CIN, COUT | May be zero |

14

VYmn

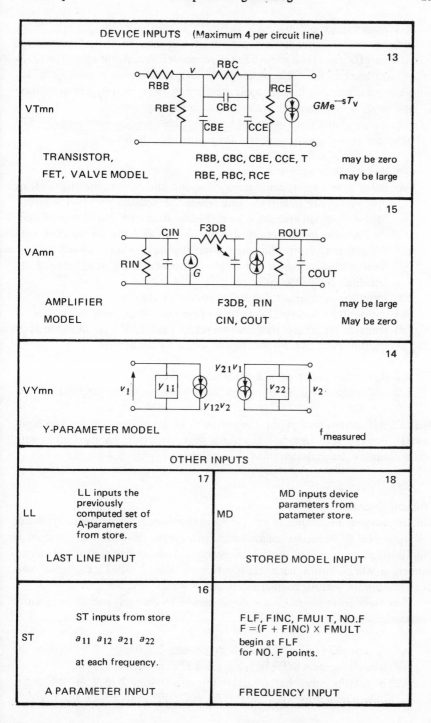

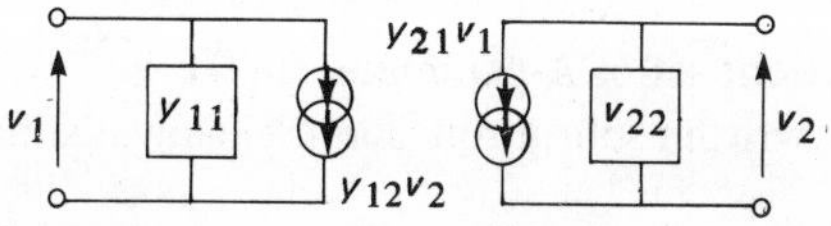


Y-PARAMETER MODEL $f_{measured}$

**OTHER INPUTS**

| | | | |
|---|---|---|---|
| LL | LL inputs the previously computed set of A-parameters from store. (17) | MD | MD inputs device parameters from patameter store. (18) |
| | LAST LINE INPUT | | STORED MODEL INPUT |
| ST | ST inputs from store $a_{11}$ $a_{12}$ $a_{21}$ $a_{22}$ at each frequency. (16) | | FLF, FINC, FMUI T, NO.F<br>F =(F + FINC) × FMULT<br>begin at FLF<br>for NO. F points. |
| | A PARAMETER INPUT | | FREQUENCY INPUT |

is the component or device number)

model parameters by outputs such as RS21?. A maximum of 15 single valued components and 4 active devices per circuit line may be used. A permanent model store MD provides input of device parameters repeatedly used. The transfer of A-parameters from line to line is through the parameter store LL which holds the parameters of the last computed line. A permanent A-parameter store ST may be filled on command and used in a similar way.

The outputs provided by 2PORT are available on the command NEXT? The nine command inputs provided are summarised as follows.

NA Begin a new analysis.
NC Enter a new set of component values for the current line of the wiring operator circuit description and repeat the analysis. Note that reference to the LL parameter store would be incorrect with this command since the last line in the modified analysis is in fact the present line with the old component values. The store ST should be used instead.
NF Repeat the analysis at a new set of frequencies. This applies only to one-line circuit description.
NL Input the next circuit line and continue the analysis.
OA Output the computed two-port A-parameters at all frequencies.
OG Output Gain, Phase, Real and Imaginary Parts of Gain at all frequencies.
OZ Output Input and Output Impedances at all frequencies.

SA Store the current set of A-parameters in ST.
US Go to a user written subroutine commencing at statement 500.

The 2PORT analysis programme contains most of the instructions for data input and is therefore very easy to use after a short period of familiarization with wiring operator circuit descriptions and reference to Tables 5.4 and 5.5.

**Program Organisation**

Circuit analysis in 2PORT is controlled entirely by a wiring operator circuit description of J1 characters length held in string array L$ (72). Interpretation of this description is carried out by conversion of successive pairs of characters entered in M$(2), into a numerical identifier K where $1 \leqslant K \leqslant 32$. Permitted pairs of characters are held for comparison in arrays W$(52) and X$(12).

The program responds to a wiring operator by changing one of two control variables A1 and A2.

A1 is normally unity but takes the values 2,3,10, or 11 to denote the unary operators RO. , RI. , ZS. , and ZP. , respectively.
A2 is initially unity but takes integer values from 2 to 7 to denote the binary operators .CC. , .BC. , .SP. , .SS. , .PP. , and .PS. , respectively.

These variables are used in Subroutine 149 for the control of parameter multiplication, addition, and conversion during the analysis.

When the characters are identified as a component type, the program calls either Subroutine 215 for component inputs, or Subroutine 254 for device and stored parameter inputs. After the A-parameters have been entered, both subroutines call Subroutine 149 to carry out the required analysis with computation controlled by A2 and A3. Parameter conversion Subroutines 110 and 131 are available to Subroutine 149 for wiring operations involving G, H,Y, or Z-parameters.

At the completion of analysis at all frequencies, the program requests the NEXT command to control further action such as output printing or continuation analysis. The list of available commands are kept in string array Y$(18) and compared to the two character input. The operation of the nine commands were summarised in the previous section.

*Component Input Subroutine 215*

All component inputs are single valued, and a maximum of 15 can be input in a single 72 character line. Array V(15) holds the component values for one circuit line in order, left to right, regardless of component type. Component values are entered into the array during an initial run of the interpretation section of the main program. Subsequent runs for each frequency then recover values from **V**.

Component type is identified by the value of K determined by the main program. On entry to the Subroutine, K selects the set of A-parameters to be entered into the analysis using the appropriate value from **V**. Available components are summarized in Table 5.5 and corresponding A-parameters were given in Section 5.3. At each frequency for analysis the A-parameters for components are entered into array B(1,8) in positions B(1,1) to B(1,8) in the order Re $a_{11}$, Im $a_{11}$, Re $a_{12}$, . . . Im $a_{22}$.

Each component is then incorporated into the analysis by calling Subroutine 149.

*Device Input Subroutine 254*

A maximum of four devices per circuit line are stored in the form of normalised Y-parameters in rows 1 to 4 of array T(5,9). Row 0 of this array is used as a permanent model store which can be entered into a program as MD. Row 5 of this array is used to store the A-parameters of the last computed line and is re-entered into the analysis as LL.

During the initial run of the interpretation section of the main program, device parameters are entered and Y-parameters of the devices stored in array T. At each frequency for analysis the A-parameters are generated from the stored Y-parameters and entered into array **B** as in Subroutine 215.

Stored A-parameters in the LL store or in the permanent store ST are similarly entered into array **B** by the device input subroutine. Each device or set of stored parameters is then incorporated into the analysis by calling Subroutine 149.

*Analysis Subroutines 149, 110, and 130*

The main numerical computation in two-port analysis is the complex multiplication, addition, and conversion of two-port parameters. This is carried out under the control of two variables A2 and A3 in Subroutine 149 which calls Subroutines 110 and 130 where necessary to convert G,H,Y, and Z-parameters to and from A-parameters.

The set of two-port parameters at each of 40 frequencies maximum is stored in the array A(40,8). Each row of **A** corresponds to an analysis frequency, and A($n$,1) to A($n$,8) stores A-parameters in the same order as array **B**. Column 0 stores the corresponding frequencies in Hz. Array C(40,8) is used as the permanent A-parameter store and is arranged identically to **A**.

Entry to Subroutine 149 is from either the component or device input subroutines with a set of A-parameters stored in array **B**. If a unary operator is in operation the A-parameters in **B** are converted to the required set provided that the entry is from the device input subroutine. Subroutine 149 then carries out the required binary operation between the sets of parameters in **A** and **B**.

Cascade and backward cascade operations are carried out internally and result in a new set of A-parameters at the given frequency stored in array **A**. Operators requiring addition of G,H,Y, or Z-parameters involve the use of Subroutines 110 and 130.

Subroutine 110 converts A-parameters held in B(1,1) to B(1,8) into the required set of parameters G.H.Y, or Z and adds them to any already stored in B(0,1), to B(0,8). The A-parameters in **A** are then transferred to B(1,1) to B(1,8) and also added. Subroutine 130 carries out the reverse process and re-enters A-parameters of the combination back into B(1,1) to B(1,8). These are then put back into array **A**. Parameter conversion follows Table 5.1, but conversions between G,H,Y, and Z-parameters are not provided because of restrictions to the program size. The control variable A2 equal to 4,5,6, or 7 denotes the conversions H, Z, Y, and G to and from A-parameters respectively, and operates almost entirely through the array **B** indices.

*Other Variables used*

| | |
|---|---|
| Control variables: | A0, A3, A4, A5, A6, J |
| Frequency generation: | F, F0, F1, F2, F3, I, P2 |
| Calculation space: | D1, D2, D3, S, R, U |

*Example* 5.4

Determine two-port wiring operator expressions for the booststrap follower circuit in Fig. 5.24. Evaluate the frequency response, input and output impedances, using 2PORT.

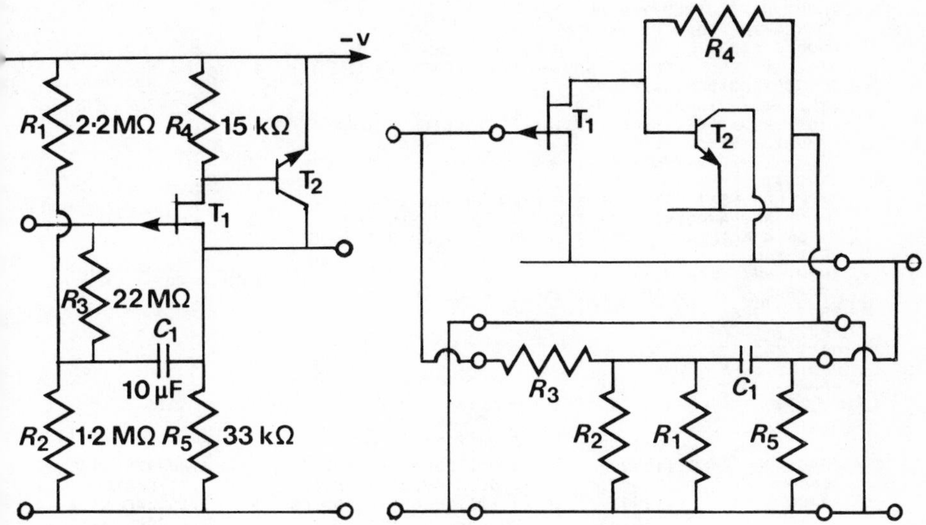


Fig. 5.24 – Boostrap follower for Example 5.4.

The two-port version of the circuit is also shown in Fig. 5.24. The wiring operator description is easily derived once the circuit has been re-drawn in this form. In order to avoid parenthesis the expression is written in two lines as

RSO4.PP.RO.VTO2.BC.VTO1

RSO3.CC.RPO2.RPO1.CSO1.RPO5.PP.RO.LL

In 2PORT the computation of the sub-network containing the two transistors and $R_4$ is analysed first, and the A-parameters of the circuit are kept in the memory storage LL reserved for the A-parameters of the previous wiring operator expression. RO has been used to convert $T_2$ to common-collector connection. The sub-network is referred to in the next line by LL which obtains A-parameters from the last line store. RO is again required to transform connections. In the second line of the expression the cascade is computed first and then connected in parallel-parallel with the sub-network containing the transistors. In 2PORT all results are finally obtained in A-parameters.

The 2PORT computer run is given in Fig. 5.25. The analysis reveals the poor high-frequency performance of the follower with regard to input impedance.

```
2-PORT ANALYSIS S
F-LOW, F+INC, F*MULT, NO.F ? 400,0,4,8
INPUT NEXT CIRCUIT LINE
 ? RS04.PP.RO.VT02.BC.VT01
RS04 ? 15E3
VT02  RBB,RBE,CBE,RBC/CBC,GM,RCE,CCE,DLY,STORE 1 OR 0
 ? 750,250,5E-12,1E6
 ? 1E-10,.2,3300,5E-12,0,0
VT01  RBB,RBE,CBE,RBC/CBC,GM,RCE,CCE,DLY,STORE 1 OR 0
 ? 0,1E8,5E-12,1E12
 ? 2E-12,.004,20E3,2E-12,0,0
.......
TYPE FOR NEXT LINE NL, NEW FREQ  NF, NEW CPTS NC
  FOR NEW ANALYSIS NA, STORE A   SA, USER 500 US
  FOR A MATRIX      OA, Z IN/OUT  OZ, GAIN     OG
NEXT ? NL
INPUT NEXT CIRCUIT LINE
 ? RS03.CC.RP02.RP01.CS01.RP05.PP.RO.LL
RS03 ? 22E6
RP02 ? 1.2E6
RP01 ? 2.2E6
CS01 ? 1E-5
RP05 ? 33E3
.......
NEXT ? OG
F HZ           V-GAIN DB       PHASE DEG       RP             IP
 400           -.125903        -4.00222E-03     .985608       -6.88467E-05
 1600          -.125911        -1.60286E-02     .985607       -2.75725E-04
 6400          -.125899        -6.41222E-02     .985608       -1.10304E-03
 25600         -.125964        -.256487         .985592       -4.41208E-03
 102400        -.126981        -1.02588         .985328       -1.76442E-02
 409600        -.143206        -4.09915         .98113        -7.03138E-02
 1.6384E+06    -.395239        -16.127          .917909       -.26541
 6.5536E+06    -3.2322         -53.3339         .411571       -.552848
NEXT ? OZ
F HZ           IN Z RP         IP              OUT Z RP       IP
 400            2.83221E+07    -1.86348E+08     5.38934        2.30579E-05
 1600           1.74627E+06    -4.75802E+07     5.38933        9.14367E-05
 6400           48043          -1.19104E+07     5.38935        3.64428E-04
 25600         -58196.7        -2.97624E+06     5.38935        1.45772E-03
 102400        -64176.5        -737722          5.38947        5.82905E-03
 409600        -55559          -162829          5.39122        .023204
 1.6384E+06    -17109.1        -17705.4         5.41774        8.60081E-02
 6.5536E+06    -1306.66        -2070.1          5.62439        .12633
NEXT ? OA
F HZ           A11 RP          IP              A12 RP         IP
               A21 RP          IP              A22 RP         IP
 400            1.0146          7.0872E-05      5.46803        4.05348E-04
                8.0845E-10      5.32179E-09     3.0324E-07     1.94671E-07
 1600           1.0146          2.83837E-04     5.46803        1.62246E-03
                7.75619E-10     2.12956E-08     3.03061E-07    7.78683E-07
 6400           1.0146          1.13549E-03     5.46804        6.48928E-03
                2.48278E-10     8.51854E-08     3.00187E-07    3.11474E-06
 25600          1.0146          4.54194E-03     5.46802        2.59571E-02
               -8.18878E-09     3.40739E-07     2.54206E-07    1.24589E-05
 102400         1.01457         1.81678E-02     5.46786        .103828
               -1.43182E-07     1.36281E-06    -4.81495E-07    4.9835E-05
 409600         1.01402         7.26712E-02     5.46514        .415316
               -2.30308E-06     5.44171E-06    -1.22528E-05    1.99294E-04
 1.6384E+06     1.00538         .290702         5.42187        1.66141
               -3.68658E-05     2.11595E-05    -2.00621E-04    7.94233E-04
 6.5536E+06     .866407         1.16381         4.72598        6.65519
               -5.90939E-04     4.55287E-05    -3.22147E-03    2.98739E-03
NEXT ?
```

Fig. 5.25 – 2PORT output for Example 5.4.

*Example* 5.5

Determine two-port wiring operator expressions for the fourth order Butterworth filter in Fig. 5.26. The operational amplifiers are Type 741 with characteristics: Input impedance = 2 M Ω // 5 pF, Gain = $10^5$, Outpu impedance = 75 Ω // 5pF, Cut-off frequency = 10 Hz. Use 2PORT to to evaluate frequency response.

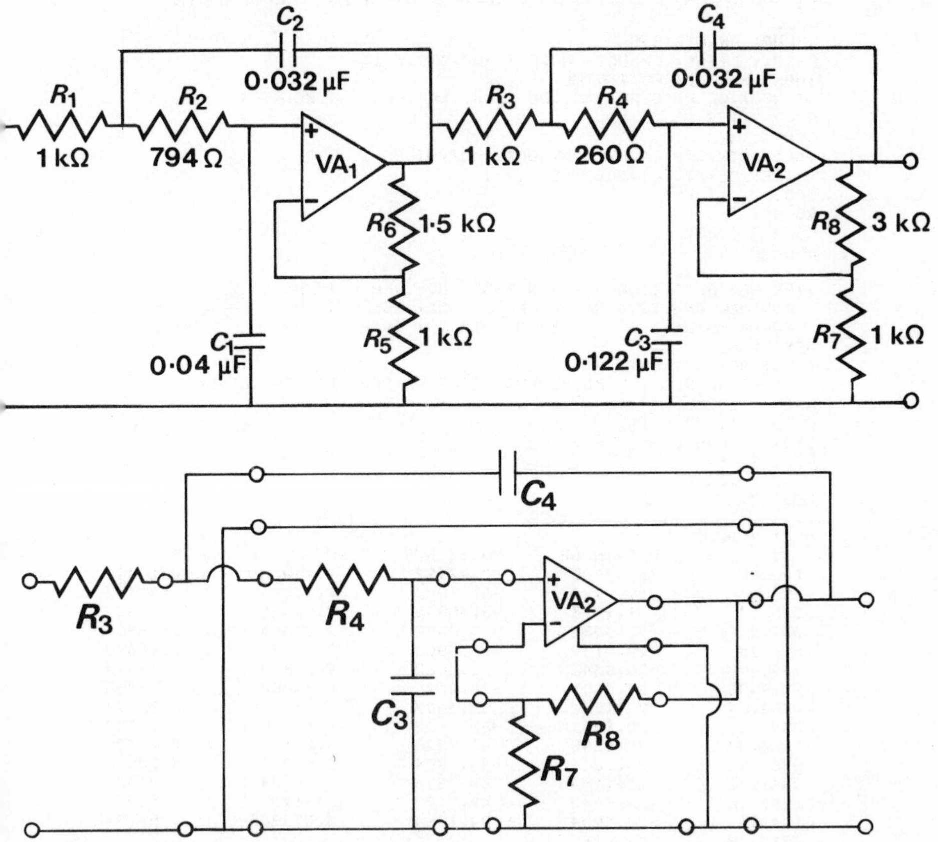


Fig. 5.26 – Butterworth filter for Example 5.5.

Fig. 5.26 also shows the two-port version of the filter circuit. The 2PORT wiring operator expression may therefore be written down by inspection as

RPO7.CC.RSO8.SP.VAO1.BC.CPO3.RSO4.PP.CSO4.BC.RSO3

RPO5.CC.RSO6.SP.MD.BC.CPO1.RSO2.PP.CSO2.BC.RSO1.CC.LL

Here the application of wiring operators is quite straightforward with LL used to refer to the last line store, and MD used to avoid repetition of the amplifier input.

The 2PORT run evaluating these expressions is given in Fig. 5.27. The most noticeable feature of the frequency response which is supposed to be maximally flat is the increase in gain by 1/3 dB as the cut-off frequency is approached. This is caused by the imperfections in the amplifier.

```
2-PORT ANALYSIS S
F-LOW, F+INC, F*MULT, NO.F ? 100,0,1.5,15
INPUT NEXT CIRCUIT LINE
 ? RP07.CC.RS08.SP.VA01.BC.CP03.RS04.PP.CS04.BC.RS03
RP07 ? 1E3
RS08 ? 3E3
VA01  RIN,CIN,GAIN,F3DB,ROUT,COUT,STORE 1 OR 0
 ? 2E6,5E-12,1E5,10,75,5E-12,1
CP03 ? .122E-6
RS04 ? 260
CS04 ? .032E-6
RS03 ? 1E3
..............
TYPE FOR NEXT LINE NL, NEW FREQ  NF, NEW CPTS NC
  FOR NEW ANALYSIS NA, STORE A   SA, USER 500 US
  FOR A MATRIX     OA, Z IN/OUT  OZ, GAIN     OG
NEXT ? NL
INPUT NEXT CIRCUIT LINE
 ? RP05.CC.RS06.SP.MD.BC.CP01.RS02.PP.CS02.BC.RS01.CC.LL
RP05 ? 1E3
RS06 ? 1.5E3
CP01 ? .04E-6
RS02 ? 794
CS02 ? .032E-6
RS01 ? 1E3
..............
NEXT ? OG
F HZ          V-GAIN DB     PHASE DEG     RP             IP
 100           19.9965      -2.97197       9.98668       -.518481
 150           19.9974      -4.45847       9.97086       -.777453
 225           19.9993      -6.68948       9.93525       -1.16527
 337.5         20.0036      -10.0402       9.85506       -1.74484
 506.25        20.0132      -15.0803       9.67439       -2.60678
 759.375       20.0345      -22.6882       9.26678       -3.87414
 1139.06       20.0801      -34.262        8.34482       -5.68435
 1708.59       20.1721      -52.1809       6.257         -8.06094
 2562.89       20.3201      -81.1132       1.60349       -10.2552
 3844.34       20.0548       46.9928      -6.86692       -7.36203
 5766.5        13.9863      -38.2704      -3.92977        3.10026
 8649.76       .461154       81.7826       .150724        1.04371
 12974.6      -13.8842       47.917        .135479        .150027
 19462        -28.1904       24.1084       3.55295E-02    1.58994E-02
 29192.9      -42.4942       1.30975       7.49537E-03    1.7137E-04
NEXT ?
```

Fig. 5.27 – 2PORT Output for Example 5.5.

## 5.6 CONCLUSION

Two-port analysis is important for its speed and accuracy when applied to filter circuits. It is also one of the few analysis techniques easily applied to distributed circuits. These two areas of application are those traditionally most used where the disadvantages of circuit description are outweighed by the advantages of computation. The value of circuit description by wiring operators lies in the more general application of two-port analysis to electronic circuits.

The ease with which circuits may be entered into the computer combined with the use of on-line time-shared terminals encourages the use of circuit analysis as a design tool through the immediate interaction between the circuit designer and the analysis results.

## FURTHER READING

Desoer, C. A. and Kuh, E. S., (1969), *Basic Circuit Theory,* Chapter 17, McGraw-Hill, New York.

Haykim, S. S., (1970), *Active Network Theory,* Chapters 4 and 5, Addison-Wesley, Reading, Mass.

Gatland, H. B., (1976), *Electronic Engineering Applications of Two-Port Networks,* Pergamon Press, Oxford.

Penfield, P., (1971), *MARTHA User's Manual,* MIT Press, Cambridge, Mass.

## PROBLEMS

5.1 Re-work Example 5.1 using two-port analysis. Why is the use of the series-parallel connection valid in this case?

5.2 Brune's two-port validity conditions for series and parallel port connections are necessary and sufficient only for three-terminal networks. Draw two active four-terminal networks to demonstrate that Brune's conditions are neither necessary nor sufficient in the general case.

5.3 Prove that the G-parameters of a transistor in common-base connection are given by

$$\begin{bmatrix} g_{11} & -1-g_{12} \\ 1-g_{21} & g_{22} \end{bmatrix} \quad \text{where} \quad \begin{bmatrix} g_{11} & g_{12} \\ g_{21} & g_{22} \end{bmatrix}$$

are the parameters for the common-emitter connection. Determine the voltage gain, input impedance, and output impedance of the amplifier in Fig. 5.28 where the transistor hybrid parameters are defined by

$$h_{ie} = 1\ \text{k}\Omega, h_{re} = 4\times10^{-4}, h_{fe} = 100, \text{and } h_{oe} = 50\ \mu S\ .$$

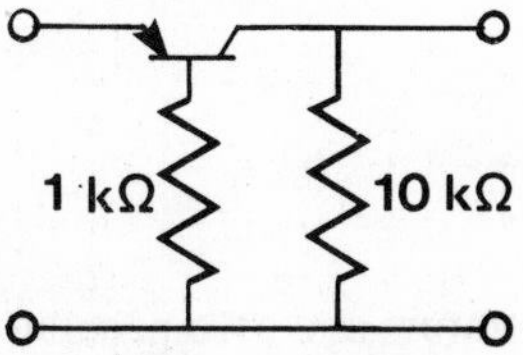


Fig. 5.28 – Problem 5.3.

5.4 Determine the two-port Y-parameters of the circuit in Fig. 5.29 with the input and output ports as shown. Use the transistor parameters from Problem 5.3 and determine the amplifier gain, input impedance, and output impedance.

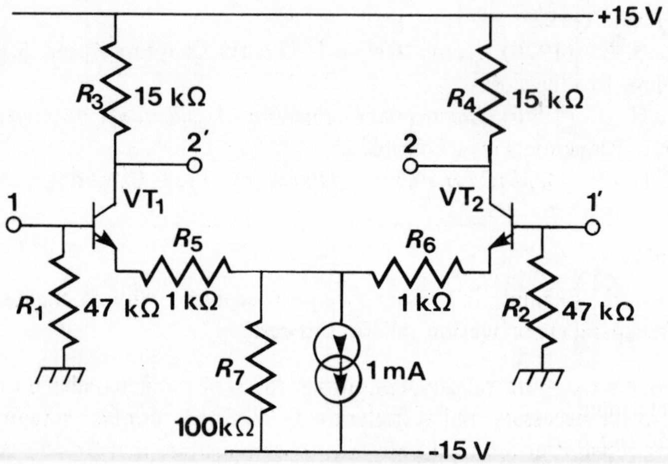


Fig. 5.29 – Problem 5.4.

5.5 Assume that the input to the amplifier in Fig. 5.29 is a voltage source connected between $R_1$ and ground. Assume that the output from the amplifier is the collector voltage of $VT_2$ which is also loaded by a 10 kΩ resistor. Determine the voltage gain, input, and output impedance of the amplifier by two-port analysis.

5.6 Show that the two circuits in Fig. 5.30 are equivalent and have the hybrid parameters

$$\begin{bmatrix} h_{11} & h_{12} \\ h_{21} & h_{22} \end{bmatrix}.$$

Derive similar one-generator and two-generator models for G, Y, and Z-parameters.

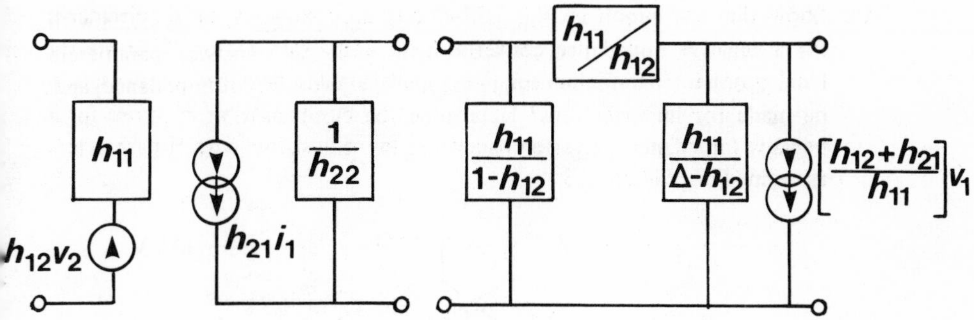


Fig. 5.30 – Problem 5.6.

5.7 Show that the A-parameters of a passive circuit can be put in the form

$$\mathbf{X}\begin{bmatrix}\lambda & 0\\ 0 & 1/\lambda\end{bmatrix}\mathbf{X}^{-1} \quad \text{where } |\mathbf{X}| = 1.$$

Determine $\lambda$ and the elements of **X**. Utilise the expressions obtained to find the A-parameters of 20 identical networks in cascade each consisting of a series resistor 1kΩ followed by a parallel capacitor 1μF.

5.8 What is the minimum value of transconductance $g_m$ required to make the circuit in Fig. 5.31(a) oscillate and what is the frequency? The VCCS is provided by a transistor with H. Parameters $\begin{bmatrix}1000 & 0\\ 500 & 2\times10^{-5}\end{bmatrix}$ and which is connected with emitter feedback resistance R as shown in Fig. 5.31(b). What is the maximum value of R for oscillation and what is the change in frequency from the ideal circuit.

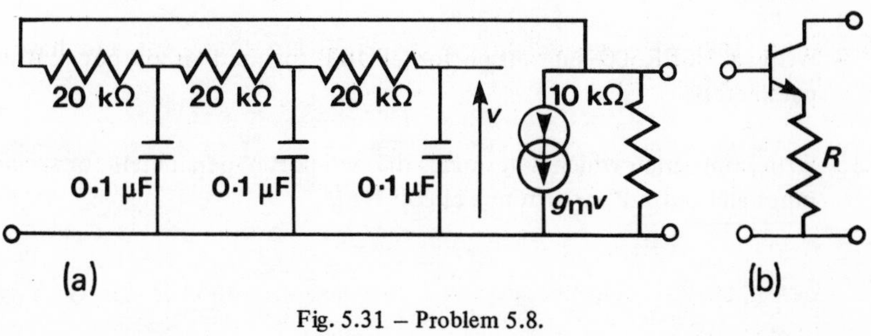


Fig. 5.31 – Problem 5.8.

5.9 Show that the circuit in Fig. 5.32 closely approximates the requirements for a negative impedance converter with unity gain factor. What are the ideal transistor parameters for this application and how can compensation be made for imperfections? Determine the compensation required for a negative impedance converter made from two transistors with H-parameters as defined in Problem 5.3.

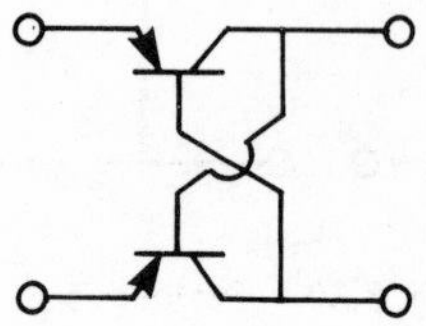

Fig. 5.32 – Problem 5.9.

5.10 Write wring operator descriptions for the circuits in Figs. 1.17, 1.20, 1.21, 4.4, 5.28, 5.29 (as in Problem 5.5), 6.15, 8.5, and 9.9.

*Computing*

5.11 Use program 2PORT to analyse familiar circuits and explore the program outputs provided. Analyse the circuits from Problem 5.10 and compare results with those obtained from program ACNLU.

5.12 Write a USER500 subroutine for 2PORT suited to the graphics facilities available to output response graphs. Plotting routines can probably be adapted from those written for Problem 4.14.

5.13 Collect the equations necessary for deriving the A-parameters of the more commonly used distributed circuit elements. Write a distributed component input subroutine called when K>32 indicates that new component identifiers in X$ have been detected.

5.14 Write a USER500 subroutine for 2PORT to input or output scattering parameters.

5.15 Write a program which determines the two-port Y-parameters for specified input and output ports from a circuit NAM.

## APPENDIX

```
0001 DIM A[40,8],B[1,8],C[40,8],T[5,9],V[15]
0002 DIM L$[72],M$[2],N$[2],W$[52],X$[12],Y$[18]
0003 PRINT '2-PORT ANALYSIS S'
0004 LET W$='RPRSCPCSLPLSVVVIIVIITDMLVTVYVASTLLMDRORICCBCSPSSPPPS'
0005 LET X$='ZSZPTRNCGYNI'
0006 LET Y$='NLNFNCNASAUSOAOZOG'
0007 LET P2=8*ATN(1)
0008 MAT C=ZER
0009 LET A0=1
0010 LET A6=0
0011 PRINT 'F-LOW, F+INC, F*MULT, NO.F';
0012 INPUT F0,F1,F2,F3
0013 ON A0 THEN GOTO 0014, 0017
0014 PRINT 'INPUT NEXT CIRCUIT LINE'
0015 INPUT L$
0016 LET A0=1
0017 LET F=F0*P2
0018 FOR I=(A0-1) TO F3
0019   LET A1=1
0020   LET A2=1
0021   LET A3=1
0022   LET A4=0
0023   LET A5=0
0024   FOR J=1 TO 8
0025     LET T[5,J]=A[I,J]
0026   NEXT J
0027   IF I<2 THEN GOTO 0030
0028   LET F=(F/P2+F1)*F2*P2
0029   PRINT '.';
0030   LET A[I,0]=F/P2
0031   LET J1=LEN(L$)
0032   FOR J=1 TO J1 STEP 3
0033     LET M$=L$[J,J+1]
0034     FOR K=1 TO 32
0035       IF K>26 THEN GOTO 0038
0036       IF W$[2*K-1,2*K]=M$ THEN GOTO 0044
0037       GOTO 0039
0038       IF X$[2*K-53,2*K-52]=M$ THEN GOTO 0042
0039     NEXT K
0040     PRINT 'UNRECOGNISED WORD AT ';J;' RE-';
0041     GOTO 0014
0042     IF K>28 THEN GOTO 0045
0043     GOTO 0061
0044     IF K>12 THEN GOTO 0050
0045     LET A4=A4+1
0046     LET J=J+2
0047     LET N$=L$[J,J+1]
0048     GOSUB 0216
0049     GOTO 0065
0050     IF K>18 THEN GOTO 0060
0051     IF K>15 THEN GOTO 0058
0052     IF A5<4 THEN GOTO 0055
0053     PRINT 'NO. MODEL INPUTS >4  RE-';
0054     GOTO 0014
0055     LET A5=A5+1
0056     LET J=J+2
0057     LET N$=L$[J,J+1]
0058     GOSUB 0255
0059     GOTO 0065
0060     IF K>20 THEN GOTO 0063
0061     LET A1=K-17
0062     GOTO 0065
0063     LET A2=K-19
0064     LET A3=1
0065   NEXT J
0066 NEXT I

*
```

```
0067 PRINT
0068 LET A6=A6+1
0069 IF A6>1 THEN GOTO 0073
0070 PRINT "TYPE FOR NEXT LINE NL, NEW FREQ  NF, NEW CPTS NC"
0071 PRINT "   FOR NEW ANALYSIS NA, STORE A    SA, USER 500 US"
0072 PRINT "   FOR A MATRIX        OA, Z IN/OUT OZ, GAIN      OG"
0073 PRINT "NEXT";
0074 INPUT N$
0075 FOR S=1 TO 17 STEP 2
0076    IF N$=Y$[S,S+1] THEN GOTO 0080
0077 NEXT S
0078 PRINT "UNRECOGNISED COMMAND   ";
0079 GOTO 0073
0080 LET A0=(S+1)/2
0081 ON A0 THEN GOTO 0014, 0011, 0016, 0003, 0105, 0108, 0098, 0084, 008
2
0082 PRINT "F HZ","V-GAIN DB","PHASE DEG","RP","IP"
0083 GOTO 0085
0084 PRINT "F HZ","IN Z RP","IP","OUT Z RP","IP"
0085 FOR I=1 TO F3
0086    LET J=A[I,1]^2+A[I,2]^2
0087    LET K=A[I,5]^2+A[I,6]^2
0088    IF A0=8 THEN GOTO 0092
0089    PRINT A[I,0],-LOG(J)/.2303,360*ATN(-A[I,2]/A[I,1])/P2,
0090    PRINT A[I,1]/J,-A[I,2]/J
0091    GOTO 0096
0092    PRINT A[I,0],(A[I,1]*A[I,5]+A[I,2]*A[I,6])/K,
0093    PRINT (A[I,2]*A[I,5]-A[I,1]*A[I,6])/K,
0094    PRINT (A[I,3]*A[I,1]+A[I,4]*A[I,2])/J,
0095    PRINT (A[I,4]*A[I,1]-A[I,3]*A[I,2])/J
0096 NEXT I
0097 GOTO 0073
0098 PRINT "F HZ","A11 RP","IP","A12 RP","IP"
0099 PRINT "        ","A21 RP","IP","A22 RP","IP"
0100 FOR I=1 TO F3
0101    PRINT A[I,0],A[I,1],A[I,2],A[I,3],A[I,4]
0102    PRINT "     ",A[I,5],A[I,6],A[I,7],A[I,8]
0103 NEXT I
0104 GOTO 0073
0105 MAT C=ZER
0106 MAT C=A
0107 GOTO 0073
0108 GOSUB 0327
0109 GOTO 0073
0110 REM                   CONVERSION A TO Y ETC
0111 LET D1=B[1,1]*B[1,7]-B[1,2]*B[1,8]
0112 LET D1=D1-B[1,5]*B[1,3]+B[1,6]*B[1,4]
0113 LET D2=B[1,1]*B[1,8]+B[1,2]*B[1,7]
0114 LET D2=D2-B[1,5]*B[1,4]-B[1,3]*B[1,6]
0115 IF A2<6 THEN GOTO 0118
0116 LET D1=-D1
0117 LET D2=-D2
0118 LET U=15-2*A2
0119 LET D3=B[1,U]^2+B[1,U+1]^2
0120 LET S=2*A2-2-8*INT(A2/4-.4)
0121 LET R=10-S
0122 LET B[0,1]=B[0,1]+(B[1,R-1]*B[1,U]+B[1,R]*B[1,U+1])/D3
0123 LET B[0,2]=B[0,2]+(B[1,R]*B[1,U]-B[1,R-1]*B[1,U+1])/D3
0124 LET B[0,3]=B[0,3]+(D1*B[1,U]+D2*B[1,U+1])/D3
0125 LET B[0,4]=B[0,4]+(D2*B[1,U]-D1*B[1,U+1])/D3
0126 LET B[0,5]=B[0,5]+B[1,U]/D3*(-1)^(A2-3)
0127 LET B[0,6]=B[0,6]-B[1,U+1]/D3*(-1)^(A2-3)
0128 LET B[0,7]=B[0,7]+(B[1,S-1]*B[1,U]+B[1,S]*B[1,U+1])/D3
0129 LET B[0,8]=B[0,8]+(B[1,S]*B[1,U]-B[1,S-1]*B[1,U+1])/D3
0130 RETURN

*
```

```
0131 REM                CONVERSION Y TO A ETC
0132 LET U=(-1)^(A2-3)
0133 LET D1=B[0,1]*B[0,7]-B[0,2]*B[0,8]
0134 LET D1=D1-B[0,5]*B[0,3]+B[0,6]*B[0,4]
0135 LET D2=B[0,1]*B[0,8]+B[0,2]*B[0,7]
0136 LET D2=D2-B[0,5]*B[0,4]-B[0,3]*B[0,6]
0137 LET D3=B[0,5]^2+B[0,6]^2
0138 LET S=2*A2-2-8*INT(A2/4-.4)
0139 LET R=10-S
0140 LET B[1,2*A2-7]=(D1*B[0,5]+D2*B[0,6])/D3*U
0141 LET B[1,2*A2-6]=(D2*B[0,5]-D1*B[0,6])/D3*U
0142 LET B[1,15-2*A2]=B[0,5]/D3*U
0143 LET B[1,16-2*A2]=-B[0,6]/D3*U
0144 LET B[1,S-1]=(B[0,7]*B[0,5]+B[0,6]*B[0,8])/D3*U
0145 LET B[1,S]=(B[0,8]*B[0,5]-B[0,6]*B[0,7])/D3*U
0146 LET B[1,R-1]=(B[0,1]*B[0,5]+B[0,2]*B[0,6])/D3*U
0147 LET B[1,R]=(B[0,2]*B[0,5]-B[0,1]*B[0,6])/D3*U
0148 RETURN
0149 REM                OPERATION CONTROL
0150 IF K<13 THEN GOTO 0179
0151 IF K>28 THEN GOTO 0179
0152 IF A1=1 THEN GOTO 0180
0153 IF A1<5 THEN GOTO 0163
0154 LET D1=B[1,45-4*A1]^2+B[1,46-4*A1]^2
0155 LET D2=(B[1,1]*B[1,5]+B[1,2]*B[1,6])/D1
0156 LET D3=(B[1,2]*B[1,5]-B[1,1]*B[1,6])/D1*(21-2*A1)
0157 MAT B=ZER
0158 LET B[1,1]=1
0159 LET B[1,7]=1
0160 LET B[1,2*A1-17]=D2
0161 LET B[1,2*A1-16]=D3
0162 GOTO 0179
0163 LET A3=A2
0164 FOR S=1 TO 8
0165   LET B[0,S]=0
0166 NEXT S
0167 LET A2=4
0168 IF A1=2 THEN GOTO 0170
0169 LET A2=7
0170 GOSUB 0111
0171 FOR S=3 TO 5 STEP 2
0172   LET B[0,S]=1-B[0,S]
0173   LET B[0,S+1]=-B[0,S+1]
0174 NEXT S
0175 LET B[0,9-2*A1]=B[0,9-2*A1]-2
0176 GOSUB 0132
0177 LET A2=A3
0178 LET A3=1
0179 LET A1=1
0180 IF A2>3 THEN GOTO 0200
0181 FOR S=1 TO 8
0182   ON A2 THEN GOTO 0211, 0183, 0186
0183   LET A[0,S]=A[I,S]
0184   LET B[0,S]=B[1,S]
0185   GOTO 0188
0186   LET A[0,S]=B[1,S]
0187   LET B[0,S]=A[I,S]
0188 NEXT S
0189 LET S=A[0,1]
0190 LET U=A[0,5]

*
```

```
0191 LET A[I,1]=S*B[0,1]-A[0,2]*B[0,2]+A[0,3]*B[0,5]-A[0,4]*B[0,6]
0192 LET A[I,2]=S*B[0,2]+A[0,2]*B[0,1]+A[0,3]*B[0,6]+A[0,4]*B[0,5]
0193 LET A[I,3]=S*B[0,3]-A[0,2]*B[0,4]+A[0,3]*B[0,7]-A[0,4]*B[0,8]
0194 LET A[I,4]=S*B[0,4]+A[0,2]*B[0,3]+A[0,3]*B[0,8]+A[0,4]*B[0,7]
0195 LET A[I,5]=U*B[0,1]-A[0,6]*B[0,2]+A[0,7]*B[0,5]-A[0,8]*B[0,6]
0196 LET A[I,6]=U*B[0,2]+A[0,6]*B[0,1]+A[0,7]*B[0,6]+A[0,8]*B[0,5]
0197 LET A[I,7]=U*B[0,3]-A[0,6]*B[0,4]+A[0,7]*B[0,7]-A[0,8]*B[0,8]
0198 LET A[I,8]=U*B[0,4]+A[0,6]*B[0,3]+A[0,7]*B[0,8]+A[0,8]*B[0,7]
0199 GOTO 0214
0200 ON A3 THEN GOTO 0201, 0209
0201 FOR S=1 TO 8
0202   LET B[0,S]=0
0203 NEXT S
0204 GOSUB 0111
0205 LET A3=2
0206 FOR S=1 TO 8
0207   LET B[1,S]=A[I,S]
0208 NEXT S
0209 GOSUB 0111
0210 GOSUB 0132
0211 FOR S=1 TO 8
0212   LET A[I,S]=B[1,S]
0213 NEXT S
0214 RETURN
0215 REM              COMPONENT INPUT
0216 IF I<>0 THEN GOTO 0220
0217 PRINT M$;N$;
0218 INPUT V[A4]
0219 GOTO 0253
0220 LET R=V[A4]
0221 MAT B=ZER
0222 LET B[1,1]=1
0223 LET B[1,7]=1
0224 IF K>6 THEN GOTO 0232
0225 ON K THEN GOTO 0226, 0226, 0228, 0230, 0230, 0228
0226 LET B[1,7-2*K]=R^(2*K-3)
0227 GOTO 0252
0228 LET B[1,8-2*K/3]=F*R
0229 GOTO 0252
0230 LET B[1,2*K-4]=-1/F/R
0231 GOTO 0252
0232 MAT B=ZER
0233 IF K>10 THEN GOTO 0236
0234 LET B[1,2*K-13]=(-1)^(K-1)/R
0235 GOTO 0252
0236 IF K=12 THEN GOTO 0243
0237 IF K>28 THEN GOTO 0246
0238 LET B[1,1]=COS(F*R)
0239 LET B[1,2]=SIN(F*R)
0240 LET B[1,7]=B[1,1]
0241 LET B[1,8]=B[1,2]
0242 GOTO 0252
0243 LET B[1,4]=-R*F
0244 LET B[1,6]=-1/R/F
0245 GOTO 0252
0246 ON K-28 THEN GOTO 0247, 0247, 0250, 0250
0247 LET B[1,1]=R
0248 LET B[1,7]=59/R-2*K/R
0249 GOTO 0252
0250 LET B[1,3]=1/R
0251 LET B[1,5]=63*R-2*K*R
0252 GOSUB 0150
0253 RETURN

*
```

```
0254 REM                    DEVICE INPUT
0255 IF I<>0 THEN GOTO 0294
0256 IF K>15 THEN GOTO 0326
0257 LET T[A5,9]=0
0258 ON K-12 THEN GOTO 0259, 0261, 0266
0259 PRINT M$;N$;"  RBB,RBE,CBE,RBC/CBC,GM,RCE,CCE,DLY,STORE 1 OR 0"
0260 GOTO 0263
0261 PRINT M$;N$;"  Y11R,Y11I,Y12R,Y12I/";
0262 PRINT "Y21R,Y21I,Y22R,Y22I,FHZ,STORE 1 OR 0"
0263 INPUT T[A5,1],T[A5,2],T[A5,3],T[A5,4]
0264 INPUT T[A5,5],T[A5,6],T[A5,7],T[A5,8],T[A5,0],R
0265 ON K-12 THEN GOTO 0275, 0285
0266 PRINT M$;N$;"  RIN,CIN,GAIN,F3DB,ROUT,COUT,STORE 1 OR 0"
0267 INPUT T[A5,1],T[A5,2],T[A5,3],T[A5,4],T[A5,7],T[A5,8],R
0268 LET T[A5,1]=1/T[A5,1]
0269 LET T[A5,5]=-T[A5,3]/T[A5,7]
0270 LET T[A5,6]=-T[A5,5]/P2/T[A5,4]
0271 LET T[A5,7]=1/T[A5,7]
0272 LET T[A5,3]=0
0273 LET T[A5,4]=0
0274 GOTO 0284
0275 LET T[A5,9]=T[A5,1]
0276 LET T[A5,1]=1/T[A5,2]+1/T[A5,4]
0277 LET T[A5,2]=T[A5,3]+T[A5,5]
0278 LET T[A5,3]=-1/T[A5,4]
0279 LET T[A5,4]=-T[A5,5]
0280 LET T[A5,5]=T[A5,6]
0281 LET T[A5,6]=T[A5,0]
0282 LET T[A5,7]=1/T[A5,7]-T[A5,3]
0283 LET T[A5,8]=-T[A5,4]+T[A5,8]
0284 LET T[A5,0]=1/P2
0285 FOR S=2 TO 8 STEP 2
0286   LET T[A5,S]=T[A5,S]/T[A5,0]
0287 NEXT S
0288 LET T[A5,0]=K
0289 IF R=0 THEN GOTO 0326
0290 FOR S=0 TO 9
0291   LET T[0,S]=T[A5,S]
0292 NEXT S
0293 GOTO 0326
0294 LET R=0
0295 ON K-12 THEN GOTO 0296, 0296, 0296, 0318, 0322, 0297
0296 LET R=A5
0297 LET A3=A2
0298 FOR S=1 TO 7 STEP 2
0299   LET B[0,S]=T[R,S]
0300   LET B[0,S+1]=T[R,S+1]*A[I,0]
0301 NEXT S
0302 ON T[A5,0]-12 THEN GOTO 0303, 0308, 0306
0303 LET B[0,5]=T[R,3]+T[R,5]*COS(A[I,0]*T[R,6])
0304 LET B[0,6]=B[0,4]-T[R,5]*SIN(A[I,0]*T[R,6])
0305 GOTO 0308
0306 LET B[0,5]=B[0,5]/(1+A[I,0]^2/(-T[R,5]/T[R,6])^2)
0307 LET B[0,6]=B[0,6]/(1+A[I,0]^2/(-T[R,5]/T[R,6])^2)
0308 LET A2=6
0309 LET A0=R
0310 GOSUB 0132
0311 LET R=A0
0312 LET A2=A3
0313 LET A3=1
0314 FOR S=1 TO 4
0315   LET B[1,S]=B[1,S]+T[R,9]*B[1,S+4]
0316 NEXT S
0317 GOTO 0325
0318 FOR S=1 TO 8
0319   LET B[1,S]=C[I,S]
0320 NEXT S
0321 GOTO 0325
0322 FOR S=1 TO 8
0323   LET B[1,S]=T[5,S]
0324 NEXT S
0325 GOSUB 0150
0326 RETURN
0327 REM USER 500 SUBROUTINE
0500 RETURN
*
```

Fig. 5.33 – Program 2PORT.

CHAPTER 6

# Transfer Function Analysis

The objective of analysis by nodal admittance matrix methods in Chapter 4 and by two-port methods in Chapter 5 was to numerically evaluate the steady-state frequency response of a linear network at a specified frequency. The complete response was then evaluated by repeated analysis. The same methods of analysis were used in Examples 1.3 and 5.2 to derive the voltage gain transfer function $F(j\omega)$ or $F(s)$ for simple filter circuits. From these transfer functions the frequency response can be rapidly evaluated without re-solving the circuit equations at each frequency. Furthermore it is possible to determine the circuit transient response to an excitation, by inverse Laplace transformation. Also, as we shall see in Chapter 8, sensitivity can be efficiently found.

## 6.1 NETWORK FUNCTIONS

The two-port admittance parameters for the bridged-T network in Fig. 6.1 were determined from Example 5.2 by substitution of $1/R$ in place of $sC_2$ giving

$$\mathbf{Y} = \begin{bmatrix} \dfrac{R+s(L_1+L_2)+s^2CL_2R+s^3CL_1L_2}{s(L_1+L_2)R+s^3CL_1L_2R} & -\dfrac{R+s(L_1+L_2)+s^3CL_1L_2}{s(L_1+L_2)R+s^3CL_1L_2R} \\ -\dfrac{R+s(L_1+L_2)+s^3CL_1L_2}{s(L_1+L_2)R+s^3CL_1L_2R} & \dfrac{R+s(L_1+L_2)+s^2L_1CR+s^3CL_1L_2}{s(L_1+L_2)R+s^3CL_1L_2R} \end{bmatrix} \tag{6.1}$$

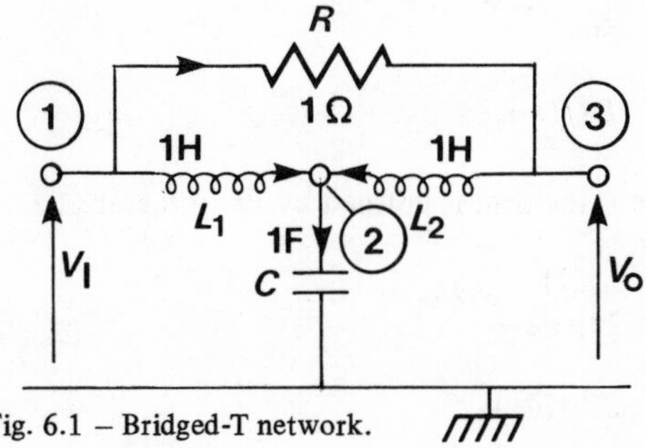


Fig. 6.1 – Bridged-T network.

We will use this result to illustrate some of the features of transfer functions, or more generally, network functions.

The matrix methods of circuit analysis in Chapters 2, 4, and 5 apply to the sinusoidal steady state when voltage and current variables represent phasors. Reactive network elements are then complex and frequency dependent. A network function in the sinusoidal steady state is therefore defined as the ratio of an output phasor and an input phasor. For example, the voltage gain transfer function $F(j\omega) = V_O(j\omega)/V_I(j\omega)$ for the bridged-T circuit is obtained from Table 5.3. Substitution of component values yields

$$F(j\omega) = \frac{(j\omega)^3 + 2j\omega + 1}{(j\omega^3) + (j\omega)^2 + 2j\omega + 1} .$$

At $\omega = 1$ rad/sec., $V_O = (1-j)V_I$, and the output voltage phasor can be evaluated for any input voltage phasor at the frequency specified.

The use of $s$ in Eq. (6.1) and previously was no more than a convenience of notation, but through the Laplace transform the definition of the network function can be generalized. In terms of the variable $s$ a network function $F(s)$ will then apply to any input waveform possessing a Laplace transform. Sinusoidal steady-state analysis is included as a special case by the substitution $s = j\omega$. Before we discuss network functions in more detail the Laplace transform must be defined, some of its properties tabulated, and its application to network analysis illustrated by an example.

**The Laplace Transform**

The Laplace transform relates the time domain behaviour of a network, for example step or impulse transient responses, to behaviour in the complex frequency or $s$-domain. The motivation for transformation is that network functions in the frequency domain are easier to manipulate because they can be algebraically determined and combined, and because they can be measured with more precision.

The unilateral Laplace transform is defined by the integral

$$F(s) = \int_{0-}^{\infty} f(t)e^{-st}dt \tag{6.2}$$

$$= \mathcal{L}\{f(t)\} .$$

The inverse Laplace transform is defined by the integral

$$f(t) = \frac{1}{2\pi j}\int_{c-j\infty}^{c+j\infty} F(s)\, e^{st}\, ds \tag{6.3}$$

$$= \mathcal{L}^{-1}\{F(s)\} .$$

The following points should be noted.

1. The value of $f(t)$ is assumed to equal zero for all $t<0$. The use of lower limit $0-$ allows $f(t)$ to be discontinuous at $t=0$. For example, a unit step at $t=0$ is unity for $t\geqslant 0$ and zero for $t<0$.
2. The variable $s$ is the complex frequency equal to $\sigma + j\omega$. The real part of $s$ must be positive and sufficiently large for integral (6.2) to be finite. If the integral is convergent for $\sigma = 0$ the Laplace transform reduces to the Fourier transform where $s = j\omega$.
3. $\sigma_a$ is called the abscissa of absolute convergence if integral (6.2) is convergent for all $\sigma > \sigma_a$. The real constant $c$ in integral (6.3) is assumed to be greater than $\sigma_a$.
4. Functions in network theory are normally Laplace transformable, although in some cases this is merely assumed for finite values of $t$ which are of interest.

Evaluation of the Laplace integral and its inverse can be directly carried out from the defining integral either numerically or by decomposition of the function into a series of simpler functions for which transforms are known. Table 6.1 gives sufficient transform pairs for application of the decomposition approach. Transforms of other elementary functions are easily derived from the general entries in the Table. (see Problem 6.1). The Laplace transform and the inverse transform are both unique except in some trivial cases. Basic properties of the transforms are summarized in Table 6.2. Tables 6.1 and 6.2 are adequate for the solution of most of the problems that we meet in circuit analysis.

**Network Analysis by the Laplace Transform**

A general network $\mathcal{N}$ driven by an input waveform $f_{IN}(t)$ and delivering an output waveform $f_{OUT}(t)$ is shown in Fig. 6.2. Analysis by the Laplace transform method for $f_{OUT}(t)$ with $f_{IN}(t)$, $\mathcal{N}$, and the initial state of $\mathcal{N}$ specified, may be summarized by the following steps.

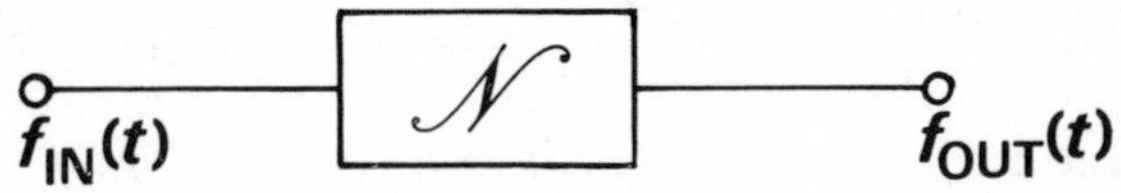


Fig. 6.2 – A General System.

STEP 1 Obtain the network differential equations and specify the initial conditions at $t=0-$.

2 Transform the differential equations to the complex frequency domain by the Laplace transform.

3 Solve for the output $F_{OUT}(s)$ as a function of $s$.

4 Transform $F_{OUT}(s)$ by the inverse Laplace transform back into the time domain to obtain $f_{OUT}(t)$.

**Table 6.1** – Laplace Transform Pairs
$\delta(t)$ is the unit impulse, $u(t)$ is the unit step.

| Time Domain $f(t)$ | $s$ – Domain $F(s)$ |
|---|---|
| $\delta(t)$ | $1$ |
| $u(t)$ | $1/s$ |
| $\dfrac{t^{n-1}e^{at}}{(n-1)!}$ | $\dfrac{1}{(s-a)^n}$ |
| $\dfrac{2t^{n-1}e^{at}}{(n-1)!}(\alpha\cos bt - \beta\sin bt)$ | $\dfrac{\alpha + j\beta}{(s-a-jb)^n} + \dfrac{\alpha - j\beta}{(s-a+jb)^n}$ |

**Table 6.2** – Laplace Transform Theorems

| Property | $f(t)$ | $F(s)$ |
|---|---|---|
| Linearity | $af_1(t) + bf_2(t)$ | $aF_1(s) + bF_2(s)$ |
| Differentiation | $\dfrac{df(t)}{dt}$ | $sF(s) - f(0-)$ |
| Integration | $\displaystyle\int_{0-}^{t} f(\tau)d\tau$ | $\dfrac{F(s)}{s}$ |
| Complex Differentiation | $t^n f(t)$ | $(-1)^n \dfrac{d^n F(s)}{ds^n}$ |
| Time delay | $f(t-\tau)u(t-\tau)$ | $e^{-s\tau}F(s)$ |
| Convolution | $\displaystyle\int_{0-}^{t} f_1(t-\tau)f_2(\tau)d\tau$ | $F_1(s)F_2(s)$ |
| Initial value | $\displaystyle\lim_{t\to 0} f(t)$ | $\displaystyle\lim_{s\to\infty} sF(s)$ |
| Final value | $\displaystyle\lim_{t\to\infty} f(t)$ | $\displaystyle\lim_{s\to 0} sF(s)$ |

The final result of the analysis is an expression for the output waveform as a function of time for $t \geqslant 0$. This response consists of two parts. The part due to the input waveform $f_{IN}(t)$ is called the **zero-state response** since it is generated by setting the initial conditions to zero. The part due to the initial conditions with the input equal to zero is called the **zero-input response**. In a linear network the zero-state response and the zero-input response may be computed separately and added by the principle of superposition to give the **complete response**. The four steps in the network analysis procedure will be illustrated by their application to the bridged-T network of Fig. 6.1.

Differential equations for the network may be obtained by summing the current leaving each node while noting that the element currents are given by

$$i_{L_1}(t) = \frac{1}{L_1}\int_{0-}^{t} e_1(t) - e_2(t)\,dt + i_{L_1}(0-)\ ,$$

$$i_{L_2}(t) = \frac{1}{L_2}\int_{0-}^{t} e_3(t) - e_2(t)\,dt + i_{L_2}(0-)\ ,$$

$$i_C = C\frac{de_2(t)}{dt},\ \text{and}\ i_R = \frac{e_1(t) - e_3(t)}{R}$$

Therefore,

$$\begin{aligned}
\frac{1}{L_1}\int_{0-}^{t} e_1(t)-e_2(t)dt + \frac{e_1(t)-e_3(t)}{R} &= I_1(t) - i_{L_1}(0-) \\
-\frac{1}{L_1}\int_{0-}^{t} e_1(t)-e_2(t)dt + C\frac{de_2(t)}{dt} \qquad & \\
-\frac{1}{L_2}\int_{0-}^{t} e_3(t)-e_2(t)dt &= I_2(t)+i_{L_1}(0-)+i_{L_2}(0-) \\
-\frac{e_1(t)-e_3(t)}{R} + \frac{1}{L_2}\int_{0-}^{t} e_3(t)-e_2(t)dt &= I_3(t)-i_{L_2}(0-)
\end{aligned} \tag{6.4}$$

Using Table 6.2, Laplace transformation of these equations into the complex frequency domain yields:

$$\begin{aligned}
\frac{e_1(s)-e_2(s)}{sL_1} + \frac{e_1(s)-e_3(s)}{R} &= I_1(s) - \frac{i_{L_1}(0-)}{s} \\
-\frac{e_1(s)-e_2(s)}{sL_1} + sCe_2(s) - \frac{e_3(s)-e_2(s)}{sL_2} &= I_2(s) + \frac{i_{L_1}(0-)}{s} + \frac{i_{L_2}(0-)}{s} + Ce_2(0-) \\
-\frac{e_1(s)-e_3(s)}{R} + sCe_2(s) - \frac{e_3(s)-e_2(s)}{sL_2} &= I_3(s) - \frac{i_{L_2}(0-)}{s}\ .
\end{aligned} \tag{6.5}$$

The customary practice of using capital letters for transformed variables must be abandoned at this stage since we are already using capital letters for fixed sources. Where necessary the function of $t$ or $s$ will be clearly indicated.

The linearity property of the Laplace transform means that the individual terms in Eqs. (6.4) were separately treated to obtain Eqs. (6.5). Thus it becomes possible to combine the first two steps of analysis into an $s$-domain modelling procedure. The fundamental characteristic equations in the time domain are

$$e(t) = \frac{L\mathrm{d}i(t)}{\mathrm{d}t}, \quad \text{and} \qquad i(t) = \frac{C\mathrm{d}e(t)}{\mathrm{d}t},$$

for the inductor and capacitor respectively. Taking the Laplace transform,

$$e(s) = sLi(s) - Li(0-) \qquad i(s) = sCe(s) - Ce(0-) \tag{6.6}$$

and rearranging yields

$$i(s) = \frac{e(s)}{sL} + \frac{i(0-)}{s} \qquad e(s) = \frac{i(s)}{sC} + \frac{e(0-)}{s}. \tag{6.7}$$

Models in the $s$-domain for inductors and capacitors derived from Eqs. (6.6) and (6.7) are shown in Fig. 6.3.

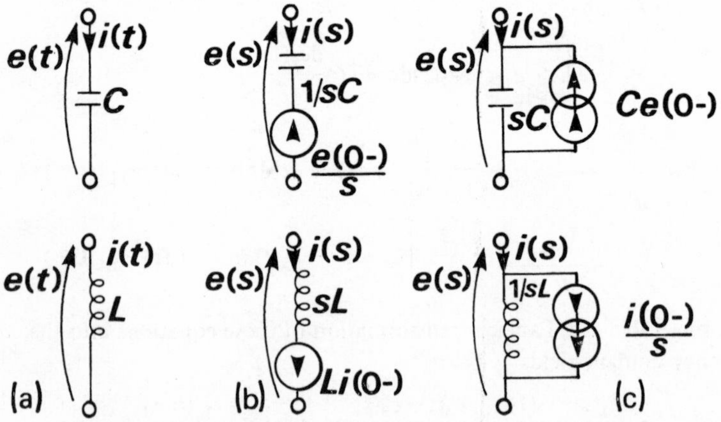


Fig. 6.3 – $s$-domain models. (a) Time domain models; (b) $s$-domain impedance models; (c) $s$-domain admittance models.

Other components involving differentiation or integration of their characteristic equations may be similarly treated. This includes mutual inductors and complex dependent sources. Resistors and other frequency independent components remain unchanged during $s$-domain modelling.

The bridged-T network is modelled in the $s$-domain for admittance matrix analysis in Fig. 6.4.

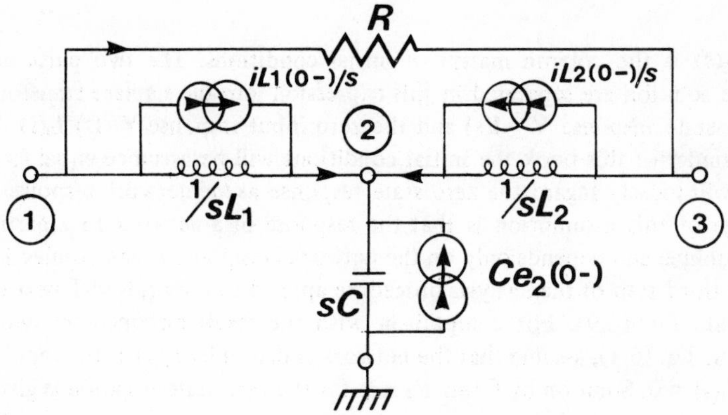


Fig. 6.4 – Bridged-T Network in $s$-domain.

The NAM in the $s$-domain may be directly written down from Fig. 6.4 to yield,

$$\begin{bmatrix} \frac{1}{R}+\frac{1}{sL_1} & -\frac{1}{sL_1} & -\frac{1}{R} \\ -\frac{1}{sL_1} & \frac{1}{sL_1}+\frac{1}{sL_2}+sC & -\frac{1}{sL_2} \\ -\frac{1}{R} & -\frac{1}{sL_2} & \frac{1}{R}+\frac{1}{sL_2} \end{bmatrix} \begin{bmatrix} e_1(s) \\ e_2(s) \\ e_3(s) \end{bmatrix}$$

$$= \begin{bmatrix} I_1(s) \\ I_2(s) \\ I_3(s) \end{bmatrix} + \begin{bmatrix} \frac{i_{L_1}(0-)}{s} \\ \frac{i_{L_1}(0-)}{s} + \frac{i_{L_2}(0-)}{s} + Ce_2(0-) \\ \frac{i_{L_2}(0-)}{s} \end{bmatrix} \quad (6.8)$$

This equation is of course identical to Eq. (6.5) but is in matrix form and has been directly obtained from the $s$-domain version of the original circuit.

The general form of the solution for node voltages from Eq. (6.8) is

$$\mathbf{e}(s) = \mathbf{Y}^{-1}(s)\,\mathbf{I}(s) + \mathbf{Y}^{-1}(s)\,\mathbf{I}_0(s) \;, \tag{6.9}$$

where $\mathbf{I}_0(s)$ is the column matrix of initial conditions. The two parts of the complete solution are separated in this expression into the Laplace transform of the zero state response $\mathbf{Y}^{-1}\,\mathbf{I}(s)$ and the zero input response $\mathbf{Y}^{-1}(s)\,\mathbf{I}_0(s)$. For the remainder of this book the initial conditions will be assumed equal to zero, and we will loosely regard the zero state response as *the* network response. The advantage of this assumption is that the response of a network to a particular input is unique and depends only on the network components and connections.

The third step of the analysis procedure applied to the bridged-T network is to evaluate Eq. (6.9). For comparison with the result obtained by two-port analysis in Eq. (6.1), assume that the network is driven by $I_1(s)$ at the input with $I_2(s) = I_3(s) = 0$. Solution by Cramer's rule for the zero state response is given by

$$e_3(s) = \frac{\Delta_{13}(s)}{\Delta(s)} I_1(s) \;,$$

and $$e_1(s) = \frac{\Delta_{11}(s)}{\Delta(s)} I_1(s) \;,$$

where $\Delta_{mn}(s)$ is the cofactor of element $y_{mn}$ of $\mathbf{Y}(s)$, and $\Delta(s)$ is the determinant of $\mathbf{Y}(s)$. The voltage gain is therefore obtained from

$$e_3(s) = \frac{\Delta_{13}(s)}{\Delta_{11}(s)} e_1(s)$$

$$= \frac{\dfrac{1}{s^2L_1L_2} + \dfrac{1}{R}\left(\dfrac{1}{sL_1} + \dfrac{1}{sL_2} + sC\right)}{\left(\dfrac{1}{R} + \dfrac{1}{sL_2}\right)\left(\dfrac{1}{sL_1} + \dfrac{1}{sL_2} + sC\right) - \dfrac{1}{s^2L_2^2}} e_1(s)$$

$$= \frac{s^3L_1L_2C + s(L_1+L_2) + R}{s^3L_1L_2C + s^2L_1CR + s(L_1+L_2) + R} e_1(s) \;. \tag{6.10}$$

This expression could also be obtained from the two-port admittance parameters in Eq. (6.1) by obtaining voltage gain from Table 5.3 as $-y_{21}/y_{22}$.

The final Step 4 of the analysis is the return to the time domain by inverse Laplace transformation. Let $e_1(t)$ be a unit step. Substitution of component values and $e_1(s) = 1/s$ into Eq. (6.10) yields

$$e_3(t) = \mathcal{L}^{-1}\left\{\frac{s^3 + 2s + 1}{s(s^3 + s^2 + 2s + 1)}\right\}.$$

Factorizing,

$$e_3(t) = \mathcal{L}^{-1}\left\{\frac{s^3 + 2s + 1}{s(s+0.6)\,(s^2 + 0.4s + 1.67)}\right\},$$

and expanding into partial fractions,

$$e_3(t) = \mathcal{L}^{-1}\left\{\frac{1}{s} + \frac{0.39}{s+0.6} - \frac{0.195 - 0.325\mathrm{j}}{s + 0.2 - 1.3\mathrm{j}} - \frac{0.195 + 0.325\mathrm{j}}{s + 0.2 + 1.3\mathrm{j}}\right\},$$

finally gives the unit step transient response through Table 6.1 as

$$e_3(t) = 1 + 0.39\mathrm{e}^{-0.6t} - 2\mathrm{e}^{-0.2t}(0.195\cos 1.3t + 0.325\sin 1.3t)\ . \tag{6.11}$$

The sinusoidal steady-state response can also be found by this method. Recall that the steady state exists as $t \to \infty$ when all transients have decayed to zero. Returning to Eq. (6.10), let $e_1(t) = \cos(\omega t + \theta)$.

From Table 6.1 with some manipulation

$$e_1(s) = \frac{s\cos\theta - \omega\sin\theta}{s^2 + \omega^2}\ .$$

Therefore,

$$e_3(s) = \frac{F(s)\,(s\cos\theta - \omega\sin\theta)}{(s - \mathrm{j}\omega)\,(s + \mathrm{j}\omega)}, \tag{6.12}$$

where $F(s)$ is the network gain function given for the bridged-T circuit by

$$F(s) = \frac{(s^3 + 2s + 1)}{(s^3 + s^2 + 2s + 1)}\ . \tag{6.13}$$

The denominator of $F(s)$ has already been factorized, and it is possible to expand $e_3(s)$ into partial fractions. However, we also know that those arising from $F(s)$ lead to exponentially decaying transient waveforms. They will be altered only in magnitude by the sinusoidal input since the exponential decays and frequency are governed by the denominators $(s^3 + s^2 + 2s + 1)$. The partial fractions due to $F(s)$ can therefore be ignored when finding the steady-state response. The remaining partial fractions due to the denominator factors $(s-j\omega)$ and $(s+j\omega)$ do not decay to zero because the exponent is equal to zero.

The steady-state partial fractions evaluated in the usual way require substitution of $s=j\omega$ for the residue of $(s-j\omega)$ and $s=-j\omega$ for the residue of $(s+j\omega)$ that is,

$$e_3(s) = \frac{F(j\omega)\,(j\omega\cos\theta - \omega\sin\theta)/2j\omega}{(s-j\omega)} + \frac{F(-j\omega)\,(-j\omega\cos\theta - \omega\sin\theta)/(-2j\omega)}{(s+j\omega)}$$

$$= \frac{\tfrac{1}{2}F(j\omega)\,(\cos\theta + j\sin\theta)}{(s-j\omega)} + \frac{\tfrac{1}{2}F(-j\omega)\,(\cos\theta - j\sin\theta)}{(s+j\omega)}$$

$$= \frac{\tfrac{1}{2}|F(j\omega)|\,e^{j\angle F(j\omega)}\,e^{j\theta}}{(s-j\omega)} + \frac{\tfrac{1}{2}|F(j\omega)|\,e^{j\angle -F(j\omega)}\,e^{-j\theta}}{(s+j\omega)},$$

$$= \tfrac{1}{2}|F(j\omega)|\left\{\frac{\cos\phi + j\sin\phi}{(s-j\omega)} + \frac{\cos\phi - j\sin\phi}{(s+j\omega)}\right\},$$

where $\phi = \theta + \angle F(j\omega)$.

The inverse Laplace transform through Table 6.1 finally yields

$$e_3(t) = |F(j\omega)|\cos\{\omega t + \theta + \angle F(j\omega)\}\ . \tag{6.14}$$

$F(j\omega)$ may therefore be interpreted as a phasor equal to the ratio of the output phasor to the input phasor. Thus the steady-state frequency response may be evaluated by making the substitution $s = j\omega$ in the network function $F(s)$. The network function has therefore been generalized as the ratio

$$\frac{\mathcal{L}\,(\text{zero-state output response})}{\mathcal{L}\,(\text{input})}\,.$$

The frequency response of the bridged-T network voltage gain is evaluated by substitution into Eq. (6.13), giving as before,

$$F(j\omega) = \frac{(j\omega)^3 + 2j\omega + 1}{(j\omega)^3 + (j\omega)^2 + 2j\omega + 1}$$

$$= \frac{1 + j(2\omega - \omega^3)}{(1 - \omega^2) + j(2\omega - \omega^3)} \ .$$

It must be emphasized that this result is correct only if the exponentials due to the partial fractions from the denominator of $F(s)$ decay to zero or a finite value. This requires the roots of the denominator polynomial to have the negative real parts and no more than one root at zero.

**Linear Network Functions**

The generalized network function is defined as the function $F(s)$ which satisfies the equation

$$\mathcal{L}\,(\text{zero-state output response}) = F(s) \times \mathcal{L}\,(\text{input}) \ .$$

Several features of network functions which apply in general were illustrated in the preceding discussion on analysis by the Laplace transform. They will be briefly summarized before application of network functions to linear circuits.

The function $F(s)$ for a lumped linear time-invariant network is a ratio of two polynomials in $s$ with real coefficients and is called a **rational function.** Therefore, $F(s) = N(s)/D(s)$, where $N(s)$ and $D(s)$ are the numerator and denominator polynomials of $F(s)$. In the example of the bridged-T network, at Eq. (6.10) each coefficient of the rational function has the additional property that it is linear in the component variables. This is true of all networks excluding those containing impedance converters or inverters and mutual inductors. This is because those elements contain two identically equal transfer admittance terms which can become multiplied together. In most practical realizations, impedance converters and inverters contain separate amplifiers for the two transfer admittances, and non-linearity of the rational function can usually be avoided.

The procedure for separating the response rational function $F(s) \times \mathcal{L}$ (input) into a sum of simple functions was carried out by expansion of the rational function into partial fractions. Inverse Laplace transformation was then easily achieved by using Table 6.1. The denominator must therefore be factorized by finding its roots, the multiple poles recognized and grouped together before the partial fraction residues can be evaluated. This is not a convenient route for computation, and direct numerical Laplace transform inversion from the rational function is preferred. However, the roots of the denominator, called the **poles** of the rational function, are often the desired result of the analysis.

The roots of numerator, called **zeroes,** affect only the residues of the partial fractions. The frequencies and decays of the exponentials in the response are solely determined by the poles of the response rational function. The poles of $F(s)$ are called the **natural frequencies** of the network.

The condition for existence of the steady-state frequency response was that the real parts of the poles of $F(s)$ must be negative with no more than one pole at zero. This is also the condition for stability of the network. The denominator polynomial can be tested for stability by Routh's criterion without finding its roots. Let $D(s) = a_0s^n + b_0s^{n-1} + a_1s^{n-2} + b_1s^{n-3} + \dots$ and put the coefficients into the array

$$\begin{array}{ccccc} a_0 & a_1 & a_2 & a_3 & \dots \\ b_0 & b_1 & b_2 & b_2 & \dots \\ c_0 & c_1 & c_2 & c_3 & \dots \\ d_0 & d_1 & d_2 & d_3 & \dots \end{array}$$

where
$$c_0 = \frac{b_0a_1 - a_0b_1}{b_0}, \quad c_1 = \frac{b_0a_2 - a_0b_2}{bp}, \dots$$

$$d_0 = \frac{c_0b_1 - b_0c_1}{c_0}, \quad d_1 = \frac{c_0b_2 - b_0c_2}{c_0}, \dots$$

$$e_0 = \frac{d_0c_1 - c_0d_1}{d_0}, \dots .$$

Continue this process until $n+1$ rows have been obtained while filling incomplete rows with zeroes at the right. Routh's criterion states that the number of sign changes as we read down the first column of the array is equal to the number of roots in $D(s)$ with positive real parts. In particular for stability all elements of the array in the first column must be greater than zero. A necessary condition for this to be possible is that all coefficients of $D(s)$ are greater than zero.

**Two-port Network Functions**

Transfer function methods of analysis are usually restricted to networks in which one port is designated as input and another as output. Therefore the general network may be treated as a two-port although one-port network functions are also defined for some purposes. Since networks in the $s$-domain are assumed linear for sinusoidal steady-state analysis, circuits with several inputs or outputs are treated by superposition and separate transfer functions derived for each signal path.

Input and output functions $\mathcal{L}\, f_{\mathrm{IN}}(t)$ and $\mathcal{L}\, f_{\mathrm{OUT}}(t)$ may be interpreted as either voltage or current port variables. Input voltage sources effectively short circuit the input port. Similarly, input current sources open circuit the port. In order that the output port is treated in the same way we will consider an output voltage to refer to an open circuit output port. An output current is assumed to refer to a short circuit output port. Thus there are four combinations of input and output configuration available. They are shown in Table 6.3.

The four sets of operating conditions for the two-port each give rise to natural frequencies which depend on the poles of the network function. Therefore there are four different denominator polynomials denoted by $p_{\mathrm{oo}}, p_{\mathrm{so}}, p_{\mathrm{os}}$, and $p_{\mathrm{ss}}$. The suffices indicate the circuit configuration, for example, $p_{\mathrm{os}}$ refers to a two-port with an open circuit input and short circuit output.

A transfer function may be defined for each circuit configuration. In terms of the port voltages and currents used throughout Chapter 5 the four transfer functions are defined as,

Open circuit forward transfer impedance $= (v_2/i_1)_{i_2=0}$

Open circuit forward voltage gain $= (v_2/v_1)_{i_2=0}$

Short circuit forward current gain $= (i_2/i_1)_{v_2=0}$

Short circuit forward transfer admittance $= (i_2/v_1)_{v_2=0}$ .

Each transfer function is simply evaluated in terms of two-port Y-parameters using Table 5.2. For example, the open circuit transfer admittance is equal to

$$\left(\frac{v_2}{i_1}\right)_{i_2=0} = \left(\frac{v_2}{v_1}\right)\left(\frac{v_1}{i_1}\right)_{Z_l=\infty} = \left(-\frac{y_{21}}{y_{22}}\right)\left(\frac{y_{22}}{\Delta}\right) = -\frac{y_{21}}{\Delta} .$$

The other three functions are entered in the summary Table 6.3.

All four transfer functions involve the same numerator $y_{21}$. The numerator polynomial of $y_{21}$ therefore contains the transmission zeroes of the network and is denoted by $q_{21}$. The reverse transmission zeroes are contained in polynomial $q_{12}$ which is the numerator polynomial of $y_{12}$. The polynomial form of the transfer functions can therefore be entered into Table 6.3 as $q_{21}/p_{\mathrm{oo}}$, $q_{21}/p_{\mathrm{so}}$ $-q_{21}/p_{\mathrm{so}}$, and $-q_{21}/p_{\mathrm{ss}}$.

Comparison of the polynomial and Y-parameter forms of the transfer function directly yields $y_{21} = -q_{21}/p_{ss}$. Therefore, $y_{11} = p_{os}/p_{ss}$, $y_{22} = p_{so}/p_{ss}$, and $\Delta = p_{oo}/p_{ss}$. Also we could expect $y_{12}$ to equal $-q_{12}/p_{ss}$. The only difficulty with the observations is that

$$\Delta = y_{11}y_{22} - y_{21}y_{12}$$

$$= \frac{p_{os}\,p_{so} - q_{12}\,q_{21}}{p_{ss}^2} = \frac{p_{oo}}{p_{ss}} \,. \tag{6.15}$$

Therefore, $p_{os}p_{so} - q_{12}q_{21}$ must contain the factor $p_{ss}$ for cancellation. This is true in general, and it is essential to make the cancellation in $s$-domain or symbolic analysis by two-ports.

Four further network functions have been added to Table 6.3. These are called driving point functions because they refer both input and output variables to the same port. For example, the input admittance of a port is defined by $(i_1/v_1)$. In a two-port this could be determined with the output port either open circuit or short circuit. The polynomials involved in the two cases are quite different. The output admittances are similarly treated, giving the four table entries. Input and output impedance network functions are of course the inverse of the corresponding admittance functions.

The complete set of two-port Y-parameters have been defined in terms of five out of the six polynomials, and the sixth arises in the determinant numerator Use of Table 5.1 and Eq. (6.15) yields the full set of two-port parameters in terms of the six polynomials as in Table 6.4.

The polynomials for the bridge-T network are easily derived from the Y parameters of Eq. (6.1).

$$p_{oo} = s^2C(L_1+L_2) + sCR$$

$$p_{so} = s^3CL_1L_2 + s^2CL_1R + s(L_1+L_2) + R$$

$$p_{os} = s^3CL_1L_2 + s^2CL_2R + s(L_1+L_2) + R$$

$$p_{ss} = s^3CL_1L_2R + sR(L_1+L_2)$$

$$q_{12} = q_{21} = s^3CL_1L_2 + s(L_1+L_2) + R \,.$$

The cancellation of $p_{os}\,p_{so} - q_{12}\,q_{21}$ by $p_{ss}$ to yield $p_{oo}$ can be readily checked. All sets of two-port parameters and the most common network functions are available from Tables 6.3 and 6.4. However, common factors in parameters and network functions are not entirely eliminated by this technique, but

**Table 6.3** – Two-Port Network Functions

$\Delta = |\mathbf{Y}|$, $\Delta_{ij}$ = Cofactor $ij$ of **Y**, $\Delta_{2,n-1}^{2,n-1}$ = Determinant of matrix consisting of rows and columns 2, 3, . . . $n-1$, of **Y**.

| Circuit | Transfer Function | 2-Port Y-parameters | Polynomials | NAM solution |
|---|---|---|---|---|
|  | o/c transfer impedance $\left(\frac{v_2}{i_1}\right)_{i_2=0}$ | $\frac{-y_{21}}{\Delta}$ | $\frac{q_{21}}{p_{oo}}$ | $\frac{\Delta_{1n}}{\Delta}$ |
| 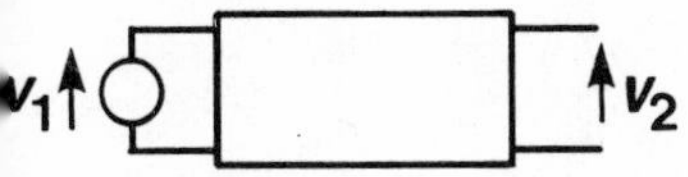 | o/c voltage gain $\left(\frac{v_2}{v_1}\right)_{i_2=0}$ | $\frac{-y_{21}}{y_{22}}$ | $\frac{q_{21}}{p_{so}}$ | $\frac{\Delta_{1n}}{\Delta_{11}}$ |
| 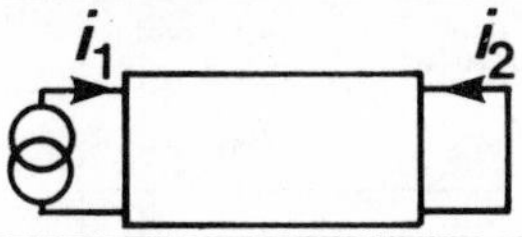 | s/c current gain $\left(\frac{i_2}{i_1}\right)_{v_2=0}$ | $\frac{y_{21}}{y_{11}}$ | $\frac{-q_{21}}{p_{os}}$ | $\frac{-\Delta_{1n}}{\Delta_{nn}}$ |
| 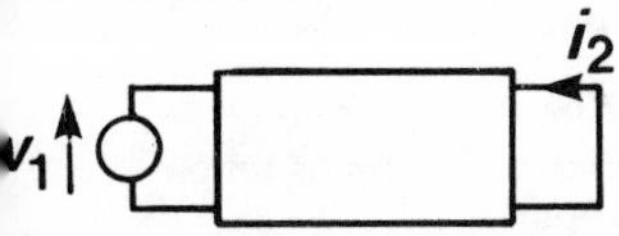 | s/c transfer admittance $\left(\frac{i_2}{v_1}\right)_{v_2=0}$ | $y_{21}$ | $\frac{-q_{21}}{p_{ss}}$ | $\frac{-\Delta_{1n}}{\Delta_{2,n-1}^{2,n-1}}$ |
| | Driving Point Function | | | |
| 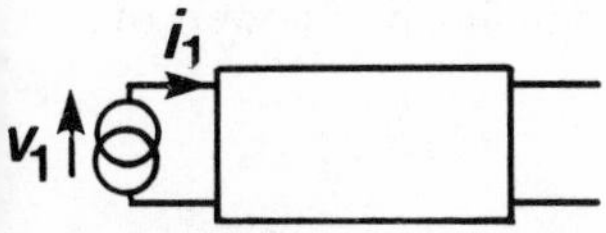 | o/c input admittance $\left(\frac{i_1}{v_1}\right)_{i_2=0}$ | $\frac{\Delta}{y_{22}}$ | $\frac{p_{oo}}{p_{so}}$ | $\frac{\Delta}{\Delta_{11}}$ |
| 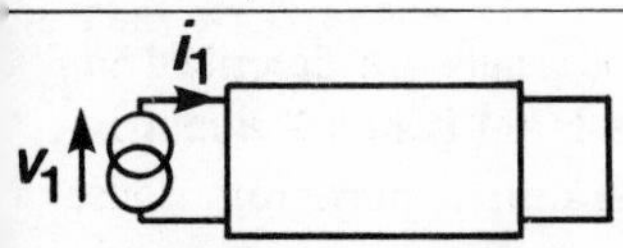 | s/c input admittance $\left(\frac{i_1}{v_1}\right)_{v_2=0}$ | $y_{11}$ | $\frac{p_{os}}{p_{ss}}$ | $\frac{\Delta_{nn}}{\Delta_{2,n-1}^{2,n-1}}$ |
| 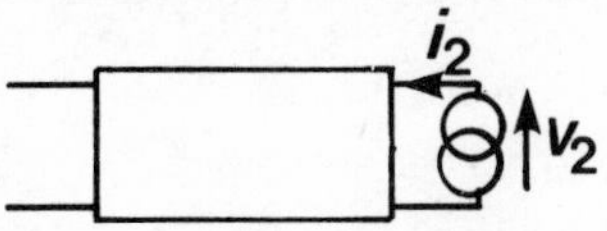 | o/c output admittance $\left(\frac{i_2}{v_2}\right)_{i_1=0}$ | $\frac{\Delta}{y_{11}}$ | $\frac{p_{oo}}{p_{os}}$ | $\frac{\Delta}{\Delta_{nn}}$ |
| 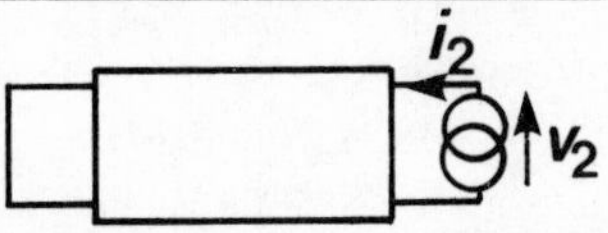 | s/c output admittance $\left(\frac{i_2}{v_2}\right)_{v_1=0}$ | $y_{22}$ | $\frac{p_{so}}{p_{ss}}$ | $\frac{\Delta_{11}}{\Delta_{2,n-1}^{2,n-1}}$ |

**Table 6.4** – Two-Port Parameters

| | 11 | 12 | 21 | 22 | $\Delta$ |
|---|---|---|---|---|---|
| A | $\frac{p_{so}}{q_{21}}$ | $\frac{p_{ss}}{q_{21}}$ | $\frac{p_{oo}}{q_{21}}$ | $\frac{p_{os}}{q_{21}}$ | $\frac{q_{12}}{q_{21}}$ |
| B | $\frac{p_{os}}{q_{12}}$ | $\frac{p_{ss}}{q_{12}}$ | $\frac{p_{oo}}{q_{12}}$ | $\frac{p_{so}}{q_{12}}$ | $\frac{q_{21}}{q_{12}}$ |
| G | $\frac{p_{oo}}{p_{so}}$ | $\frac{-q_{12}}{p_{so}}$ | $\frac{q_{21}}{p_{so}}$ | $\frac{p_{ss}}{p_{so}}$ | $\frac{p_{os}}{p_{so}}$ |
| H | $\frac{p_{ss}}{p_{os}}$ | $\frac{q_{12}}{p_{os}}$ | $\frac{-q_{21}}{p_{os}}$ | $\frac{p_{oo}}{p_{os}}$ | $\frac{p_{so}}{p_{os}}$ |
| Y | $\frac{p_{os}}{p_{ss}}$ | $\frac{-q_{12}}{p_{ss}}$ | $\frac{-q_{21}}{p_{ss}}$ | $\frac{p_{so}}{p_{ss}}$ | $\frac{p_{oo}}{p_{ss}}$ |
| Z | $\frac{p_{so}}{p_{oo}}$ | $\frac{q_{12}}{p_{oo}}$ | $\frac{q_{21}}{p_{oo}}$ | $\frac{p_{os}}{p_{oo}}$ | $\frac{p_{ss}}{p_{oo}}$ |

the polynomial manipulations corresponding to the wiring operators of Table 5.4 can be fully cancelled. Considerable reduction in the build-up of powers of $s$ and uncancelled factors can therefore be achieved.

### Nodal Admittance Matrix Network Functions

The same set of network functions can be obtained for a network described in terms of its nodal admittance matrix. Assume that the NAM is an $n \times n$ matrix with the inout port from node ① to reference and the output port from node ⓝ to reference. Then since $\mathbf{Y}'\,\mathbf{e}' = \mathbf{I}'$,

$$\begin{bmatrix} e'_1 \\ \vdots \\ e'_n \end{bmatrix} = \begin{bmatrix} \Delta_{11}/\Delta & \cdots & \Delta_{n1}/\Delta \\ \vdots & & \vdots \\ \Delta_{1n}/\Delta & \cdots & \Delta_{nn}/\Delta \end{bmatrix} \begin{bmatrix} I'_1 \\ \vdots \\ I'_n \end{bmatrix}.$$

If the intervening nodes are not used as inputs, $I_2' = I_3' = \ldots I_{n-1}' = 0$, and

$$\begin{bmatrix} e_1' \\ \\ e_n' \end{bmatrix} = \begin{bmatrix} \Delta_{11}/\Delta & \Delta_{n1}/\Delta \\ \\ \Delta_{1n}/\Delta & \Delta_{nn}/\Delta \end{bmatrix} \begin{bmatrix} I_1' \\ \\ I_n' \end{bmatrix} . \tag{6.16}$$

Thus the cofactors $\Delta_{11}, \Delta_{n1}, \Delta_{1n}, \Delta_{nn}$, the determinant $\Delta$ of $\mathbf{Y}'$ correspond closely to the polynomials $p_{so}$, $q_{12}$, $q_{21}$, $p_{os}$, and $p_{oo}$ respectively, differing by powers of $s$ and multiplicative factors.

The set of transfer and driving point functions are found by solution of Eq. (6.16) under the various open and short circuit port conditions. For example, $I_n' = 0$ for open circuit output network functions, and Eq. (6.16) reduces to

$$e_1' = \frac{\Delta_{11}}{\Delta} I_1' \tag{6.17}$$

$$e_n' = \frac{\Delta_{1n}}{\Delta} I_1' . \tag{6.18}$$

The open circuit transfer impedance equals $(v_2/i_1)_{i_2=0}$, where $v_2$ corresponds to $e_n'$, and from Eq. (6.18) we obtain the network function as $\Delta_{1n}/\Delta$.

Short circuit network functions are slightly more difficult since the condition $v_2 = e_n' = 0$ leads to the equations

$$e_1' = \frac{\Delta_{11}}{\Delta} I_1' + \frac{\Delta_{n1}}{\Delta} I_n' \tag{6.19}$$

$$0 = \frac{\Delta_{1n}}{\Delta} I_1' + \frac{\Delta_{nn}}{\Delta} I_n' . \tag{6.20}$$

Solution for the short circuit transfer admittance yields

$$\left(\frac{i_2}{v_1}\right)_{v_2=0} = \frac{-\Delta_{1n}\Delta}{\Delta_{11}\Delta_{nn} - \Delta_{1n}\Delta_{n1}} .$$

Cancellation of common factors is clearly a necessity. In order that we end up with an admittance, the denominator must contain the factor $\Delta$ and cancel to the determinant of an $(n-2) \times (n-2)$ matrix. This matrix can be empirically shown to be derived from $\mathbf{Y}$ by deleting rows 1 and $n$, and columns 1 and $n$. This is denoted by the symbol for a minor $M_{2,3,\ldots(n-1)}^{2,3,\ldots(n-1)}$ showing rows and columns present. The determinant $\Delta_{2,3,\ldots(n-1)}^{2,3,\ldots(n-1)}$ has a positive sign in this case. From the theory of determinants, therefore, we have the factorization

$$\Delta_{11}\Delta_{nn}-\Delta_{1n}\Delta_{n1} = \Delta\Delta_{2,\ldots(n-1)}^{2,\ldots(n-1)} .$$

The set of network functions derived from the nodal admittance matrix is entered in Table 6.3. Evaluating ths short circuit transfer admittance from $-q_{21}/p_{ss}$ yields

$$\frac{-(s^3CL_1L_2 + sL_1 + sL_2 + R)}{(s^3CL_1L_2R + sRL_1 + sRL_2)}$$

The same result can be obtained from $-\Delta_{13}/\Delta_2^2$ applied to the NAM in Eq. (6.8), noting that $\Delta_2^2 = sC + 1/sL_1 + 1/sL_2$.

## 6.2 *s*-DOMAIN ANALYSIS

Network functions in Table 6.3 involve evaluation of the determinants of the nodal admittance matrix **Y** and of sub-matrices of **Y**. Since solutions are required in the *s*-domain the elements of **Y** are either rational functions of *s* or rational functions of *s* and the symbolic component variables. The methods described in this section are applicable to both situations.

The available techniques for *s*-domain analysis are limited to much smaller networks than numerical NAM inversion at a specified frequency. Nevertheless analysis in the *s*-domain is important, as much can be learnt from network functions, for example, about active networks with feedback. Symbolic network analysis is also important as the first stage of symbolic sensitivity analysis which will be considered in Chapter 8.

Three different analysis methods are described, each of which can be successfully programmed on a digital computer. Topological methods based on the incidence matrix rely on the fact that the full network description is contained within the matrix and a component symbol list. Modifications cater for active networks. Methods based on the NAM rely on the cancellation of common terms and factors during inversion. Signal flow graph techniques also lead to symbolic network functions although computer programs based on this approach are considered less successful. However, signal flow graph methods are needed to reduce block diagrams thus allowing *s*-domain analysis of networks in stages.

### Topological Methods

Symbolic analysis by topological methods will be derived in this section in terms of the reduced incidence matrix. The relationship between nodal analysis methods in previous chapters and symbolic methods in this and the following section will then be clearly set out. Parallel development of the theory in terms of cut-set or tie-set matrices is also possible.

Recall from Chapter 2 the definition of the reduced incidence matrix as the matrix of elements equal to +1, −1, or 0 indicating in columns the branches connected to each node. Recall also that $n_N$, $n_B$, $n_T$, and $n_L$ are the numbers of nodes, branches, tree branches, and link branches in a network graph, where $n_T = n_N - 1$.

If we choose a tree in a graph and lable the tree branches 1, 2, . . ., $n_T$, and the link branches $n_T + 1 \ldots n_B$, we can partition the reduced incidence matrix **A** into two parts $\mathbf{A}_T$ and $\mathbf{A}_L$ where

$$\mathbf{A} = \begin{bmatrix} \mathbf{A}_T \\ \mathbf{A}_L \end{bmatrix} .$$

$\mathbf{A}_T$ is an $n_T \times n_T$ sub-matrix of **A** containing elements due to the tree branches. $\mathbf{A}_L$ is an $n_L \times n_T$ matrix of elements due to the link branches. Branch voltages **e** are given in terms of node voltages $\mathbf{e}'$ by Eq. (2.35) as

$$\mathbf{e} = \mathbf{A}\,\mathbf{e}' .$$

Partitioning **A** and similarly partitioning **e**,

$$\begin{bmatrix} \mathbf{A}_T \\ \mathbf{A}_L \end{bmatrix} \mathbf{e}' = \begin{bmatrix} \mathbf{e}_T \\ \mathbf{e}_L \end{bmatrix} . \tag{6.21}$$

The upper half of Eq. (6.21) relates tree branch voltages to node voltages by

$$\mathbf{A}_T \mathbf{e}' = \mathbf{e}_T . \tag{6.22}$$

Since both the tree branch voltages and the node voltages are sets of independent voltages for the network, $\mathbf{A}_T$ must have an inverse and

$$\mathbf{e}' = \mathbf{A}_T^{-1}\,\mathbf{e}_T . \tag{6.23}$$

We know by definition that the elements of **A** are ±1 or 0. The elements of the inverse $\mathbf{A}_T^{-1}$ are also ±1 or 0 since node voltages can be expressed as a simple combination of tree branch voltages. In fact the matrix $\mathbf{B}_T$ defined by $\mathbf{B}_T^t = \mathbf{A}_T^{-1}$ and known as the datum node matrix may be derived from the directed graph. Each column of $\mathbf{B}_T$ corresponds to a node and lists the tree branches in the path from the node to the reference node. Elements are +1 if the corresponding branch traversed is directed towards the reference, −1 if directed away from the reference, and 0 if the branch is not traversed.

Since the elements of $\mathbf{B_T}$ are derived from $\mathbf{A_T}$ as a ratio of a cofactor and the determinant of $\mathbf{A_T}$, then $\mathbf{A_T}$ and all non-zero cofactors of $\mathbf{A_T}$ must equal the same integer.Non-zero terms of $\mathbf{B_T}$ are then equal to $\pm 1$. Similarly the determinant and all non-zero cofactors of $\mathbf{B_T}$ are all equal to an integer. Both integers must be equal to either $+1$ or $-1$ since $|\mathbf{A_T}| = 1/|\mathbf{B_T}|$. We conclude that $|\mathbf{A_T}| = \pm 1$.

We noted in Chapter 2 that in general a network graph contains a number of different trees. By rearranging the branch numbering they could in turn be entered in $\mathbf{A_T}$, and $|\mathbf{A_T}|$ would then equal $\pm 1$. Any set of $n_T$ branches which do not constitute a tree if entered into $\mathbf{A_T}$ would yield the result $|\mathbf{A_T}| = 0$.

Thus the value of the determinant $|\mathbf{A_T}|$ will indicate whether a chosen set of branches form a tree. It is not necessary to renumber branches in order to perform this test. The set of $n_T$ rows of $\mathbf{A}$ corresponding to a chosen set of branches form an $n_T \times n_T$ sub-matrix of $\mathbf{A}$ called a **major** of $\mathbf{A}$. If its determinant equals $\pm 1$ the set of branches form a tree. If the determinant equals zero the branches do not form a tree.

*The Binet-Cauchy Theorem*

A further property of the incidence matrix is derived from the Binet-Cauchy theorem. Consider two matrixes $\mathbf{A}$ and $\mathbf{B}$ where $\mathbf{A}$ is $m \times n$ and $\mathbf{B}$ is $n \times m$ with $m<n$.

*The Binet-Caucy theorem states that* $|\mathbf{AB}| = \Sigma$ *(products of corresponding majors of* $\mathbf{A}$ *and* $\mathbf{B}$*).*

For example, if $\mathbf{A} = \begin{bmatrix} 2 & 1 & 7 & 2 \\ 1 & 2 & 1 & 0 \end{bmatrix}$ and $\mathbf{B}^t = \begin{bmatrix} 1 & 0 & 2 & 4 \\ 2 & 1 & 1 & 2 \end{bmatrix}$, then

$$\mathbf{AB} = \begin{bmatrix} 2 & 1 & 7 & 2 \\ 1 & 2 & 1 & 0 \end{bmatrix} \begin{bmatrix} 1 & 2 \\ 0 & 1 \\ 2 & 1 \\ 4 & 2 \end{bmatrix} = \begin{bmatrix} 24 & 16 \\ 3 & 5 \end{bmatrix},$$

and $|\mathbf{AB}| = 72$.

By the Binet-Cauchy theorem,

$$|\mathbf{AB}| = \begin{vmatrix} 2 & 1 \\ 1 & 2 \end{vmatrix} \begin{vmatrix} 1 & 0 \\ 2 & 1 \end{vmatrix} + \begin{vmatrix} 2 & 7 \\ 1 & 1 \end{vmatrix} \begin{vmatrix} 1 & 2 \\ 2 & 1 \end{vmatrix} + \begin{vmatrix} 2 & 2 \\ 1 & 0 \end{vmatrix} \begin{vmatrix} 1 & 4 \\ 2 & 2 \end{vmatrix} +$$

$$\begin{vmatrix} 1 & 7 \\ 2 & 1 \end{vmatrix}\begin{vmatrix} 0 & 2 \\ 1 & 1 \end{vmatrix} + \begin{vmatrix} 1 & 2 \\ 2 & 0 \end{vmatrix}\begin{vmatrix} 0 & 7 \\ 1 & 2 \end{vmatrix} + \begin{vmatrix} 7 & 2 \\ 1 & 0 \end{vmatrix}\begin{vmatrix} 2 & 4 \\ 1 & 2 \end{vmatrix}$$

$$= 3 + 15 + 12 + 26 + 16 + 0$$

$$= 72.$$

The results agree as expected. The example also illustrates how majors of **A** and the corresponding majors of **B** are selected in sequence.

If we now consider the reduced incidence matrix, the determinant $|\mathbf{A}^t\mathbf{A}|$ can be evaluated. Since majors in $\mathbf{A}^t$ and **A** correspond they have identical determinants. We have also established that the determinant of a major of **A** equals $\pm 1$ if the corresponding set of branches form a tree and 0 if not. Non-zero products of corresponding majors of $\mathbf{A}^t$ and **A** therefore correspond to trees of the network and equal +1.

*The sum of the products for all majors of* **A** *is therefore equal to the total number of trees in the network*

The Binet-Cauchy theorem can also be applied to the evaluation of the NAM determinant $|\mathbf{A}^t\mathbf{Y}\mathbf{A}|$ from

$$|\mathbf{A}^t\mathbf{Y}\mathbf{A}| = \Sigma\,(\text{products of corresponding majors of } \mathbf{A}^t\mathbf{Y} \text{ and } \mathbf{A}).$$

Consider the limited case where **Y** is a diagonal matrix containing no trans-admittance terms. Each column of $\mathbf{A}^t$ is multiplied by the corresponding admittance from **Y** to yield $\mathbf{A}^t\mathbf{Y}$. The value of the determinant of a major of $\mathbf{A}^t\mathbf{Y}$ therefore equals that of the same major of $\mathbf{A}^t$ multiplied by the product of the corresponding branch admittances. The product of corresponding majors of $\mathbf{A}^t\mathbf{Y}$ and **A** is equal to the product of the corresponding tree branch admittances provided that the branches form a tree and zero otherwise.

*The determinant of a network NAM is therefore equal to the sum for all trees of the products of tree branch admittances.*

*Cofactor Evaluation*

The network function is a ratio of two cofactors of $\mathbf{A}^t\mathbf{Y}\,\mathbf{A}$ or of one cofactor and the determinant. The cofactor $\Delta_{ij}$ of element $ij$ of the NAM is obtained by evaluating the determinant of $\mathbf{A}^t\mathbf{Y}\,\mathbf{A}$ after deleting row $i$ and column $j$. Therefore row $i$ must be deleted from $\mathbf{A}^t$ and column $j$ from **A**. Additionally the cofactor must be allocated the sign $(-1)^{i+j}$. The physical interpretation of deleting a row of $\mathbf{A}^t$ or a column of **A** is that the corresponding node in the network is connected to the reference node by a short circuit. Two circuits are therefore required for cofactor generation when $i \neq j$.

The effect on a typical network tree of shorting nodes ⓘ and ⓙ to the reference node, is illustrated in Fig. 6.5.

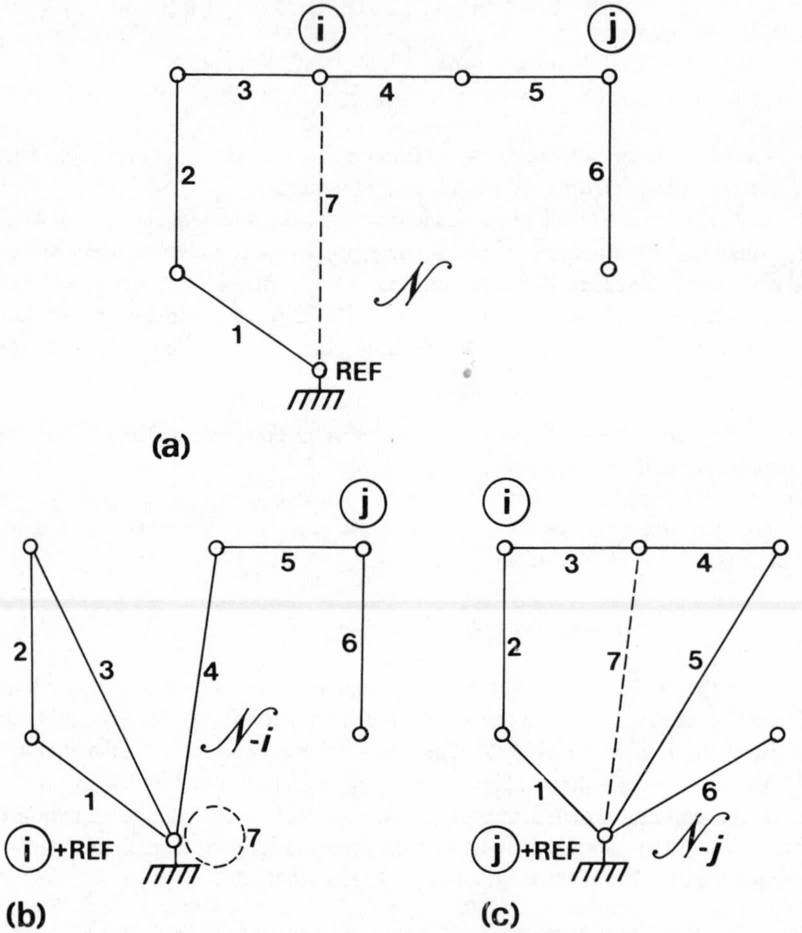


Fig. 6.5 – Trees for Cofactor Evaluation
(a) Original tree; (b) Node ⓘ shorted; (c) Node ⓙ shorted.

Denote the original network by $\mathcal{N}$, the network with node ⓘ short circuit to reference by $\mathcal{N}_{-i}$, and with node ⓙ short circuit to reference by $\mathcal{N}_{-j}$. All trees in $\mathcal{N}$ become trees of $\mathcal{N}_{-i}$ in Fig. 6.5(a) if any one of the branches in the created loop of branches 1,2, and 3 is removed. In $\mathcal{N}_{-j}$ a different loop is created consisting of branches 1,2,3,4, and 5, and any one of them can be removed to create trees in $\mathcal{N}_{-j}$. When branches 1, 2, or 3 are removed, trees are created in both networks, but when branches 4 or 5 are removed, trees are created only in network $\mathcal{N}_{-j}$.

Now in terms of matrices, let $\mathbf{A}_{-i}$ and $\mathbf{A}_{-j}$ be reduced incidence matrices of $\mathcal{N}_{-i}$ and $\mathcal{N}_{-j}$ respectively. If trees in $\mathcal{N}_{-i}$ and $\mathcal{N}_{-j}$ are common to both networks, corresponding majors of $\mathbf{A}_{-i}$ and $\mathbf{A}_{-j}$ will equal +1 or −1 and the product of the determinants of correponding majors of $\mathbf{A}^t_{-i}\mathbf{Y}$ and $\mathbf{A}_{-j}$ will equal +1 or −1 multiplied by the product of tree branch admittances. If the tree in one network is not a tree of the other network, one major will equal zero, and the product of the majors will also equal zero.

*Cofactor $\Delta_{ij}$ of the NAM of a network $\mathcal{N}$ is therefore equal to the sum for all trees common to networks $\mathcal{N}_{-i}$ and $\mathcal{N}_{-j}$ of the products of tree branch admittances multiplied by $(-1)^{i+j}\ |\mathbf{A}_{-i}||\mathbf{A}_{-j}|$.*

*Example* 6.1

Find the open-circuit transfer impedance of the bridged-T network from Fig. 6.1.

The network graphs of the original network $\mathcal{N}$, and the shorted node networks $\mathcal{N}_{-1}$ and $\mathcal{N}_{-3}$ were derived from Fig. 6.1. as shown in Fig. 6.6. From Table 6.1 the required transfer function is $\Delta_{13}/\Delta$.

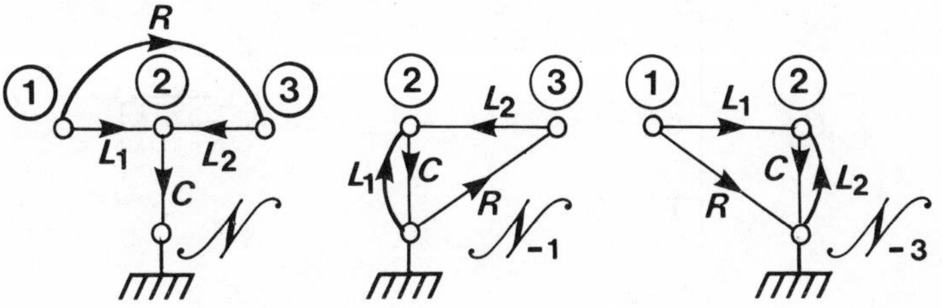


Fig. 6.6 – Network graph fro Example 6.1.

The reduced incidence matrices corresponding to $\mathcal{N}$, $\mathcal{N}_{-1}$, and $\mathcal{N}_{-3}$ are given by

$$\mathbf{A} = \begin{matrix} L_1 \\ L_2 \\ C \\ R \end{matrix}\begin{bmatrix} 1 & -1 & 0 \\ 0 & -1 & 1 \\ 0 & 1 & 0 \\ 1 & 0 & -1 \end{bmatrix}, \quad \mathbf{A}_{-1} = \begin{bmatrix} -1 & 0 \\ -1 & 1 \\ 1 & 0 \\ 0 & -1 \end{bmatrix}, \quad \mathbf{A}_{-3} = \begin{bmatrix} 1 & -1 \\ 0 & -1 \\ 0 & 1 \\ 1 & 0 \end{bmatrix}.$$

Results can now be tabulated.

*Determinant evaluation.*

| Branches | Major | \|Major\| | Graph $\mathcal{N}$ |
|---|---|---|---|
| $L_1,L_2,C$ | $\begin{bmatrix} 1 & -1 & 0 \\ 0 & -1 & 1 \\ 0 & 1 & 0 \end{bmatrix}$ | $-1$ | 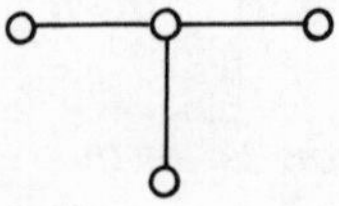 |
| $L_1,L_2,R$ | $\begin{bmatrix} 1 & -1 & 0 \\ 0 & -1 & 1 \\ 1 & 0 & -1 \end{bmatrix}$ | $0$ | 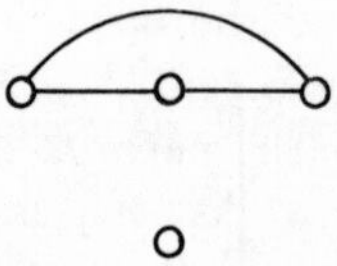 |
| $L_1,C,R$ | $\begin{bmatrix} 1 & -1 & 0 \\ 0 & 1 & 0 \\ 1 & 0 & -1 \end{bmatrix}$ | $-1$ | 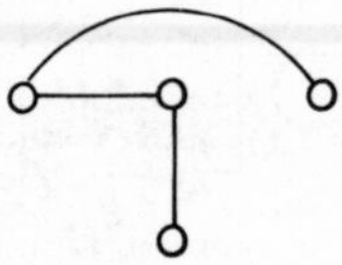 |
| $L_2,C,R$ | $\begin{bmatrix} 0 & -1 & 1 \\ 0 & 1 & 0 \\ 1 & 0 & -1 \end{bmatrix}$ | $-1$ | 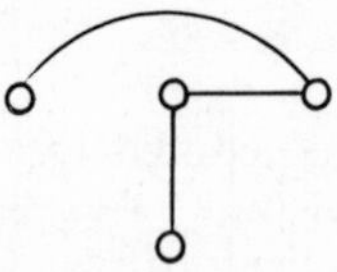 |

Therefore

$$|\mathbf{A}^t\mathbf{YA}| = \frac{sC}{s^2L_1L_2} + \frac{sC}{sL_1R} + \frac{sC}{sL_2R}$$

$$= (s^2CL_2 + s^2CL_1 + sCR)/s_2L_1L_2R \ .$$

*Cofactor $\Delta_{13}$ evaluation*

| Branches | $\mathbf{A}_{-1}$ Majors | $\mathbf{A}_{-3}$ Majors | $\mathbf{A}_{-1}$ \|Majors\| | \|Majors\| $\mathbf{A}_{-3}$ |
|---|---|---|---|---|
| $L_1,L_2$ | $\begin{bmatrix} -1 & 0 \\ -1 & 1 \end{bmatrix}$ | $\begin{bmatrix} 1 & -1 \\ 0 & -1 \end{bmatrix}$ | $-1$ | $-1$ |
| $L_1,C$ | $\begin{bmatrix} -1 & 0 \\ 1 & 0 \end{bmatrix}$ | $\begin{bmatrix} 1 & -1 \\ 0 & 1 \end{bmatrix}$ | $0$ | $1$ |
| $L_1,R$ | $\begin{bmatrix} -1 & 0 \\ 0 & -1 \end{bmatrix}$ | $\begin{bmatrix} 1 & -1 \\ 1 & 0 \end{bmatrix}$ | $1$ | $1$ |
| $L_2,C$ | $\begin{bmatrix} -1 & 1 \\ 1 & 0 \end{bmatrix}$ | $\begin{bmatrix} 0 & -1 \\ 0 & 1 \end{bmatrix}$ | $-1$ | $0$ |
| $L_2,R$ | $\begin{bmatrix} -1 & 1 \\ 0 & -1 \end{bmatrix}$ | $\begin{bmatrix} 0 & -1 \\ 1 & 0 \end{bmatrix}$ | $1$ | $1$ |
| $C,R$ | $\begin{bmatrix} 1 & 0 \\ 0 & -1 \end{bmatrix}$ | $\begin{bmatrix} 0 & 1 \\ 1 & 0 \end{bmatrix}$ | $-1$ | $-1$ |

$\mathcal{N}_{-1}$ Graphs $\mathcal{N}_{-3}$

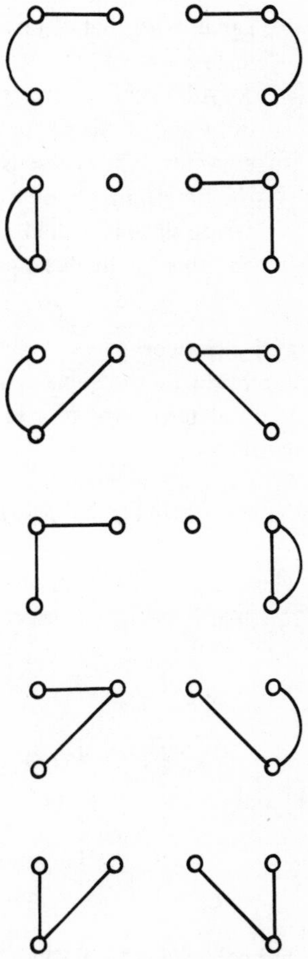

Therefore $\Delta_{13}$ of $\mathbf{A}^t\mathbf{Y}\mathbf{A}$ is given by

$$\Delta_{13} = \frac{1}{s^2 L_1 L_2} + \frac{1}{sL_1R} + \frac{1}{sL_2R} + \frac{sC}{R} ,$$

$$= (s^3 L_1 L_2 C + sL_1 + sL_2 + R)/s^2 L_1 L_2 R .$$

The transfer function,

$$\frac{\Delta_{13}}{\Delta} = \frac{s^3L_1L_2C + sL_1 + sL_2 + R}{s^2CL_1 + s^2CL_2 + sCR}$$

agrees with that obtained from two-port polynomial evaluation as $q_{21}/p_{oo}$ in Section 6.1.

All terms in the transfer function are positive because they contain only passive elements and because input and output ports have one node in common. Thus there is no necessity to evaluate the sign from the product of the determinants of two majors. This observation is helpful during computation since it is then necessary only to check majors for a zero determinant, and this can be done without evaluation.

**Active Networks**

Extension of the topological method to active networks is based on expansion of the determinant of a sum of two square matrices. If **P** and **Q** are two $n \times n$ matrices then

$$|\mathbf{P}+\mathbf{Q}| = R_0+R_1+R_2 \ldots +R_n$$

where $R_k$ is the sum of determinants obtained by replacing the columns (or rows) of **P** by the corresponding columns (or rows) of **Q**, $k$ at a time. Therefore,

$$R_0 = |\mathbf{P}| ,$$

$$R_1 = |[q_1p_2p_3 \ldots p_n]| + |[p_1q_2p_3 \ldots p_n]| + \ldots + |[p_1p_2 \ldots q_n]| ,$$

$$R_2 = |[q_1q_2p_3 \ldots p_n]| + |[q_1p_2q_3 \ldots p_n]| + \ldots + |[p_1p_2 \ldots q_{n-1}q_n]| ,$$

and continuing until finally

$$R_n = |\mathbf{Q}| ,$$

where $p_i$ and $q_j$ represent columns of **P** and **Q**.

Matrices **P** and **Q** may be identified with the modified branch admittance matrix $\mathbf{Y}_m$ of an active network. $\mathbf{Y}_m$ was first introduced in Section 2.3, and fully discussed in relation to many different active network elements. The diagonal matrix of passive element terms in $\mathbf{Y}_m$ are separated from terms due to dependent sources and mutual inductances and entered into the $n_B \times n_B$ matrix **P**. The remaining terms are entered into matrix **Q**. $\mathbf{Y}_m$ is therefore equal to the sum **P**+**Q** and the NAM may be put into the form $\mathbf{A}^t\mathbf{P}\mathbf{A} + \mathbf{A}^t\mathbf{Q}\mathbf{A}$.

Topological methods for evaluating the terms of the determinant and cofactors of this sum are based on expansion followed by use of the Binet-Cauchy theorem. As a result of a somewhat lengthy discussion (see *Further Reading*), the substitution of columns may be interpreted as the appearance of transfer admittances in groups, $k$ at a time for $k$=1,2 . . . $n_G$ where $n_G$ is the number of VCCS, as tree branch admittances across the voltage sending branches of the VCCS, and the substitution of VCCS current receiving branches for corresponding voltage sending branches. Trees which include voltage sending branches are accepted as active network trees only if they remain trees after substitution by current receiving branches. Purely passive element trees from $R_0$ are obtained as before, and terms added to the active tree terms.

Determinants and cofactors can therefore be evaluated for networks containing VCCS dependent sources by an active tree selection procedure which is additional to that previously described for passive component networks. Usually the number of VCCS is small in comparison with passive components, and in matrix form the test for active trees is quite straightforward. Topologically the procedure gets out of hand when the number of VCCS exceeds two or three. The procedure is illustrated in the following example.

*Example* 6.2

Find the open circuit transfer impedance of the bridged-T network from Fig. 6.1 with the addition of mutual inductance $M < \sqrt{(L_1 L_2)}$ between the two inductors.

The circuit together with an admittance model of the mutual inductor derived from Table 2.2 are shown in Fig. 6.7. The use of $M_1'$ and $M_2'$ is convenient since we need to distinguish the two coupling terms.

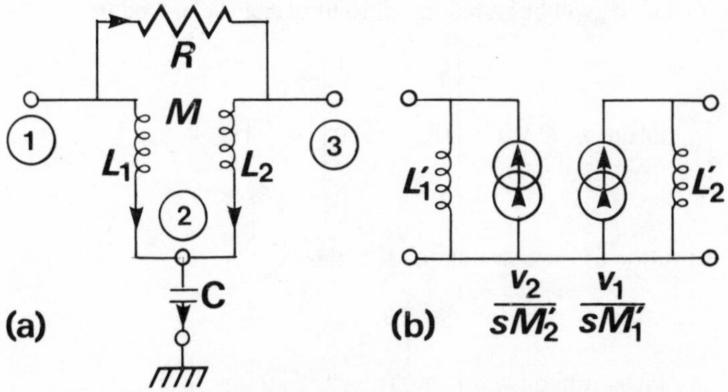


Fig 6.7 – (a) Circuit for Example 6.2; (b) Mutual inductor model.
$L_1' = L_1(1 - M^2/L_1L_2)$, $L_2' = L_2(1 - M^2/L_1L_2)$
$M_1' = M_2' = M(L_1L_2/M^2 - 1)$

All passive component trees were found in Example 6.1, and they apply to this example with the modification that $L_1'$ and $L_2'$ now replace $L_1$ and $L_2$. Here we will find the active element terms and add them to the earlier result. The solution required is $\Delta_{13}/\Delta$.

The reduced incidence matrix is augmented by the addition of rows labelled $M_1'$ and $M_2'$ which correspond to the voltage sending branches of the VCCS. Therefore,

$$\mathbf{A} = \begin{matrix} L_1' \\ L_2' \\ C \\ R \\ M_1' \\ M_2' \end{matrix} \begin{bmatrix} 1 & -1 & 0 \\ 0 & -1 & 1 \\ 0 & 1 & 0 \\ 1 & 0 & -1 \\ 1 & -1 & 0 \\ 0 & -1 & 1 \end{bmatrix}, \quad \mathbf{A}_{-1} = \begin{bmatrix} -1 & 0 \\ -1 & 1 \\ 1 & 0 \\ 0 & -1 \\ -1 & 0 \\ -1 & 1 \end{bmatrix}, \quad \mathbf{A}_{-3} = \begin{bmatrix} 1 & -1 \\ 0 & -1 \\ 0 & 1 \\ 1 & 0 \\ 1 & -1 \\ 0 & -1 \end{bmatrix}.$$

Also for substitution we will need rows corresponding to current receiving branches of the VCCS.

$$\begin{matrix} M_1' \\ M_2' \end{matrix} \begin{bmatrix} 0 & 1 & -1 \\ -1 & 1 & 0 \end{bmatrix} \qquad \begin{bmatrix} 1 & -1 \\ 1 & 0 \end{bmatrix}.$$

Results for trees containing active elements can now be tabulated, but owing to the number, only the branches will be listed together with the result of evaluating the determinants. Before tabulation the set of branches $L_2'$, $C$, and $M_1'$ will be tested in full to illustrate the procedure.

$$|\text{Major } L_2' \, C M_1'| = \begin{vmatrix} 0 & -1 & 1 \\ 0 & 1 & 0 \\ 1 & -1 & 0 \end{vmatrix} = -1$$

Substitution of the row due to $M_1'$ yields

$$|\text{Substituted major}| = \begin{vmatrix} 0 & -1 & 1 \\ 0 & 1 & 0 \\ 0 & 1 & -1 \end{vmatrix} = 0 .$$

Therefore the product of admittances $sC/s^2L_2'M_1'$ is not a term of the NAM determinant.

Results will now be tabulated, noting that branch sets containing $L_1'$ and $M_1'$ or $L_2'$ and $M_2'$ cannot be trees and they are not tested.

*Determinant Evaluation*

| Branches | $L_1'CM_2'$ | $L_1'RM_2'$ | $L_2'CM_1'$ | $L_2'RM_1'$ | $CRM_1'$ | $CRM_2'$ | $CM_1'M_2'$ | $RM_1'M_2'$ |
|---|---|---|---|---|---|---|---|---|
| $\lvert$Major of $\mathbf{A}\rvert$ | 1 | 0 | −1 | 0 | −1 | −1 | −1 | 0 |
| $\lvert$Substituted Major of $\mathbf{A}\rvert$ | 0 | 0 | 0 | 0 | 1 | 1 | 1 | 0 |

The additional terms in $\Delta$ due to the mutual admittance coupling are therefore given by

$$-\frac{sC}{sM_1'R}-\frac{sC}{sM_2'R}-\frac{sC}{s^2M_1'M_2'}\,.$$

Signs are obtained from the product of evaluated determinants.

*Cofactor $\Delta_{13}$ Evaluation*

| Branches | $L_1'M_2'$ | $L_2'M_1'$ | $CM_1'$ | $CM_2'$ | $RM_1'$ | $RM_2'$ | $M_1'M_2'$ |
|---|---|---|---|---|---|---|---|
| $\lvert$Substituted Major of $\mathbf{A}_{-1}\rvert$ | 0 | 0 | −1 | 0 | 1 | 1 | 1 |
| $\lvert$Major of $\mathbf{A}_{-3}\rvert$ | −1 | 1 | −1 | 0 | −1 | −1 | −1 |

The additional terms in $\Delta_{13}$ due to the coupling are therefore given by

$$\frac{sC}{sM_1'}-\frac{1}{sRM_1'}-\frac{1}{sRM_2'}-\frac{1}{s^2M_1'M_2'}\,.$$

The important feaures of topological analysis illustrated by this example are the addition of rows to the incidence matrix (which is for convenience when using matrix methods), use of the appropriate matrix together with row substitution, and the evaluation of sign. Results obtained can be checked against those directly derived from the circuit NAM. If this check is made by Cramer's rule the cancellation of terms involved, from 44 to 6 for $\Delta$, will show some of the difficulties which must be overcome by the direct NAM inversion method.

**Topological Analysis by Computer**

Realization of the topological analysis method as an efficient computer program is not particularly easy, and many of the available programs are slow and inconvenient to use. The objective of computation must first be established. Symbolic analysis is used to evaluate network functions in the $s$-domain. Much more importantly, it is then possible to differentiate with respect to all components and find sensitivities. Frequently there are hundreds of symbolic terms in a network function, and the information cannot be stored for evaluation. Even a printout is of little use. If we accept that transfer functions in the form of rational functions in $s$ are the main objective and that a symbolic printout is not necessary, the program can be made more efficient. However, it must be realized that any algorithm which separately enumerates the contribution of all tree admittance products to the network function, is inherently slow. Related methods which avoid this feature and also the problem of sign determination are available in the research literature, and are essential further reading in the study of topological analysis methods. With the knowledge that a straightforward computer program would be limited to the analysis of small circuits, the program described in Section 6.4 was written to demonstrate the attraction of the topological approach for transfer function analysis. The main advantage is that network functions are generated without the cancelling terms which predominate, and therefore cause numerical inaccuracy in other methods. As we shall see later in this section, signal flow graphs are particularly poor in this respect.

There are four major obstacles to the programming of an efficient algorithm.

*Network trees must be generated without duplication*

All network functions require tree admittance products for both numerator and denominator. Preferably one set of trees should be generated and then checked to see if it also provides a term of the other set. In some network functions both sets cannot be uniquely and simultaneously generated. Trees of $\mathcal{N}$, for example, are also trees of $\mathcal{N}_{-i}$ if they contain a branch from node ⓘ to the reference node. However, several trees of $\mathcal{N}$ may yield the same tree of $\mathcal{N}_{-i}$, and the set of trees for $\mathcal{N}_{-i}$ is not then uniquely determined. Network functions of the same order in numerator and denominator, for example voltage gain, avoid this problem.

*Branch sets must be tested rapidly*

The selection of branch sets as possible trees will in many cases result in a zero determinant when the appropriate major is tested. The evaluation of determinants by normal algorithms is too slow, and much faster methods must be utilized. It is worth recalling the special characteristic of majors of the incidence matrix that all elements and the determinant are ±1 or zero. This allows the determinant to be evaluated without any multiplications, and Gauss matrix reduction then requires $n^2 - n/2$ additions only.

If the sign of the determinant can be separately predicted, it becomes possible for simpler tests to be applied. For example, the presence of a loop, which produces a zero determinant, also implies separation of the branch set into two disconnected groups, provided that we can assume that all nodes in the network are included. It is then necessary only to check that all branches of the set are connected together. This type of test is efficient; the more so because it is fully carried out only in the minority of cases when a tree is present.

*Branch sets predictably forming a loop must not be generated*

The generation of branch sets is based on some form of cyclic selection. It is possible to avoid a large proportion, typically 30% to 90%, of possible selections by prediction and thus avoid time-wasting loop tests.

Consider the example of a network with 5 nodes plus reference node and 15 branches. The number of possible selections of 5 branches from 15 is ${}_5C_{15} = 3003$. Another method of making selections is to choose one branch from a list of those connected to each node. In this case there are a maximum of $5^5 = 3125$ selections assuming that all nodes join 5 branches. In general there are more possible selections by this method since the same branch could be chosen twice. However, networks do not usually have the same number of branches connected to each node. If this network were to have 7 connections to the reference node and 2, 6, 8, 4, and 3 connections to nodes ① to ⑤ respectively, the number of possible selections is reduced to 1152. Furthermore the avoidance of duplicate branch selection and parallel branch selection can easily reduce this figure by a factor of at least 2:1. Of the 500 or so remaining selections a high proportion will in fact be trees.

A tree generation algorithm based on selection of branches from a node-branch list is summarized by the following steps.

STEP 1 List the branches connected to each node, ensuring that the set of branches at the top of the list for each node is unique and does not contain parallel branches.

2 Cyclically step down the list from the previously selected branch set changing node Ⓝ most rapidly and node ① least rapidly.

3 If a duplicate or parallel branch is introduced, continue cyclic generation at the highest numbered node for that branch until a selection is obtained.

It is quite easy to introduce a parallel branch detector as well as duplicate branch detector in Step 3. This is based on a list of pairs of parallel branches and does not fully cater for the less common case of three or more parallel branches. Each test involves just one IF-GOTO statement since the organization of the list of Step 1 guarantees that the only possible duplicate or parallel branch is the one newly introduced. Adjustment of the cyclic operation at Step 3 entirely avoids further selection of branch sets containing that branch combination.

Consider the following branch-node list as an example of this procedure. The graph is easily drawn.

| Node | 1 | 2 | 3 | 4 | 5 | 6 |
|---|---|---|---|---|---|---|
| Branches | 1 | 2 | 8 | 3 | 5 | 9 |
| | 4 | 4 | 3 | 5 | 7 | 8 |
| | 7 | 6 | 1 | 6 | | |
| | 10 | 10 | | | | |

Tree generation proceeds as follows

| *Generated branch set* | *Action* | *Test Result* |
|---|---|---|
| 1 2 8 3 5 9 | Select | Tree |
| 1 2 8 3 5 8 | Reject | |
| 1 2 8 3 7 9 | Select | Tree |
| 1 2 8 3 7 8 | Reject | |
| 1 2 8 5 5 9 | Reject | |
| 1 2 8 5 7 9 | Select | Tree |
| 1 2 8 5 7 8 | Reject | |
| 1 2 8 6 5 9 | Select | Tree |
| 1 2 8 6 7 8 | Reject | |
| 1 2 3 3 5 9 | Reject | |
| 1 2 3 5 5 9 | Reject | |
| 1 2 3 5 7 9 | Select | Loop |

| *Generated branch set* | *Action* | | *Test Result* |
|---|---|---|---|
| 1 2 3 5 7 8 | | Select | Loop |
| 1 2 3 6 5 9 | | Select | Tree |
| 1 2 3 6 5 8 | | Select | Tree |
| 1 2 3 6 7 9 | | Select | Tree |
| 1 2 3 6 7 8 | | Select | Tree |
| 1 2 1 3 5 9 | Reject | | |
| 1 4 8 3 5 9 | | Select | Tree |
| . | | | |
| . | | | |
| . | | | |

This example contains ${}_6C_{10} = 210$ combinations of six branches selected from ten. Out of 576 possible selections from the branch-node list, 170 do not contain duplicate branches and 129 do not contain either duplicate or parallel branches. Of the 129 loop tests performed, 77 result in trees.

*Simple tests must be devised to simultaneously obtain both numerator and denominator.*

The tree generation and testing procedure is efficient, fast, and requires almost no storage. One further test is required to fully determine the voltage gain of a passive circuit.

If trees are generated for network $\mathcal{N}_{-i}$ in order to determine $\Delta_{ii}$, the trees in common with the network $\mathcal{N}_{-j}$ must also be determined in order to obtain $\Delta_{ij}$. Topologically the test may be states as follows:

Trees of $\mathcal{N}_{-i}$ are also trees of $\mathcal{N}_{-j}$ if node ⓙ is connected by tree branches to the reference node via node ⓘ.

All signs are positive for passive networks with one node in common to the input and output ports of the amplifier.

Similar succinct statement cannot be so easily made for active networks, although the topological situation can be understood and a method of sign determination established. Programming becomes involved, and an efficient topological test has not yet been devised to replace determinant evaluation for the active element trees. Recall that multiplications are not involved. Also evaluations need not be completed unless the factor appears in the network function.

Networks tested for evaluation of $\Delta$, $\Delta_{ii}$, and $\Delta_{ij}$ are summarized in the following table.

| | *Passive element terms* | *Active element terms* |
|---|---|---|
| $\Delta$ | $\mathcal{N}$ | $\mathcal{N}$(sub) and |
| $\Delta_{ii}$ | $\mathcal{N}_{-i}$ | $\mathcal{N}_{-i}$ (sub) and $\mathcal{N}_{-i}$ |
| $\Delta_{ij}$ | $\mathcal{N}_{-i}$ and $\mathcal{N}_{-j}$ | $\mathcal{N}_{-i}$ (sub) and $\mathcal{N}_{-j}$ . |

$\mathcal{N}$(sub) for example, indicates the substitution of current receiving branches in $\mathcal{N}$ for voltage sending branches of coupled branches. Treatment of the VCCS input and output as circuit branches in their own right considerably simplifies tree generation by separating the passive part of the circuit from the active. It also automatically ensures that the correct substitution of branches $k$ at a time for all required $k$, are all carried out.

**NAM Inversion Methods**

Symbolic and $s$-domain inversion of the nodal admittance matrix can be carried out with careful attention to common factor cancellation so that the transfer functions are obtained with the lowest possible polynomial order. The general solution for all node voltages by Gauss matrix reduction was first discussed in Section 4.1. Later in that section, network reduction was shown to be numerically equivalent, but cancellation of common terms in Gauss reduction was necessary to make the techniques symbolically equivalent. The general expression (4.10) for pivotal reduction was used in Chapter 4 to remove or suppress unwanted nodes from the NAM. Application of expression (4.10) to the evaluation of determinants required for the network functions in Table 6.3. will be closely examined for common factor cancellation.

We should perhaps first recall from Section 4.2. that the NAM of an $n$ node circuit with node ① as input and node ② as output can be converted to a 2X2 matrix of admittance parameters by node suppression. Diagonal elements $y_{n,n}, y_{n-1,n-1}, \ldots y_{3,3}$ are used in turn as pivots for application to Eq. (4.10). The use of node ② as output and the suppression of nodes ⓝ to ③ simplifies the suffixes without loss of generality. The reduced NAM is 2X2, and network functions can be evaluated from Table 6.3. as before. When the NAM determinant is required, pivoted reduction on element (2,2) will produce $\Delta$ in position (1,1) of the fully reduced NAM before final division by the pivot.

An element $y_{ij}$ in the NAM is changed every time a node is suppressed. Expression (4.10) may therefore be written in the iterative form,

$$y_{ij}^{(k)} = y_{ij}^{(k-1)} - y_{im}^{(k-1)}\, y_{mj}^{(k-1)} \Big/ y_{mm}^{(k-1)} \,, \tag{6.24}$$

where $(k)$ indicates the iteration and $m$ is the pivot used at iteration $k$. Reduction of an $n \times n$ NAM to $2 \times 2$ involves use of pivots $m=n, n-1, \ldots, 3$ for $k=1, 2, \ldots, m-2$. Various cancellations arising from NAM inversion by pivotal reduction can now be checked.

*Pivot cancellation*

Application of Eq. (6.24) at iteration $k=1$ yields,

$$y_{ij}^{(1)} = \frac{y_{ij}^{(0)} y_{nn}^{(0)} - y_{in}^{(0)} y_{nj}^{(0)}}{y_{nn}^{(0)}} . \tag{6.25}$$

Similarly at iteration $k=2$,

$$y_{ij}^{(2)} = \frac{y_{ij}^{(1)} y_{n-1\,n-1}^{(1)} - y_{i\,n-1}^{(1)} y_{n-1\,j}^{(1)}}{y_{n-1\,n-1}^{(1)}} . \tag{6.26}$$

Substitution from Eq. (6.25) into Eq. (6.26) is lengthy but worthwhile, giving finally,

$$y_{ij}^{(2)} = \frac{\left\{\begin{matrix} y_{ij}^{(0)} y_{nn}^{(0)} y_{n-1\,n-1}^{(0)} - y_{in}^{(0)} y_{nj}^{(0)} y_{n-1\,n-1}^{(0)} \\ - y_{ij}^{(0)} y_{n-1\,n}^{(0)} y_{n\,n-1}^{(0)} - y_{i\,n-1}^{(0)} y_{nn}^{(0)} y_{n-1\,j}^{(0)} \\ + y_{in}^{(0)} y_{n\,n-1}^{(0)} y_{n-1\,j}^{(0)} + y_{i\,n-1}^{(0)} y_{n-1\,n}^{(0)} y_{nj}^{(0)} \end{matrix}\right\}}{y_{n-1\,n-1}^{(1)} y_{nn}^{(0)}} .$$

The common denominator $y_{nn}^{(0)^2}$ of the numerator terms in Eq. (6.26) has cancelled once with the numerator. This becomes possible because terms not containing $y_{nn}^{(0)}$ cancelled out. Since similar equations apply for each iteration in turn, cancellation of the previous pivot will be exact.

*Denominator cancellation*

Consider now the rational function form of the NAM with $y_{ij}=N_{ij}/D_{ij}$

From Eq. (6.25),

$$\frac{N_{ij}^{(1)}}{D_{ij}^{(1)}} = \frac{N_{ij}^{(0)} N_{nn}^{(0)} D_{in}^{(0)} D_{nj}^{(0)} - N_{in}^{(0)} N_{nj}^{(0)} D_{ij}^{(0)} D_{nn}^{(0)}}{D_{in}^{(0)} D_{nj}^{(0)} D_{ij}^{(0)} D_{nn}^{(0)}} \Bigg/ \frac{N_{nn}^{(0)}}{D_{nn}^{(0)}}$$

$$= \frac{N_{ij}^{(0)} N_{nn}^{(0)} - N_{in}^{(0)} N_{nj}^{(0)} \dfrac{D_{ij}^{(0)} D_{nn}^{(0)}}{D_{in}^{(0)} D_{nj}^{(0)}}}{D_{ij}^{(0)} D_{nn}^{(0)}} \Bigg/ \frac{N_{nn}^{(0)}}{D_{nn}^{(0)}}$$

Since diagonal elements of the NAM include all the denominator factors from the same row and the same column, cancellation of $D_{ij}D_{nn}$ with $D_{in}D_{nj}$ will be exact. If $i \neq j$, both terms will be included in $D_{nn}$, and if $i=j$, one term will be included in each of $D_{nn}$ and $D_{ij}$. We may therefore write,

$$\frac{N_{ij}^{(1)}}{D_{ij}^{(1)}} = \frac{N_{ij}^{(1)}}{D_{ij}D_{nn}} \Bigg/ \frac{N_{nn}^{(0)}}{D_{nn}} . \tag{6.27}$$

At the second iteration, substituting into Eq. (6.26),

$$\frac{N_{ij}^{(2)}}{D_{ij}^{(2)}} = \frac{\dfrac{1}{N_{nn}^{(0)}}\left[N_{ij}^{(1)}N_{n-1\ n-1}^{(1)} - N_{in-1}^{(1)}N_{n-1\,j}^{(1)}\dfrac{D_{ij}D_{n-1\ n-1}}{D_{i\ n-1}D_{n-1\ j}}\right]}{D_{ij}D_{n-1\ n-1}D_{nn}} \Bigg/ \frac{N_{n-1\ n-1}^{(1)}}{D_{n-1\ n-1}} \tag{6.28}$$

We know from pivot cancellation that the division by $N_{nn}^{(0)}$ is exact. The denominator at any iteration $k$ is explicitly known as $D_{ij} \prod_{r=n-k+1}^{n} D_{rr}$. Therefore we only need to iterate the numerator making the cancellations indicated which are exact at each iteration.

Division by the pivot $N_{n-1\ n-1}^{(1)}/D_{n-1\ n-1}$ in Eq. (6.28) is not carried out at this stage since it cannot be fully cancelled until the next iteration.

One further cancellation of the denominator is required when determinants or cofactors are evaluated. The denominator in the form $D_{ij} \prod D_{rr}$ will contain several diagonal element denominators. Since any term not on the matrix diagonal is contained in two diagonal elements because of the method of filling the NAM, some terms of the denominator will be squared. These terms are easily listed and must cancel with the numerator.

*Diagonal element cancellations*

Diagonal elements of the reduced matrix found by this procedure contain common terms which must cancel before division by the previous pivot can be carried out. This is easily seen for the case of reciprocal networks where the NAM is symmetric. Thus all off-diagonal numerators are negative, all diagonal term numerators and all denominators are positive. Therefore, off-diagonal terms $N_{ij}\ N_{n-k+1\ n-k+1}$ at iteration $k$ are negative because the numerators are of opposite sign. Both terms in the numerator expression are then negative and no cancellation can take place. However, when $i=j$ for diagonal terms $N_{ij}N_{n-k+1\ n-k+1}$ is positive and opposite in sign to the second numerator term. Cancellation therefore takes place.

This effect was clearly demonstrate in the description of network reduction in Section 4.1 where the Gauss reduction was shown to require cancellation and star-to-delta transformation did not. The presence of cancellation leads to numerical inaccuracy and to difficulties with symbolic analysis. Removal of the cancellation uses the property that rows sum to zero in the INAM. Therefore the diagonal term in row $i$ can be replaced by

$$y_{ij} = - \sum_{j=1\, j \neq i}^{n} y_{ij} \ .$$

Then all terms involved in the diagonal are of the same sign, and cancellation is impossible. When the INAM is partially evaluated the upper limit of summation can be reduced from $n$ to $m$, where $m$ is the pivot in use, since terms with higher column numbers have already been reduced to zero. Use of the summation still requires only original denominators. However, slight modification to the basic method is required since the common denominator of the sum is not identical to that expected in Eq. (6.28).

Only the main points of term and factor cancellation have been described. Conversion of the method into an efficient computer program involves many more finer points of technique which will be found in *Further Reading*. Programs have been written for full matrix conversion so that sensitivity can be evaluated by the methods of Chapter 8.

*Example* 6.3

Find the NAM determinant of the bridged-T network of Fig. 6.1.

In order to conform with our notation for NAM inversion methods, node numbers ② and ③ of the circuit will be interchanged. The circuit is redrawn in Fig. 6.8.

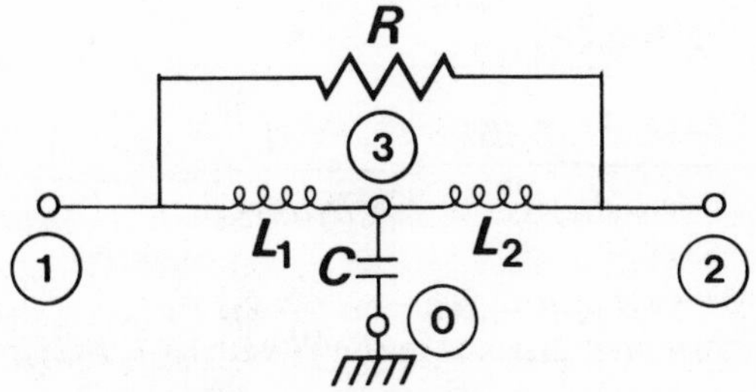


Fig. 6.8 – Bridged-T network with nodes renumbered.

The INAM of the circuit is given by

$$
\begin{array}{c} \\ 0 \\ \\ 1 \\ \\ 2 \\ \\ 3 \end{array}
\begin{array}{c}
\begin{array}{cccc} \quad 0 \quad & \quad 1 \quad & \quad 2 \quad & \quad 3 \quad \end{array} \\
\left[\begin{array}{cccc}
\dfrac{sC}{1} & \dfrac{0}{1} & \dfrac{0}{1} & \dfrac{-sC}{1} \\[2ex]
\dfrac{0}{1} & \dfrac{R+sL_1}{sL_1R} & \dfrac{-1}{R} & \dfrac{-1}{sL_1} \\[2ex]
\dfrac{0}{1} & \dfrac{-1}{R} & \dfrac{R+sL_2}{sL_2R} & \dfrac{-1}{sL_2} \\[2ex]
\dfrac{-sC}{1} & \dfrac{-1}{sL_1} & \dfrac{-1}{sL_2} & \dfrac{s^3CL_1L_2+sL_1+sL_2}{s^2L_1L_2}
\end{array}\right],
\end{array}
$$

where numerators and denominators have been clearly defined. Rows and columns are numbered from zero to conform to the node numbering.

Application to Eq. (6.27) with pivot $y_{33}^{(0)}$ yields, for $n=3, i=1, j=0$,

$$y_{10}^{(1)} = \frac{-sC\, s^2L_1L_2/sL_1}{s^2L_1L_2\, y_{33}^{(0)}} = \frac{-s^2CL_2}{s^2L_1L_2\, y_{33}^{(0)}}\,.$$

for $n=3, i=2, j=1$,

$$y_{21}^{(1)} = \frac{-(s^3CL_1L_2 + sL_1 + sL_2) - R\, s^2L_1L_2/sL_2sL_1}{R\, s^2L_1L_2\, y_{33}^{(0)}}\,,$$

$$= \frac{-(s^3CL_1L_2 + sL_1 + sL_2 + R)}{s^2L_1L_2R\, y_{33}^{(0)}}\,,$$

and for $n=3, i=2, j=0$,

$$y_{20}^{(1)} = \frac{-sC\, s^2L_1L_2/sL_2}{s^2L_1L_2\, y_{33}^{(0)}} = \frac{-s^2CL_1}{s^2L_1L_2\, y_{33}^{(0)}}\,.$$

Since the INAM is symmetric $y_{01} = y_{10}$ and $y_{12} = y_{21}$.

Direct calculation of diagonal term $y_{22}^{(1)}$ with $n=3, i=2, j=2$, gives,

$$y_{22}^{(1)} = \frac{(R+sL_2)(s^3CL_1L_2 + sL_1 + sL_2) - sL_2Rs^2L_1L_2/sL_2sL_2}{sL_2R\, s^2L_1L_2\, y_{33}^{(0)}}\,.$$

The term $sL_1R$ cancels in the numerator to yield

$$y_{22}^{(1)} = \frac{s^4CL_1L_2^2 + s^3CL_1L_2R + s^2L_1L_2 + s^2L_2^2 + sL_2R}{s^2L_1L_2\, sL_2R\, y_{33}^{(0)}} \; .$$

Calculation of $y_{22}^{(1)}$ as the negative of the sum of off-diagonal elements in the same row from

$$y_{22}^{(1)} = -y_{21}^{(1)} - y_{20}^{(1)}$$

gives
$$y_{22}^{(1)} = \frac{s^3CL_1L_2 + s^2CL_1R + sL_1 + sL_2 + R}{s^2L_1L_2R\, y_{33}^{(0)}} \; .$$

This expression cannot be directly used in Eq. (6.28) without either reinstating missing denominator factors or cancelling them also from the second numerator term of Eq. (6.28). Cancellation by the previous pivot is thereby maintained, and the need to cancel numerator terms in direct calculation of diagonal terms is avoided.

Finally for this pivot,

$$y_{11}^{(1)} = -y_{10}^{(1)} - y_{12}^{(1)} \; .$$

$$= \frac{(s^3cL_1L_2 + s^2CL_2R + sL_1 + sL_2 + R)sL_1}{s^2L_1L_2R\, y_{33}^{(0)}\, sL_1}$$

Since this is a reciprocal network $y_{12}^{(1)}$ has been replaced by $y_{21}^{(1)}$ in computing the diagonal element and, in general, matrix reduction can be similarly limited to matrix elements below the diagonal. For active networks the original expression must be used because symmetry is lost.

Application of Eq. (6.28) with pivot $y_{22}^{(1)}$ yields, for $n$=2, $i$=1, $j$=0

$$y_{10}^{(2)} = \frac{\dfrac{1}{s^3CL_1L_2+sL_1+sL_2}\begin{bmatrix}-sCL_2(s^4CL_1L_2^2+s^3CL_1L_2R+s^2L_1L_2+s^2L_2^2+sL_2R) \\ -(s^3CL_1L_2+sL_1+sL_2+R)\,(s^2CL_1)sL_2R/R\end{bmatrix}}{sL_2R\, s^2L_1L_2\, y_{22}^{(1)}}$$

The numerator of the previous pivot cancels exactly to yield

$$y_{11}^{(2)} = -y_{10}^{(2)} = \frac{s^3CL_1L_2+s^3CL_2^2+s^2CL_2R}{s^2L_1L_2RsL_2\, y_{22}^{(1)}} \; .$$

If the INAM were to be further reduced the denominator should be increased to $s^4L_1^2L_2^2R$, but the process is ended, so we may instead cancel any extra factors.

The determinant of the NAM is given by $y_{11}$ of the reduced matrix before division by the final pivot, that is,

$$\Delta = \frac{s^2CL_1 + s^2CL_2 + sCR}{s^2L_1L_2R},$$

The cancellations avoided by use of the equality $y_{11}^{(2)} = -y_{10}^{(2)}$ in place of direct calculation of $y_{11}^{(2)}$ from Eq. (6.28) were all terms of $(s^3CL_1L_2 + sL_1 + sL_2 + R)^2$ which occur with both positive and negative signs in the numerator.

## Signal Flow Graph Methods

Circuit and system analysis based on transfer functions is of most value when results from the analysis must be interpreted to gain insight into circuit behaviour. Circuits are treated as a connected group of non-interacting sub-circuits or blocks. The contribution of individual blocks can then be examined in isolation as well as in relation to the remainder of the circuit. Signal flow graph methods are used to describe a network consisting of interconnected blocks and to analyse the network to find its transfer function as a rational function in $s$. Simultaneous equations and electric circuits can also be put into signal flow graph form, and the analysis method can therefore be used to solve linear equations, for example the NAM equations, or to perform network analysis at the component level. Signal flow graph methods are therefore quite versatile and are frequently used in the analysis of small circuits. However, the method lacks the organization of equation generation by matrix methods and, as we shall see, it involves the cancellation of terms by subtraction. Analysis based on signal flow graphs cannot therefore be expected to perform as well as methods which avoid cancellation. Signal flow graphs are best regarded as suitable for reducing block diagrams. They are also useful for graphically showing signal flow in small circuits where the number of feedback loops is small and analysis relatively straightforward.

An inductor in the $s$-domain is governed by the equation, $e(s) = sLi(s)$. A signal flow graph representing this equation is shown in Fig. 6.9, where as usual $i$ and $e$ are understood to be functions of $s$.

Fig. 6.9 – Signal Flow Graph of $e = sLi$.

Variables $e$ and $i$ are each represented by a node, and the transmission function $sL$ is represented by a directed branch. The function of a node is to add signals which terminate there and to distribute signals through branches to other nodes. The function of a branch is to process the signal. In examples from circuit theory that we will consider, the branch will represent a transfer function as a rational function in $s$.

*Signal flow graph construction*

Signal flow graphs representing block diagrams of interconnected transfer functions functions are easily constructed, differing from block diagrams of feedback control systems only by the mode of signal addition. The example in Fig. 6.10 illustrates the main point of difference.

The summing points in a block diagram must be separated from any signal take-off points by an introduced unity gain branch. The branches from P to A and from B to C have both been introduced for this reason. The isolation of the feedback loop between nodes A and B is an important consequence of the extra branches. An input or **source** node is a node with only outgoing branches, and an output or **sink** node is a node with only incoming branches. Unity gain branches must often be added to the output variable node to make it into a sink node. In Fig. 6.10, R is a source node and C is a sink node.

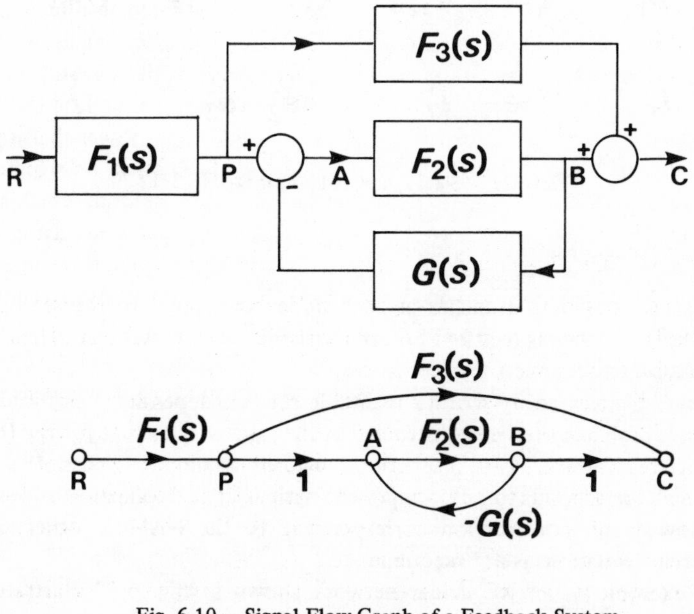


Fig. 6.10 – Signal Flow Graph of a Feedback System.

Signal flow graphs may be constructed from a set of simultaneous equations by rewriting them in a suitable form. For example the NAM equations for a three-node network given by $\mathbf{Y}'\mathbf{e}' = \mathbf{I}'$ may be rewritten in the form

$$e_1' = (I_1' - y_{12}e_2' - y_{13}e_3')/y_{11}$$

$$e_2' = (I_2' - y_{21}e_1' - y_{23}e_3')/y_{22}$$

$$e_3' = (I_3' - y_{31}e_1' - y_{32}e_2')/y_{33} \; . \quad (6.29)$$

The signal flow graph is shown in Fig. 6.11.

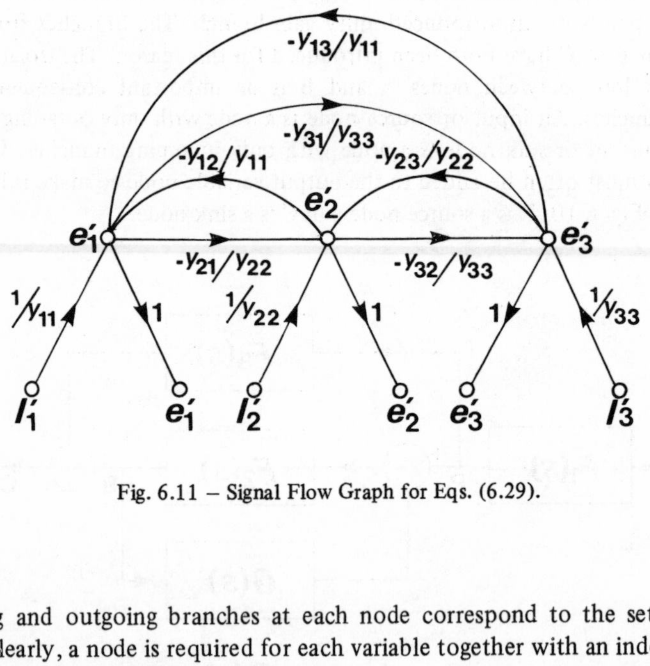


Fig. 6.11 – Signal Flow Graph for Eqs. (6.29).

Incoming and outgoing branches at each node correspond to the set of Eqs. (6.29). Clearly, a node is required for each variable together with an independent equation for all nodes except source nodes.

Circuits are treated by writing a suitable set of independent equations. The signal flow graph can then be constructed in the same way as that for Eq. (6.29). The objective, however, is to keep the equations as simple as possible so that each branch corresponds to one component with as little duplication as possible. Signal flow graph construction corresponding to the NAM or other similar sets of circuit equations is not recommended.

The example of an RC ladder network shown in Fig. 6.12 illustrates the technique.

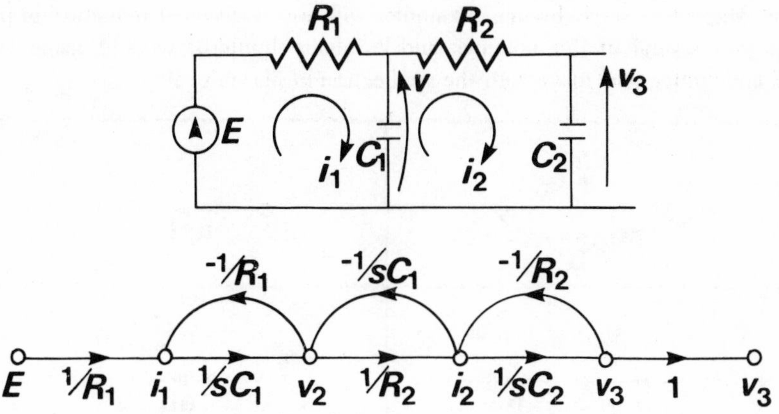


Fig. 6.12 – Signal Flow Graph of an RC Ladder Network.

The set of circuit equations used to construct the signal flow graph were

$$i_1 = \frac{1}{R_1} E - \frac{1}{R_1} v_2$$

$$v_2 = \frac{1}{sC} i_1 - \frac{1}{sC} i_2$$

$$i_2 = \frac{1}{R_2} v_2 - \frac{1}{R_2} v_3$$

$$v_3 = \frac{1}{sC} i_2 \ .$$

Another construction for a passive circuit is given in Example 6.4, and several are requested in the Problems.

*Transfer Function Evaluation*

Signal flow graphs may be reduced to a single branch between a source node and a sink node by the process of node removal, using the techniques of block diagram reduction from feedback and control system theory. Alternatively the transfer function can be written down by inspection, using Mason's rule.

Reduction methods based on branch combination and node removal are not usually necessary for the size of block diagram or circuit for which signal flow graphs are generally used. Therefore we will do no more than summarize the basic operations in Table 6.5 which are used to progressively reduce a signal flow graph to a single branch. Examples will not be given of reduction in practice, but several of the examples and Problems should be worked, using Table 6.5 and comparison made with the application of Mason's rule.

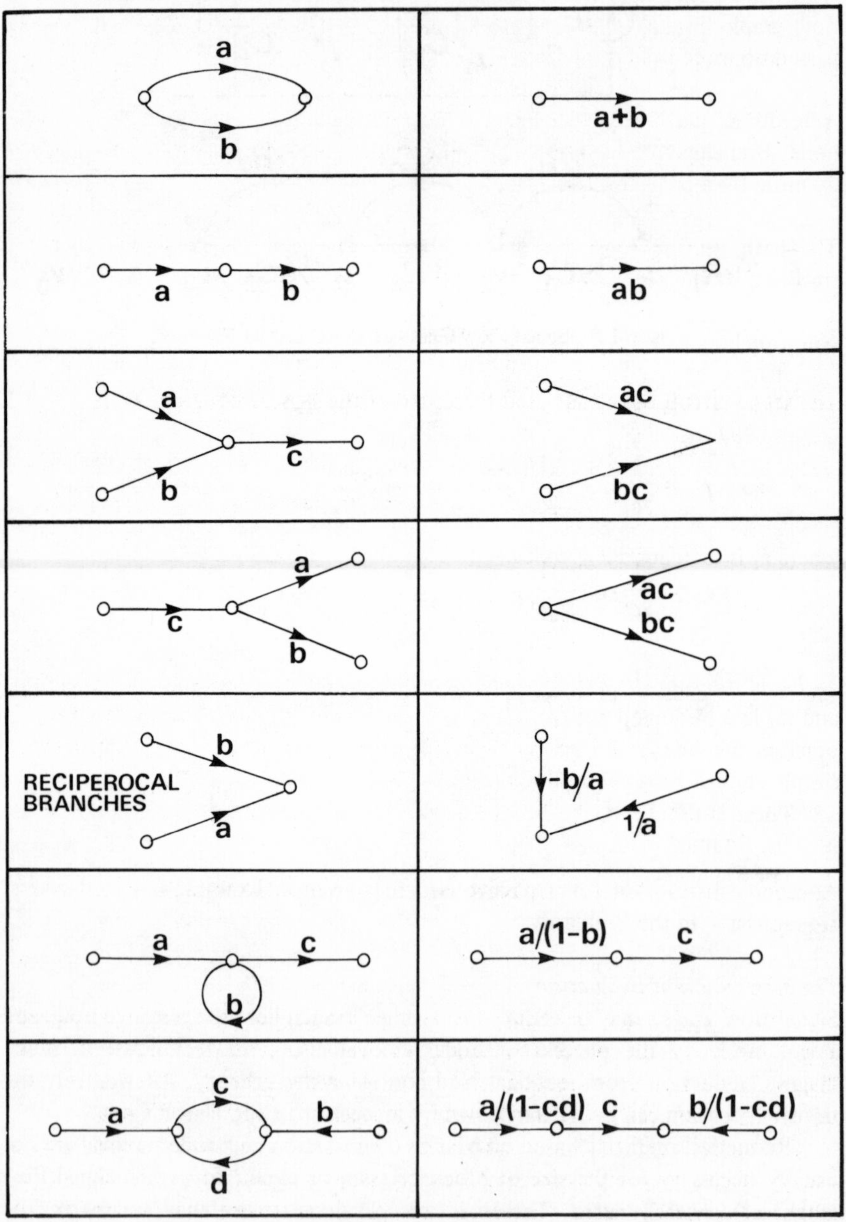


Table 6.5 – Signal Flow Graph Reduction Operations

Some of the terms used in the statement of Mason's rule must first be defined.

A **forward path** is any path from the input node to the output node of a signal flow graph. Branches are traversed in the forward direction, and any node is passed no more than once.

A **feedback path** or **loop** is any path originating and terminating at the same node. Branches are traversed in the forward direction, and any node is passed no more than once.

The **path gain** or **loop gain** is the product of transfer functions of branches traversed by the path.

**Non-touching** paths or loops do not share any common nodes.

Mason's rule states that the transfer function $T$ of a signal flow graph is given by

$$T = \Sigma_i P_i \Delta_i / \Delta ,$$

where $P_i$ is the path gain of forward path $i$,

$$\Delta = 1 - \Sigma_j P_{j1} + \Sigma_j P_{j2} - \Sigma_j P_{j3} + \ldots ,$$

$\Sigma_j P_{jk}$ is the sum of loop gain products of non-touching loops taken $k$ at a time, and $\Delta_i$ is $\Delta$ evaluated with all gains of loops touching $P_i$ set to zero. After a little practice the rule is quite easy to apply provided that the signal flow graph is simple enough for paths and loops to be listed.

The transfer functions for the signal flow graphs of Figs. 6.10 and 6.12 can now be obtained.

For Fig. 6.10,

Forward paths $P_1 = F_1(s)F_3(s), \ P_2 = F_1(s)F_2(s),$

Loops $P_{11} = -F_2(s)G(s)$, non-touching to $P_1$.

Therefore, $\Delta = 1 + F_2(s)G(s), \ \Delta_1 = 1 + F_2(s)G(s), \ \Delta_2 = 1,$

and $$T = \frac{C}{R} = \frac{F_1(s)F_3(s)\,(1+F_2(s)G(s)) + F_1(s)F_2(s)}{1 + F_2(s)G(s)} .$$

For Fig. 6.12,

$$\text{Forward path} \quad P_1 = 1/s^2C_1C_2R_1R_2 \;,$$

$$\text{Loops} \quad P_{11} = -1/s\,C_1R_1,\; P_{21} = -1/s\,C_1R_2,$$
$$P_{31} = -1/s\,C_2R_2 \;,$$

and since $P_{11}$ and $P_{31}$ are non-touching, $P_{12} = 1/s^2\,C_1C_2R_1R_2$.

Therefore, $\Delta = 1 + 1/sC_1R_1 + 1/sC_1R_2 + 1/sC_2R_2 + 1/s^2C_1C_2R_1R_2$ ,
$\Delta_1 = 1$,

and
$$T = \frac{v_3(s)}{E(s)} = \frac{1}{s^2C_1C_2R_1R_2 + sC_1R_1 + sC_2R_1 + sC_2R_2 + 1} \,.$$

The following example for the bridged-T network shows the increase in difficulty of locating all loops as networks become slightly more complicated. The solution should be compared to those of the previous Examples in this chapter from the viewpoint of the realization of the method as a computer program.

*Example* 6.4

Find the open circuit transfer impedance of the bridged-T network in Fig. 6.1 by signal flow graphs.

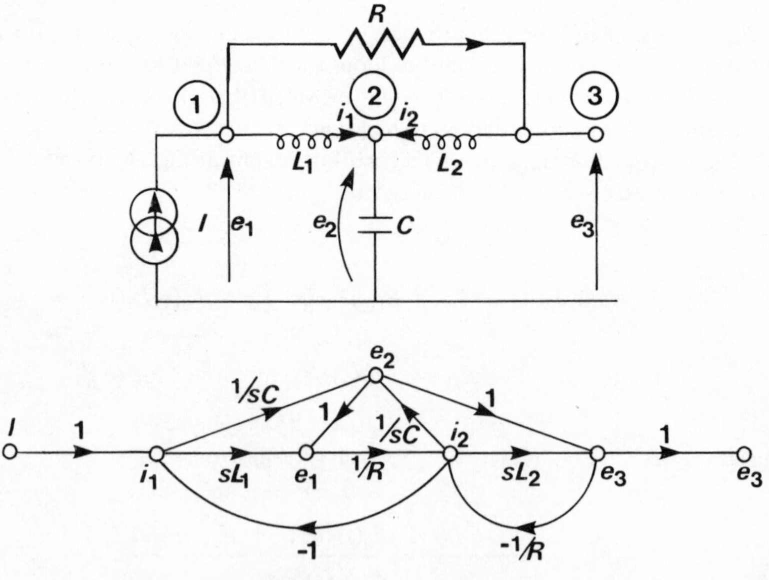


Fig. 6.13 – Bridged-T network for signal flow graph analysis.

The circuit is redrawn in Fig. 6.13 with additional variables for deriving the following set of circuit equations.

$$e_2 = \frac{1}{sC}(i_1 + i_2)$$

$$e_3 = e_2 + sL_2 i_2$$

$$i_2 = \frac{1}{R}(e_1 - e_3)$$

$$e_1 = e_2 + sL_1 i_1$$

$$i_1 = I - i_2 \, .$$

Forward path gains $P_1$ to $P_4$ are

$$\frac{1}{sC}, \frac{sL_1}{sCR}, \frac{s^2L_1L_2}{R}, \frac{1}{s^2L_2CR} \, .$$

Feedback loop gains $P_{11}$ to $P_{61}$ are

$$\frac{-sL_2}{R}, \frac{-1}{sCR}, \frac{1}{sCR}, \frac{-sL_1}{R}, \frac{-1}{sCR}, \frac{1}{sCR} \, .$$

Since all feedback loops touch each other and all forward paths, and all $1/sCR$ terms cancel,

$$\Delta = 1 + \frac{sL_2}{R} + \frac{sL_1}{R} \, ,$$

and $\Delta_1 = \Delta_2 = \Delta_3 = \Delta_4 = 1$ .

Therefore,

$$\Sigma_{i=1}^{4} P_i\Delta_i = (s^3CL_1L_2 + sL_1 + sL_2 + R)/sCR$$

$$\Delta = (s^2L_1C + s^2L_2C + sCR)/sCR$$

and $$T = \frac{e_3}{I} = \frac{s^3CL_1L_2 + sL_1 + sL_2 + R}{s^2L_1C + s^2L_2C + sCR} \, .$$

For comparison, this example will be worked from the NAM circuit equations derived from Eq. (6.8) and rewritten in a form suitable for construction of a signal flow graph as

$$e_1 = \frac{sL_1R}{R+sL_1}\left(\frac{1}{sL_1}e_2 + \frac{1}{R}e_3 + I_1\right) .$$

$$e_2 = \frac{s^2L_1L_2}{s^3CL_1L_2+sL_1+sL_2}\left(\frac{1}{sL_1}e_1 + \frac{1}{sL_2}e_3\right)$$

$$e_3 = \frac{sL_2R}{R+sL_2}\left(\frac{1}{R}e_1 + \frac{1}{sL_2}e_2\right) .$$

The signal flow graph, which is a simplified version of Fig. 6.11, is shown in Fig. 6.14.

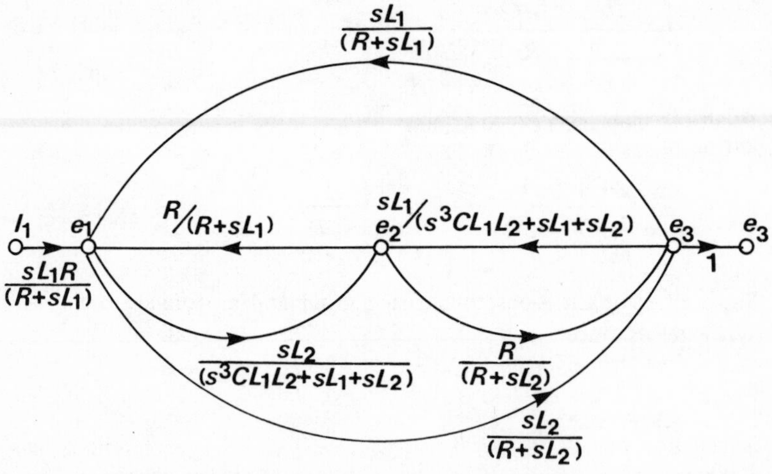


Fig. 6.14 – Signal flow graph for NAM equations of bridged-T network.

The forward path gains are,

$$P_1 = \frac{s^2L_1L_2R^2}{(R+sL_1)\,(R+sL_2)\,(s^3CL_1L_2+sL_1+sL_2)},$$

$$P_2 = \frac{s^2L_1L_2R}{(R+sL_1)\,(R+sL_2)} .$$

The feedback loop gains are,

$$P_{11} = \frac{s^2L_1L_2}{(R+sL_1)(R+sL_2)}$$

$$P_{21} = P_{31} = \frac{s^2L_2L_1R}{(R+sL_2)(R+sL_1)(s^3CL_1L_2+sL_1+sL_2)}$$

$$P_{41} = \frac{sL_2R}{(R+sL_1)(s^3CL_1L_2+sL_1+sL_2)}$$

$$P_{51} = \frac{sL_1R}{(R+sL_2)(s^3CL_1L_2+sL_1+sL_2)} \,.$$

All feedback loops touch each other and all forward paths, and $\Delta_1 = \Delta_2 = 1$. The transfer function is there given by

$$T = \frac{P_1 + P_2}{1-P_{11}-P_{21}-P_{31}-P_{41}-P_{51}} \,.$$

Putting all path gains over a common factor

$$T = \frac{s^2L_1L_2R(s^3CL_1L_2+sL_1+sL_2+R)}{\left\{\begin{array}{l}(R+sL_1)(R+sL_2)(s^3CL_1L_2+sL_1+sL_2) \quad - \\ s^2L_1L_2(s^3CL_1L_2+sL_1+sL_2) - 2s^2L_1L_2R \; - \\ sL_2R(R+sL_2) - sL_1R(R+sL_1)\end{array}\right\}}$$

Cancellation of all negative denominator terms with terms from expansion of the positive term brackets, and cancellation of $s^2L_1L_2R$ from numerator and denominator, will yield the transfer function obtained earlier in the example.

This cancellation of positive and negative terms, from 21 terms to 3 in this example using the NAM signal flow graph, and from 7 to 3 using carefully chosen circuit equations, is a feature of the signal flow graph method. The two previously described methods in this chapter which avoid cancellation are therefore preferred to circuit analysis. Signal flow graphs are, however, of considerable use in transfer function analysis for the reduction of block diagrams which arise when circuits are analysed as isolated sub-circuit blocks.

## 6.3 RESPONSE

The two main responses available from transfer function analysis have already been found. In Section 6.1 we proved by the derivation of Eq. (6.14) that the steady-state frequency response could be evaluated by making the substitution $s=j\omega$ into the network function. The resulting complex number is a phasor, and for various substituted values of frequency the response can be tabulated. More generally we proved that the transient response of a network could be obtained by taking the inverse Laplace transform of the product.

Network function $F(s) \times$ Laplace transform of the input.

Thus the output waveform can be evaluated for various input waveforms for a series of values of time and transient response tabulated.

Numerical complication arises in both response computations because network function from the analysis in this chapter consist of ratios of two unfactorized polynomials in $s$. Nevertheless, evaluation of frequency or transient response is relatively straightforward. Unfortunately, computer output in the form of tabulated numerical results is not very informative since the plethora of numbers diguises features of the response which are important to the designer. However, many different graphical presentations are available which very much ássist the interpretation of results in particular design situations. In this section we will list the most common graphical outputs relevant to frequency and transient response. The computer program should be able to supply any relevant graph on demand.

**Frequency Response**

The general form of the unfactorized rational function is usually obtained as

$$F(s) = \Sigma_{i=0}^{m} a_i s^i / \Sigma_{j=0}^{n} b_j s^j \tag{6.30}$$

Factorization yields

$$F(s) = K\Pi_{i=1}^{n_Z} (s-z_i)^{m_i} / \Pi_{j=1}^{n_P} (s-p_j)^{m_j} \ . \tag{6.31}$$

Since the coefficients in Eq. (6.30) are real, poles $p_j$ and zeroes $z_i$ in Eq. (6.31) are either real, or complex and occurring in complex pairs. $n_Z$ and $n_P$ are the total number of distinct zeroes and poles respectively, and the power $m$ is the multiplicity. The separation of real roots from complex roots and the recognition of equal roots after solution of the numerator and denominator polynomials is one of the more difficult aspects of rational function factorization, owing to numerical inaccuracy in locating roots. Fortunately the process is necessary only for the denominator polynomial in transient response evaluation.

Substitution of $s=j\omega$ in Eq. (6.30) yields a complex number $\alpha+j\beta$ representing a phasor applicable at the given frequency. The following responses are all easily evaluated.

| | | |
|---|---|---|
| Real part | $\alpha$ | |
| Imaginary part | $\beta$ | |
| Magnitude | $\sqrt{(\alpha^2+\beta^2)}$ | |
| Magnitude dB | $10 \log_{10}(\alpha^2+\beta^2)$ | |
| Phase $\theta°$ | $\tan^{-1}\beta/\alpha$ | $-180° < \theta \leqslant 180°$ . |

Several different graphical presentations of these frequency responses are in common use.

*Polar plots*

A polar plot in rectangular coordinates is a graph of $Im\{F(j\omega)\}$ versus $\text{Re}\{F(j\omega)\}$ for $-\infty < \omega > \infty$. The polar plot may also be generated in polar coordinates by plotting $|F(j\omega)|$ and $\angle F(j\omega)$. The graphs obtained are of course identical whichever set of coordinates are used.

Rectangular coordinates are usually used for plotting by computer, and the polar plot is constructed from tabulated values of real and imaginary ports of $F(j\omega)$ for values of $\omega$ covering the range of frequencies of interest. Negative frequencies are not plotted since it can easily be shown that the value of $F(-j\omega)$ is the complex conjugate of $F(j\omega)$. The section of the locus for $-\infty < \omega < 0$ is the mirror image about the real axis of the locus for $\infty > \omega > 0$.

The Nyguist plot is a polar plot of the open-loop transfer function of a feedback or control system. It is used in stability investigations because the open-loop gain can be stable and therefore possess a frequency response when the closed loop is unstable.

*Bode plots*

Bode plots consist of two separate graphs which are sometimes superimposed on a common frequency axis. The gain of a network function in dB given by $20 \log_{10}|F(j\omega)|$ is plotted against $\log_{10}\omega$. The phase in degrees is also plotted against $\log_{10}\omega$.

One minor difficulty arises since phase obtained from $\tan^{-1} \beta/\alpha$ is restricted to the range $-180° < \theta \leqslant 180°$ by the use of real and imaginary parts. Before plotting can commence segments of the phase response must be corrected by adding or subtracting increments of 360° to avoid discontinuities in the graph. Also the phase, usually at a high or low frequency, must be separately established in order to avoid an overall increment.

When the factorized form of the rational function is available, the contribution of individual pole and zero factors in Eq. (6.31) can be separately evaluated for both magnitude dB and phase. The total gain or phase is then obtained by addition, and phase is therefore without ambiguity. This separation of pole and zero contributions to gain and phase is also the basis of Bode plot sketching using the asymptotes and known response shapes for real and quadratic factors of the rational function.

*Nichols plot*

The Nichols plot is a graph of magnitude dB versus phase. It is often used in place of a Nyguist plot because the logarithmic magnitude scale satisfactorily covers a greater range than the linear Nyguist scale.

**Transient Response**

Transient response was determined in Section 6.1 by inverse Laplace transformation of the factorized form of the rational function. This involved expansion of the rational function into partial fractions. Since the numerator is merely evaluated during expansion it is necessary only to factorize the denominator polynomial when the rational function is in the more usual unfactorized form. Even so, the procedure is a little tedious and prone to numerical error both in solving for denominator roots and in solving for residues. When the rational function denominator is not factorized it is usually preferable to use the techniques of state-space analysis in Chapter 7 for evaluation of transient response.

In general, factorization of the network function denominator will yield multiple quadratic factors. Evaluation of partial fraction residues in the form required for inverse Laplace transformation using Table 6.1 is not quite straightforward. Most available techniques are not sufficiently general and treat multiple roots by modification of basic methods. However, it is possible to solve for all residues in one step using the solution of linear simultaneous equations by any standard method. The equations are set up by equating the coefficients of the rational function numerator with those obtained by putting the partial fractions back over a common denominator.

For example, expansion of $F(s)=s^2/\{(s+1)^2+2^2\}^2$ is needed in the form

$$F(s) = \frac{\alpha+j\beta}{(s+1-j2)} + \frac{\alpha-j\beta}{(s+1+j2)} + \frac{\gamma+j\delta}{(s+1-j2)^2} + \frac{\gamma-j\delta}{(s+1+j2)^2}$$

If the numerator is obtained by recombination of the partial fractions and equated to the required numerator $s^2$, we obtain

$$\begin{aligned} s^2 = {} & (\alpha+j\beta)(s+1-j2)(s+1+j2)^2 + (\alpha-j\beta)(s+1+j2)(s+1-j2)^2 + \\ & + (\gamma+j\delta)(s+1+j2)^2 + (\gamma-j\delta)(s+1-j2)^2 \end{aligned}$$

The coefficients of $s^3$, $s^2$, $s^1$, and $s^0$ are collected, and the resulting simultaneous equations may be put into matrix form as

$$\begin{matrix} s^3 \\ s^2 \\ s^1 \\ s^0 \end{matrix} \begin{bmatrix} 2 & 0 & 0 & 0 \\ 6 & -4 & 2 & 0 \\ 14 & -8 & 4 & -8 \\ 10 & -20 & -6 & -8 \end{bmatrix} \begin{bmatrix} \alpha \\ \beta \\ \gamma \\ \delta \end{bmatrix} = \begin{bmatrix} 0 \\ 1 \\ 0 \\ 0 \end{bmatrix}. \tag{6.32}$$

Solution yields $\alpha = 0, \beta = -5/32, \gamma = 6/32$, and $\delta = 8/32$.
Generation of the columns of the matrix in Eq. (6.32) is in fact quite straightforward in the general case. Since the matrix contains real coefficients and all residues are determined as real numbers, the equality and recognition of complex conjugate residue is guaranteed and the number of equations for solution is minimized.

Consider a real pole factor of $F(s)$ given by $(s-p_1)^m$. $F(s)$ may therefore be written as

$$F(s) = \frac{N(s)}{(s-p_1)^m D_1(s)},$$

where $D(s) = (s-p_1)^m D_1(s)$. Note that $D_1(s)$ may also contain factors $(s-p_1)$ since residues are required over all powers of $(s-p_1)$ up to the maximum. When separated into partial fractions

$$F(s) = \frac{\lambda_1}{(s-p_1)^m} + \frac{N_1(s)}{D_1(s)}.$$

The column of the coefficient matrix corresponding to $\lambda_1$ therefore contains the coefficients of the polynomial $D_1(s)$.

Now consider the quadratic factor case where

$$F(s) = \frac{N(s)}{\{(s-p_2)^2 + p_3^2\}^m D_2(s)}$$

$$= \frac{\lambda_2 + j\lambda_3}{(s-p_2-jp_3)^m} + \frac{\lambda_2 - j\lambda_3}{(s-p_2+jp_3)^m} + \frac{N_2(s)}{D_2(s)}.$$

The real and imaginary parts of the residues $\lambda_2$ and $\lambda_3$ are treated separately.

The column of the coefficient matrix corresponding to $\lambda_2$ contains the coefficients of

$$[(s-p_2+jp_3)^m + (s-p_2-jp_3)^m]D_2(s) \ ,$$

or more concisely

$$2D_2(s) \times \text{ Real Part of } \{s-p_2+jp_3\}^m \ .$$

Similarly the column of the coefficient matrix corresponding to $\lambda_3$ contains the coefficients of

$$-2D_2(s) \times \text{ Imaginary part of } \{s-p_2+jp_3\}^m \ .$$

Thus partial fraction expansion may be accomplished by solution of simultaneous equations once the coefficient matrix has been generated. Uncancelled factors common to numerator and denominator of the rational function have no effect on the solution and yield zero residues where appropriate. Evaluation of transient response using Table 6.1 involves substitution of poles and residues into one of the two expressions given, followed by summation of constituent transits.

Response to any input waveform can be found when the input is known in the form of its Laplace transform as a rational function in $s$. When inputs are tabulated functions of time, however, the output response must usually be obtained by convolution. Impulse and step responses are normally available from transient response programs and can usually be presented as graphs of response amplitude versus time.

## 6.4 COMPUTER PROGRAMMING

The use of transfer function coefficients as an intermediate set of parameters between circuit components and frequency response is numerically efficient and gives additional insight into circuit behaviour. However, the full symbolic listing of coefficients is only of limited use and owing to the vast number of terms for even quite small circuits. Computer programs must therefore be primarily intended for analysis yielding numerical functions of the Laplace variable $s$. Topological methods of transfer function analysis are attractive because they require very little involved programming, easily provide sensitivities to all circuit components, and do not produce cancelling terms.

### Transfer Function Analysis using TOPSEN

TOPSEN is a topological transfer function analysis program written in BASIC for on-line computation from time-shared terminals. A five-node, twenty-branch circuit will run in approximately 12K bytes of memory using Data General Basic 3.8 on the NOVA 1220 computer. The program listing is given in the Appendix to this chapter.

The data structure of this program is based on that of ACNLU from Chapter 4 and contains identical program statements and data storage for entry and modification of the circuit components. After reorganizing the connection data into a form suitable for tree generation the program calls subroutines in succession to carry out the various steps of the analysis process. Separate subroutines are provided for tree generation (3000), loop test (4000), active loop test (5000), transfer function and sensitivity update (7000), and frequency response evaluation (9000).

PROGRAM INPUTS

*Components – Statement numbers 160 to 440*

Components are entered in the form of a list of component types, values, and node connections. Resistors must be numbered from 1 to R9, capacitors from 1 to C9, inductors from 1 to L9, and VCCS from 1 to G9. Resistor values and node connections are entered into matrix S(R9+1,2). Inductors are similarly entered into matrix M(L9+1,2), and capacitors entered into matrix A(C9+1,2). VCCS have four node connection, value, and a type number, and are stored in matrix G(G9+1,5).

*Frequencies – Statement number 450 to 470*

Frequencies for transfer function evaluation are entered as F0,F1,F2 and F3, and analysis is carried out at F3 frequencies beginning at F0 Hz. Linear or logarithmic scales are generated from the expression $F_{n+1} = (F_n + F1)* F2$, and converted to rad/sec. using P2 $= 2\pi$.

*Output selection – Statement numbers 490 to 580*

The following outputs contained in M$(6) may be selected.

1. LT List trees.
2. TF Voltage Gain Transfer Function
3. SE Voltage Gain Transfer Function and sensitivities to all components.

These commands are decoded as integers 1 to 3 in T7 and control the calling of subroutines to provide the required output.

*Input and output nodes – Statement Numbers 600 to 620*

N1 is the input node and N2 is the output node. N0 is set equal to N1 and controls the generation of trees with the input node grounded.

## ORGANIZATION OF THE BRANCH-NODE CONNECTION DATA

*Filling the branch data matrix – Statement Numbers 650 to 1190*

All circuit components including the VCCS current generators are entered into matrix U(T9,6) where T9 is the total number of components. The separate component data arrays **S**, **M**, **A**, and **G** are not essential since matrix **U** could contain the same data. However, the convenience of programming for circuit modification from ACNLU was retained as the storage requirements are not too great.

Various component counts of use in the remainder of the program are

- T9 Total number of components (R9+L9+C9+G9)
- P9 Total number of passive components (R9+L9+C9)
- G6 Number of type 2 VCCS, i.e. GM/s.
- G7 Number of type 3 VCCS, i.e. sGM.

The columns of **U** are filled as follows.

Column 0 Component type: 1,2,3 for R,L,C,
and −3, −2, −1 for sGM, GM/s, and GM.
1 Component number.
2 Component first node or VCCS receive − node.
3 Component second node or VCCS receive + node.
4 Row number in **U** of a parallel branch to this branch.
5 Calculation space for transfer function update.
6 Component value.

*Filling the Node-Branch matrix – Statement Numbers 1200 to 1940*

Matrix V (T8,N), where N is the number of nodes, and T8 is 2+ Maximum number of branches connected to any one node, is used in tree generation. Each column from 0 to N lists the branches connected to the corresponding node. Row 0 is used as calculation space but normally carries the row number of **V** containing the branch number of a possible tree. Row T8–1 contains the total number of branches in the corresponding column, and row T8 is used for calculation space in the loop test subroutine.

Matrix **V** is initially filled from **U** after determination of T8. The order of branches in the columns of **V** is then rearranged so that row 1 contains a possible tree. Row 0 elements are than all set to unity except for column N0 set to zero. Matrices **U** and **V** are filled whenever component data is input or modified.

## OPERATION CONTROL

*Commands – Statements 2790 to 2920*

The following commands contained in N$(16) govern the action of the program when input data has been organized for analysis and also after each command has been executed.

1. ER End run.
2. EV Evaluate frequency response.

3. GO Analysis run.
4. NC Modify component list.
5. NN Modify input and output nodes.
6. NO Modify output required for analysis.
7. NF Enter frequencies for frequency response.
8. MP Print matrices **U**, **V**.

These commands are decoded as integers 1 to 8 in I6 which controls program operation.

ANALYSIS AND OUTPUTS

*Tree generation – Subroutine 3000*

Statements 3010 to 3180 generate branch sets forming possible trees by the cyclic method described in Section 6.2. Statements 3110 and 3120 find the other node of a newly introduced branch. Statment 3150 directly checks if the branch is parallel to the one already listed, and 3160 checks if the branch is already listed at its other node. Matrix **V** is specifically defined so that these checks do not involve a search. Statment 3170 alters the cyclic generation to omit all intervening branch sets in which the parallel or duplicate branch would have been included.

Statments 3190 to 3320 test the branch set, first calling subroutine 4000 to test for loops. Branch sets involving active elements are also tested by subroutine 5000. The value of V(0,0) indicates the status of the tree as follows.

V(0,0) = 1 No trees left to generate.
2 $\mathcal{N}_{-N1}$ passive tree.
3 $\mathcal{N}_{-N1}$ and $\mathcal{N}_{-N2}$ passive common tree.
4 $\mathcal{N}_{-N1}$ (sub) and $\mathcal{N}_{-N1}$ active common tree.
5 $\mathcal{N}_{-N1}$ (sub) and $\mathcal{N}_{-N2}$ active common tree.
6 Conditions 4 and 5 simultaneously.

P1 and Q1 then contain the sign of the denominator term and numerator term respectively for updating the transfer function. T5 counts the number of branch sets generated.

This subroutine also operates with NO=0 and generates trees of $\mathcal{N}$. Outputs with V(0,0) equal to 2 or 4 are valid. Other possibilities cannot be utilized since trees of $\mathcal{N}_{-N2}$, for example, are produced with duplication in some networks. TOPSEN is limited for this reason to voltage gain although some of the subroutines are more general.

*Loop test – Subroutine 4000*

This subroutine is used for passive branch sets since it does not generate the sign of the term involved. If NO=N1 the network $\mathcal{N}_{-N1}$ is tested, and if NO=0 the network $\mathcal{N}$ is tested. The subroutine returns V(0,0) equal to 1 if a loop is detected, 2 for a tree to $\mathcal{N}_{-N1}$, and 3 for a common tree to $\mathcal{N}_{-N1}$ and $\mathcal{N}_{-N2}$.

Statements 4020 to 4150 select the first branch, enter its nodes into array W(N+1), and note the branch is connected by entering 1 into the appropriate column of row T8 in **V**. Branches are successively attached to the tree while checking that a loop is not created. When all possible branches have been connected, statements 4370 to 4390 check to see if the full set of N or N–1 have been connected. Statements 4410 to 4680 check for a common tree.

T4 counts the number of loop tests. This subroutine is in fact also used as a pre-test for the active branch sets of $\mathcal{N}_{-N1}$(sub.) Subroutine 5000 is then only used when there is a chance that an active tree exists.

*Active loop test – Subroutine 5000*

Statements 5040 to 5160 fill matrix B(N,N) with rows of the incidence matrix corresponding to the branch set being tested. On first pass through with C8=1, current receiving node numbers of VCCS are used, and on subsequent passes, voltage sending node numbers are used. Statements 5170 to 5240 replace row N1 of **B** with row N2 for the test network $\mathcal{N}_{-N2}$ so that both row and column N2 can be deleted for evaluation of $\Delta_{N1N2}$. Statements 5250 to 5550 evaluate the determinant of **B** with row and column N0 equal to N1 or N2 deleted. Gaussian reduction is used with column interchange. P1 and Q1 return the sign of the product of determinants and V(0,0) returns 1 if no tree is detected or 4, 5, or 6 as previously noted. T6 counts the number of active loop tests. If N0=0, the subroutine tests network $\mathcal{N}$(sub.) and $\mathcal{N}$.

This subroutine must be regarded as a provisional version since it is slow and has not been exhaustively tested. It does, however, work on all networks that have been analysed so far. The problem of speed is best illustrated by analysis of the circuit from Example 4.3, Fig. 4.23. This circuit contains 202 passive trees and only 6 active trees. Program TOPSEN performed 214 passive tree tests and 197 active tree tests. A considerable amount of redundant testing is clearly carried out by this subroutine, whereas passive tree generation is remarkably efficient. Improvements to efficiency are currently being incorporated. These involve topological tests to rapidly detect the more commonly occurring causes of failure to generate an active tree (for example isolation of a node when changing from $\mathcal{N}_{-N1}$ (sub) to $\mathcal{N}_{-N1}$). The program at present takes approximately 15 minutes of terminal time on a busy time-shared NOVA 1220 system to obtain the transfer function for this circuit.

*Transfer function update – Subroutine 7000*

Statements 7010 to 7070 generate the set of components identified by unity in column 5 of matrix **U**, which normalize the tree admittance products. S1 contains the power of $s$ due to normalizing. Statements 7090 to 7110 add or remove from the list, components due to the tree. The normalizing factor is given by R1 × R2 × ..... × L1 × ..... × $s^{L9+G6}$. Tree resistors and inductors must therefore be deleted and tree capacitors and $g_m$ added to obtain the normalized transfer function term cleared of denominator factors. Statements 7120 to 7190 evaluate and add this function to the numerator and denominator of the voltage gain transfer function stored as D(2,S3), where S3 = C9+L9+G6+G7 and is the maximum power of $s$. Row 1 of **D** contains the numerator and row 2 the denominator coefficients.

Statements 7230 to 7310 update the sensitivities of the numerator and denominator to all components. Matrix E(T9,S3) contains the numerator sensitivities in rows to each of the T9 components. Denominator sensitivities are similarly stored in F(T9,S3). Since numerator and denominator symbolic terms are linear in components, each differentiation merely involves deletion of each component in turn from the factor added to the transfer function.

*Transfer function evaluation – Subroutine 9000*

Statements 9010 to 9240 directly evaluate and print the frequency response from the transfer function stored in D(2,S3). Statements 9260 to 9390 similarly and separately evaluate the numerator and denominator sensitivities ∂(Num)/∂(Cpt). and ∂(Denom)/∂(Cpt), as the complex numbers S1 +j T1 and S2 +j T2 respectively. Statements 9400 to 9460 process these separate sensitivities to produce gain, phase, real part, and imaginary part sensitivities. Sensitivity is fully treated in Chapter 8. However, the part of this program concerned with sensitivity is small and not easily separated from the test of the program. A more detailed description is not therefore given in Chapter 8, but the program is broadly similar to the sensitivity sections of ACNLU.

*Outputs*

Statements 2040 to 2780 organize the calling of subroutines using the controls T6 and T7 which are derived from the OUTPUT and NEXT commands. Statements 2090 to 2170 initialize matrices containing the transfer function and sensitivities. Statements 2190 to 2420 call subroutine 3000 to generate trees, and lists trees as sets of branch numbers. If the transfer function is required, tree listing is omitted and statements 2450 to 2470 call subroutine 7000 to update the transfer function and sensitivities. Statments 2490 to 2640 print the voltage gain transfer function and sensitivities. Statements 2670 to 2770 call subroutine 9000 to evaluate the frequency response and sensitivities. Statements 1960 to 2020 print matrices **U** and **V**.

*Other program variables*

| | |
|---|---|
| Control variables | T3, I6, I7 |
| Calculation space | N$(2), K$(28), J$(8), C8, D1, D2, F4, F5, G8, L5, L8, I1, I2, I9, J1, J2, P3, R5, R8, S1, S2, S9, T1, T2, V1, V2, W(N+1), B(N,N). |

**Example 6.5**

List the trees, determine the transfer function, and evaluate the frequency response of the circuit in Fig. 6.15.

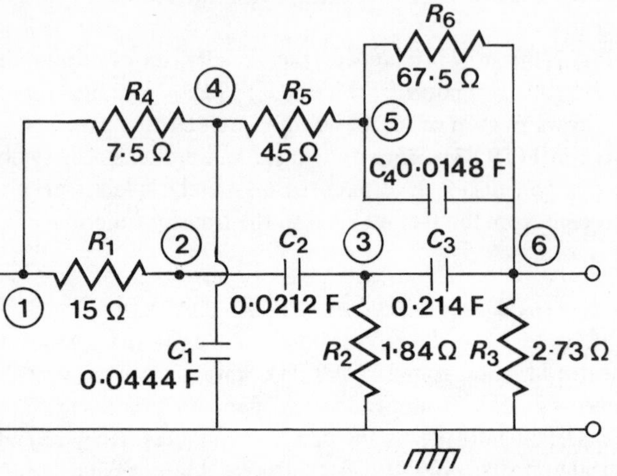


Fig. 6.15 – Circuit for Example 6.5.

This filter is derived from a Guillemin parallel ladder synthesis for prescribed driving point and transfer functions. The voltage gain is given by

$$F(s) = \frac{1}{15}\left(\frac{s^2 + s + 1}{s^2 + 4s + 3}\right)$$

for exact component values. Those used have been rounded to three significant figures, and inaccuracy is about 0.1%. The computer run reproduced in Fig. 6.16 was obtained using TOPSEN. Considerably more output is available for this example, including all transfer function sensitivities in the form ∂(Numerator)/∂(Component) and ∂(Denominator)/∂(Component), and frequency response sensitivities in the same form as those obtained from ACNLU (see Example 4.3 continued, in Section 8.6).

The cyclic tree generation method produced 84 branch sets which were then tested by the loop test subroutine. All were in fact trees. This compares very well with the number of possible selections of 5 branches from 10, that is, ${}_5C_{10} = 252$, and with the selection of one branch at each node except the reference node, that is $2 \times 3 \times 3 \times 3 \times 4 = 216$. Common trees of $\mathcal{N}_{-1}$ and $\mathcal{N}_{-6}$ contributed 18 numerator terms.

The transfer function obtained is fourth order in both numerator and denominator and must contain a common factor which numerically cancels when component values are exact. Voltage gain was evaluated from 0 to

1 Hz using command EV. These values agree fairly well against those obtained from programs 2PORT and ACNLU, and also against those directly computed from the exact transfer function.

```
TOPOLOGICAL TRANSFER FUNCTION ANALYSIS
RLC INPUTS: COMPONENT NUMBER,VALUE,2 NODE CONNECTIONS
GM  INPUTS: GM NUMBER,VALUE,TYPE,4 NODE CONNECTIONS
      TYPE: 1 FOR GM, 2 FOR GM/S, 3 FOR SGM
     NODES: SEND +, - , RECEIVE +, -
NO.NODES         ? 6
NO. R,L,C,GM     ? 6,0,4,0
R ? 1,15,1,2
R ? 2,1.84,3,0
R ? 3,2.73,6,0
R ? 4,7.5,1,4
R ? 5,45,4,5
R ? 6,67.5,5,6
C ? 1,.0444,4,0
C ? 2,.0212,2,3
C ? 3,.214,3,6
C ? 4,.0148,5,6
OUTPUTS: LT LIST TREES, TF VOLTAGE GAIN F(S)
       : SE VOLTAGE GAIN AND SENSITIVITIES
OUTPUT ? LT
INPUT NODE,OUTPUT NODE          ? 1,6
NEXT COMMAND: ER END RUN        EV EVALUATE RESPONSE
MODIFICATION: NC COMPONENTS     NF NEW FREQUENCIES
            : NN IN/OUT NODES   NO OUTPUTS
      OUTPUT: MP MATRICES U V   GO ANALYSE
NEXT ? GO
BRANCH - COMPONENT LIST
 1   1/R 1      2   1/R 2      3   1/R 3      4   1/R 4      5   1/R 5
 6   1/R 6      7    SC 1      8    SC 2      9    SC 3      10   SC 4

 D             8  2  7  5  3
 D             8  2  7  5  6
 D             8  2  7  5  9
 D             8  2  7  5  10
 D             8  2  7  6  3
 D             8  2  7  6  9
 D             8  2  7  10  3
 D             8  2  7  10  9
 D             8  2  5  6  3
 D             8  2  5  6  9
 D             8  2  5  10  3
 D             8  2  5  10  9
 D             8  2  4  5  3
 N D           8  2  4  5  6
 D             8  2  4  5  9
 N D           8  2  4  5  10
 D             8  2  4  6  3
 D             8  2  4  6  9
 D             8  2  4  10  3
 D             8  2  4  10  9
 D             8  9  7  5  3
 D             8  9  7  5  6
 D             8  9  7  5  10
 D             8  9  7  6  3
 D             8  9  7  10  3
 D             8  9  5  6  3
 D             8  9  5  10  3
 D             8  9  4  5  3
 N D           8  9  4  5  6
 N D           8  9  4  5  10
 D             8  9  4  6  3
 D             8  9  4  10  3
```

Fig. 6.16 – TOPSEN computer run for Example 6.5.

Fig. 6.16 (*continued*)

```
   D              1  2  7  5  3
   D              1  2  7  5  6
   D              1  2  7  5  9
   D              1  2  7  5  10
   D              1  2  7  6  3
   D              1  2  7  6  9
   D              1  2  7  10  3
   D              1  2  7  10  9
   D              1  2  5  6  3
   D              1  2  5  6  9
   D              1  2  5  10  3
   D              1  2  5  10  9
   D              1  2  4  5  3
   N D            1  2  4  5  6
   D              1  2  4  5  9
   N D            1  2  4  5  10
   D              1  2  4  6  3
   D              1  2  4  6  9
   D              1  2  4  10  3
   D              1  2  4  10  9
   D              1  8  7  5  3
   D              1  8  7  5  6
   N D            1  8  7  5  9
   D              1  8  7  5  10
   D              1  8  7  6  3
   N D            1  8  7  6  9
   D              1  8  7  10  3
   N D            1  8  7  10  9
   D              1  8  5  6  3
   N D            1  8  5  6  9
   D              1  8  5  10  3
   N D            1  8  5  10  9
   D              1  8  4  5  3
   N D            1  8  4  5  6
   N D            1  8  4  5  9
   N D            1  8  4  5  10
   D              1  8  4  6  3
   N D            1  8  4  6  9
   D              1  8  4  10  3
   N D            1  8  4  10  9
   D              1  9  7  5  3
   D              1  9  7  5  6
   D              1  9  7  5  10
   D              1  9  7  6  3
   D              1  9  7  10  3
   D              1  9  5  6  3
   D              1  9  5  10  3
   D              1  9  4  5  3
   N D            1  9  4  5  6
   N D            1  9  4  5  10
   D              1  9  4  6  3
   D              1  9  4  10  3
   84 TREES       84 PASSIVE    0 ACTIVE LOOP TESTS
NEXT ? NO
OUTPUT ? TF
NEXT ? GO
VOLTAGE GAIN  NUMERATOR      DENOMINATOR
S^ 0          2.73            122.73
S^ 1          4.77687         255.794
S^ 2          5.1241          179.033
S^ 3          2.39048         51.1327
S^ 4          .341155         5.11051
NEXT ? NF
```

Fig. 6.16 (*continued*)

```
F-LOW,F+INC,F*MULT,NO.F
 ? 0,.05,1,20
NEXT ? EV
HZ.         GAIN DB        PHASE DEG.     REAL PART       IMAG.PART
 0          -33.0558       0              2.22439E-02     0
 .05        -33.9195       -4.21383       .020084         -1.47975E-03
 .1         -35.8811       2.12675        1.60563E-02     5.96265E-04
 .15        -36.6816       22.5863        .013529         5.62779E-03
 .2         -35.0441       40.6069        1.34322E-02     1.15156E-02
 .25        -32.8503       47.939         1.52585E-02     .01691
 .3         -31.0426       49.3853        1.82571E-02     2.12899E-02
 .35        -29.6633       48.3918        2.18287E-02     2.45792E-02
 .4         -28.6108       46.4395        2.55715E-02     2.68897E-02
 .45        -27.7959       44.1573        2.92405E-02     2.83928E-02
 .5         -27.1545       41.8238        3.27001E-02     2.92616E-02
 .55        -26.6419       39.5639        3.58848E-02     2.96484E-02
 .6         -26.2267       37.4318        3.87728E-02     2.96782E-02
 .65        -25.8863       35.4472        4.13674E-02     2.94496E-02
 .7         -25.6041       33.6127        4.36856E-02     2.90385E-02
 .75        -25.3678       31.9225        4.57504E-02     2.85021E-02
 .8         -25.1684       30.3674        4.75875E-02     .027883
 .85        -24.9986       28.9364        .049222         2.72128E-02
 .9         -24.8529       27.6185        5.06774E-02     2.65143E-02
 .95        -24.7272       26.4034        5.19753E-02     2.58045E-02
NEXT ?
```

The program TOPSEN contains many facilities not explored by computed examples simply because space is limited. In common with most circuit analysis programs, vast amounts of numerical output can be generated, and it is preferable for the user to gain practical experience of the program on familiar circuit examples with known behaviour. It soon becomes obvious that graphic output is essential in this type of program. However, such programs are usually specific to one computer system or installation and are not readily transferred to any other system. This is the reason why none of the programs in this book contain graphic facilities, and similarly do not use disc files, or other less standardized aspects of BASIC.

## 6.5 CONCLUSION

Transfer function analysis in the $s$-domain is one of the more difficult analysis requirements to satisfy for large circuits. However, recent progress in research has made transfer function analysis important in applications where the speed of response evaluation outweighs the inaccuracy caused by modelling in terms of amplifiers. In many applications, the transfer function is a required output from the analysis, and in RC-active network design where active elements are in fact amplifiers, transfer function analysis is particularly advantageous.

## FURTHER READING

Desoer, C. A. and Kuh, E. S., (1969), *Basic Circuit Theory*, Chs. 13 to 15, McGraw-Hill, New York.

Haykin, S. S., (1970), *Active Network Theory,* Chs. 5, 7 and 8, Addison Wesley, Reading Mass.

Holbrook, J. B., (1966), *Laplace Transforms for Electronic Engineers,* Pergamon Press, Oxford, 2nd ed.

*Topological Analysis*

Chan, S. P., (1969), *Introductory Topological Analysis of Electrical Networks,* Holt, Rinehart, and Winston, New York.
(see also subsequent papers by S. P. Chan *et al* from 1970 in *IEEE Trans. on Circuit Theory or Circuits and Systems*).

Murdock, J. B., (1970). *Network Theory,* Ch. 6, McGraw-Hill, New York.

*Signal Flow Graphs*

Haykin, S. S., *ibid.,* Ch. 10.

Mason, S. J. and Zimmerman, H. J., (1960), *Electronic Circuits, Signals, and Systems,* Chs. 4 and 5, John Wiley, New York.

*NAM Inversion*

Downs, T., (1970), Inversion of the Nodal Admittance Matrix in Symbolic Form, *Electronics Letters,* **6**, 74-76 and 690-691.

## PROBLEMS

6.1 Determine Laplace transforms of the following time functions.

a. $f(t) = te^{-at} \sin \omega t$

b. $f(t) = u(t) - u(t-1)$

c. $f(t) = \Sigma_{n=0}^{\infty} \{u(t-nT) - u(t-a-nT)\}$

d. $f(t) = 5e^{-t} - 4e^{-2t} - 18te^{-2t} + 3e^{-t} \cos(t-60°)$

6.2 Determine inverse Laplace transforms of the following functions of $s$.

a. $1/(s+2)^3$

b. $1/s(s+2)^3$

c. $1(s^2+2s+2)^3$

d. $(3s^3+4s^2+3s+2)/(s+2)^2\,(s^2+2s+2)^2\,(s+1)$

e. $e^{-2s}/(s^2+2s+2)$.

6.3 Determine initial and final values of $f(t)$ and $\mathrm{d}[f(t)]/\mathrm{d}t$ where $f(t)$ is given by the inverse Laplace transform of the following functions of $s$.

a. $1/s(s+2)^3$

b. $2(s+1)/(s^2+2s+5)$

6.4 The circuit in Fig. 6.17 has zero initial conditions at $t=0-$. The switch is closed at $t=0$ and opened at $t=1$ sec.
Find an expression for the output voltage $v(t)$.

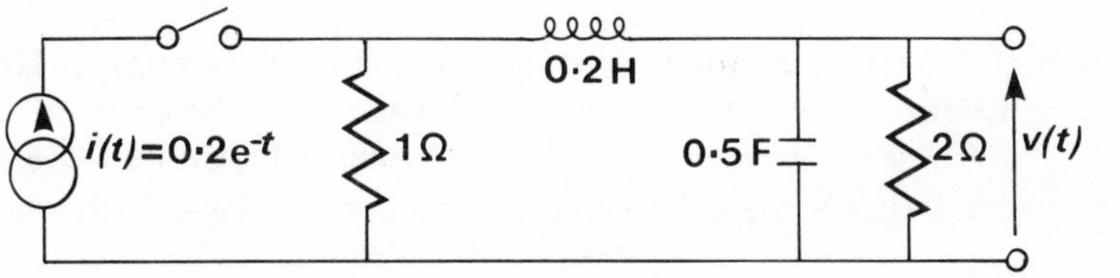


Fig. 6.17 – Problem 6.4.

6.5 Compile a list of expressions for each wiring operator in Table 5.4 which give the six polynomials of the connected networks in terms of the six (for unary operators) or twelve (for binary operators) polynomials of the original networks. For example, the cascade connection of networks A and B yields network AB in which

$$p_{so}^{AB} = p_{so}^{A}\, p_{so}^{B} + p_{ss}^{A}\, p_{oo}^{B} \,.$$

6.6 Determine the two-port Y and A-parameters of the bridged-T network in Fig. 6.1 using the polynomial expressions from Problem 6.5 and Table 6.4. Examine the build-up of powers of $s$ in comparison with normal two-port analysis where cancellation is unpractical by computer.

6.7 Determine the topological conditions for a tree of $\mathcal{N}_{-i}$(sub) to also be a tree of $\mathcal{N}_{-i}$, or $\mathcal{N}_{-j}$, or $\mathcal{N}_{-i}$ and $\mathcal{N}_{-j}$ simultaneously.

6.8 List the tree of the network $\mathcal{N}$ in Fig. 8.9 by the cyclic method of Section 6.2. List also those trees common to the network $\mathcal{N}_{-3}$ and note that trees of $\mathcal{N}_{-3}$ are not uniquely generated.

6.9 List the active trees of the filter in Fig. 4.23 common to $\mathcal{N}_{-1}$(sub) and $\mathcal{N}_{-1}$, and those common to $\mathcal{N}_{-1}$(sub) and $\mathcal{N}_{-6}$. Examine the trees by matrix methods as well as in terms of the answer to Problem 6.7.

6.10 Determine the voltage gain transfer function of the circuits in Figs. 1.20, 1.21, 4.28, and 6.15 by the signal flow graph method.

6.11 Apply the NAM inversion technique in the $s$-domain and symbolically to the NAM obtained for the circuit in Fig. 6.15.

*Computing*

6.12 Use program TOPSEN on a number of familiar networks and explore the facilities available. Compare the responses obtained with those from other computer programs including ACNLU from Chapter 4 and 2PORT from Chapter 5.

6.13 It was noted in Section 6.4 that Subroutine 5000 in program TOPSEN is slow and needs either complete replacement using a topological test, or a rapid pre-test which detects the most commonly occurring tree failure modes. Examine various pre-tests as a means of speeding-up the program.

6.14 Add graph plotting subroutines to program TOPSEN suited to the computing facilities available.

6.15 Write a subroutine to give the inverse Laplace transform of the transfer function $F(s)$ obtained from program TOPSEN as a ratio of two polynomials in $s$.

6.16 Write a program to give the inverse Laplace transform of transfer functions $F(s)$ obtained with the denominator factorized into multiple real and quadratic poles.

6.17 Explore the possibility of programming two-port analysis in the $s$-domain using the polynomial expression obtained in answer to Problem 6.5.

6.18 Write a subroutine for inversion of the NAM in the $s$-domain and construct an analysis program around it.

## APPENDIX

```
0010 PRINT "TOPOLOGICAL TRANSFER FUNCTION ANALYSIS"
0020 PRINT "RLC INPUTS: COMPONENT NUMBER,VALUE,2 NODE CONNECTIONS"
0030 PRINT "GM  INPUTS: GM NUMBER,VALUE,TYPE,4 NODE CONNECTIONS"
0040 PRINT "       TYPE: 1 FOR GM, 2 FOR GM/S, 3 FOR SGM"
0050 PRINT "     NODES: SEND +, - , RECEIVE +, - "
0060 DIM N$[2],M$[6],L$[16],K$[28],J$[8]
0070 LET I6=0
0080 LET F3=0
0090 LET P2=8*ATN(1)
0100 LET M$="LTTFSE"
0110 LET L$="EREVGONCNNNONFMP"
0120 LET K$=" SGMGM/S  GM      1/R1/SL  SC"
0130 LET J$=" R L CGM"
0140 PRINT "NO.NODES",
0150 INPUT N
0160 PRINT "NO. R,L,C,GM",
0170 INPUT R9,L9,C9,G9
0180 DIM S[R9+1,2],M[L9+1,2],A[C9+1,2],G[G9+1,5]
0190 LET R8=R9
0200 LET L8=L9
0210 LET C8=C9
0220 LET G8=G9
0230 GOTO 0270
0240 PRINT "NO. MODIFIED R,L,C,GM";
0250 INPUT R8,L8,C8,G8
0260 LET T3=0
0270 FOR S9=1 TO R8
0280   PRINT "R";
0290   INPUT S8,S[S8,0],S[S8,1],S[S8,2]
0300 NEXT S9
0310 FOR S9=1 TO L8
0320   PRINT "L";
0330   INPUT S8,M[S8,0],M[S8,1],M[S8,2]
0340 NEXT S9
0350 FOR S9=1 TO C8
0360   PRINT "C";
0370   INPUT S8,A[S8,0],A[S8,1],A[S8,2]
0380 NEXT S9
0390 FOR S9=1 TO G8
0400   PRINT "GM";
0410   INPUT S8,G[S8,0],G[S8,1],G[S8,2],G[S8,3],G[S8,4],G[S8,5]
0420 NEXT S9
0430 IF I6=4 THEN GOTO 0660
0440 GOTO 0480
0450 PRINT "F-LOW,F+INC,F*MULT,NO.F"
0460 INPUT F0,F1,F2,F3
0470 GOTO 2830
0480 IF I6>0 THEN GOTO 0510
0490 PRINT "OUTPUTS: LT LIST TREES, TF VOLTAGE GAIN F(S)"
0500 PRINT "        : SE VOLTAGE GAIN AND SENSITIVITIES"
0510 PRINT "OUTPUT";
0520 INPUT N$
0530 FOR I9=1 TO 3
0540   LET I7=I9
0550   IF N$=M$[2*I7-1,2*I7] THEN GOTO 0590
0560 NEXT I9
0570 PRINT "UNRECOGNISED OUTPUT RETYPE";
0580 GOTO 0520
0590 IF I6>1 THEN GOTO 2770
0600 PRINT "INPUT NODE,OUTPUT NODE",
0610 INPUT N1,N2
0620 LET N0=N1
0630 LET T3=0
0640 IF I6=5 THEN GOTO 1820
0650 REM FILL BRANCH DATA MATRIX U"
0660 LET T9=R9+L9+C9+G9
0670 LET P9=T9-G9
```

*

```
0680 MAT U=ZER[T9,6]
0690 FOR I9=1 TO R9
0700   LET U[I9,0]=1
0710   LET U[I9,1]=I9
0720   LET U[I9,2]=S[I9,1]
0730   LET U[I9,3]=S[I9,2]
0740   LET U[I9,6]=S[I9,0]
0750 NEXT I9
0760 FOR I9=1 TO L9
0770   LET U[I9+R9,0]=2
0780   LET U[I9+R9,1]=I9
0790   LET U[I9+R9,2]=M[I9,1]
0800   LET U[I9+R9,3]=M[I9,2]
0810   LET U[I9+R9,6]=M[I9,0]
0820 NEXT I9
0830 LET S9=R9+L9
0840 FOR I9=1 TO C9
0850   LET U[I9+S9,0]=3
0860   LET U[I9+S9,1]=I9
0870   LET U[I9+S9,2]=A[I9,1]
0880   LET U[I9+S9,3]=A[I9,2]
0890   LET U[I9+S9,6]=A[I9,0]
0900 NEXT I9
0910 LET G6=0
0920 LET G7=0
0930 LET S9=S9+C9
0940 FOR I9=1 TO G9
0950   LET U[I9+S9,0]=-G[I9,1]
0960   LET U[I9+S9,1]=I9
0970   LET U[I9+S9,2]=G[I9,5]
0980   LET U[I9+S9,3]=G[I9,4]
0990   LET U[I9+S9,6]=G[I9,0]
1000   ON G[I9,1] THEN GOTO 1040, 1010, 1030
1010   LET G6=G6+1
1020   GOTO 1040
1030   LET G7=G7+1
1040 NEXT I9
1050 LET S3=L9+C9+G6+G7
1060 REM PARALLEL BRANCH LIST
1070 FOR I9=1 TO T9-1
1080   IF U[I9,4]>0 THEN GOTO 1190
1090   FOR S9=I9+1 TO T9
1100     IF U[S9,4]>0 THEN GOTO 1180
1110     IF U[S9,2]<>U[I9,2] THEN GOTO 1160
1120     IF U[S9,3]<>U[I9,3] THEN GOTO 1180
1130     LET U[I9,4]=S9
1140     LET U[S9,4]=I9
1150     GOTO 1190
1160     IF U[I9,2]<>U[S9,3] THEN GOTO 1180
1170     IF U[I9,3]=U[S9,2] THEN GOTO 1130
1180   NEXT S9
1190 NEXT I9
1200 REM FILL NODE BRANCH MATRIX V
1210 LET T8=2
1220 DIM W[N+1]
1230 FOR I9=1 TO N+1
1240   LET W[I9]=0
1250   FOR S9=1 TO T9
1260     IF U[S9,2]<>I9-1 THEN GOTO 1290
1270     LET W[I9]=W[I9]+1
1280     GOTO 1300
1290     IF U[S9,3]=I9-1 THEN GOTO 1270
1300   NEXT S9
1310   IF W[I9]>T8-2 THEN LET T8=W[I9]+2
1320 NEXT I9

*
```

```
1330 MAT V=ZER[T8,N]
1340 FOR S9=0 TO N
1350   LET V[T8-1,S9]=W[S9+1]
1360   LET V[0,S9]=0
1370   LET V[T8,S9]=0
1380 NEXT S9
1390 FOR S9=1 TO T9
1400   LET V1=U[S9,2]
1410   LET V2=U[S9,3]
1420   LET V[T8,V1]=V[T8,V1]+1
1430   LET V[T8,V2]=V[T8,V2]+1
1440   LET V[V[T8,V1],V1]=S9
1450   LET V[V[T8,V2],V2]=S9
1460 NEXT S9
1470 REM RE-ARRANGE BRANCH ORDER IN V
1480 LET R8=0
1490 FOR S9=1 TO V[T8-1,0]
1500   LET V1=U[V[S9,0],2]
1510   LET V2=U[V[S9,0],3]
1520   IF V1=0 THEN LET V1=V2
1530   FOR I9=1 TO V[T8-1,V1]
1540     IF V[I9,V1]<>V[S9,0] THEN GOTO 1580
1550     IF V[0,V1]>0 THEN GOTO 1580
1560     LET V[0,V1]=I9
1570     LET R8=R8+1
1580   NEXT I9
1590 NEXT S9
1600 FOR I9=1 TO N
1610   IF V[0,I9]=0 THEN GOTO 1770
1620   IF V[T8-1,I9]=1 THEN GOTO 1770
1630   FOR S9=1 TO V[T8-1,I9]
1640     IF V[0,I9]=S9 THEN GOTO 1760
1650     LET V1=U[V[S9,I9],2]
1660     LET V2=U[V[S9,I9],3]
1670     IF V1=I9 THEN LET V1=V2
1680     IF V1=0 THEN GOTO 1760
1690     IF V[0,V1]>0 THEN GOTO 1760
1700     FOR C8=1 TO V[T8-1,V1]
1710       IF V[C8,V1]<>V[S9,I9] THEN GOTO 1730
1720       LET V[0,V1]=C8
1730     NEXT C8
1740     LET R8=R8+1
1750     GOTO 1600
1760   NEXT S9
1770 NEXT I9
1780 IF R8=N THEN GOTO 1860
1790 PRINT "SUITABLE V MATRIX NOT FOUND"
1800 PRINT "INPUT V(";T8;",";N;")"
1810 MAT INPUT V
1820 FOR I9=1 TO N
1830   LET V[0,I9]=1
1840 NEXT I9
1850 GOTO 1940
1860 FOR I9=1 TO N
1870   IF V[0,I9]=1 THEN GOTO 1930
1880   LET V1=V[1,I9]
1890   LET V2=V[V[0,I9],I9]
1900   LET V[V[0,I9],I9]=V1
1910   LET V[1,I9]=V2
1920   LET V[0,I9]=1
1930 NEXT I9
1940 LET V[0,N0]=0
1950 GOTO 2770
1960 PRINT " MATRIX U"
1970 MAT PRINT U
1980 PRINT
```

*

```
1990 PRINT "MATRIX V"
2000 MAT PRINT V
2010 PRINT
2020 GOTO 2830
2030 REM TREE LIST
2040 LET T6=0
2050 LET V[0,0]=0
2060 LET T4=0
2070 LET T5=0
2080 ON I7 THEN GOTO 2190, 2150, 2090
2090 MAT E=ZER[T9,S3]
2100 MAT F=ZER[T9,S3]
2110 FOR S9=0 TO T9
2120   LET E[S9,0]=0
2130   LET F[S9,0]=0
2140 NEXT S9
2150 MAT D=ZER[2,S3]
2160 LET D[1,0]=0
2170 LET D[2,0]=0
2180 ON I7 THEN GOTO 2190, 2240, 2240
2190 PRINT "BRANCH - COMPONENT LIST"
2200 FOR I9=1 TO T9
2210   PRINT I9;K$[U[I9,0]*4+13,U[I9,0]*4+16];U[I9,1],
2220 NEXT I9
2230 PRINT
2240 GOSUB 3000
2250 ON I7 THEN GOTO 2260, 2450, 2450
2260 LET N$="++"
2270 ON V[0,0] THEN GOTO 2420, 2290, 2280, 2340, 2310, 2310
2280 PRINT " N";
2290 PRINT " D";
2300 GOTO 2360
2310 IF Q1*(-1)^(N1+N2)<1 THEN LET N$[1,1]="-"
2320 PRINT N$[1,1];"N";
2330 IF V[0,0]=5 THEN GOTO 2360
2340 IF P1<1 THEN LET N$[2,2]="-"
2350 PRINT N$[2,2];"D";
2360 PRINT " ",
2370 FOR I9=1 TO N
2380   IF V[0,I9]=0 THEN GOTO 2400
2390   PRINT V[V[0,I9],I9];
2400 NEXT I9
2410 GOTO 2230
2420 PRINT T5;"TREES",T4;"PASSIVE",T6;"ACTIVE LOOP TESTS"
2430 GOTO 2830
2440 REM TRANSFER FUNCTION GENERATION
2450 IF V[0,0]=1 THEN GOTO 2490
2460 GOSUB 7000
2470 GOTO 2240
2480 REM PRINT TRANSFER FUNCTIONS
2490 PRINT "VOLTAGE GAIN","NUMERATOR","DENOMINATOR"
2500 FOR I1=0 TO S3
2510   PRINT "S^";I1,D[1,I1],D[2,I1]
2520 NEXT I1
2530 LET T3=1
2540 ON I7-1 THEN GOTO 2830, 2550
2550 PRINT "SENSITIVITY","D(NUM)/D(CPT)","D(DEN)/D(CPT)"
2560 FOR I1=1 TO T9
2570   LET V1=U[I1,0]
2580   IF V1<0 THEN LET V1=4
2590   PRINT "COMPONENT";J$[2*V1-1,2*V1];U[I1,1]
2600   FOR J1=0 TO S3
2610     PRINT "S^";J1,E[I1,J1],F[I1,J1]
2620   NEXT J1
2630 NEXT I1
```

*

```
2640 LET T3=2
2650 GOTO 2830
2660 REM EVALUATION
2670 IF T3>0 THEN GOTO 2700
2680 PRINT "ANALYSE FIRST"
2690 GOTO 2830
2700 IF F3=0 THEN GOTO 0450
2710 IF I7-1>T3 THEN GOTO 2680
2720 PRINT "HZ.","GAIN DB","PHASE DEG.","REAL PART","IMAG.PART"
2730 IF I7=2 THEN GOTO 2750
2740 PRINT "CPT.NO.","DB SEN","DEG SEN","RP SEN","IP SEN"
2750 GOSUB 9000
2760 GOTO 2830
2770 IF I6>0 THEN GOTO 2830
2780 LET T3=0
2790 PRINT "NEXT COMMAND: ER END RUN        EV EVALUATE RESPONSE"
2800 PRINT "MODIFICATION: NC COMPONENTS     NF NEW FREQUENCIES"
2810 PRINT "            : NN IN/OUT NODES   NO OUTPUTS"
2820 PRINT "      OUTPUT: MP MATRICES U V   GO ANALYSE"
2830 PRINT "NEXT";
2840 INPUT N$
2850 FOR I9=1 TO 8
2860   LET I6=I9
2870   IF N$=L$[2*I6-1,2*I6] THEN GOTO 2910
2880 NEXT I9
2890 PRINT "UNRECOGNISED COMMAND RETYPE";
2900 GOTO 2840
2910 ON I6 THEN GOTO 2920, 2670, 2030, 0240, 0600, 0480, 0450, 1960
2920 STOP

*


3000 REM SUBROUTINE TO GENERATE THE NEXT TREE
3010 IF V[0,0]=0 THEN GOTO 3200
3020 LET V[0,0]=1
3030 LET I1=N+1
3040 LET I1=I1-1
3050 IF I1=0 THEN GOTO 3340
3060 IF I1=NO THEN GOTO 3040
3070 IF V[0,I1]<V[T8-1,I1] THEN GOTO 3100
3080 LET V[0,I1]=1
3090 GOTO 3040
3100 LET V[0,I1]=V[0,I1]+1
3110 IF I1=U[V[V[0,I1],I1],2] THEN LET I2=U[V[V[0,I1],I1],3]
3120 IF I1=U[V[V[0,I1],I1],3] THEN LET I2=U[V[V[0,I1],I1],2]
3130 IF I2=0 THEN GOTO 3200
3140 IF I2=NO THEN GOTO 3200
3150 IF V[V[0,I1],I1]=U[V[V[0,I2],I2],4] THEN GOTO 3170
3160 IF V[V[0,I1],I1]<>V[V[0,I2],I2] THEN GOTO 3200
3170 IF I2>I1 THEN LET I1=I2
3180 GOTO 3070
3190 REM TEST POSSIBLE TREE
3200 LET V[0,0]=1
3210 LET P1=1
3220 LET Q1=1
3230 GOSUB 4000
3240 IF V[0,0]=1 THEN GOTO 3030
3250 IF G9=0 THEN GOTO 3330
3260 LET J1=0
3265 LET J1=J1+1
3270 IF V[V[0,J1],J1]>P9 THEN GOTO 3300
3280 IF J1<N THEN GOTO 3265
3290 GOTO 3330
3300 LET V[0,0]=1
3310 GOSUB 5000
3320 IF V[0,0]=1 THEN GOTO 3030
3330 LET T5=T5+1
3340 RETURN

*
```

```
4000 REM SUBROUTINE FOR LOOP TEST
4010 LET T4=T4+1
4020 FOR J1=1 TO N
4030   LET V[T8,J1]=0
4040   LET W[1+J1]=0
4050 NEXT J1
4060 LET V1=1
4070 ON V[0,0] THEN GOTO 4080, 4430
4080 IF N0=1 THEN LET V1=2
4090 LET V[T8,V1]=1
4100 LET W[1]=U[V[V[0,V1],V1],2]
4110 IF W[1]=N0 THEN LET W[1]=0
4120 LET W[2]=U[V[V[0,V1],V1],3]
4130 IF W[2]=N0 THEN LET W[2]=0
4140 LET V1=1
4150 LET V1=V1+1
4160 FOR J1=1 TO N
4170   IF J1=N0 THEN GOTO 4360
4180   IF V[T8,J1]=1 THEN GOTO 4360
4190   LET C8=U[V[V[0,J1],J1],2]
4200   LET L8=U[V[V[0,J1],J1],3]
4210   IF C8=N0 THEN LET C8=0
4220   IF L8=N0 THEN LET L8=0
4230   FOR J2=1 TO V1
4240     LET V2=L8
4250     IF W[J2]=C8 THEN GOTO 4300
4260     LET V2=C8
4270     IF W[J2]=L8 THEN GOTO 4300
4280   NEXT J2
4290   GOTO 4360
4300   FOR J2=1 TO V1
4310     IF W[J2]=V2 THEN GOTO 4720
4320   NEXT J2
4330   LET W[1+V1]=V2
4340   LET V[T8,J1]=1
4350   GOTO 4150
4360 NEXT J1
4370 LET V2=1
4380 IF N0=0 THEN LET V2=0
4390 IF V1=N+1-V2 THEN LET V[0,0]=2
4400 IF N0=0 THEN GOTO 4720
4410 REM COMMON TREE TEST -N0 AND -N2
4420 ON V[0,0] THEN GOTO 4720, 4020
4430 FOR J1=1 TO V[T8-1,N0]
4440   LET V2=U[V[J1,N0],2]
4450   IF V2=N0 THEN LET V2=U[V[J1,N0],3]
4460   IF V[V[0,V2],V2]<>V[J1,N0] THEN GOTO 4510
4470   IF V2=N2 THEN GOTO 4700
4480   LET V[T8,V2]=1
4490   LET W[V1]=V2
4500   LET V1=V1+1
4510 NEXT J1
4520 IF V1=1 THEN GOTO 4720
4530 FOR J1=1 TO N
4540   IF J1=N0 THEN GOTO 4680
4550   IF V[T8,J1]=1 THEN GOTO 4680
4560   FOR J2=1 TO V1-1
4570     LET V2=U[V[V[0,J1],J1],3]
4580     IF W[J2]=U[V[V[0,J1],J1],2] THEN GOTO 4630
4590     LET V2=U[V[V[0,J1],J1],2]
4600     IF W[J2]=U[V[V[0,J1],J1],3] THEN GOTO 4630
4610   NEXT J2
4620   GOTO 4680
4630   IF V2=N2 THEN GOTO 4700
4640   LET W[V1]=V2
4650   LET V1=V1+1
4660   LET V[T8,J1]=1
4670   GOTO 4520
4680 NEXT J1
4690 GOTO 4720
4700 LET V[0,0]=3
4710 LET Q1=(-1)^(N1+N2)
4720 RETURN
```

*

```
5000 REM SUBROUTINE FOR ACTIVE LOOP TEST
5010 LET T6=T6+1
5020 LET T4=T4-1
5030 LET C8=1
5040 MAT B=ZER[N,N]
5050 FOR J1=1 TO N
5060   IF V[0,J1]=0 THEN GOTO 5160
5070   ON C8 THEN GOTO 5080, 5110, 5110
5080   LET V1=U[V[V[0,J1],J1],2]
5090   LET V2=U[V[V[0,J1],J1],3]
5100   GOTO 5140
5110   IF V[V[0,J1],J1]<=P9 THEN GOTO 5080
5120   LET V1=G[U[V[V[0,J1],J1],1],2]
5130   LET V2=G[U[V[V[0,J1],J1],1],3]
5140   LET B[J1,V1]=1
5150   LET B[J1,V2]=-1
5160 NEXT J1
5170 IF C8<3 THEN GOTO 5250
5180 LET N0=N2
5190 FOR J1=1 TO N
5200   LET B[N1,J1]=B[N2,J1]
5210 NEXT J1
5220 LET R8=1
5230 IF ABS(N2-N1)>1 THEN LET R8=(-1)^(ABS(N2-N1)+1)
5240 GOTO 5260
5250 LET R8=1
5260 LET I1=N+1
5265 LET I1=I1-1
5270 IF I1=N0 THEN GOTO 5550
5280 IF B[I1,I1]<>0 THEN GOTO 5410
5290 IF I1=1 THEN GOTO 5680
5300 LET J1=I1
5305 LET J1=J1-1
5310 IF J1=N0 THEN GOTO 5330
5320 IF B[I1,J1]<>0 THEN GOTO 5350
5330 IF J1>1 THEN GOTO 5305
5340 GOTO 5680
5350 FOR I2=1 TO I1
5360   LET L8=B[I2,I1]
5370   LET B[I2,I1]=B[I2,J1]
5380   LET B[I2,J1]=L8
5390 NEXT I2
5400 LET R8=-R8
5410 IF I1=1 THEN GOTO 5540
5420 LET I2=I1
5425 LET I2=I2-1
5430 IF I2=N0 THEN GOTO 5530
5440 IF B[I2,I1]=0 THEN GOTO 5530
5450 IF B[I2,I1]=B[I1,I1] THEN GOTO 5500
5460 FOR J1=1 TO I1-1
5470   LET B[I2,J1]=B[I2,J1]+B[I1,J1]
5480 NEXT J1
5490 GOTO 5530
5500 FOR J1=1 TO I1-1
5510   LET B[I2,J1]=B[I2,J1]-B[I1,J1]
5520 NEXT J1
5530 IF I2>1 THEN GOTO 5425
5540 IF B[I1,I1]<0 THEN LET R8=-R8
5550 IF I1>1 THEN GOTO 5265
5560 ON C8 THEN GOTO 5570, 5610, 5650
5570 LET P1=R8
5580 LET Q1=R8
5590 LET C8=2
5600 GOTO 5040
5610 LET P1=P1*R8
5620 LET V[0,0]=4
5630 LET C8=3
5640 GOTO 5040
5650 LET Q1=Q1*R8
5660 IF V[0,0]=1 THEN LET V[0,0]=5
5670 IF V[0,0]=4 THEN LET V[0,0]=6
5680 ON C8 THEN GOTO 5700, 5630, 5690
5690 LET N0=N1
5700 RETURN

*
```

```
7000 REM SUBROUTINE TO UPDATE TRANSFER FUNCTION
7010 LET J1=1
7020 FOR I1=1 TO R9+L9
7030   LET U[I1,5]=1
7040 NEXT I1
7050 FOR I1=R9+L9+1 TO T9
7060   LET U[I1,5]=0
7070 NEXT I1
7080 LET S1=L9+G6
7090 FOR I1=1 TO N
7100   LET U[V[V[0,I1],I1],5]=1-U[V[V[0,I1],I1],5]
7110 NEXT I1
7120 FOR I1=1 TO T9
7130   IF U[I1,5]=0 THEN GOTO 7180
7140   LET J1=J1*U[I1,6]
7150   IF ABS(U[I1,0])=3 THEN LET S1=S1+1
7160   IF U[I1,0]=-2 THEN LET S1=S1-1
7170   GOTO 7190
7180   IF U[I1,0]=2 THEN LET S1=S1-1
7190 NEXT I1
7200 IF V[0,0]=5 THEN GOTO 7220
7210 LET D[2,S1]=D[2,S1]+P1*J1
7220 ON V[0,0]-1 THEN GOTO 7240, 7230, 7240, 7230, 7230
7230 LET D[1,S1]=D[1,S1]+Q1*J1*(-1)^(N1+N2)
7240 IF I7=2 THEN GOTO 7320
7250 FOR I1=1 TO T9
7260   IF U[I1,5]=0 THEN GOTO 7310
7270   IF V[0,0]=5 THEN GOTO 7290
7280   LET F[I1,S1]=F[I1,S1]+P1*J1/U[I1,6]
7290   ON V[0,0]-1 THEN GOTO 7310, 7300, 7310, 7300, 730
7300   LET E[I1,S1]=E[I1,S1]+Q1*J1*(-1)^(N1+N2)/U[I1,6]
7310 NEXT I1
7320 RETURN
```

*

```
9000 REM FREQUENCY RESPONSE EVALUATION
9010 LET F4=F0
9015 LET J1=0
9020 LET J1=J1+1
9030 IF J1=1 THEN GOTO 9050
9040 LET F4=(F4+F1)*F2
9050 LET F5=F4*P2
9060 LET V1=0
9070 LET V2=0
9080 LET D1=0
9090 LET D2=0
9100 FOR I1=0 TO S3 STEP 2
9110   LET V1=V1+D[1,I1]*(-1)^(2+I1/2)*F5^I1
9120   LET D1=D1+D[2,I1]*(-1)^(2+I1/2)*F5^I1
9130 NEXT I1
9140 FOR I1=1 TO S3 STEP 2
9150   LET V2=V2+D[1,I1]*(-1)^(1.5+I1/2)*F5^I1
9160   LET D2=D2+D[2,I1]*(-1)^(1.5+I1/2)*F5^I1
9170 NEXT I1
9180 LET P1=D1*D1+D2*D2
9190 LET C8=(V1*D1+V2*D2)/P1
9200 LET G8=(V2*D1-D2*V1)/P1
9210 LET L8=57.2958*ATN(G8/C8)
9220 LET P3=C8*C8+G8*G8
9230 LET R8=4.34295*LOG(P3)
9240 PRINT F4,R8,L8,C8,G8
9250 REM SENSITIVITY EVALUATION
9260 IF I7=2 THEN GOTO 9480
9265 LET S9=0
9270 LET S9=S9+1
9280 LET S1=0
9290 LET S2=0
9300 LET T1=0
9310 LET T2=0
9320 FOR I1=0 TO S3 STEP 2
9330   LET S1=S1+E[S9,I1]*(-1)^(2+I1/2)*F5^I1
9340   LET T1=T1+F[S9,I1]*(-1)^(2+I1/2)*F5^I1
9350 NEXT I1
9360 FOR I1=1 TO S3 STEP 2
9370   LET S2=S2+E[S9,I1]*(-1)^(1.5+I1/2)*F5^I1
9380   LET T2=T2+F[S9,I1]*(-1)^(1.5+I1/2)*F5^I1
9390 NEXT I1
9400 LET R5=((S1+T2*G8-T1*C8)*D1+(S2-T2*C8-T1*G8)*D2)/P1
9410 LET L5=((S2-T2*C8-T1*G8)*D1-(S1+T2*G8-T1*C8)*D2)/P1
9420 LET R8=(R5*C8+L5*G8)/P3*8.68589
9430 LET L8=(L5*C8-R5*G8)/P3*57.2958
9440 LET S1=U[S9,0]
9450 IF S1<0 THEN LET S1=4
9460 PRINT J$[2*S1-1,2*S1];U[S9,1],R8,L8,R5,L5
9470 IF S9<T9 THEN GOTO 9270
9480 IF J1<F3 THEN GOTO 9020
9490 RETURN

*
```

Fig. 6.18 – Program TOPSEN.

CHAPTER 7

# State Variable Analysis

Transfer function analysis in Chapter 6 provided a concise method of writing and solving network differential equations which relate input and output functions of time. The zero-input response is a necessary part of the general solution which is caused by the initial condition or *state* of a network at time $t = 0$. Initial conditions depend on the entire history of the network for $-\infty < t < 0$, and their specification at the instant $t = 0$ accounts for the effect of past history on the future network response for $t > 0$. The idea of state may be generalized as a property of the network by recognizing that any time $t = \tau$ may be chosen as a reference point in time. At any time $\tau$ the effect of the past history of the system may be taken into account by knowledge of the values of a set of state variables. In network theory these are chosen to be the inductor currents and capacitor voltages.

In this chapter we consider circuit analysis in terms of state variables by rewriting network equations as simultaneous first-order differential equations. Many numerical procedures from the theory of matrices, eigenvalues, and eigenvectors may then be applied to the solution of these equations in the time and frequency domains. In a much wider context the concept of state may be applied to any physical system, and state-space analysis is an important unifying mathematical technique.

## 7.1 GENERATION OF STATE EQUATIONS

The standard form of the state equations is given in matrix form by

$$\dot{\mathbf{x}} = \mathbf{A}\mathbf{x} + \mathbf{B}\mathbf{u} , \tag{7.1}$$

where $\mathbf{x}$ is a column vector of state variables, and $\mathbf{u}$ is a column vector of input variables. Since the required solution variables may not be state variables an output equation is also required and is given in matrix form by

$$\mathbf{y} = \mathbf{C}\mathbf{x} + \mathbf{D}\mathbf{u} , \tag{7.2}$$

where $\mathbf{y}$ is a column vector of output variables.

Note that matrices **A**, **B**, **C**, and **D** are not the topological matrices defined in Chapters 2 and 6. However, they are matrices which describe circuit components and their connections. The first requirement of state variable analysis is for an algorithm which determines the **A**, **B**, **C**, and **D** matrices for any circuit. Secondly the state equations must be solved to yield steady-state frequency and transient response to any input.

Methods for finding state equations simple circuits will be illustrated by the following two examples

**Example 7.1**

Find the state and output equations for the circuit in Fig. 7.1.

The capacitor voltages and inductor current, $e_1$, $e_2$, and $i_3$ respectively are selected as state variables and are marked on Fig. 7.1. Since the standard form of the state equations involves $\dot{e}_1$, $\dot{e}_2$, and $\dot{i}_3$, the circuit equations should be determined with $i_1$, $i_2$, and $e_3$ expressed in terms of $e_1$, $e_2$, $i_3$ and $E$ only, The component equations can then be used to convert to the required form. This example will be solved by two separate methods, each of which can be generalized and applied to active networks.

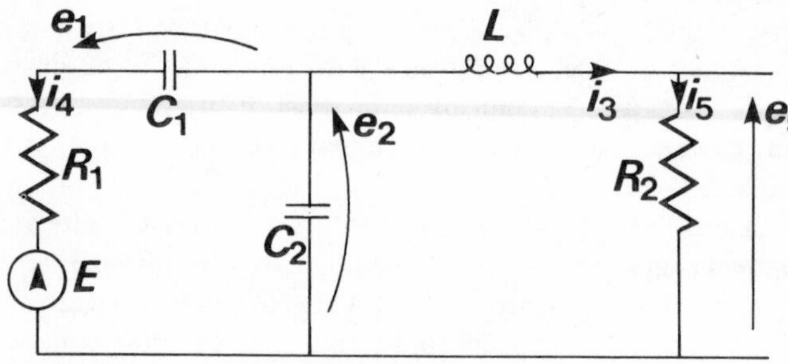


Fig. 7.1 – Circuit for Example 7.1.

*Solution by superposition*

The circuit is prepared for analysis by:

I replacement of all capacitors by fixed voltage sources equal in this example to $e_1$ and $e_2$,

II replacement of all inductors by fixed current sources equal here to $i_3$.

The circuit with sources inserted is shown in Fig. 7.2.

Solution is required for the state variables $i_1$, $i_2$, and $e_3$ and the output variable $e_5$. By the principle of superposition the contribution of individual

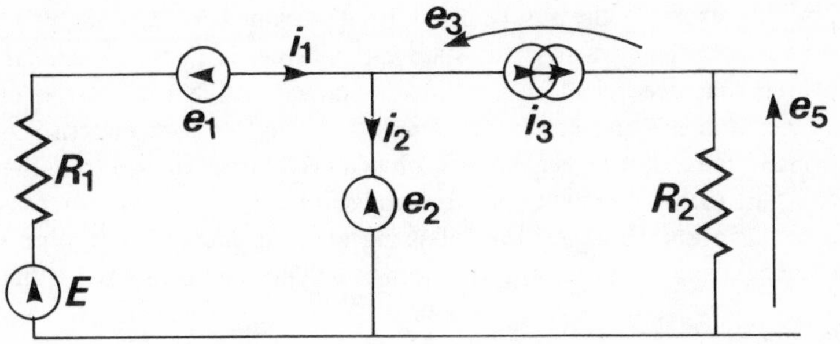


Fig. 7.2 – Circuit of Fig. 7.1 redrawn for analysis by superposition.

sources can be calculated separately and summed. Results can be easily tabulated in this case.

| Source | $e_1$ | $e_2$ | $i_3$ | $E$ |
|---|---|---|---|---|
| Solution variable | | | | |
| $i_1$ | $-e_1/R_1$ | $-e_2/R_1$ | 0 | $E/R_1$ |
| $i_2$ | $-e_1/R_1$ | $-e_2/R_1$ | $-i_3$ | $E/R_1$ |
| $e_3$ | 0 | $e_2$ | $-i_3R_2$ | 0 |
| $e_5$ | 0 | 0 | $i_3R_2$ | 0 |

Since $i_1 = C_1\dot{e}_1$, $i_2 = C_2\dot{e}_2$, and $e_3 = L\dot{i}_3$, the state and output equations may be obtained by summing the contributions tabulated above, and expressed in matrix form as

$$\begin{bmatrix}\dot{e}_1\\ \dot{e}_2\\ \dot{i}_3\end{bmatrix} = \begin{bmatrix}-1/C_1R_1 & -1/C_1R_1 & 0\\ -1/C_2R_1 & -1/C_2R_1 & -1/C_2\\ 0 & 1/L & -R_2/L\end{bmatrix}\begin{bmatrix}e_1\\ e_2\\ i_3\end{bmatrix} + \begin{bmatrix}1/R_1C_1\\ 1/R_1C_2\\ 0\end{bmatrix}[E]$$

$$[e_5] = [\ 0 \quad 0 \quad R_2]\begin{bmatrix}e_1\\ e_2\\ i_3\end{bmatrix} + [0]\ [E]$$

In this example the solution has been obtained by d.c. analysis, and, provided that the chosen set of state variables are in fact independent, and provided that dependent sources are contained within the resistive circuit branches, the method can be generalized through $N$-port analysis. Organization of the solution begins with an $N$-port matrix for the resistive and active part of the circuit. Capacitors and inductors appear as sources across the circuit ports. However, the following method of solution is more easily generalized, and the superposition method will not be taken any further.

*Solution by cut-set and tie-set analysis*

The circuit of Fig. 7.1 is redrawn in Fig. 7.3 to show a tree in heavy type, a cut-set, and a tie-set.

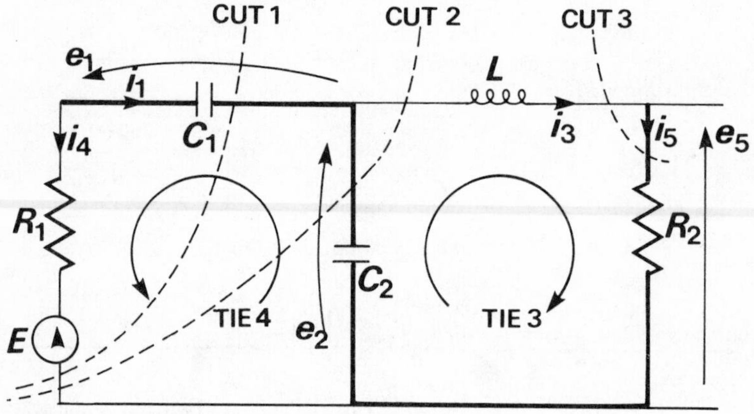


Fig. 7.3 – Circuit of Fig. 7.1 redrawn to show a tree, a cut-set, and a tie-set.

KCL applied to cuts ① and ② yields

$$i_1 + i_4 = 0$$

$$i_2 + i_3 + i_4 = 0 \ .$$

KVL applied to tie ① yields

$$e_2 - e_3 - e_5 = 0 \ .$$

Variables $e_1$, $e_2$ and $i_3$ are the required state variables. $i_1$, $i_2$, and $e_3$ may be replaced as before by $C_1\dot{e}_1$, $C_2\dot{e}_2$, and $L\dot{i}_3$, Therefore,

$$C_1\dot{e}_1 = -i_4$$

$$C_2\dot{e}_2 = -i_3 - i_4$$

$$L\dot{i}_3 = e_2 - e_5 \; .$$

Variables $i_4$ and $e_5$ are not state variables and must be eliminated by subsidiary equations involving the resistors. Use of the remaining equations in the cut-set and tie-set, together with Ohm's law for the resistors, yields

$$e_5/R_2 - i_3 = 0$$

$$E + i_4R_1 - e_1 - e_2 = 0 \; .$$

On substitution

$$C_1\dot{e}_1 = (-e_1 - e_2 + E)/R_1$$

$$C_2\dot{e}_2 = (-e_1 - e_2 + E)/R_1 - i_3$$

$$L\dot{i}_3 = e_2 - i_3R_2 \; ,$$

8

and the state equations can be obtained in the same matrix form as before.

The essential features of solution by combined cut-set and tie-set analysis which must be incorporated in the generalized procedure may be summarised as follows:

1. A tree must be drawn to include all capacitors, some resistors, and no inductors. Such a tree is called a **proper** tree.
2. Cut-set equations for the capacitive tree branches and tie-set equations for the inductive link branches form the basis of the set of state equations.
3. Tree resistor currents and link resistor voltages which appear as variables must be eliminated from the set of equations using further cut-set and tie-set equations.
4. Similar procedures are required to express the output variables in terms of state variables.

Before moving on to a formal development of the general procedure for writing state equations we will take a final look at the bridged-T network from the previous chapter. Further difficulties not covered by the solution steps above will be demonstrated.

**Example 7.2**

Find the state and output equations for the bridged-T network in Fig. 6.1.

The circuit is redrawn in Fig. 7.4 showing state variables $i_1$, $i_2$, and $e_3$.

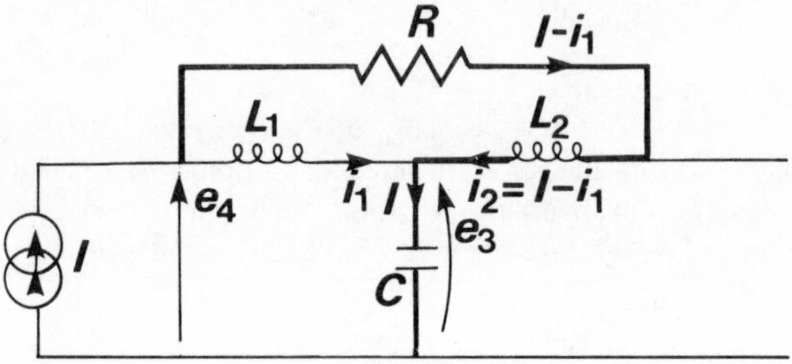


Fig. 7.4 – Circuit of Fig. 6.1 redrawn to show state variables.

Since the capacitor current is constrained by the source to $I$, the sum of the inductor currents also equals $I$. Therefore we must conclude that one inductor current cannot be independent and the number of state variables is equal to 2 and not 3 as expected. A second difficulty arises because the source $I$ is not associated with a component. Our definition of a proper tree is therefore inadequate for this circuit. We will find a set of state equations as best we can and later discuss the modifications necessary for these **excess element** networks.

Consider KVL for the loop consisting of $L_1$, $L_2$, and R. Summing voltages and using Ohm's law we obtain

$$e_1 - R(I - i_1) - e_2 = 0 \; .$$

Since $e_1 = L_1\dot{i}_1$, and $e_2 = L_2(\dot{I} - \dot{i}_1)$,

$$(L_1 + L_2)\,\dot{i}_1 = -Ri_1 + RI + L_2\dot{I} \; . \tag{7.3}$$

For the capacitor

$$C\dot{e}_2 = I \; . \tag{7.4}$$

The output is given by

$$e_3 = e_2 + L_1\dot{i}_1 - (1 - i_1)R \; . \tag{7.5}$$

Rearrangement of Eqs. (7.3) to (7.5) together with substitution for $\dot{i}_1$ in Eq. (7.5) yields the state and output equations,

$$\begin{bmatrix} \dot{e}_2 \\ \dot{i}_1 \end{bmatrix} = \begin{bmatrix} 0 & 0 \\ 0 & -R/(L_1+L_2) \end{bmatrix} \begin{bmatrix} e_2 \\ i_1 \end{bmatrix} + \begin{bmatrix} 1/C & 0 \\ R/(L_1+L_2) & L_2/(L_1+L_2) \end{bmatrix} \begin{bmatrix} I \\ \dot{I} \end{bmatrix}$$

$$[e_3] = [1 \quad RL_2/(L_1+L_2)] \begin{bmatrix} e_2 \\ i_1 \end{bmatrix} + [-RL_2/(L_1+L_2) \quad L_1L_2/(L_1+L_2)] \begin{bmatrix} I \\ \dot{I} \end{bmatrix}$$

In this solution it is necessary to consider the excess inductor $L_2$ as a tree branch and the current source $I$ as a link branch if Eqs. (7.3) and (7.4) are to have the same interpretation as in the solution for Example 7.1. The inclusion of excess inductors as tree branches (and similarly excess capacitors as link branches) within the proper tree leads to the appearance of derivatives of sources in the state equations.

**Linear Time Invariant Networks**

In order to conform with various conflicting standards of terminology, state equations will be derived in terms of the tie-set matrix **C** and cut-set matrix **D** even though we have defined the state equation standard form in terms of **A**, **B**, **C**, and **D**, matrices. Almost all variables throughout the remainder of Section 7.1 including the topological **C** and **D** matrices will have suffixes such as T or L denoting tree or link, or C, R, or L denoting component type. The matrices in use should therefore be quite clear.

Let us make the following assumptions so that the standard branch and dependent sources from Chapter 2 may be retained for the present.

I Assume that the network contains a proper tree.

II Assume that the branches include fixed sources as in Fig. 2.7.

III Assume that the dependent sources can be isolated into groups so that admittance matrix **Y** and impedance matrix **Z** can be partitioned as:

$$\mathbf{Y} = \begin{bmatrix} \mathbf{Y}_C & \mathbf{0} & \mathbf{0} & \mathbf{0} \\ \mathbf{0} & \mathbf{Y}_G & \mathbf{0} & \mathbf{0} \\ \mathbf{0} & \mathbf{0} & \mathbf{Y}_L & \mathbf{0} \\ \mathbf{0} & \mathbf{0} & \mathbf{0} & \mathbf{Y}_R \end{bmatrix} \qquad \mathbf{Z} = \begin{bmatrix} \mathbf{Z}_C & \mathbf{0} & \mathbf{0} & \mathbf{0} \\ \mathbf{0} & \mathbf{Z}_G & \mathbf{0} & \mathbf{0} \\ \mathbf{0} & \mathbf{0} & \mathbf{Z}_L & \mathbf{0} \\ \mathbf{0} & \mathbf{0} & \mathbf{0} & \mathbf{Z}_R \end{bmatrix}$$

where the suffices C, G, L, and R refer to tree capacitors, tree resistors, link inductors, and link resistors, respectively.

The branch voltage vector **e** and branch current vector **i** may be similarly partitioned first into tree branch and link branch variables with suffixes T and L respectively, and then into the component groups using the same suffixes as above, that is:

$$[\mathbf{e}] = \left[\begin{array}{c} \mathbf{e}_T \\ \hdashline \mathbf{e}_L \end{array}\right] = \left[\begin{array}{c} \mathbf{e}_{TC} \\ \mathbf{e}_{TG} \\ \hdashline \mathbf{e}_{LL} \\ \mathbf{e}_{LR} \end{array}\right]$$

$$[\mathbf{i}] = \left[\begin{array}{c} \mathbf{i}_T \\ \hdashline \mathbf{i}_L \end{array}\right] = \left[\begin{array}{c} \mathbf{i}_{TC} \\ \mathbf{i}_{TG} \\ \hdashline \mathbf{i}_{LL} \\ \mathbf{i}_{LR} \end{array}\right]$$

The source vectors **E** and **I** are partitioned in the same way. All sub-vector variables are defined by the following table

| | Branch voltages | Branch currents | Source voltages | Source currents |
|---|---|---|---|---|
| Tree branch capacitors | $\mathbf{e}_{TC}$ | $\mathbf{i}_{TC}$ | $\mathbf{E}_{TC}$ | $\mathbf{I}_{TC}$ |
| Tree branch resistors | $\mathbf{e}_{TG}$ | $\mathbf{i}_{TG}$ | $\mathbf{E}_{TG}$ | $\mathbf{I}_{TG}$ |
| Link branch inductors | $\mathbf{e}_{LL}$ | $\mathbf{i}_{LL}$ | $\mathbf{E}_{LL}$ | $\mathbf{I}_{LL}$ |
| Link branch resistors | $\mathbf{e}_{LR}$ | $\mathbf{i}_{LR}$ | $\mathbf{E}_{LR}$ | $\mathbf{I}_{LR}$ |

The branch equations from Chapter 2, Eqs. (2.7) and (2.8) are

$$\mathbf{i} = \mathbf{Ye} + \mathbf{I} - \mathbf{YE} \tag{7.6}$$

$$\mathbf{e} = \mathbf{Zi} + \mathbf{E} - \mathbf{ZI}\ . \tag{7.7}$$

Partitions may be inserted in all matrices provided that a proper tree has been drawn for the circuit and branches numbered in correct order. Equations for tree branch currents and link branch voltages may be selected from Eqs. (7.6) and (7.7) and written separately as

$$\mathbf{i}_{TC} = \mathbf{Y}_C \frac{d}{dt} \mathbf{e}_{TC} + \mathbf{I}_{TC} - \mathbf{Y}_C \frac{d}{dt} \mathbf{E}_{TC} \tag{7.8}$$

$$\mathbf{i}_{TG} = \mathbf{Y}_G \mathbf{e}_{TG} + \mathbf{I}_{TG} - \mathbf{Y}_G \mathbf{E}_{TG} \tag{7.9}$$

$$\mathbf{e}_{LL} = \mathbf{Z}_L \frac{d}{dt} \mathbf{i}_{LL} + \mathbf{E}_{LL} - \mathbf{Z}_L \frac{d}{dt} \mathbf{I}_{LL} \tag{7.10}$$

$$\mathbf{e}_{LR} = \mathbf{Z}_R \mathbf{i}_{LR} + \mathbf{E}_{LR} - \mathbf{Z}_R \mathbf{I}_{LR} \; . \tag{7.11}$$

The partitions in **Y** and **Z** are not restrictive for passive circuits with diagonal admittance or impedance matrices. However, the formulation of Eqs. (7.8) to (7.11) restricts the use of dependent source elements to real values in the resistive partitions, to inductive transfer impedance in $\mathbf{Z}_L$, and to capacitance transfer admittance in $\mathbf{Y}_C$. One more restriction which we can accept for convenience, is that tree branches contain only voltage sources. The remaining sources will be omitted from the following analysis, but they can easily be reinstated into the final solution. All of these restrictions will later be removed.

KCL for the cut-set and KVL for the tie-set were expressed in matrix form by Eqs. (2.11) and (2.12) as

$$\mathbf{D}^t \mathbf{i} = 0 \quad \text{and} \quad \mathbf{C}^t \mathbf{e} = 0 \; , \tag{7.12}$$

where $\mathbf{D}^t$ and $\mathbf{C}^t$ are the transposed cut-set and tie-set matrices respectively. Partitions may be inserted into Eqs. (7.12) to separate tree and link variables to yield

$$[\mathbf{D}_T^t \;\; \mathbf{D}_L^t] \begin{bmatrix} \mathbf{i}_T \\ \mathbf{i}_L \end{bmatrix} = 0 \quad \text{and} \; [\mathbf{C}_T^t \;\; \mathbf{C}_L^t] \begin{bmatrix} \mathbf{e}_T \\ \mathbf{e}_L \end{bmatrix} = 0 \; , \tag{7.13}$$

where suffices T and L signify tree and link partitions. Since $\mathbf{D}_T$ and $\mathbf{C}_L$ are both unit matrices, tree and link branch variables are related by

$$\mathbf{i}_T = -\mathbf{D}_L^t \, \mathbf{i}_L \; \text{and} \; \mathbf{e}_L = -\mathbf{C}_T^t \, \mathbf{e}_T \; . \tag{7.14}$$

Eqs. (7.14) may be further partitioned showing the variables required by Eqs. (7.8) to (7.11) in the form

$$\begin{bmatrix} \mathbf{i}_{TC} \\ \mathbf{i}_{TG} \end{bmatrix} = -\begin{bmatrix} \mathbf{D}_{CL} & \mathbf{D}_{CR} \\ \mathbf{D}_{GL} & \mathbf{D}_{GR} \end{bmatrix} \begin{bmatrix} \mathbf{i}_{LL} \\ \mathbf{i}_{LR} \end{bmatrix} \tag{7.15}$$

$$\begin{bmatrix} \mathbf{e}_{LL} \\ \mathbf{e}_{LR} \end{bmatrix} = -\begin{bmatrix} \mathbf{C}_{LC} & \mathbf{C}_{LG} \\ \mathbf{C}_{RC} & \mathbf{C}_{RG} \end{bmatrix} \begin{bmatrix} \mathbf{e}_{TC} \\ \mathbf{e}_{TG} \end{bmatrix} . \tag{7.16}$$

The first suffix of the partitioned $\mathbf{D}_L^t$ and $\mathbf{C}_L^t$ matrices indicates the dependent variable type and the second suffix indicates the independent variable type.

The required state variables are tree branch capacitor voltages $\mathbf{e}_{TC}$ and link branch inductor currents $\mathbf{i}_{LL}$. From Eqs. (7.15) and (7.16) we may obtain by rearrangement,

$$\begin{bmatrix} \mathbf{i}_{TC} \\ \mathbf{e}_{LL} \end{bmatrix} = -\begin{bmatrix} \mathbf{0} & \mathbf{D}_{CL} \\ \mathbf{C}_{LC} & \mathbf{0} \end{bmatrix} \begin{bmatrix} \mathbf{e}_{TC} \\ \mathbf{i}_{LL} \end{bmatrix} - \begin{bmatrix} \mathbf{0} & \mathbf{D}_{CR} \\ \mathbf{C}_{LG} & \mathbf{0} \end{bmatrix} \begin{bmatrix} \mathbf{e}_{TG} \\ \mathbf{i}_{LR} \end{bmatrix} . \tag{7.17}$$

Variables $\mathbf{e}_{TG}$ and $\mathbf{i}_{LR}$ due to resistive elements must now be eliminated. From Eqs. (7.15) and (7.16) we also have

$$\begin{bmatrix} \mathbf{e}_{LR} \\ \mathbf{i}_{TG} \end{bmatrix} = -\begin{bmatrix} \mathbf{C}_{RC} & \mathbf{0} \\ \mathbf{0} & \mathbf{D}_{GL} \end{bmatrix} \begin{bmatrix} \dot{\mathbf{e}}_{TC} \\ \dot{\mathbf{i}}_{LL} \end{bmatrix} - \begin{bmatrix} \mathbf{C}_{RG} & \mathbf{0} \\ \mathbf{0} & \mathbf{D}_{GR} \end{bmatrix} \begin{bmatrix} \mathbf{e}_{TG} \\ \mathbf{i}_{LR} \end{bmatrix} . \tag{7.18}$$

We must now involve the branch equations (7.8) to (7.11) in order to proceed. Eqs. (7.11) and (7.9) with sources $\mathbf{I}_{LR}$ and $\mathbf{E}_{TG}$ omitted may be rearranged to yield

$$\begin{bmatrix} \mathbf{e}_{LR} \\ \mathbf{i}_{TG} \end{bmatrix} = \begin{bmatrix} \mathbf{0} & \mathbf{Z}_R \\ \mathbf{Y}_G & \mathbf{0} \end{bmatrix} \begin{bmatrix} \mathbf{e}_{TG} \\ \mathbf{i}_{LR} \end{bmatrix} + \begin{bmatrix} \mathbf{E}_{LR} \\ \mathbf{I}_{TG} \end{bmatrix} . \tag{7.19}$$

Similarly from Eqs. (7.8) and (7.10), omitting sources $\mathbf{I}_{LL}$ and $\mathbf{E}_{TC}$ we can obtain

$$\begin{bmatrix} \mathbf{i}_{TC} \\ \mathbf{e}_{LL} \end{bmatrix} = \begin{bmatrix} \mathbf{Y}_C & \mathbf{0} \\ \mathbf{0} & \mathbf{Z}_L \end{bmatrix} \begin{bmatrix} \dot{\mathbf{e}}_{TC} \\ \dot{\mathbf{i}}_{LL} \end{bmatrix} + \begin{bmatrix} \mathbf{I}_{TC} \\ \mathbf{E}_{LL} \end{bmatrix} . \tag{7.20}$$

Solution of Eqs. (7.18) and (7.19) for $\mathbf{e}_{TG}$ and $\mathbf{i}_{LR}$ yields

$$\begin{bmatrix} \mathbf{e}_{TG} \\ \mathbf{i}_{LR} \end{bmatrix} = -\begin{bmatrix} \mathbf{C}_{RG} & \mathbf{Z}_R \\ \mathbf{Y}_G & \mathbf{D}_{GR} \end{bmatrix}^{-1} \begin{bmatrix} \mathbf{C}_{RC} & \mathbf{0} \\ \mathbf{0} & \mathbf{D}_{GL} \end{bmatrix} \begin{bmatrix} \mathbf{e}_{TC} \\ \mathbf{i}_{LL} \end{bmatrix} - \begin{bmatrix} \mathbf{C}_{RG} & \mathbf{Z}_R \\ \mathbf{Y}_G & \mathbf{D}_{GR} \end{bmatrix} \begin{bmatrix} \mathbf{E}_{LR} \\ \mathbf{I}_{TG} \end{bmatrix} . \tag{7.21}$$

$$\begin{bmatrix} \mathbf{i}_{TC} \\ \mathbf{e}_{LL} \end{bmatrix} = \left[ \begin{bmatrix} \mathbf{0} & \mathbf{D}_{CR} \\ \mathbf{C}_{LG} & \mathbf{0} \end{bmatrix} \begin{bmatrix} \mathbf{C}_{RG} & \mathbf{Z}_R \\ \mathbf{Y}_G & \mathbf{D}_{GR} \end{bmatrix}^{-1} \begin{bmatrix} \mathbf{C}_{RC} & \mathbf{0} \\ \mathbf{0} & \mathbf{D}_{GL} \end{bmatrix} - \begin{bmatrix} \mathbf{0} & \mathbf{D}_{CL} \\ \mathbf{C}_{LC} & \mathbf{0} \end{bmatrix} \right] \begin{bmatrix} \mathbf{e}_{TC} \\ \mathbf{i}_{LL} \end{bmatrix} + \begin{bmatrix} \mathbf{0} & \mathbf{D}_{CR} \\ \mathbf{C}_{LG} & \mathbf{0} \end{bmatrix} \begin{bmatrix} \mathbf{C}_{RG} & \mathbf{Z}_R \\ \mathbf{Y}_G & \mathbf{D}_{GR} \end{bmatrix}^{-1} \begin{bmatrix} \mathbf{E}_{LR} \\ \mathbf{I}_{TG} \end{bmatrix} \tag{7.22}$$

$$= \mathbf{p} \begin{bmatrix} \mathbf{e}_{TC} \\ \mathbf{i}_{LL} \end{bmatrix} + \mathbf{R} \begin{bmatrix} \mathbf{E}_{LR} \\ \mathbf{I}_{TG} \end{bmatrix} . \tag{7.23}$$

Substitution from Eq. (7.20) yields finally

$$\begin{bmatrix} \dot{\mathbf{e}}_{TC} \\ \dot{\mathbf{i}}_{LL} \end{bmatrix} = \begin{bmatrix} \mathbf{Y}_C & \mathbf{0} \\ \mathbf{0} & \mathbf{Z}_L \end{bmatrix}^{-1} \left[ \mathbf{P} \begin{bmatrix} \mathbf{e}_{TC} \\ \mathbf{i}_{LL} \end{bmatrix} + \mathbf{R} \begin{bmatrix} \mathbf{E}_{LR} \\ \mathbf{I}_{TG} \end{bmatrix} - \mathbf{U} \begin{bmatrix} \mathbf{I}_{TC} \\ \mathbf{E}_{LL} \end{bmatrix} \right] , \tag{7.24}$$

where $\mathbf{U}$ is a unit matrix. In standard form Eq. (7.24) may be written

$$\begin{bmatrix} \dot{\mathbf{e}}_{TC} \\ \dot{\mathbf{i}}_{LL} \end{bmatrix} = \mathbf{A} \begin{bmatrix} \mathbf{e}_{TC} \\ \mathbf{i}_{LL} \end{bmatrix} + \mathbf{B} \begin{bmatrix} \mathbf{E}_{LR} \\ \mathbf{I}_{TG} \\ \mathbf{I}_{TC} \\ \mathbf{E}_{LL} \end{bmatrix} ,$$

where the state equation matrices **A** and **B** are given by

$$\mathbf{A} = \begin{bmatrix} \mathbf{Y_C} & 0 \\ 0 & \mathbf{Z_L} \end{bmatrix}^{-1} \mathbf{P}, \text{ and } \mathbf{B} = \begin{bmatrix} \mathbf{Y_C} & 0 \\ 0 & \mathbf{Z_L} \end{bmatrix}^{-1} [\mathbf{R} \quad -\mathbf{U}] \ .$$

Within the three limitations listed at the beginning of this section which primarily affect active networks, the state equations for any network with a proper tree can be found. The output equation for a network must also be determined by a similar solution procedure which expresses output variables in terms of state variables.

*Example* 7.3

Determine the state equation for the transitor amplifier in Fig. 7.5(a).

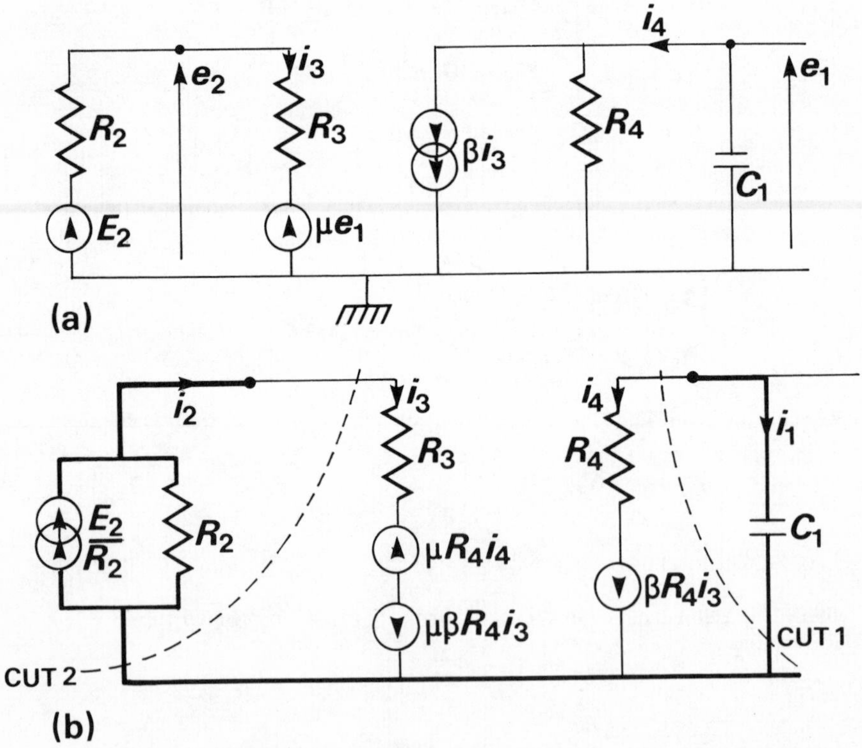


Fig. 7.5 – (a) Amplifier for Example 7.3; (b) Amplifier prepared for analysis.

The circuit is redrawn for analysis in Fig. 7.5(b) showing a proper tree and all dependent sources as transfer resistances between link resistor branches. The cut-set and tie-set matrices may be filled using the tree shown and partitioned, that is:

$$\mathbf{C} = \begin{bmatrix} 0 & -1 \\ 1 & 0 \\ \hdashline 1 & 0 \\ 0 & 1 \end{bmatrix}, \qquad \mathbf{D} = \begin{bmatrix} 1 & 0 \\ 0 & 1 \\ \hdashline 0 & -1 \\ 1 & 0 \end{bmatrix}.$$

Therefore,

$$\mathbf{C}_{\mathrm{T}}^{\mathrm{t}} = \left[\begin{array}{c:c} 0 & 1 \\ -1 & 0 \end{array}\right], \qquad \mathbf{D}_{\mathrm{L}}^{\mathrm{t}} = \begin{bmatrix} 0 & 1 \\ \hdashline -1 & 0 \end{bmatrix},$$

$$= \begin{bmatrix} \mathbf{C}_{\mathrm{LC}} & \mathbf{C}_{\mathrm{LG}} \\ \mathbf{C}_{\mathrm{RC}} & \mathbf{C}_{\mathrm{RG}} \end{bmatrix}, \qquad = \begin{bmatrix} \mathbf{D}_{\mathrm{CL}} & \mathbf{D}_{\mathrm{CR}} \\ \mathbf{D}_{\mathrm{GL}} & \mathbf{D}_{\mathrm{GR}} \end{bmatrix}.$$

The admittance and impedance matrix partitions are given by

$$\begin{bmatrix} \mathbf{Y}_{\mathrm{C}} & \mathbf{0} \\ \mathbf{0} & \mathbf{Z}_{\mathrm{L}} \end{bmatrix} = [C_1], \qquad \mathbf{Y}_{\mathrm{G}} = [1/R_2],$$

and $$\mathbf{Z}_{\mathrm{R}} = \begin{bmatrix} R_3 - \mu\beta R_4 & \mu R_4 \\ -\beta R_4 & R_4 \end{bmatrix}.$$

Substitution into Eq. (7.22) yields

$$\mathbf{P} = [0 \quad 0 \quad 1] \begin{bmatrix} 1 & R_3 - \mu\beta R_4 & \mu R_4 \\ 0 & -\beta R_4 & R_4 \\ 1/R_2 & -1 & 0 \end{bmatrix}^{-1} \begin{bmatrix} 0 \\ -1 \\ 0 \end{bmatrix}$$

$$= \left[\left(\frac{\mu\beta}{R_2 + R_3} - \frac{1}{R_4}\right)\right],$$

$$\text{and,}\quad \mathbf{R} = [0 \quad 0 \quad 1]\begin{bmatrix} 1 & R_3-\mu\beta R_4 & \mu R_4 \\ 0 & -\beta R_4 & R_4 \\ 1/R_2 & -1 & 0 \end{bmatrix}^{-1}$$

$$= [* \quad * \quad -\beta R_2/(R_2+R_3)] \ ,$$

where terms marked * are not needed and are therefore omitted. Finally, substitution into Eq. (7.24) yields,

$$[\dot{e}_1] = [1/C_1]\left[\left(\frac{\mu\beta}{R_2+R_3} - \frac{1}{R_4}\right)\right][e_1] - [1/C_1]\left[\left(\frac{\beta}{R_2+R_3}\right)\right][E_2]$$

The output equation is trivial since the output required is a state variable.

The main difficulty with this analysis of a quite simple amplifier was that dependent sources had to be reorganised. A more general solution is preferred in which this organization is included in the matrix manipulation procedure. The generalization will be described in the following section.

**Excess Element Networks**

The matrix procedure just described for generation of state equations follows the scheme summarized at the end of Example 7.1. By using the standard circuit branch from Chapter 2 for the inclusion of fixed sources, and by manipulating partitioned matrices in a similar way to that used for the generation of the *N*-port matrix in Chapter 4, the state equations of a large class of linear active networks can be easily found. The limitations introduced are not fundamental to the method used but are included to keep the mathematics lucid.

Two main modifications are necessary before the state equations for a general active network can be generated. The two difficulties which must be overcome were briefly illustrated by Example 7.2 where we found that our definition of a proper tree was going to be inadequate.

In some networks there is an excess of circuit components for which state variables are normally assigned. This situation arises when a node exists at which only inductors and current sources terminate. One inductor current must then be selected as dependent on the sum of the remaining node currents. Similarly in the dual situation a loop of capacitive and voltage source branches means that one capacitor voltage must be selected as dependent on the sum of the remaining loop voltages. The proper tree must therefore be modified to include excess inductive elements, and link branches may contain excess capacitative elements.

The inclusion of dependent sources within the admittance and impedance matrices makes them non-diagonal and unnecessarily complicates both storage and inversion as well as introducing restrictions on the type and grouping of sources. Also, as we saw in Example 7.2, fixed sources are not always easily incorporated within our standard branch definition. It is therefore convenient to redefine the standard branch as a single element which may be a component, a fixed source, or a dependent source. The definition of a proper tree must therefore be extended to include all voltage sources. All current sources are therefore link branches. Some limitations still exist on source connections; for example, loops consisting only of voltage sources, and voltage sources in parallel with tree capacitors, are not permitted.

As a consequence of these extensions to the definition of the proper tree, further partitions are required in the tree branch voltage, link branch current, cut-set, and tie-set matrices. Eqs. (7.15) and (7.16) are rewritten as

$$\mathbf{i}_T = \begin{bmatrix} \mathbf{i}_{TC} \\ \mathbf{i}_{TG} \\ \mathbf{i}_{TL} \\ \mathbf{i}_{TE} \\ \mathbf{i}_{TV} \end{bmatrix} = -\mathbf{D}_L^t \begin{bmatrix} \mathbf{i}_{LL} \\ \mathbf{i}_{LR} \\ \mathbf{i}_{LC} \\ \mathbf{i}_{LI} \\ \mathbf{i}_{LJ} \end{bmatrix}, \text{ and } \mathbf{e}_L = \begin{bmatrix} \mathbf{e}_{LL} \\ \mathbf{e}_{LR} \\ \mathbf{e}_{LC} \\ \mathbf{e}_{LI} \\ \mathbf{e}_{LJ} \end{bmatrix} = -\mathbf{C}_T^t \begin{bmatrix} \mathbf{e}_{TC} \\ \mathbf{e}_{TG} \\ \mathbf{e}_{TL} \\ \mathbf{e}_{TE} \\ \mathbf{e}_{TV} \end{bmatrix} \tag{7.25}$$

Variables additional to those previously used are defined by the following table.

| | *Branch voltages* | *Branch currents* |
|---|---|---|
| Tree branch inductors | $\mathbf{e}_{TL}$ | $\mathbf{i}_{TL}$ |
| Fixed voltage sources | $\mathbf{e}_{TE}$ | $\mathbf{i}_{TE}$ |
| Dependent voltage sources | $\mathbf{e}_{TV}$ | $\mathbf{i}_{TV}$ |
| Link branch capacitors | $\mathbf{e}_{LC}$ | $\mathbf{i}_{LC}$ |
| Fixed current sources | $\mathbf{e}_{LI}$ | $\mathbf{i}_{LI}$ |
| Dependent current sources | $\mathbf{e}_{LJ}$ | $\mathbf{i}_{LJ}$ |

Partitions in $\mathbf{D}_L^t$ and $\mathbf{C}_T^t$ are denoted as before by two suffixes.

Branch equations (7.8) to (7.11) can now be simplified by the omission of sources, giving

$$\begin{aligned} \mathbf{i}_{TC} &= \mathbf{Y}_{TC}\,\dot{\mathbf{e}}_{TC} & \mathbf{e}_{LL} &= \mathbf{Z}_{LL}\,\dot{\mathbf{i}}_{LL} \\ \mathbf{e}_{TG} &= \mathbf{Y}_G^{-1}\,\mathbf{i}_{TG} & \mathbf{i}_{LR} &= \mathbf{Z}_R^{-1}\,\mathbf{e}_{LR} \\ \mathbf{e}_{TL} &= \mathbf{Z}_{TL}\,\dot{\mathbf{i}}_{TL} & \mathbf{i}_{LC} &= \mathbf{Y}_{LC}\,\mathbf{e}_{LC} \end{aligned} \tag{7.26}$$

Dependent sources are incorporated by a matrix equation defining the dependent variables as a function of all branch voltages and currents in the network. The dependence may be represented by a real number, for example voltage gain, or it may involve derivatives, for example the transfer impedance of a mutual inductance. In general, it is impossible to predict when dependent sources introduce linear dependence in the network state variables. Also some assumptions will be made here in order to obtain a solution without introducing too many complications.

Dependent source voltages $\mathbf{e}_{TV}$ and $\mathbf{i}_{LJ}$ depend on $\mathbf{e}$ and $\mathbf{i}$. However, use of Eqs. (2.14) and (2.24) which relate branch variables to tree voltages and link currents by $\mathbf{e} = \mathbf{D}\,\mathbf{e}_T$ and $\mathbf{i} = \mathbf{C}\,\mathbf{i}_L$ enables us to write

$$\begin{bmatrix} \mathbf{e}_{TV} \\ \mathbf{i}_{LJ} \end{bmatrix} = \begin{bmatrix} \mathbf{S}_{VV} & \mathbf{S}_{IV} \\ \mathbf{S}_{VI} & \mathbf{S}_{II} \end{bmatrix} \begin{bmatrix} \mathbf{e} \\ \mathbf{i} \end{bmatrix} = \begin{bmatrix} \mathbf{S}_{VV}\,\mathbf{D} & \mathbf{S}_{IV}\,\mathbf{C} \\ \mathbf{S}_{VI}\,\mathbf{D} & \mathbf{S}_{IZ}\,\mathbf{C} \end{bmatrix} \begin{bmatrix} \mathbf{e}_T \\ \mathbf{i}_L \end{bmatrix} .$$

The notation is fairly obvious with $\mathbf{S}_{IV}$, for example, containing all CCVS. Inserting partitions yields

$$\begin{bmatrix} \mathbf{e}_{TV} \\ \mathbf{i}_{LJ} \end{bmatrix} = \mathbf{S} \begin{bmatrix} \mathbf{e}_T \\ \mathbf{i}_L \end{bmatrix} = \mathbf{S}_1 \begin{bmatrix} \mathbf{e}_{TC} \\ \mathbf{i}_{LL} \end{bmatrix} + \mathbf{S}_2 \begin{bmatrix} \mathbf{e}_{TG} \\ \mathbf{i}_{LR} \end{bmatrix} + \mathbf{S}_3 \begin{bmatrix} \mathbf{e}_{TL} \\ \mathbf{i}_{LC} \end{bmatrix} + \mathbf{S}_4 \begin{bmatrix} \mathbf{e}_{TE} \\ \mathbf{i}_{LI} \end{bmatrix} . \tag{7.27}$$

$\mathbf{S}_1$ to $\mathbf{S}_4$ may involve derivative terms in the form $L\,\mathrm{d}/\mathrm{d}t$ or $C\,\mathrm{d}/\mathrm{d}t$ where $L$ and $C$ arise from complex tranfer impedances or admittances. In order to separate these from the non-derivative terms, Eq. (7.27) will be rewritten as

$$\mathbf{e}_5 = \mathbf{S}_1\mathbf{e}_1 + \mathbf{S}_2\mathbf{e}_2 + \mathbf{S}_3\mathbf{e}_3 + \mathbf{S}_4\mathbf{e}_4 + \mathbf{S}_5\dot{\mathbf{e}}_1 + \mathbf{S}_6\dot{\mathbf{e}}_4 , \tag{7.28}$$

where $\mathbf{e}_1 = [\mathbf{e}_{TC}\ \mathbf{i}_{LL}]^t$, $\dot{\mathbf{e}}_1 = [\dot{\mathbf{e}}_{TC}\ \dot{\mathbf{i}}_{LL}]^t$, etc., and $\mathbf{S}_5$ and $\mathbf{S}_6$ contain the terms of $\mathbf{S}_1$ and $\mathbf{S}_4$ involving derivatives. $\mathbf{S}_2$ and $\mathbf{S}_3$ have been restricted to non-derivative terms by assumption.

Extraction of state variable terms and others of use from Eq. (7.25) combined with the branch equations (7.26) yields

$$\begin{bmatrix} \mathbf{Y}_{TC} & \mathbf{0} \\ \mathbf{0} & \mathbf{Z}_{LL} \end{bmatrix} \begin{bmatrix} \dot{\mathbf{e}}_{TC} \\ \dot{\mathbf{i}}_{LL} \end{bmatrix} = \mathbf{P}_1\mathbf{e}_1 + \mathbf{P}_2\mathbf{e}_2 + \mathbf{P}_3\mathbf{e}_3 + \mathbf{P}_4\mathbf{e}_4 + \mathbf{P}_5\mathbf{e}_5 \, , \tag{7.29}$$

$$\begin{aligned} \mathbf{e}_2 &= \begin{bmatrix} \mathbf{0} & \mathbf{Y}_G^{-1} \\ \mathbf{Z}_R^{-1} & \mathbf{0} \end{bmatrix} \begin{bmatrix} \mathbf{e}_{LR} \\ \mathbf{i}_{TG} \end{bmatrix} \\ &= \mathbf{Q}_1\mathbf{e}_1 + \mathbf{Q}_2\mathbf{e}_2 + \mathbf{Q}_3\mathbf{e}_3 + \mathbf{Q}_4\mathbf{e}_4 + \mathbf{Q}_5\mathbf{e}_5 \, , \end{aligned} \tag{7.30}$$

and

$$\begin{aligned} \mathbf{e}_3 &= \begin{bmatrix} \mathbf{0} & \mathbf{Z}_{TL} \\ \mathbf{Y}_{LC} & \mathbf{0} \end{bmatrix} \begin{bmatrix} \dot{\mathbf{e}}_{LC} \\ \dot{\mathbf{i}}_{TL} \end{bmatrix} . \\ &= \mathbf{R}_1\dot{\mathbf{e}}_1 + \mathbf{R}_4\dot{\mathbf{e}}_4 \, . \end{aligned} \tag{7.31}$$

In Eq. (7.25), excess element variables are expressed in terms of all other variables available, that is, $\mathbf{e}_1$, $\mathbf{e}_2$, $\mathbf{e}_3$, $\mathbf{e}_4$, and $\mathbf{e}_5$. However, each capacitive loop, for example, requires removal of one capacitor only, and its voltage can be expressed in terms of state and source variables. $\mathbf{R}_2$ and $\mathbf{R}_3$ are not therefore required and they are assumed to be zero matrices. Omission of $\mathbf{R}_5\,\dot{\mathbf{e}}_5$ in Eq. (7.31) is by assumption. A less restrictive assumption would be that $\mathbf{R}_5\,\dot{\mathbf{e}}_5$ involved first derivatives $\dot{\mathbf{e}}_1$, $\dot{\mathbf{e}}_2$, $\dot{\mathbf{e}}_3$, and $\dot{\mathbf{e}}_4$ only, by substitution from Eq. (7.28).

Reduction to standard form is now straightforward. Dependent source variables are eliminated by substitution from Eq. (7.28) into Eqs. (7.29) and (7.30) yielding

$$\dot{\mathbf{e}}_1 = \mathbf{P}_1'\mathbf{e}_1 + \mathbf{P}_2'\mathbf{e}_2 + \mathbf{P}_3'\mathbf{e}_3 + \mathbf{P}_4'\mathbf{e}_4' + \mathbf{P}_5'\dot{\mathbf{e}}_4 \, , \tag{7.32}$$

where

$$\mathbf{P}_1' = \left[ \begin{bmatrix} \mathbf{Y}_{TC} & \mathbf{0} \\ \mathbf{0} & \mathbf{Z}_{LL} \end{bmatrix} - \mathbf{P}_5\mathbf{S}_5 \right]^{-1} (\mathbf{P}_1 + \mathbf{P}_5\mathbf{S}_1) \text{ , etc.,}$$

and

$$\mathbf{e}_2 = \mathbf{Q}_1'\mathbf{e}_1 + \mathbf{Q}_3'\mathbf{e}_3 + \mathbf{Q}_4'\mathbf{e}_4 + \mathbf{Q}_5'\dot{\mathbf{e}}_1 + \mathbf{Q}_6'\dot{\mathbf{e}}_4 \, , \tag{7.33}$$

where $\mathbf{Q}_1' = [\mathbf{U} - \mathbf{Q}_2]^{-1}(\mathbf{Q}_1 + \mathbf{Q}_5\mathbf{S}_1)$, etc. and $\mathbf{U}$ is a unit matrix.

The standard form is obtained by substitution of $\mathbf{e}_3$ from Eq. (7.31) into

Eqs. (7.32) and (7.33) followed by substitution of $\mathbf{e}_2$ from Eq. (7.33) into Eqs. (7.31). We finally obtain

$$\dot{\mathbf{e}}_1 = (\mathbf{U} - \mathbf{P}_2'\mathbf{Q}_3'\mathbf{R}_1 - \mathbf{Q}_5' - \mathbf{P}_3'\mathbf{R}_1)^{-1}\ [(\mathbf{P}_1' + \mathbf{P}_2'\mathbf{Q}_1')\mathbf{e}_1$$

$$+ (\mathbf{Q}_4' + \mathbf{P}_4')\mathbf{e}_4 + (\mathbf{P}_2'\mathbf{Q}_3'\mathbf{R}_4 + \mathbf{Q}_6' + \mathbf{P}_3'\mathbf{R}_4 + \mathbf{P}_5')\dot{\mathbf{e}}_4]\ . \quad (7.34)$$

The output equation involves elimination of variables in a similar procedure to that used above. The out put variables $\mathbf{y}$ are assumed to be a combination of branch variables $\mathbf{e}$ and $\mathbf{i}$ defined by a matrix expression which can be put in the form

$$\mathbf{y} = \mathbf{T}\begin{bmatrix}\mathbf{e}_T \\ \mathbf{i}_L\end{bmatrix}$$

using Eqs. (2.14) and (2.24) as before, This can be expanded to

$$\mathbf{y} = \mathbf{T}_1\mathbf{e}_1 + \mathbf{T}_2\mathbf{e}_2 + \mathbf{T}_3\mathbf{e}_3 + \mathbf{T}_4\mathbf{e}_4 + \mathbf{T}_5\mathbf{e}_5$$

and substitution made from Eqs. (7.28), (7.31), and (7.33). The output equation is finally obtained by substitution for $\dot{\mathbf{e}}_1$ from Eq. (7.34).

The state and output equations generated here are more general than those in Eq. (7.24) and involve derivatives of the fixed sources. The restrictions introduced to obtain a solution can be avoided in practice, and generalized procedures which simultaneously process the state and output equation are available. However, dependent sources can still be arranged to prevent the generation of state variables in the expected standard form.

*Example* 7.4

Determine the state equation for the circuit in Fig. 7.6.

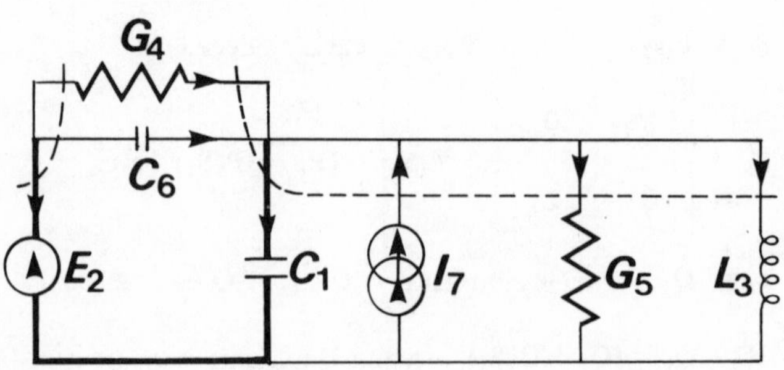


Fig. 7.6 – Circuit for Example 7.4.

A proper tree is marked on the figure together with current directions and cuts ① and ②. One point of immediate importance is the allocation of sign to fixed source branches. If we are to use **E** and **I** as branch voltages and currents then sign conventions in the standard branch of Chapter 2 must be reversed. This is because $\mathbf{e}+\mathbf{E}=\mathbf{0}$ and $\mathbf{i}+\mathbf{I}=\mathbf{0}$ in the absence of finite branch admittance elements. Standard current directions for voltage and current sources including dependent sources are shown in Fig. 7.7.

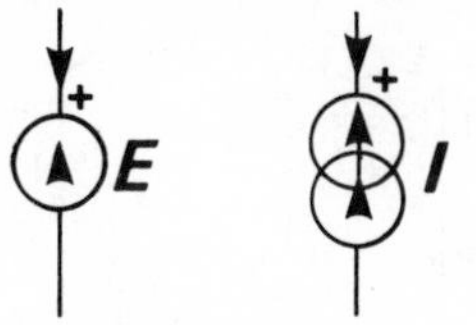


Fig. 7.7 – Standard source branch current directions.

Dependent sources are of course already treated by the standard of Fig. 7.7, as reference to Fig. 2.19 will confirm.

Tree and link branch variables are related by the topological matrices from Eq. 7.25 as

$$\begin{bmatrix} i_{TC1} \\ i_{TE2} \end{bmatrix} = \begin{bmatrix} -1 & 1 & -1 & 1 & 1 \\ 0 & -1 & 0 & -1 & 0 \end{bmatrix} \begin{bmatrix} i_{LL3} \\ i_{LG4} \\ i_{LG5} \\ i_{LC6} \\ i_{LI7} \end{bmatrix}$$

and

$$\begin{bmatrix} e_{LL3} \\ e_{LG4} \\ e_{LG5} \\ e_{LC6} \\ e_{LI7} \end{bmatrix} = \begin{bmatrix} 1 & 0 \\ -1 & 1 \\ 1 & 0 \\ -1 & 1 \\ -1 & 0 \end{bmatrix} \begin{bmatrix} e_{TC1} \\ e_{TE2} \end{bmatrix}$$

At this point it is worth noting that $\mathbf{C}_T = -\mathbf{D}_L^t$, and it is only necessary to derive one matrix from the directed graph. The second equation then requires the negative of the transposed matrix.

Since there are no dependent sources, $\mathbf{e}_5$ for Eq. (7.27) is not required. $\mathbf{e}_1$ to $\mathbf{e}_4$ are given by

$$\mathbf{e}_1 = \begin{bmatrix} e_{TC1} \\ i_{LL3} \end{bmatrix}, \quad \mathbf{e}_2 = \begin{bmatrix} i_{LG4} \\ i_{LG5} \end{bmatrix}, \quad \mathbf{e}_3 \ [i_{LC6}], \text{ and } \mathbf{e}_4 = \begin{bmatrix} e_{TE2} \\ i_{LI7} \end{bmatrix}.$$

Selecting the equation containing state variables as in Eq. (7.29),

$$\begin{bmatrix} C_1 & 0 \\ 0 & L_3 \end{bmatrix} \dot{\mathbf{e}}_1 = \begin{bmatrix} i_{TC1} \\ e_{LL3} \end{bmatrix}$$

$$= \begin{bmatrix} 0 & -1 \\ 1 & 0 \end{bmatrix} \mathbf{e}_1 + \begin{bmatrix} 1 & -1 \\ 0 & 0 \end{bmatrix} \mathbf{e}_2 + \begin{bmatrix} 1 \\ 0 \end{bmatrix} \mathbf{e}_3 + \begin{bmatrix} 0 & 1 \\ 0 & 0 \end{bmatrix} \mathbf{e}_4 .$$

Selecting resistor and link capacitor variables as in Eqs. (7.30) and (7.31)

$$\mathbf{e}_2 = \begin{bmatrix} G_4 & 0 \\ 0 & G_5 \end{bmatrix} \begin{matrix} e_{LG4} \\ e_{LG5} \end{matrix} = \begin{bmatrix} -G_4 & 0 \\ G_5 & 0 \end{bmatrix} \mathbf{e}_1 + \begin{bmatrix} G_4 & 0 \\ 0 & 0 \end{bmatrix} \mathbf{e}_4 .$$

$$\mathbf{e}_3 = C_6 \dot{e}_{LC6} = [-C_6 \quad 0]\ \dot{\mathbf{e}}_1 + [C_6 \quad 0]\ \dot{\mathbf{e}}_4 .$$

Substitution for $\mathbf{e}_2$ and $\mathbf{e}_3$ in the state variable equation yields

$$\begin{bmatrix} C_1 + C_6 & 0 \\ 0 & L_3 \end{bmatrix} \dot{\mathbf{e}}_1 = \begin{bmatrix} -(G_4 + G_5) & -1 \\ 1 & 0 \end{bmatrix} \mathbf{e}_1 + \begin{bmatrix} G_4 & 1 \\ 0 & 0 \end{bmatrix} \mathbf{e}_4 + \begin{bmatrix} C_6 & 0 \\ 0 & 0 \end{bmatrix} \dot{\mathbf{e}}_4 .$$

In standard form we finally obtain

$$\begin{bmatrix} \dot{e}_{TC1} \\ \dot{i}_{LL3} \end{bmatrix} = \begin{bmatrix} -(G_4+G_5)/(C_1+C_6) & -1/(C_1+C_6) \\ 1/L_3 & 0 \end{bmatrix} \begin{bmatrix} e_{TC1} \\ i_{LL3} \end{bmatrix}$$

$$+ \begin{bmatrix} G_4/C_1+C_6) & 1/(C_1+C_6) \\ 0 & 0 \end{bmatrix} \begin{bmatrix} e_{TE2} \\ i_{LI7} \end{bmatrix} + \begin{bmatrix} C_6/(C_1+C_6) & 0 \\ 0 & 0 \end{bmatrix} \begin{bmatrix} \dot{e}_{TE2} \\ \dot{i}_{LI7} \end{bmatrix}$$

**Network Functions**

Network functions from Chapter 6 for circuits with a single input and a single output are rational functions in the Laplace variable $s$. The general form of the unfactorized rational function from Eq. (6.30) relates the network output to the input by

$$F_{\mathrm{OUT}}(s) = \frac{\sum_{i=0}^{m} a_i s^i}{\sum_{j=0}^{n} b_j s^j} F_{\mathrm{IN}}(s) , \tag{7.35}$$

where $m \leqslant n$. Eq. (7.35) may be rewritten,

$$\sum_{j=0}^{n} b_j s^j F_{\mathrm{OUT}}(s) = \sum_{i=0}^{m} a_i s^i F_{\mathrm{IN}}(s) . \tag{7.36}$$

Each term of the form $s^k F(s)$ may be expressed with the help of Table 6.2, and assuming zero initial conditions, in the time domain as $\mathrm{d}^k f(t)/\mathrm{d}t^k$. Therefore,

$$\sum_{j=0}^{n} b_j \frac{\mathrm{d}^j}{\mathrm{d}t^j} f_{\mathrm{OUT}}(t) = \sum_{i=0}^{m} a_i \frac{\mathrm{d}^i}{\mathrm{d}t^i} f_{\mathrm{IN}}(t) . \tag{7.37}$$

Non-zero initial conditions will be reinstated during the solution of this equation in state variable form. However, we must first put Eq. (7.37) into standard form by defining a suitable set of state variables.

Assume initially that $a_0=1$, and $a_1=a_2=a_3 \ldots = a_m=0$. Derivatives of the input are not then required. Define a set of state variables by the equations

$$x_1 = f_{\mathrm{OUT}}(t)$$

$$x_2 = \dot{x}_1 = \frac{\mathrm{d}}{\mathrm{d}t} f_{\mathrm{OUT}}(t)$$

$$x_3 = \dot{x}_2 = \frac{\mathrm{d}^2}{\mathrm{d}t^2} f_{\mathrm{OUT}}(t)$$

$$\vdots$$

$$x_n = \dot{x}_{n-1} = \frac{\mathrm{d}^{n-1}}{\mathrm{d}t^{n-1}} f_{\mathrm{OUT}}(t) .$$

An expression for $\dot{x}_n$ can be obtained from Eq. (7.37), remembering that the right-hand side equals $f_{\mathrm{IN}}(t)$, or in terms of a standard variable name $u_1$. Solving Eq. (7.37) yields

$$\dot{x}_n = -\sum_{j=0}^{n-1} b_j \frac{\mathrm{d}^j}{\mathrm{d}t^j} f_{\mathrm{OUT}}(t) + f_{\mathrm{IN}}(t) .$$

We may therefore write state equations in standard form as

$$\begin{bmatrix} \dot{x}_1 \\ \dot{x}_2 \\ \dot{x}_3 \\ \cdot \\ \cdot \\ \cdot \\ \cdot \\ \cdot \\ \cdot \\ \dot{x}_n \end{bmatrix} = \begin{bmatrix} 0 & 1 & & & & 0 \\ & 0 & 1 & & & \\ & & 0 & & & \\ & & & \cdot & & \\ & & & & \cdot & \\ & & & & \cdot \; 1 & \\ 0 & & & & 0 & 1 \\ -b_0 & -b_1 & -b_2 & & -b_{n-2} & -b_{n-1} \end{bmatrix} \begin{bmatrix} x_1 \\ x_2 \\ x_3 \\ \cdot \\ \cdot \\ \cdot \\ \cdot \\ \cdot \\ \cdot \\ x_n \end{bmatrix} + \begin{bmatrix} 0 \\ 0 \\ 0 \\ \cdot \\ \cdot \\ \cdot \\ \cdot \\ \cdot \\ \cdot \\ 1 \end{bmatrix} [u_1] \tag{7.38}$$

The output equation is given by $y=x_1$ for $a_0=1$ and by $y=a_0x_1$ for $a_0 \neq 1$. Therefore in standard matrix from the output is given by

$$y = [a_0 \quad 0 \quad 0 \quad \ldots . \quad 0] \begin{bmatrix} x_1 \\ \cdot \\ \cdot \\ \cdot \\ x_n \end{bmatrix} . \tag{7.39}$$

The general case for which $a_1, a_2, \ldots a_m$ are non-zero involves derivatives of the input. However, the proper form of the state equation not involving input derivatives can easily be found. This step is of course essential for input signals known as a table of numerical values, because differentiation is then a very inaccurate process. Elimination of input derivatives is based on the property that an input $\dot{f}_{IN}(t)$ to a linear time invariant network will produce an output $\dot{f}_{OUT}(t)$. Inputs of derivatives due to terms $a_1\dot{u}_1$, $a_2\ddot{u}_2$, etc., in Eq. (7.37) give rise to outputa as $a_1\dot{x}_1$, $a_2\ddot{x}_1$, etc. In terms of state variables the output equation becomes

$$y = [a_0\, a_1\, a_2 \ldots a_m\; 0 \ldots 0] \begin{bmatrix} x_1 \\ \cdot \\ \cdot \\ \cdot \\ x_n \end{bmatrix} \tag{7.40}$$

in the general case for $m<n$.

If $m=n$, the final term involving $a_m\dot{x}_n$ must be obtained from Eq. (7.38) as

$$[-b_0a_m \quad -b_1a_m \quad \ldots \quad \ldots \quad -b_{n-1}a_m] \begin{bmatrix} x_1 \\ \cdot \\ \cdot \\ \cdot \\ x_n \end{bmatrix} + a_m u_1 \ .$$

Therefore,

$$y = [a_0-b_0a_m \quad a_1-b_1a_m \quad \ldots\ldots \quad a_m-b_{n-1}a_m] \begin{bmatrix} x_1 \\ \cdot \\ \cdot \\ \cdot \\ x_n \end{bmatrix} + [a_m]\ [u_1] \tag{7.41}$$

Thus the network functions from Chapter 6 may be put into standard state equation form and transient response evaluated by the solution methods of the following section. More complex procedures may be used to generate similar state equations for networks with many inputs and outputs.

## 7.2 SOLUTION OF STATE EQUATIONS

The standard form of the state equation and output equation from Section 7.1 is

$$\dot{\mathbf{x}} = \mathbf{A}\mathbf{x} + \mathbf{B}\mathbf{u} \tag{7.42}$$

$$\mathbf{y} = \mathbf{C}\mathbf{x} + \mathbf{D}\mathbf{u} \ . \tag{7.43}$$

In this section matrices **A**, **B**, **C**, and **D** will without exception be the state matrices. The input vector **u** will in general contain derivatives of the input although they can be removed from the state equation but not the output equation by redefining state variables. Matrix **A** is called the characteristic matrix, **B** couples the input to the state variables, **C** couples the state variables to the output, and **D** is due to the direct coupling from input to output.

### Solution by Laplace Transform

Application of the Laplace transform to Eqs. (7.42) and (7.43) yields

$$s\mathbf{X}(s) - \mathbf{x}(0) = \mathbf{A}\mathbf{X}(s) + \mathbf{B}\mathbf{U}(s)$$

$$\mathbf{Y}(s) = \mathbf{C}\mathbf{X}(s) + \mathbf{D}\mathbf{U}(s) \ .$$

Therefore,

$$(s\mathbf{I}-\mathbf{A})\mathbf{X}(s) = \mathbf{x}(0) + \mathbf{BU}(s) ,$$

where **I** is the unit matrix, and

$$\mathbf{X}(s) = (s\mathbf{I}-\mathbf{A})^{-1}\mathbf{x}(0) + (s\mathbf{I}-\mathbf{A})^{-1}\mathbf{BU}(s) \tag{7.44}$$

$$\mathbf{Y}(s) = \mathbf{C}(s\mathbf{I}-\mathbf{A})^{-1}\mathbf{x}(0) + [\mathbf{C}(s\mathbf{I}-\mathbf{A})^{-1}\mathbf{B} + \mathbf{D}]\,\mathbf{U}(s) . \tag{7.45}$$

The first part of the solution for $\mathbf{X}(s)$ in Eq. (7.44) due to the initial conditions $\mathbf{x}(0)$ is called the **zero input response**, and the second part due to the input $\mathbf{U}(s)$ is called the **zero state response**. Solution in the time domain is finally obtained by inverse Laplace transformation of Eqs. (7.44) and (7.45).

Taking the inverse Laplace transform of Eq. (7.44),

$$\mathbf{x}(t) = \mathcal{L}^{-1}\{(s\mathbf{I}-\mathbf{A})^{-1}\}\mathbf{x}(0) + \mathcal{L}^{-1}\{(s\mathbf{I}-\mathbf{A})^{-1}\mathbf{BU}(s)\} .$$

If $\mathcal{L}^{-1}\{(s\mathbf{I}-\mathbf{A})^{-1}\}$ is denoted by $\phi(t)$ and $\mathcal{L}^{-1}\mathbf{BU}(s)$ equals $\mathbf{Bu}(t)$, then

$$\mathbf{x}(t) = \phi(t)\mathbf{x}(0) + \int_0^t \phi(t-\tau)\,\mathbf{Bu}(\tau)\mathrm{d}\tau , \tag{7.46}$$

where the zero state response is in the form of a convulution integral. $\phi(t)$ is called the **state transition matrix** since it converts an initial state at $t$=0 into the state at any time $t$>0.

In order to define the state transition matrix more fully we must consider the direct solution of Eq. (7.42). The zero input response is obtained from the solution of the simultaneous first-order differential equations

$$\dot{\mathbf{x}} = \mathbf{Ax} \tag{7.47}$$

A solution of the form $\mathrm{x} = \mathrm{e}^{\mathbf{A}t}x_0$, where

$$\mathrm{e}^{\mathbf{A}t} = \mathbf{I} + \mathbf{A}t + \frac{\mathbf{A}^2t^2}{2!} + \frac{\mathbf{A}^3t^3}{3!} \ldots \tag{7.48}$$

may be proposed by analogy with the solution of the single equation $\dot{x} = ax$. Substitution of the assumed solution in Eq. (7.47) yields

$$\mathrm{e}^{\mathbf{A}t}\mathbf{Ax}_0 = \mathbf{A}\mathrm{e}^{\mathbf{A}t}\mathbf{x}_0 .$$

Provided that we can show

$$\frac{d}{dt}e^{At} = e^{At}A = Ae^{At},$$

the solution is verified. Both equalities can easily be proven by use of the series expansion (7.48). Thus we can identify the state transition matrix $\phi(t)$ with $e^{At}$.

Eq. (7.46) may therefore be rewritten as

$$\mathbf{x}(t) = e^{\mathbf{A}t}\mathbf{x}_0 + \int_0^t e^{\mathbf{A}(t-\tau)}\,\mathbf{B}\mathbf{u}(\tau)d\tau \tag{7.49}$$

$$= e^{\mathbf{A}t}\mathbf{x}_0 + e^{\mathbf{A}t}\int_0^t e^{-\mathbf{A}\tau}\,\mathbf{B}\mathbf{u}(\tau)d\tau\ . \tag{7.50}$$

Efficient computation of the state transition matrix is therefore an essential prerequisite to solution of the state equation, and much of the remainder of this chapter is concerned with the problem.

*Example* 7.5

Solve the state equation derived in Example 7.4 by the Laplace transform method assuming $C_1 = 1.0$ F, $C_6 = 0.5$ F, $G_4 = 2$ S, $G_5 = 4$ S, and $L_3 = 0.2$ H. Assume that the input $E_2$ is a unit step, $I_7$ is not excited, and that the initial conditions are $e_{TC1} = 2$ V and $i_{LI7} = 0.5$ A.

Substitution of numerical values into the state equation yields

$$\dot{\mathbf{x}} = \begin{bmatrix} -4 & -2/3 \\ 9/2 & 0 \end{bmatrix} \mathbf{x} + \begin{bmatrix} 4/3 & 2/3 & 1/3 \\ 0 & 0 & 0 \end{bmatrix} \mathbf{u}$$

where $\mathbf{x} = \begin{bmatrix} e_{TC1} \\ i_{LL3} \end{bmatrix}$ and $\mathbf{u} = \begin{bmatrix} e_{TE2} \\ i_{LI7} \\ \dot{e}_{TE2} \end{bmatrix}$ .

Taking Laplace transforms,

$$\begin{bmatrix} (s+4) & 2/3 \\ -9/2 & s \end{bmatrix} \mathbf{X} = \begin{bmatrix} 2 \\ 1/2 \end{bmatrix} + \begin{bmatrix} 4/3 & 2/3 & 1/3 \\ 0 & 0 & 0 \end{bmatrix} \begin{bmatrix} 1/s \\ 0 \\ 1 \end{bmatrix},$$

where $\mathbf{x}(0) = \begin{bmatrix} 2 \\ 1/2 \end{bmatrix}$ and $\mathbf{U}(s) = \begin{bmatrix} 1/s \\ 0 \\ 1 \end{bmatrix}$.

Therefore,

$$\mathbf{X}(s) = \begin{bmatrix} (s+4) & 2/3 \\ -9/2 & s \end{bmatrix}^{-1} \left[ \begin{bmatrix} 2 \\ 1/2 \end{bmatrix} + \begin{bmatrix} 4/3s + 1/4 \\ 0 \end{bmatrix} \right],$$

and the zero input response is given by

$$\mathbf{x}(t) = \mathcal{L}^{-1} \left\{ \begin{bmatrix} (s+4) & 2/3 \\ -9/2 & s \end{bmatrix}^{-1} \begin{bmatrix} 2 \\ 1/2 \end{bmatrix} \right\}$$

$$= \mathcal{L}^{-1} \left\{ \begin{bmatrix} s/(s+1)(s+3) & -2/3(s+1)(s+3) \\ 9/2(s+1)(s+3) & (s+4)/(s+1)(s+3) \end{bmatrix} \begin{bmatrix} 2 \\ 1/2 \end{bmatrix} \right\}$$

$$= \begin{bmatrix} (-\frac{1}{2}e^{-t} + \frac{3}{2}e^{-3t}) & (-\frac{1}{3}e^{-t} + \frac{1}{3}e^{-3t}) \\ (\frac{9}{4}e^{-2} - \frac{9}{4}e^{-3t}) & (\frac{3}{2}e^{-t} - \frac{1}{2}e^{-3t}) \end{bmatrix} \begin{bmatrix} 2 \\ \frac{1}{2} \end{bmatrix}$$

$$= \begin{bmatrix} -\frac{7}{6}e^{-t} + \frac{19}{6}e^{-3t} \\ \frac{21}{4}e^{-t} - \frac{19}{4}e^{-3t} \end{bmatrix}.$$

The zero state unit step response is given by

$$\mathbf{x}(t) = \mathcal{L}^{-1} \left\{ \begin{bmatrix} (s+4) & 2/3 \\ -9/2 & s \end{bmatrix}^{-1} \begin{bmatrix} 1/3 + 4/3s \\ 0 \end{bmatrix} \right\}$$

$$= \mathcal{L}^{-1} \left\{ \begin{bmatrix} (s+4)/3(s+1)(s+3) \\ 3(s+4)/2s(s+1)(s+3) \end{bmatrix} \right\}$$

$$= \begin{bmatrix} (\frac{1}{2}e^{-t} - \frac{1}{6}e^{-3t}) \\ (2 - \frac{9}{4}e^{-t} + \frac{1}{4}e^{-3t} \end{bmatrix}.$$

The total response is of course the sum of zero input and zero state responses.

We have also obtained the state transition matrix as a by-product of the solution for zero input response in the exponential form

$$e^{\mathbf{A}t} = \begin{bmatrix} (-\frac{1}{2}e^{-t} + \frac{3}{2}e^{-3t}) & (-\frac{1}{3}e^{-t} + \frac{1}{3}e^{-3t}) \\ (\frac{9}{4}e^{-t} - \frac{9}{4}e^{-3t}) & (\frac{3}{2}e^{-t} - \frac{1}{2}e^{-3t}) \end{bmatrix} .$$

**Solution by Eigenvalues**

The *eigenvalues* of an $n \times n$ matrix $\mathbf{A}$ are defined as the roots of the matrix characteristic equation $|\mathbf{A} - \lambda\mathbf{I}| = 0$, where $\mathbf{I}$ is the unit matrix. Since $\lambda$ only appears on the diagonal of $[\mathbf{A}-\lambda\mathbf{I}]$, evaluation of the determinant results in a polynomial of degree $n$ in the variable $\lambda$. The elements of $\mathbf{A}$ are real and the eigenvalues are therefore real or occur in complex conjugate pairs. In general, eigenvalues will be multiple, but for most of this chapter they will be assumed to be distinct. The theory developed here can be extended to the multiple eigenvector case.

The eigenvalues of

$$\mathbf{A} = \begin{bmatrix} -4 & -2/3 \\ 9/2 & 0 \end{bmatrix}$$

from Example 7.5 are obtained from evaluation of the determinant

$$\begin{bmatrix} (-4-\lambda) & -2/3 \\ 9/2 & -\lambda \end{bmatrix} ,$$

that is $\quad \lambda^2 + 4\lambda + 3 = 0 .$

Therefore $\lambda_1 = -1$ and $\lambda_2 = -3$. These are of course the same values as those obtained for the poles of transfer functions formed from $[s\mathbf{I}-\mathbf{A}]^{-1}$ since the inverse equals adjoint $[s\mathbf{I}-\mathbf{A}]/|s\mathbf{I}-\mathbf{A}|$. Thus the eigenvalues are an important property of $\mathbf{A}$ which define the natural frequencies of the network.

The theory of eigenvalues and eigenvectors may be applied to the problem of direct determination of the state transition matrix $e^{\mathbf{A}t}$ from $\mathbf{A}$. Initially this will be carried out through the Cayley-Hamilton theorem.

*The Cayley-Hamilton theorem states that a matrix satisfies its own characteristic equation.*

This may be confirmed for the matrix above by substitution of **A** into the characteristic equation giving

$$\mathbf{A}^2 + 4\mathbf{A} + 3\mathbf{I}$$

Substituting values,

$$\begin{bmatrix} -4 & -2/3 \\ 9/2 & 0 \end{bmatrix}\begin{bmatrix} -4 & -2/3 \\ 9/2 & 0 \end{bmatrix} + 4\begin{bmatrix} -4 & -2/3 \\ 9/2 & 0 \end{bmatrix} + 3\begin{bmatrix} 1 & 0 \\ 0 & 1 \end{bmatrix},$$

therefore,

$$\begin{bmatrix} 13 & 8/3 \\ -36/2 & -3 \end{bmatrix} + \begin{bmatrix} -16 & -8/3 \\ 18 & 0 \end{bmatrix} + \begin{bmatrix} 3 & 0 \\ 0 & 3 \end{bmatrix} = \begin{bmatrix} 0 & 0 \\ 0 & 0 \end{bmatrix}.$$

Proof of this theorem will not be given here, but a short proof for the case of distinct eigenvalues will be given in the following section on diagonalization. The Cayley-Hamilton theorem will be applied to the evaluation of a matrix raised to a power.

The $n \times n$ matrix **A** may be substituted into its characteristic equation to yield the polynomial

$$c_0\mathbf{I} + c_1\mathbf{A} + c_2\mathbf{A}^2 \;\ldots\ldots\; c_n\mathbf{A}^n = \mathbf{0}\,. \tag{7.51}$$

$\mathbf{A}^n$ can therefore be expressed in terms of powers of **A** up to $(n-1)$ by solution of Eq. (7.51), yielding

$$\mathbf{A}^n = -\frac{c_0}{c_n}\mathbf{I} - \frac{c_1}{c_n}\mathbf{A} - \frac{c_2}{c_n}\mathbf{A}^2 \ldots - \frac{c_{n-1}}{c_n}\mathbf{A}^{n-1}$$

Pre-multiplication by **A** followed by substitution for $\mathbf{A}^n$ gives

$$\mathbf{A}^{n+1} = \frac{c_{n-1}\,c_0}{c_n^2}\mathbf{I} + \left(\frac{c_{n-1}\,c_1}{c_n^2} - \frac{c_0}{c_n}\right)\mathbf{A} \ldots \left(\frac{c_{n-1}^2}{c_n^2} - \frac{c_{n-2}}{c_n}\right)\mathbf{A}^{n-1}\,.$$

This process may be repeated for all higher powers of **A** than $n$. An important conclusion based on this result derived from the Cayley-Hamilton theorem is that a power series expansion of a matrix function may be replaced by a finite series of $n$ terms.

The state transition matrix was defined by the expansion

$$e^{\mathbf{A}t} = \mathbf{I} + \mathbf{A}t + \frac{\mathbf{A}^2t^2}{2!} + \frac{\mathbf{A}^3t^3}{3!} \;\ldots\ldots,$$

and it can now be replaced by the expression

$$e^{\mathbf{A}t} = k_0\mathbf{I} + k_1\mathbf{A} + k_2\mathbf{A}^2 + k_3\mathbf{A}^3 \;\ldots\; k_{n-1}\mathbf{A}^{n-1} \tag{7.52}$$

The coefficients $k_0, k_1 \ldots k_{n-1}$ must be functions of time, and since they each involve an infinite series of terms from an exponential function we might reasonably expect them to be sums of exponential functions of time.

Since Eq. (7.51) is satisfied by the eigenvalues of **A**, we may use the same argument to show that Eq. (7.52) is also satisfied by the eigenvalues. Substitution of each eigenvalue in turn therefore gives a set of simultaneous equations which can be solved for the values $k_0, k_1 \ldots k_{n-1}$, that is,

$$\begin{aligned} e^{\lambda_1 t} &= k_0 + k_1\lambda_1 + k_2\lambda_1^2 \;\ldots\; k_{n-1}\lambda_1^{n-1} \\ &\vdots \\ e^{\lambda_n t} &= k_0 + k_1\lambda_n + k_2\lambda_n^3 \;\ldots\; k_{n-1}\lambda_n^{n-1} \end{aligned} \tag{7.53}$$

Therefore,

$$\begin{bmatrix} k_0 \\ \cdot \\ \cdot \\ \cdot \\ k_{n-1} \end{bmatrix} = \begin{bmatrix} 1 & \lambda_1 & \lambda_1^2 & \ldots\ldots & \lambda_1^{n-1} \\ \cdot & & & & \\ \cdot & & & & \\ \cdot & & & & \\ 1 & \lambda_n & \lambda_n^2 & \ldots\ldots & \lambda_n^{n-1} \end{bmatrix}^{-1} \begin{bmatrix} e^{\lambda_1 t} \\ \cdot \\ \cdot \\ \cdot \\ e^{\lambda_n t} \end{bmatrix}. \tag{7.54}$$

The matrix inversion can be carried out provided that the eigenvalues are distinct. When multiple eigenvalues are present there are insufficient equations for solution and the procedure is modified. Assuming, for example, that $\lambda_1$ was a multiple eigenvalue of order 3. The first equation of Eqs. (7.53) is obtained as before by substitution of $\lambda_1$ giving

$$e^{\lambda_1 t} = k_0 + k_1\lambda_1 + k_2\lambda_1^2 \;\ldots\; k_{n-1}\lambda_1^{n-1} .$$

Differentiation is then used to obtain the two necessary further equations,

$$\begin{aligned} t\,e^{\lambda_1 t} &= k_1 \;+ 2k_2\lambda_1 + 3k_3\lambda_1^2 \;\ldots\; (n-1)k_{n-1}\lambda_1^{n-2} \\ t^2 e^{\lambda_1 t} &= 2k_2 + 6k_3\lambda_1 + 12k_4\lambda_1^2 \;\ldots\; (n-1)(n-2)k_{n-2}\lambda_1^{n-3} . \end{aligned}$$

The resulting time multiplied exponentials are to be expected from our previous work on partial fraction expansion.

When the eigenvalues arise in complex conjugate pairs, Eq. (7.54) involves complex numbers and the exponential functions of time are also complex. The solutions for $k_0 \ldots k_{n-1}$ are therefore oscillatory with exponential damping due to the real parts of the eigenvalues.

*Example* 7.6

Determine the state transition matrix for the state equation of Example 7.5 by the eigenvalue method. Complete the solution for zero state response by the convolution integral.

The **A** matrix in Example 7.5 was given by

$$\begin{bmatrix} -4 & -2/3 \\ 9/2 & 0 \end{bmatrix}$$

and the eigenvalues are obtained as solutions of $(-4-\lambda)(-\lambda) + 3 = 0$. Therefore $\lambda^2 + 4\lambda + 3 = 0$ and $\lambda_1 = -1, \lambda_2 = -3$.

Substitution in Eq. (7.53) yields

$$e^{-t} = k_0 + k_1(-1)$$

$$e^{-3t} = k_0 + k_1(-3) \ ,$$

therefore,

$$\begin{bmatrix} k_0 \\ k_1 \end{bmatrix} = \begin{bmatrix} 1 & -1 \\ 1 & -3 \end{bmatrix}^{-1} \begin{bmatrix} e^{-t} \\ e^{-3t} \end{bmatrix}$$

$$= -\tfrac{1}{2} \begin{bmatrix} -3 & 1 \\ -1 & 1 \end{bmatrix} \begin{bmatrix} e^{-t} \\ e^{-3t} \end{bmatrix}$$

$$= \begin{bmatrix} \tfrac{3}{2}\,e^{-t} - \tfrac{1}{2}\,e^{-3t} \\ \tfrac{1}{2}\,e^{-t} - \quad e^{-3t} \end{bmatrix} .$$

Substitution into Eq. (7.52) gives

$$e^{\mathbf{A}t} = (\tfrac{3}{2}e^{-t} - \tfrac{1}{2}e^{-3t})\begin{bmatrix} 1 & 0 \\ 0 & 1 \end{bmatrix} + (\tfrac{1}{2}e^{-t} - \tfrac{1}{2}e^{-3t})\begin{bmatrix} -4 & -2/3 \\ 9/2 & 0 \end{bmatrix}$$

$$= \begin{bmatrix} (-\tfrac{1}{2}e^{-t} + \tfrac{3}{2}e^{-3t}) & (-\tfrac{1}{3}e^{-t} + \tfrac{1}{3}e^{-3t}) \\ (\tfrac{9}{4}e^{-t} - \tfrac{9}{4}e^{-3t}) & (\tfrac{3}{2}e^{-t} - \tfrac{1}{2}e^{-3t}) \end{bmatrix}$$

This state transition matrix is the same as that derived in Example 7.5.

The convolution integral for the zero state response is given from Eq. (7.49) by

$$x(t) = \int_0^t e^{\mathbf{A}(t-\tau)}\,\mathbf{B}u(\tau)\,d\tau$$

$$= \int_0^t \begin{bmatrix} -\tfrac{1}{2}e^{-t+\tau} + \tfrac{3}{2}e^{-3t+3\tau} & u(t-\tau) & * \\ \tfrac{9}{4}e^{-t+\tau} - \tfrac{9}{4}e^{-3t+3\tau} & u(t-\tau) & * \end{bmatrix} \begin{bmatrix} \tfrac{4}{3} & \tfrac{2}{3} & \tfrac{1}{3} \\ 0 & 0 & 0 \end{bmatrix} \begin{bmatrix} u(\tau) \\ 0 \\ u'(\tau) \end{bmatrix} d\tau$$

where the asterisk has been inserted for clarity in place of the two terms of $e^{\mathbf{A}t}$ which are not needed as they are eventually multiplied by zero.

The step functions $u(t-\tau)$ are needed to define the values of $\tau$ for which the exponentials are present. The excitation function may be multiplied to yield

$$\begin{bmatrix} \tfrac{4}{3}u(\tau) + \tfrac{1}{3}u'(\tau) \\ 0 \end{bmatrix}$$

and products $u(t-\tau)\ u(\tau)$ and $u(t-\tau)u'(\tau)$ arise from integration. Now $u(t-\tau)u(\tau)$ is unity for $0<\tau<t$ and zero elsewhere and has no effect on integration between 0 and $t$. However, the product $u(t-\tau)u'(\tau)$ only exists as an impulse $u'(\tau)$ at $\tau=0$. This has the effect of sampling the exponentials $e^{\tau}$ and $e^{3\tau}$ at $\tau=0$ when they are both unity. The impulse area, in this case 1/3, is therefore the value of both

$$\int_0^t \tfrac{1}{3}u'(\tau)e^{\tau}d\tau \quad \text{and} \quad \int_0^t \tfrac{1}{3}u'(\tau)e^{3\tau}\,d\tau\ .$$

We may now perform the integration

$$x(t) = \begin{bmatrix} \frac{4}{3}(-\frac{1}{2}e^{-t}e^{\tau} + \frac{1}{2}e^{-3t}e^{3\tau}) \\ \frac{4}{3}(\frac{9}{4}e^{-t}e^{\tau} \quad - \frac{3}{4}e^{-3t}e^{3\tau}) \end{bmatrix}_0^t + \begin{bmatrix} \frac{1}{3}(-\frac{1}{2}e^{-t} + \frac{3}{2}e^{-3t}) \\ \frac{1}{3}(\frac{9}{4}e^{-t} \quad - \frac{9}{4}e^{-3t}) \end{bmatrix}$$

$$= \begin{bmatrix} (\frac{1}{2}e^{-t} - \frac{1}{6}e^{-3t}) \\ (2 - \frac{9}{4}e^{-t} + \frac{1}{4}e^{-3t}) \end{bmatrix}.$$

This result was previously obtained in Example 7.5.

**Sylvester' Theorem**

Sylvester's theorem is also based on the solution of eigenvalues and expresses the exponential function of a matrix as the summation given by

$$e^{\mathbf{A}t} = \Sigma_{i=1}^{n} e^{\lambda_i t} \prod_{\substack{j=1 \\ j \neq i}}^{n} \frac{[\lambda_j \mathbf{I} - \mathbf{A}]}{(\lambda_j - \lambda_i)}, \tag{7.55}$$

for distinct eigenvalues $\lambda_1, \ldots \lambda_n$.

In our previous example

$$\mathbf{A} = \begin{bmatrix} -4 & -2/3 \\ 9/2 & 0 \end{bmatrix},$$

$\lambda_1 = -1, \lambda_2 = -3$, and $n=2$. Therefore,

$$e^{\mathbf{A}t} = e^{-t}\left\{\frac{-3\begin{bmatrix} 1 & 0 \\ 0 & 1 \end{bmatrix} - \begin{bmatrix} -4 & -2/3 \\ 9/2 & 0 \end{bmatrix}}{-2}\right\} +$$

$$e^{-3t}\left\{\frac{-1\begin{bmatrix} 1 & 0 \\ 0 & 1 \end{bmatrix} - \begin{bmatrix} -4 & -2/3 \\ 9/2 & 0 \end{bmatrix}}{2}\right\}.$$

This expression may be simplified to yield $e^{\mathbf{A}t}$ as before

**Matrix Diagonalization**

Solution of the state equation $\mathbf{x} = \mathbf{Ax} + \mathbf{Bu}$ is only partially completed by evaluation of the state transition matrix. The convolution integral must also be evaluated before the zero state response can be determined. The two methods so far used, first in the frequency domain by Laplace transformation and later by convolution, are not practical for large systems in the form used in the examples. In both frequency and time domains, more efficient evaluation of the zero state response results from the prior diagonalization of the state transition matrix.

The equation $\mathbf{Ax} = \lambda\mathbf{x}$ has non-trivial solutions only if $|\mathbf{A}-\lambda\mathbf{I}| = 0$, and it was by solving the polynomials in $\lambda$ obtained from this determinant that the eigenvalues of $\mathbf{A}$ were found. The resulting non-zero solutions for $\mathbf{x}$ consist of column vectors each of which corresponds to an eigenvalue $\lambda_i$. and satisfies $\mathbf{Ax} = \lambda_i\mathbf{x}$. If a vector $\mathbf{Q}_i$ satisfies this equation then clearly so does any scaler multiple $k\mathbf{Q}_i$. Vectors $\mathbf{Q}_i$ are called **eigenvectors** and are usually scaled to unit length. In the following, however, scaling is not carried out since the examples used are obscured by significant figures after the decimal point.

Returning once again to the $\mathbf{A}$ matrix of our previous example, the eigenvalues of the matrix

$$\begin{bmatrix} -4 & -2/3 \\ 9/2 & 0 \end{bmatrix}$$

were found to be $\lambda_1 = -1$ and $\lambda_2 = -3$. Substitution of the eigenvalues into $\mathbf{Ax}-\lambda_i\mathbf{x} = 0$ yields the two equations

$$\begin{bmatrix} -3 & -2/3 \\ 9/2 & 1 \end{bmatrix} \begin{bmatrix} x_1 \\ x_2 \end{bmatrix} = \mathbf{0} \quad \text{and} \quad \begin{bmatrix} -1 & -2/3 \\ 9/2 & 3 \end{bmatrix} \begin{bmatrix} x_1 \\ x_2 \end{bmatrix} = \mathbf{0}$$

Non-trivial solutions for $x_1$ and $x_2$ must satisfy the relationships

$$x_2 = -4.5x_1 \text{ for } \lambda_1 \text{ and } x_2 = -1.5x_1 \text{ for } \lambda_2 .$$

Eigenvectors may therefore be written as

$$\mathbf{Q}_1 = \begin{bmatrix} 1 \\ -4.5 \end{bmatrix} \quad \text{and} \quad \mathbf{Q}_2 = \begin{bmatrix} 1 \\ -1.5 \end{bmatrix},$$

and when scaled to unit length,

$$\mathbf{Q}_1 = \begin{bmatrix} 0.217 \\ -0.976 \end{bmatrix} \quad \text{and} \quad \mathbf{Q}_2 = \begin{bmatrix} 0.555 \\ -0.832 \end{bmatrix} .$$

Throughout most of this section eigenvalues are assumed distinct. One important property of the corresponding set of eigenvectors is that they are linearly independent. Thus the determinant $|\mathbf{Q}| = |[\mathbf{Q}_1\ \mathbf{Q}_2 \ldots \mathbf{Q}_n]| \neq 0$ and $\mathbf{Q}$ therefore possesses an inverse. Substitution of each eigenvector into $\mathbf{Ax} = \lambda_i \mathbf{x}$ yields the set of equations

$$\mathbf{AQ}_1 = \lambda_1 \mathbf{Q}_1 \qquad \mathbf{AQ}_2 = \lambda_2 \mathbf{Q}_2 \ \ldots . \ \mathbf{AQ}_n = \lambda_n \mathbf{Q}_n \ .$$

Combination of these equations into one matrix expression yields

$$\mathbf{AQ} = \mathbf{Q}\Lambda \ ,$$

where
$$\Lambda = \begin{bmatrix} \lambda_1 & & & & 0 \\ & \lambda_2 & & & \\ & & \cdot & & \\ & & & \cdot & \\ 0 & & & & \lambda_n \end{bmatrix} .$$

Since $\mathbf{Q}^{-1}$ exists for distinct eigenvalues

$$\mathbf{A} = \mathbf{Q}\Lambda\mathbf{Q}^{-1} \tag{7.56}$$

and
$$\Lambda = \mathbf{Q}^{-1}\mathbf{AQ} \tag{7.57}$$

The transformation $\mathbf{Q}^{-1}\mathbf{AQ}$ has reduced the matrix $\mathbf{A}$ to the **diagonal canonical form** $\Lambda$. This may be checked for the eigenvectors derived in the previous paragraph by the substitution of numerical values, that is,

$$\Lambda = \mathbf{Q}^{-1}\mathbf{AQ} = \begin{bmatrix} -1/2 & -1/3 \\ 3/2 & 1/3 \end{bmatrix} \begin{bmatrix} 4 & -2/3 \\ 9/2 & 0 \end{bmatrix} \begin{bmatrix} 1 & 1 \\ -9/2 & -3/2 \end{bmatrix}$$

$$= \begin{bmatrix} -1 & 0 \\ 0 & -3 \end{bmatrix} . \tag{7.58}$$

When the eigenvalues are not distinct, it is sometimes possible that a set of linearly independent eigenvectors can still be found and diagonalization achieved. In general, by the use of generalized eigenvectors it will be possible only to generate the **Jordan canonic form** which has some elements of the super-diagonal equal to unity instead of zero.

Utilization of diagonalization begins by expressing a matrix power in simplified form. Since

$$\mathbf{A} = \mathbf{Q}\Lambda\mathbf{Q}^{-1} \, ,$$

$$\mathbf{A}^i = \mathbf{Q}\Lambda\mathbf{Q}^{-1}\,\mathbf{Q}\Lambda\mathbf{Q}^{-1} \ldots \mathbf{Q}\Lambda\mathbf{Q}^{-1}$$

$$= \mathbf{Q}\Lambda^i\mathbf{Q}^{-1} \, .$$

We have already shown that any function $F(\mathbf{A})$ can be expanded as the power series

$$F(\mathbf{A}) = \Sigma_{i=1}^{n-1} k_i \mathbf{A}^i \, ,$$

therefore, $F(\mathbf{A}) = \mathbf{Q}\Sigma_{i=1}^{n-1} k_i \Lambda^i \mathbf{Q}^{-1}$

$$= \mathbf{Q}F(\Lambda)\mathbf{Q}^{-1} \, .$$

Particular expressions of use in the solution of state equations are,

$$e^{\mathbf{A}t} = \mathbf{Q}\, e^{\Lambda t}\, \mathbf{Q}^{-1} \tag{7.59}$$

and

$$[s\mathbf{I}-\mathbf{A}]^{-1} = \mathbf{Q}[s\mathbf{I}-\Lambda]^{-1}\mathbf{Q}^{-1} \, . \tag{7.60}$$

For example, using the values of **A**, **Q**, and $\Lambda$ from Eq. (7.58)

$$e^{\mathbf{A}t} = \begin{bmatrix} 1 & 1 \\ -9/2 & -3/2 \end{bmatrix} \begin{bmatrix} e^{-t} & 0 \\ 0 & e^{-3t} \end{bmatrix} \begin{bmatrix} -1/2 & -1/3 \\ 3/2 & 1/3 \end{bmatrix}$$

and,

$$[s\mathbf{I}-\mathbf{A}]^{-1} = \begin{bmatrix} 1 & 1 \\ -9/2 & -3/2 \end{bmatrix} \begin{bmatrix} 1/(s-1) & 0 \\ 0 & 1/(s-3) \end{bmatrix} \begin{bmatrix} -1/2 & -1/3 \\ 3/2 & 1/3 \end{bmatrix} .$$

Evaluation yields the state transition matrix equal to that derived in Example 7.6.

Finally we are now able to prove the Cayley-Hamilton theorem for the case of distinct eigenvalues. The theorem states that if

$$|\mathbf{A}-\lambda\mathbf{I}| = c_0 + c_1\lambda + c_2\lambda^2 \ldots + c_n\lambda^n = 0$$

then $$F(\mathbf{A}) = c_0\mathbf{I} + c_1\mathbf{A} + c_2\mathbf{A}^2 \ldots + c_n\mathbf{A}^n = \mathbf{0} .$$

Factorizing $F(\mathbf{A})$ while assuming distinct eigenvalues,

$$F(\mathbf{A}) = c_n(\mathbf{A}-\lambda_1\mathbf{I})(\mathbf{A}-\lambda_2\mathbf{I}) \ldots (\mathbf{A}-\lambda_n\mathbf{I}) ,$$

therefore, $$\mathbf{Q}^{-1}F(\mathbf{A})\mathbf{Q} = c_n\mathbf{Q}^{-1}(\mathbf{A}-\lambda_1\mathbf{I})\mathbf{Q}\mathbf{Q}^{-1}(\mathbf{A}-\lambda_2\mathbf{I})\mathbf{Q} \ldots \mathbf{Q}^{-1}(\mathbf{A}-\lambda_n\mathbf{I})\mathbf{Q}$$

$$= c_n(\mathbf{\Lambda}-\lambda_1\mathbf{I})(\mathbf{\Lambda}-\lambda_2\mathbf{I}) \ldots (\mathbf{\Lambda}-\lambda_n\mathbf{I}) .$$

Since $\mathbf{\Lambda}$ has diagonal elements $\lambda_1, \lambda_2, \ldots \lambda_n$ and zeroes elsewhere, $(\mathbf{\Lambda}-\lambda_i\mathbf{I})$ will contain row $i$ with all zero elements. Therefore the product

$$\prod_{i=1}^{n} (\mathbf{\Lambda}-\lambda_i\mathbf{I}) = \mathbf{0} ,$$

and $$\mathbf{Q}^{-1} F(\mathbf{A})\mathbf{Q} = \mathbf{0} .$$

Since $\mathbf{Q} \neq 0$, $F(\mathbf{A}) = 0$ and the Cayley-Hamilton theorem is proven for the case of distinct eigenvalues.

**Solution by Diagonalization**

The sinusoidal steady-state solution to the state equations in the frequency domain from Eq. (7.45) is given by

$$\mathbf{Y}(s) = [\mathbf{C}(s\mathbf{I}-\mathbf{A})^{-1}\mathbf{B} + \mathbf{D}]\ \mathbf{U}(s) . \tag{7.61}$$

In the most usual case, analysis is for a single input and single output with $\mathbf{D} = \mathbf{0}$ because the input is not directly coupled to the output. Therefore,

$$\mathbf{Y}(s) = \mathbf{C}\,(s\mathbf{I}-\mathbf{A})^{-1}\,\mathbf{B}\mathbf{U}(s)$$

where $\mathbf{C}$ is a row vector and $\mathbf{BU}(s)$ is a column vector. Substitution from Eq. (7.60) yields

$$\mathbf{Y}(s) = \mathbf{C}\mathbf{Q}\,[s\mathbf{I}-\mathbf{\Lambda}]^{-1}\mathbf{Q}^{-1}\mathbf{B}\mathbf{U}(s) . \tag{7.62}$$

Frequency response can be rapidly evaluated since $(s\mathbf{I}-\Lambda)$ is diagonal. Although the frequency response solution is a relatively minor application of state variable analysis it is advantageous compared to nodal analysis, particularly for the single input–single output system evaluated at many different frequencies.

The matrix inversion $(s\mathbf{I}-\mathbf{A})^{-1}$ may also be directly carried out by the Souriau-Frame Algorithm which generates the $n^2$ rational transfer functions in $n^4$ computer operations. The NAM inversion method from Chapter 6 could also be employed for the same purpose.

Solution of Eq. (7.61) in the time domain by inverse Laplace transformation is relatively straightforward for simple excitation functions, for example step or impulse inputs. However, much of the advantage of the factorization and partial fraction expansion inherent in the solution for eigenvectors is lost if $\mathbf{U}(s)$ is not also known in factorized form in the frequency domain. Computationally there is little advantage in diagonalization compared to partial fraction expansion involving solution for polynomial roots. In general the following numerical solutions are preferred in which the time domain convolution integral is generated by means of a recurrence formula at successive time increments.

The solution of the non-homogeneous state equation was given in Eq. (7.49) by

$$\mathbf{x}(t) = \mathrm{e}^{\mathbf{A}t}\mathbf{x}_0 + \int_0^t \mathrm{e}^{\mathbf{A}(t-\tau)}\mathbf{B}\mathbf{u}(\tau)\mathrm{d}\tau \ . \tag{7.63}$$

Since any point in time may be designated as the time reference, Eq. (7.63) may be used to update the response at time $nT$ to yield the response at time $(n+1)T$ by the recurrence relationship,

$$\mathbf{x}[(n+1)T] = \mathrm{e}^{\mathbf{A}nT}\,\mathbf{x}[nT] + \int_{nT}^{(n+1)T} \mathrm{e}^{\mathbf{A}[(n+1)T-\tau]}\,\mathbf{B}\mathbf{u}(\tau)\mathrm{d}\tau \ , \tag{7.64}$$

where the output is generated successively at $0, T, 2T, \ldots. \ nT, (n+1)T, \ldots T_{\mathrm{MAX}}$. Diagonalization of the exponentials using eigenvectors yields

$$\mathbf{x}[(n+1)T] = \mathbf{Q}\mathrm{e}^{\Lambda nT}\mathbf{Q}^{-1}\mathbf{x}[nT] + \int_{nT}^{(n+1)T} \mathbf{Q}\mathrm{e}^{\Lambda[(n+1)T-\tau]}\,\mathbf{Q}^{-1}\mathbf{B}\mathbf{u}(\tau)\mathrm{d}\tau \ .$$

Assuming that the input excitation remains constant during the period of integration,

$$\mathbf{x}[(n+1)T] = \mathbf{Q}\mathrm{e}^{\Lambda nT}\mathbf{Q}^{-1}\mathbf{x}[nT] + \mathbf{Q}\,[\textstyle\int_0^T \mathrm{e}^{\Lambda(T-\tau)}\,\mathrm{d}\tau]\,\mathbf{Q}^{-1}\mathbf{B}\mathbf{u}(nT) \, ,$$

where the limits of integration have been changed by change of variable $\tau$ to $\tau - nT$.

Integration of the diagonal matrix $e^{\Lambda(T-\tau)}$ is carried out element by element to yield

$$\mathbf{G} = \Lambda^{-1}(e^{\Lambda T}-\mathbf{I}) = \begin{bmatrix} \frac{e^{\lambda_1 T}-1}{\lambda_1} & & & 0 \\ & \frac{e^{\lambda_2 T}-1}{\lambda_2} & & \\ & & \ddots & \\ 0 & & & \frac{e^{\lambda_n T}-1}{\lambda_n} \end{bmatrix} . \tag{7.65}$$

The recursion formula giving the value of $\mathbf{x}[(n+1)T]$ in terms of $\mathbf{x}[nT]$ may therefore be written in the form

$$\mathbf{x}[(n+1)T] = \mathbf{Q}e^{\Lambda T}\mathbf{Q}^{-1}\,\mathbf{x}[nT] + \mathbf{QGQ}^{-1}\mathbf{B}\,\mathbf{u}(nT) . \tag{7.66}$$

Computation proceeds from time $t = 0$, $n = 0$, in increments of $T$ sec. Prior to integration the eigenvalues $\Lambda$, eigenvector matrix $\mathbf{Q}$, its inverse $\mathbf{Q}^{-1}$, and $\mathbf{G}$ must be determined. Also Eq. (7.63) must be evaluated for $t$=0+ to find the value of $\mathbf{x}[nT]$ for $n$=0 which does not entirely depend on initial conditions, for example when impulses are present at $t$=0.

Direct integration of Eq. (7.64) under the same assumptions yields the recursion formula

$$\mathbf{x}[(n+1)T] = e^{\mathbf{A}T}\mathbf{x}[nT] + \mathbf{A}^{-1}(e^{\mathbf{A}T}-\mathbf{I})\mathbf{B}\,\mathbf{u}(nT) . \tag{7.67}$$

This equation forms the basis of Liu's method of transient response evaluation in which

$$e^{\mathbf{A}T} = \Sigma_{i=0}^{\infty} \frac{\mathbf{A}^i T^i}{i!}$$

and $$\mathbf{A}^{-1}(e^{\mathbf{A}T}-\mathbf{I}) = T\,\Sigma_{i=0}^{\infty} \frac{\mathbf{A}^i T^i}{(i+1)!}$$

are evaluated to prescribed accuracy from the truncated power series. The recursion formula is used from $t$=0 in increments of $T$ sec. until periodic checks reveal errors which propagate with increasing time. The integration is then restarted with recomputed matrices from the corrected current state.

*Example* 7.7

Use Branin's recursion formula, Eq. (7.66), to generate the transient response previously determined exactly in Example 7.5 for time increment $T$=0.1 sec.

We have already determined the following matrices for this circuit by using its **A** matrix throughout this section. To summarize, therefore,

$$\mathbf{Q} = \begin{bmatrix} 1 & 1 \\ -9/2 & -3/2 \end{bmatrix}, \quad \mathbf{Q}^{-1} = \begin{bmatrix} -1/2 & -1/3 \\ 3/2 & 1/3 \end{bmatrix},$$

$$e^{\Lambda T} = \begin{bmatrix} e^{-0.1} & 0 \\ 0 & e^{-0.3} \end{bmatrix}, \quad \mathbf{G} = \begin{bmatrix} (1-e^{-0.1}) & 0 \\ 0 & \frac{1}{3}(1-e^{-0.3}) \end{bmatrix}$$

and $$\mathbf{Bu}(t) = \begin{bmatrix} 4/3 & 2/3 & 1/3 \\ 0 & 0 & 0 \end{bmatrix} \begin{bmatrix} u(t) \\ 0 \\ u'(t) \end{bmatrix},$$

where the excitation is a unit step $u(t)$ and an impulse $u'(t)$ is generated by the circuit configuration.

Substitution of numerical values in Eq. (7.66) yields

$$\mathbf{Q}e^{\Lambda T}\mathbf{Q}^{-1} = \begin{bmatrix} 0.6588 & -0.0547 \\ 0.3690 & 0.9868 \end{bmatrix},$$

and $$\mathbf{QGQ}^{-1}\mathbf{B}\,\mathbf{u}(nT) = \begin{bmatrix} 0.1093 & 0.02733 \\ 0.02631 & 0.00658 \end{bmatrix} \begin{bmatrix} u(t) \\ u'(t) \end{bmatrix}.$$

From Eq. (7.63) the value of the output at $t$=0+ is given by

$$\mathbf{x}(0+) = \mathbf{x}(0) + \int_0^{0+} \begin{bmatrix} \frac{4}{3}u(\tau) + \frac{1}{3}u'(\tau) \\ 0 \end{bmatrix} d\tau$$

$$= \begin{bmatrix} 2 \\ 1/2 \end{bmatrix} + \begin{bmatrix} 1/3 \\ 0 \end{bmatrix} = \begin{bmatrix} 2.333 \\ 0.5 \end{bmatrix}.$$

The recursion formula from Eq. (7.66) is therefore

$$\mathbf{x}[(n+1)T] = \begin{bmatrix} 0.6588 & -0.0547 \\ 0.369 & 0.9868 \end{bmatrix} \mathbf{x}[nT] + \begin{bmatrix} 0.1093 & 0.0273 \\ 0.0263 & 0.00658 \end{bmatrix} \begin{bmatrix} u(t) \\ u'(t) \end{bmatrix}$$

with the initial value $\mathbf{x}[0] = [2.333 \quad 0.5]^t$

Exact values were computed from the solution to Example 7.5 as

$$\mathbf{x}(t) = \begin{bmatrix} -\frac{2}{3}\,e^{-t} + 3e^{-3t} \\ 2 + 3e^{-t} - \frac{9}{2}\,e^{-3t} \end{bmatrix}.$$

Results may now be tabulated.

| $nT$ | $x[nT]$ | $x(t)$ exact |
|---|---|---|
| 0 | $[2.333 \quad 0.5]^t$ | $[2.333 \quad 0.5]^t$ |
| $T$ | $[1.6464 \quad 1.3872]^t$ | $[1.6192 \quad 1.3808]^t$ |
| $2T$ | $[1.1181 \quad 2.0027]^t$ | $[1.1006 \quad 1.9865]^t$ |
| $3T$ | $[0.7364 \quad 2.4151]^t$ | $[0.7258 \quad 2.3929]^t$ |
| $4T$ | $[0.4623 \quad 2.6813]^t$ | $[0.4567 \quad 2.6555]^t$ |
| $5T$ | $[0.2672 \quad 2.8428]^t$ | $[0.2650 \quad 2.8155]^t$ |

This solution should be repeated by computer for various values of $T$ and for many more steps. The tabulated results correspond to a relatively large increment and are not particularly accurate.

**Solution by Numerical Integration**

Most computer centres have quite sophisticated computer programs available which are apparently capable of integrating the state equations. Unfortunately the popular methods in use are not well suited to state equation solutions arising from electronic circuit analysis. This is because the wide spread of eigenvalues causes numerical instability in which the computed response diverges from the actual response with increased time. Nevertheless we will examine the basic methods of numerical integration since the advanced methods used in practice arise from the same principles.

The two recursion formulae given in Eqs. (7.66) and (7.67) require computed approximations to the state transition matrix $e^{\mathbf{A}T}$ which are derived either directly from the expansion or through the eigenvalues and eigenvectors of **A**. We will show that the simple integration techniques correspond to the use of approximations to $e^{\mathbf{A}T}$ consisting of the first few terms only.

Euler's method of integration makes a first-order approximation to the derivative $\dot{\mathbf{x}}$ at $t=nT$, where $T$ is a time increment for the integration given by

$$\dot{\mathbf{x}}[nT] = \frac{1}{T}\{\mathbf{x}[(n+1)T] - \mathbf{x}[nT]\}$$

$$= \mathbf{Ax}[nT] + \mathbf{Bu}[nT] .$$

The Euler recursion formula is therefore obtained as

$$\mathbf{x}[(n+1)T] = (\mathbf{I} + T\mathbf{A})\,\mathbf{x}[nT] + T\mathbf{Bu}[nT] . \tag{7.68}$$

This equation could be directly obtained from Eq. (7.67) by using two terms of the expansion for $e^{\mathbf{A}T}$ and one term only for the expansion for $\mathbf{A}^{-1}(e^{\mathbf{A}T}-\mathbf{I})$. The recursion formula Eq. (7.68) cannot therefore be expected to give very good accuracy.

Trapezoidal integration approximates the area of the function from $nT$ to $(n+1)T$ by the area of a trapezium. Thus

$$\int_{nT}^{(n+1)T} \mathbf{Ax}(t) + \mathbf{Bu}(t)\,\mathrm{d}t \approx$$

$$\frac{T}{2}\{\mathbf{Ax}[nT] + \mathbf{Bu}[nT] + \mathbf{Ax}[(n+1)T] + \mathbf{Bu}[(n+1)T]\} .$$

Substitution for $\mathbf{Ax}\,[(n+1)T]$ from Eq. (7.68), together with the assumption that $\mathbf{Bu}[nT]$ remains constant until $(n+1)T$, yields the increae in $\mathbf{x}$ as

$$\mathbf{x}[(n+1)T] - \mathbf{x}[nT] = \frac{T}{2}\{\mathbf{Ax}[nT] + (\mathbf{I}+T\mathbf{A})\mathbf{Ax}[nT] +$$

$$\mathbf{A}T\mathbf{Bu}[nT] + 2\mathbf{Bu}[nT]\} .$$

$$\mathbf{x}[(n+1)T] = \left(\mathbf{I} + T\mathbf{A} + \frac{T^2\mathbf{A}^2}{2}\right)\mathbf{x}[nT] + \left(T\mathbf{I} + \frac{T^2\mathbf{A}}{2}\right)\mathbf{Bu}[nT] . \tag{7.69}$$

This equation involves only one more term of the expansions for $e^{\mathbf{A}T}$ and $\mathbf{A}^{-1}(e^{\mathbf{A}T}-\mathbf{I})$ than the Euler method in Eq. (7.68) and is therefore a minor improvement.

The popular and widely available fourth order Runge-Kutta recursion formula involves terms of the expansion for $e^{\mathbf{A}T}$ up to $T^4\mathbf{A}^4/24$. Here we will do no more than quote the recursion equation in the form

$$\mathbf{x}[(n+1)T] = \left(\mathbf{I} + T\mathbf{A} + \frac{T^2\mathbf{A}^2}{2} + \frac{T^3\mathbf{A}^3}{6} + \frac{T^4\mathbf{A}^4}{24}\right)\mathbf{x}[nT] +$$

$$\left(\frac{\mathbf{I}}{6} + \frac{T\mathbf{A}}{6} + \frac{T^2\mathbf{A}^2}{12} + \frac{T^3\mathbf{A}^3}{24}\right)\mathbf{Bu}[nT] +$$

$$\left(\frac{2\mathbf{I}}{3} + \frac{T\mathbf{A}}{3} + \frac{T^2\mathbf{A}^2}{12}\right)\mathbf{Bu}[(n+\tfrac{1}{2})T] + \mathbf{Bu}[(n+1)T] \ . \tag{7.70}$$

Note that $\mathbf{Bu}[nT]$ has now been assumed to vary over the period $nT$ to $(n+1)T$, and separate values have appeared in the formula for $\mathbf{u}[nT]$, $\mathbf{u}[(n+\tfrac{1}{2})T]$, and $\mathbf{u}[(n+1)T]$. The terminology for the mid-point evaluation is obvious.

The integration formulae given in Eqs. (7.66) to (7.70) have in common the property that they predict a value for $\mathbf{x}[(n+1)T]$ when given the values of $\mathbf{x}[nT]$ and $\mathbf{u}[nT]$ to $\mathbf{u}[(n+1)T]$, all of which are known at the time of evaluation. The formulae are therefore known as **predictors.** They are also said to be **single step** integrators since they require information only one step back from the current time step. Accuracy can be improved if the use of a predictor expression is followed by the iterative application of a **corrector** expression in which $\mathbf{x}[(n+1)T]$ appears on both sides of the equals sign.

The procedure can be illustrated using the Euler expression (7.68) as predictor and the trapezoidal expression (7.69) before substitution for $\mathbf{x}[(n+1)T]$ as corrector. The predictor–corrector pair are therefore given by

$$\mathbf{x}[(n+1)T] = (\mathbf{I} + T\mathbf{A})\ \mathbf{x}[nT] + T\mathbf{Bu}[nT] \tag{7.71}$$

$$\mathbf{x}[(n+1)T] = \left(\mathbf{I} + \frac{T\mathbf{A}}{2}\right)\mathbf{x}[nT] + \frac{T\mathbf{A}}{2}\mathbf{x}[(n+1)T]$$

$$+ \frac{T\mathbf{B}}{2}\mathbf{u}[nT] + \frac{T\mathbf{B}}{2}\mathbf{u}[(n+1)T] \ . \tag{7.72}$$

In use, Eq. (7.71) predicts the value of $\mathbf{x}[(n+1)T]$, and the value obtained is used as the first estimated value for substitution on the right-hand side of Eq. (7.72). The left-hand side of Eq. (7.72) is then evaluated to provide an improved value for $\mathbf{x}[(n+1)T]$ and the procedure is iterated until convergence is obtained. In practice a fixed small number of iterations is usually employed since it is otherwise better to reduce $T$ and use the predictor alone. It is of course necessary to check the convergence of the corrector, and dependent on the relative accuracies of predictor and corrector it is possible to vary the step size to optimise both speed and accuracy for an efficient integration.

The popular Adams-Bashforth predictor–corrector utilizes the formulae

$$\begin{aligned}\mathbf{x}[(n+1)T] = \left(\mathbf{I} + \frac{55T\mathbf{A}}{24}\right)\mathbf{x}[nT] + \frac{55T}{24}\mathbf{Bu}[nT] \\ \frac{-59T}{24}(\mathbf{Ax}[n-1)T] + \mathbf{Bu}[(n-1)T]) \\ \frac{+37T}{24}(\mathbf{Ax}[(n-2)T] + \mathbf{Bu}[(n-2)T]) \\ \frac{-9T}{24}(\mathbf{Ax}[(n-3)T] + \mathbf{Bu}[(n-3)T])\end{aligned} \tag{7.73}$$

$$\begin{aligned}\mathbf{x}[(n+1)T] = \left(\mathbf{I} + \frac{19T\mathbf{A}}{24}\right)\mathbf{x}[nT] + \frac{19T}{24}\mathbf{Bu}[nT] \\ \frac{-5T}{24}(\mathbf{Ax}[(n-1)T] + \mathbf{Bu}[(n-1)T]) \\ \frac{+T}{24}(\mathbf{Ax}[(n-2)T] + \mathbf{Bu}[(n-2)T]) \\ \frac{+9T}{24}(\mathbf{Ax}[(n+1)T] + \mathbf{Bu}[(n+1)T])\end{aligned} \tag{7.74}$$

The first two points $\mathbf{x}[T]$ and $\mathbf{x}[2T]$ must be generated by another method before the Adams-Bashforth predictor–corrector can be utilised. This is because the integration requires information from previously computed time steps in order to incorporate the function gradient into the solution. Previously in one-step integration it was necessary to generate this information in the form of the Taylor series expansion of $e^{\mathbf{A}T}$.

At this point it is a worthwhile exercise to test the various integration formulae with various step lengths on several known networks by means of simple computer programs. In general all integration methods so far described are severely tested when applied to networks with widely separated eigenvalues. The networks and resulting state equations are then called **stiff**. Algorithms for determination of eigenvalues and eigenvectors are also prone to inaccuracy in these circumstances. We must therefore turn to the considerably more sophisticated integration methods recently available for application in general purpose state variable analysis programs. For a detailed discussion of methods suitable for stiff equations the reader must refer to the *Further Reading* listed at the end of this chapter and to numerous research papers in the literature.

## 7.3 CONCLUSION

In this chapter we have empahsized the eigenvalue–eigenvector approach to the solution of state equations for the insight that this gives into the nature of response. There is, however, little doubt that modern numerical integration algorithms are superior both in speed and accuracy. This is particularly true for network problems which usually result in stiff equations. Unfortunately the application of state space analysis to non-linear dynamic networks is very much complicated by the problem of state equation generation which is extraordinarily difficult in comparison with filling the nodal admittance matrix. Overall computational efficiency is therefore difficult to achieve and it is preferable at present to apply the same numerical integration techniques to circuit equations set up on the nodal basis.

## FURTHER READING

Kuo, F. F. and Kaiser, J. F., eds., (1966), *System analysis by Digital Computer,* Chapters 3 and 4. J. Wiley, New York.

Rohrer, R. A., (1970), *Circuit Theory – An Introduction to the State Variable Approach,* McGraw-Hill, New York.

*Numerical Integration*

Calahan, D. A., (1972), *Computer-Aided Network Design,* Chapter 9, McGraw-Hill, New York, rev. ed.

Hamming, R. W., (1962), *Numerical Methods for Scientists and Engineers,* Chapters 13 to 16, McGraw-Hill, New York.

*Non-Linear Networks*

Calahan, D. A., MacNee, A. B. and McMahon, E. L., (1974), *Introduction to Modern Circuit Analysis,* Chapter 14, Holt Rinehart and Winston, New York.

Haykin, S. S., (1970), *Active Network Theory,* Chapter 9, Addison-Wesley, Reading, Mass.

## PROBLEMS

7.1 Determine the eigenvalues and eigenvectors of the matrix

$$\mathbf{A} = \begin{bmatrix} 8 & -8 & -2 \\ 4 & -3 & -2 \\ 3 & -4 & 1 \end{bmatrix}.$$

7.2 Determine the state transition matrix $e^{\mathbf{A}t}$ for **A** from Problem 7.1, using each of the methods described in Section 7.2.

7.3 Repeat Problem 7.2 using the matrix

$$\mathbf{A} = \begin{bmatrix} -1 & 0 & 0 \\ 0 & -4 & 4 \\ 0 & -1 & 0 \end{bmatrix}.$$

7.4 Write standard form state equation for the rational function

$$F(s) = \frac{s^2 + 6s + 10}{s^4 + 6s^3 + 13s^2 + 14s + 6}.$$

7.5 For a single input–single output network in which $\mathbf{D}^{-1} = 1/D$, show that the zeroes of the transfer function are the eigenvalues of $\mathbf{A} - (1/D)\,\mathbf{B}\,\mathbf{C}$.

7.6 Write state and output equations for the circuits in Figs. 1.20, 1.21 (with $M$=0.75), 1.21 (with $M$=1), 6.17 and 7.8.

7.7 Find the natural frequencies of the networks in Fig. 7.8.

7.8 Write state and output equations for the excess element networks in Fig. 7.9.

7.9 Determine the output voltage $v_2(t)$ for the circuit of Fig. 7.8(a) if the initial conditions of the state variables are $i_{L1}(0) = 10^{-3}$, $i_{L2}(0) = 0.5 \times 10^{-3}$, $e_{C1}(0) = 1$ V, and the input current is $10^{-3}u(t)$.

7.10 Solve Problem 7.9 by means of a simple numerical integration algorithm and compare numerical results obtained as the time step is changed, against the precise results.

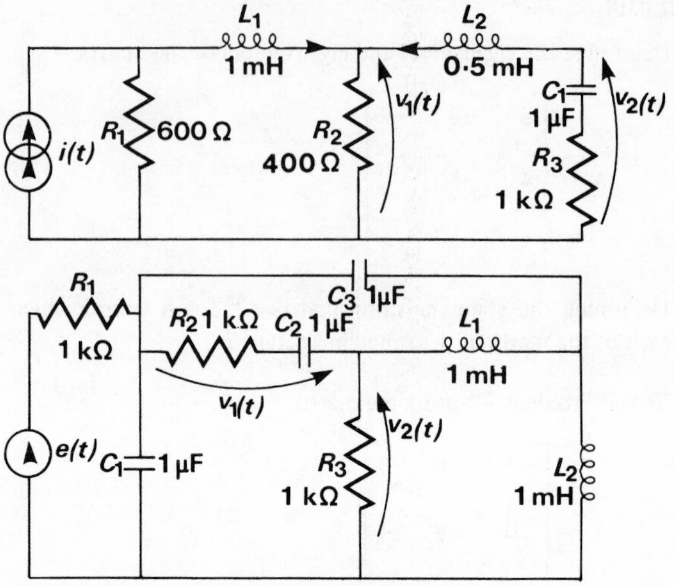


Fig. 7.8 – Problems 7.6 and 7.7.
(Outputs $v_1(t)$ and $v_2(t)$).

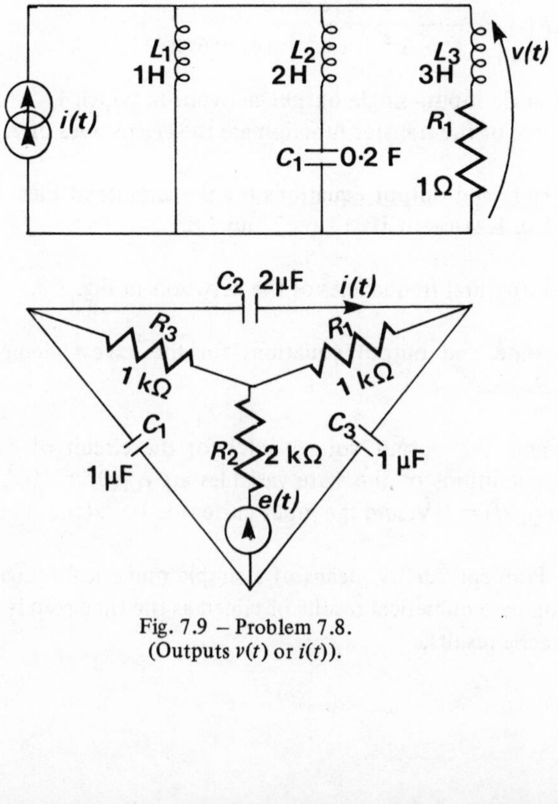


Fig. 7.9 – Problem 7.8.
(Outputs $v(t)$ or $i(t)$).

*Computing*

7.11 Write a circuit analysis program based on the formulation of state equations.

7.12 Write a numerical integration subroutine to give transient response. Pay particular attention to algorithms for stiff equations.

7.13 Compare various simple integration formulae for numerical integration of state equations. Try different time steps and modification of step length during integration.

7.14 Write a program specifically for inverse Laplace transformation of a rational function $F(s)$ expressed as a ratio of unfactorized polynomials in $s$.

7.15 Write a subroutine to give the steady-state frequency response from the state equations.

CHAPTER 8

# Sensitivity Analysis

The first seven chapters of this book have been concerned with the analysis of a fixed circuit. The circuit diagram, the component values, and the device models employed have been defined from the outset, and a unique d.c. frequency, or transient response has normally been obtained. The response obtained by analysis of the circuit with the **nominal** component values, that is, those predicted by the circuit design process, is called the **nominal** response. It is normally a close approximation to the **required** response given as part of the design specification.

In practice, component values are not accurate. Manufacturing tolerances of resistors and capacitors, for example, are usually between ±0.1% and ±20%. Transistor current gain is subject to large variation: typically +100%, −50%. Most integrated circuit components vary in the range ±5% to ±20%. Moreover, components vary owing to environmental effects, such as temperature and humidity. The final two chapters of this book are concerned with evaluating the effect of these variations on the nominal response of the circuit. The general term **sensitivity** is applied to all measures of the effect on circuit response of variation in individual component values. The collective effect of simultaneous variation in all component values is called the **tolerance**, which will be considered in the next chapter.

## 8.1 SENSITIVITY MEASURES

### Absolute Sensitivity

Consider the value of a circuit response $y$ which is a function of the $n$ component values $p_1, p_2, \ldots p_n$,

that is, $$y = y(p_1, p_2, \ldots p_n) \, .$$

For the purpose of the definitions which follow, the response function $y$ need not be specified. It would normally be expected to be the frequency response at a given frequency $\omega_0$ or the transient response at a given time $t_0$, and the value of $\omega_0$ or $t_0$ would appear in the function $y$ as an independent variable. Other response functions, such as d.c. voltages in the circuit or the $Q$ of a tuned amplifier, are equally valid. Incremental changes in the component values $p_1, \ldots p_n$, denoted by $\Delta p_1, \ldots \Delta p_n$, cause an incremental change $\Delta y$ in $y$.

The Taylor expansion of a function about a point gives the change in value of the function caused by variation of its parameters in terms of the function gradients at the point and the parameter increments, assuming that the function can be differentiated. Expanding about the nominal value of response $y$, the increment $\Delta y$ is given by

$$\Delta y = \Sigma_{i=1}^{n} \frac{\partial y}{\partial p_i} \Delta p_i + \frac{1}{2!} \Sigma_{i=1}^{n} \Sigma_{j=1}^{n} \frac{\partial^2 y}{\partial p_i \partial p_j} \Delta p_i \, \Delta p_j$$

$$+ \frac{1}{3!} \Sigma_{i=1}^{n} \Sigma_{j=1}^{n} \Sigma_{k=1}^{n} \frac{\partial^3 y}{\partial p_i \partial p_j \partial p_k} \Delta p_i \Delta p_j \Delta p_k + \ldots \quad (8.1)$$

This equation involving the partial derivations of the function $y$ with respect to the component values is an infinite series and is valid if all derivatives exist. In all cases considered in this book, terms involving derivatives of third and higher order will be neglected on the assumption that changes in the component value are sufficiently small for accuracy to be maintained. Indeed, in most cases the second-order terms will be neglected giving the approximate relationship.

$$\Delta y = \Sigma_{i=1}^{n} \frac{\partial y}{\partial p_i} \Delta p_i \; . \quad (8.2)$$

*The absolute sensitivity $S_i$ of the response $y = y(p_1 \ldots p_n)$ to the component $p_i$ is defined as the first-order partial derivative $\partial y / \partial p_i$*, that is,

$$S_i = \partial y / \partial p_i = \underset{\Delta p_i \to 0}{\text{Limit}} \{\Delta y / \Delta p_i\} \; . \quad (8.3)$$

Vector $\mathbf{G}$ is the transpose of the **gradient** vector $\nabla y$ which is a row vector of first-order partial derivatives, that is,

$$\nabla y = \mathbf{G}^t = \left[ \frac{\partial y}{\partial p_1} \; \frac{\partial y}{\partial p_2} \; \ldots\ldots \; \frac{\partial y}{\partial p_n} \right],$$

$$= [S_1 \;\; S_2 \;\; \ldots\ldots \;\; S_n] \; . \quad (8.4)$$

When the response is evaluated at $m$ sample points, $t_1, t_2, \ldots t_m$ of an independent parameter $t$, (usually either frequency or time) the response $y_i$ is given by

$$y_i = y(p_1, \ldots p_n, t_i) = y_i(p_1, \ldots p_n) ,$$

recognizing that the functions $y$ are different at each value of the independent parameter and are labelled accordingly.

The gradient vector **G** is most usefully replaced by the $m \times n$ gradient matrix **S** where,

$$\mathbf{S} = \begin{bmatrix} \frac{\partial y_1}{\partial p_1} & \frac{\partial y_1}{\partial p_2} & \cdots & \frac{\partial y_1}{\partial p_n} \\ \frac{\partial y_2}{\partial p_1} & & & \vdots \\ \vdots & & & \vdots \\ \frac{\partial y_m}{\partial p_1} & & \cdots & \frac{\partial y_m}{\partial p_n} \end{bmatrix} . \tag{8.5}$$

This matrix is often referred to as the **sensitivity matrix** or the **first-order sensitivity matrix.**

When second-order derivatives in Eq. (8.1) contribute significantly to the computation of the variation $\Delta y$, they are called second-order sensitivity coefficients.

The derivatives

$$\frac{\partial^2 y}{\partial p_i \partial p_j} \quad i=1, \ldots n, \; j=1, \ldots n$$

are usually used in the form of the $n \times n$ symmetric matrix **H** where

$$\mathbf{H} = \begin{bmatrix} \frac{\partial^2 y}{\partial p_1^2} & \frac{\partial^2 y}{\partial p_1 \partial p_2} & \cdots & \frac{\partial^2 y}{\partial p_1 \partial p_n} \\ \frac{\partial^2 y}{\partial p_2 \partial p_1} & & & \vdots \\ \vdots & & & \vdots \\ \frac{\partial^2 y}{\partial p_n \partial p_1} & & \cdots & \frac{\partial^2 y}{\partial p_n^2} \end{bmatrix} . \tag{8.6}$$

This matrix is known as the **Hessian matrix** and is also referred to as the **second-order sensitivity matrix**. Owing to the larger number of terms in the Hessian, it is not normally used for responses at a large number of sample points unless it can be obtained as a symbolic transfer function.

The matrix form of Eq. (8.1) may be written

$$\Delta y = \mathbf{G}^t \Delta \mathbf{p} + \frac{1}{2} \Delta \mathbf{p}^t \mathbf{H} \, \Delta \mathbf{p} , \tag{8.7}$$

where $\Delta \mathbf{p}$ is the vector containing the incremental changes in component valve defined by

$$\Delta \mathbf{p} = [\Delta p_1 \, \Delta p_2 \ldots \Delta p_n]^t$$

**Relative and Semi-relative Sensitivity**

*The relative sensitivity $S_i^r$ of the response $y = y(p_1, p_2, \ldots p_n)$ to the component $p_i$ is defined by*

$$S_i^r = \frac{\partial(\log_n y)}{\partial(\log_n p_i)} , \tag{8.8}$$

$$= \frac{p_i}{y} \frac{\partial y}{\partial p_i} = \frac{p_i}{y} S_i . \tag{8.9}$$

*The semi-relative sensitivities $Q_i$ and $Q_i^r$ of the response $y = y\,(p_1, \ldots p_n)$ to the component $p_i$ are defined by*

$$Q_i = \frac{\partial(\log_n y)}{\partial p_i} = \frac{1}{y} S_i , \tag{8.10}$$

$$Q_i^r = \frac{\partial y}{\partial(\log p_i)} = p_i \, S_i \tag{8.11}$$

Relative and semi-relative sensitivities apply particularly to RC active network design where sensitivity to different components must be compared and measures of overall circuit sensitivity must be minimized. The following theorem is important in this respect since it sets a constraint on the sum of the relative sensitivities which must hold even when, for example, the sum of the squared sensitivities is minimized.

**Theorem**

*The sum of the relative sensitivities $\Sigma_{i=1}^{n} S_i^{\mathrm{r}}$ in a network is unity for impedance or admittance functions and zero for gain functions*

**Proof**

*Admittance functions*

Consider any admittance function

$$Y = Y(G_1, \ldots G_i, C_1, \ldots C_j, L_1^{-1}, \ldots L_k^{-1}, g_{m_1}, \ldots g_{m_l}, \omega) \ ,$$

where $i$, $j$, $k$ and $l$ are the numbers of the conductances, capacitances, inverse inductors, and voltage controlled current sources, respectively.

Recall from Chapter 6 that admittance functions are linear homogeneous functions of the variables $G$, $C$, $L$, and $g_\mathrm{m}$.

If all admittances are scaled by a factor $\lambda$, then $Y$ is also scaled by $\lambda$, that is,

$$\lambda Y = Y(\lambda G_1, \ldots \lambda G_i, \lambda C_1, \ldots \lambda C_j, \lambda L_1^{-1}, \ldots \lambda L_k^{-1}, \lambda g_{\mathrm{m}_1}, \ldots \lambda g_{\mathrm{m}_l}, \omega) \, .$$

Differentiating with respect to $\lambda$ yields

$$Y = \Sigma_{h=1}^{i} \frac{\partial Y}{\partial \lambda G_h} \frac{\partial \lambda G_h}{\partial \lambda} + \Sigma_{h=1}^{j} \frac{\partial Y}{\partial \lambda C_h} \frac{\partial \lambda C_h}{\partial \lambda} + \Sigma_{h=1}^{k} \frac{\partial Y}{\partial \lambda L_h^{-1}} \frac{\partial \lambda L_h^{-1}}{\partial \lambda}$$

$$+ \Sigma_{h=1}^{l} \frac{\partial Y}{\partial \lambda g_{\mathrm{m}_h}} \frac{\partial \lambda g_{\mathrm{m}_h}}{\partial \lambda}$$

Setting $\lambda = 1$ and dividing through $Y$ gives

$$1 = \Sigma_{h=1}^{i} \frac{G_h}{Y} \frac{\partial Y}{\partial G_h} + \Sigma_{h=1}^{j} \frac{C_h}{Y} \frac{\partial Y}{\partial C_h} + \Sigma_{h=1}^{k} \frac{L_h^{-1}}{Y} \frac{\partial Y}{\partial L_h^{-1}}$$

$$+ \Sigma_{h=1}^{l} \frac{g_{\mathrm{m}_h}}{Y} \frac{\partial Y}{\partial g_{\mathrm{m}_h}} \ ,$$

$$= \Sigma_{i=1}^{n} S_i^{\mathrm{r}} \ .$$

*Impedance functions*

The proof for impedance functions follows that for admittance functions beginning with the function $Z = Z(R_1, \ldots R_i, C_1^{-1}, \ldots C_j^{-1}, L_1, \ldots L_k, r_1, \ldots r_l, \omega)$ where $i$, $j$, $k$, and $l$ are the numbers of resistors, inverse capacitors, inductors, and current controlled voltage sources respectively.

*Gain functions*

The proof for voltage or current gain functions is similar to that for admittance functions. Consider the gain function

$$K = K(G_1, \ldots G_i, C_1, \ldots C_j, L_1^{-1}, \ldots L_k^{-1}, g_{m_1}, \ldots g_{m_l}, \omega) \;,$$

in terms of admittances as before.

Again, recalling from Chapter 6, gain functions are independent of any admittance scaling factor $\lambda$, therefore,

$$K = K(\lambda G_1, \ldots \lambda G_i, \lambda C_1, \ldots \lambda C_j, \lambda L_1^{-1}, \ldots \lambda L_k^{-1}, \lambda g_{m_1}, \ldots \lambda g_{m_l}, \omega) \;.$$

Differentiate with respect to $\lambda$, set $\lambda = 1$, and divide through by $K$ to obtain

$$0 = \Sigma_{h=1}^{i} \frac{G_h}{K} \frac{\partial K}{\partial G_h} + \Sigma_{h=1}^{j} \frac{C_h}{K} \frac{\partial K}{\partial C_h} + \Sigma_{h=1}^{k} \frac{L_h^{-1}}{K} \frac{\partial K}{\partial L_h^{-1}}$$

$$+ \Sigma_{h=1}^{l} \frac{g_{m_h}}{K} \frac{\partial K}{\partial g_{m_h}}$$

$$= \Sigma_{i=1}^{n} S_i^{\mathrm{r}}$$

When all admittance variables are changed to impedance variables, the identical result is obtained.

*Example* 8.1

Determine the absolute and relative sensitivities of the voltage gain to variation in the circuit admittances for the amplifier in Fig. 8.1 at $\omega = 10^7$rad/sec. Show that the sum of the relative sensitivities is zero for voltage gain and unity for input impedance.

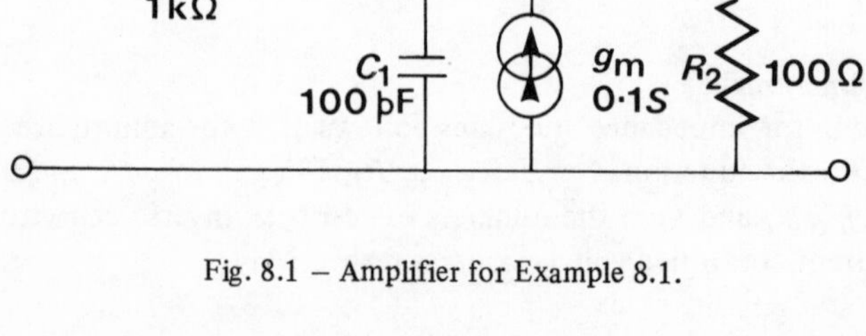

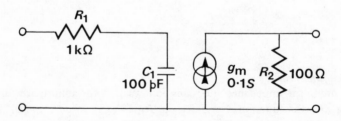

Fig. 8.1 – Amplifier for Example 8.1.

By symbolic analysis, voltage gain is given by

$$K = \frac{g_m R_2}{1+sC_1R_1} = \frac{10}{1+10^{-7}s} = 5-5\mathrm{j}$$

at $\omega = 10^{-7}$ rad/sec. In terms of admittances,

$$K = \frac{g_m G_1}{G_2(G_1+sC_1)} ,$$

where $G_1 = 1/R_1$ and $G_2 = 1/R_2$.

Absolute sensitivities $S_i$ are therefore obtained by differentiation as

$$\frac{\partial K}{\partial G_1} = \frac{sC_1 g_m}{G_2(G_1+sC_1)^2} = 0.5 \times 10^4$$

$$\frac{\partial K}{\partial C_1} = \frac{-sG_1 g_m}{G_2(G_1+sC_1)^2} = 0.5 \times 10^{11}$$

$$\frac{\partial K}{\partial g_m} = \frac{G_1}{G_2(G_1+sC_1)} = 50\,(1-\mathrm{j})$$

$$\frac{\partial K}{\partial G_2} = \frac{-g_m G_1}{G_2^2(G_1+sC_1)} = 500\,(1-\mathrm{j}) .$$

Relative sensitivities $S_i^r$ are therefore given by

$$\frac{G_1}{K}\frac{\partial K}{\partial G_1} = \frac{sC_1}{(G_1+sC_1)} = 0.5\,(1+\mathrm{j})$$

$$\frac{C_1}{K}\frac{\partial k}{\partial C_1} = \frac{-sC_1}{(G_1+sC_1)} = -0.5\,(1+\mathrm{j})$$

$$\frac{g_m}{K}\frac{\partial K}{\partial g_m} = 1 = 1$$

$$\frac{G_2}{K}\frac{\partial K}{\partial G_2} = -1 = -1 .$$

The sum of the relative sensitivities $S_i^r$ is clearly zero, both symbolically for any value of frequency and numerically at $\omega = 10^7$rad/sec. The gradient vector **G** is given by

$$\mathbf{G} = [0.5\times10^4 \quad 0.5\times10^{11} \quad 50(1-j) \quad 500(1-j]^t$$

The amplifier input impedance $Z$ in terms of impedances is

$$Z = R_1 + C_1^{-1}/s = 10^3 - 10^3 j\ , \quad \text{at } \omega = 10^7 \text{ rad/sec.}$$

Relative sensitivities $S_i^r$ are

$$\frac{R_1}{Z}\frac{\partial Z}{\partial R_1} = \frac{sR_1}{C_1^{-1} + sR_1} = 0.5\,(1 + j)$$

$$\frac{C_1^{-1}}{Z}\frac{\partial Z}{\partial C_1^{-1}} = \frac{C_1^{-1}}{C_1^{-1} + sR_1} = 0.5\,(1 - j)\ .$$

Sensitivities to $g_m$ and $R_2$ are zero and have been omitted. Clearly the sum of relative sensitivities $S_i^r$ is unity.

The numerical results in this example illustrate the enormous range of values obtained for sensitivity even in quite simple electronic circuits. The importance of relative sensitivity in the assessment of computed results is obvious.

## 8.2 SENSITIVITY ANALYSIS METHODS

**Differentiation**

When a circuit can be analysed, either by hand or by computer, to give symbolic transfer or impedance functions, sensitivity of any order may be obtained by direct differentiation. Example 8.1 was solved using symbolic analysis, and it illustrates the technique. Sensitivity through the transfer function can be extended to give gain, phase, transient, and other response sensitivities. These will be treated in detail in Section 8.4.

In cases where sensitivity is required at a number of points on a frequency or transient response graph the symbolic technique is efficient since the analysis is performed only once. Analysis through transfer functions is particularly recommended for tolerance analysis and computer-aided design of RC-active networks.

**Differences**

Standard difference formulae from numerical analysis can be utilized for approximating derivates in terms of first and higher order differences. Consider the effect of a change $\Delta p_i$ in component value $p_i$. If $\Delta p_i$ is sufficiently small (<1% for most circuits),

$$\frac{\partial y}{\partial p_i} \approx \frac{\Delta y}{\Delta p_i} = \frac{y(p_1, \ldots, p_i + \Delta p_i, \ldots, p_n) - y(p_i, \ldots, p_i, \ldots p_n)}{\Delta p_i} \tag{8.12}$$

Each component sensitivity evaluation requires one circuit analysis with an increment added to each component in turn. The nominal response is also required. More elaborate and more accurate difference formulae require more than one analysis for each component and they are not generally used. For $n$ components, therefore $n + 1$ analyses are required, each followed by subtraction of the nominal response. The difference between the two responses is small and leads to numerical inaccuracy.

Second-order sensitivity can also be obtained by differences, using the relationship

$$\frac{\partial^2 y}{\partial p_i \partial p_j} \approx \left\{ \left(\frac{\partial y}{\partial p_j}\right)_{p_i + \Delta p_i} - \left(\frac{\partial y}{\partial p_j}\right)_{p_i} \right\} \Big/ \Delta p_i \,, \tag{8.13}$$

which is a first order approximation expressing second-order derivatives in terms of differences in gradient at nearby points. Substitution from Eq. (8.12) into (8.13) yields

$$\begin{aligned} \frac{\partial^2 y}{\partial p_i \partial p_j} = \frac{1}{\Delta p_i \Delta p_j} \Big\{ & y(p_1, \ldots, p_i + \Delta p_i, p_j + \Delta p_j, \ldots p_n) \\ & - y(p_i, \ldots, p_i + \Delta p_i, p_j, \ldots p_n) \\ & - y(p_1, \ldots, p_i, p_j + \Delta p_j, \ldots p_n) \\ & + y(p_1, \ldots, p_i, p_j, \ldots p_n) \Big\}. \end{aligned} \tag{8.14}$$

The terms in Eq. (8.14) involving only one increment were also involved in computing the first derivatives. In order to compute the $n(n+1)/2$ second-order derivatives, one further analysis for each derivative is required with both increments added simultaneously to the component values.

The use of difference techniques for sensitivity evaluation is not recommended, because of the excessive number of analyses required and the possible problem of poor numerical accuracy. However, in some networks difference techniques cannot be avoided.

**Nodal Analysis**

Nodal analysis was developed in Chapter 2, giving the solution for node voltages

$$\mathbf{e}' = (\mathbf{A}^t\mathbf{Y}_m\mathbf{A})^{-1}\,\mathbf{A}^t(\mathbf{I}-\mathbf{Y}_m\mathbf{E}) \;, \tag{8.15}$$

in terms of the modified diagonal admittance matrix $\mathbf{Y}_m$ including dependent sources and other mutual coupling elements, the reduced incidence matrix **A**, and the branch current and voltage source matrices **I** and **E**. In Chapter 4 the solution was developned further in terms of the nodal admittance matrix $\mathbf{Y}'$ and nodal current sources $\mathbf{I}'$ as

$$\mathbf{e}' = (\mathbf{Y}')^{-1}\,\mathbf{I}' = \mathbf{Z}\mathbf{I}' \;. \tag{8.16}$$

The dash in $\mathbf{Y}'$ will be dropped for convenience in the following analysis.

On the assumption of linearity, increments to any fixed branch current source in **I** or voltage source in **E** are additive in effect. In linearized circuits increments must be assumed sufficiently small not to significantly change admittances or sources in the linearization. If $\Delta\mathbf{e}'$, $\Delta\mathbf{I}$, and $\Delta\mathbf{E}$ are the increments to $\mathbf{e}'$, **I**, and **E** respectively,

$$\Delta\mathbf{e}' = (\mathbf{A}^t\mathbf{Y}_m\mathbf{A})^{-1}\,\mathbf{A}^t\,(\Delta\mathbf{I}-\mathbf{Y}_m\,\Delta\mathbf{E}). \tag{8.17}$$

This equation merely restates one of the consequences of linearity (see Chapter 1, Section 1.5).

When the modified admittance matrix $\mathbf{Y}_m$ changes by a small increment $\Delta\mathbf{Y}_m$, node voltage variation may be obtained from

$$(\mathbf{A}^t\,\Delta_m\,\mathbf{A})\mathbf{e}' = \mathbf{A}^t(\mathbf{I}-\mathbf{Y}_m\,\mathbf{E}) \tag{8.18}$$

which is a version of Eq. (8.17) avoiding the matrix inverse. Since increments $\Delta\mathbf{Y}_m$ cause changes $\Delta\mathbf{e}'$ in the node voltages $\mathbf{e}'$ the total increment is required, and

$$(\mathbf{A}^t\,\Delta\mathbf{Y}_m\,\mathbf{A})\mathbf{e}' + (\mathbf{A}^t\,\mathbf{Y}_m\,\mathbf{A})\Delta\mathbf{e}' = \mathbf{A}^t(-\Delta_m\mathbf{E}) \;.$$

Rearranging yields

$$\Delta\mathbf{e}' = -(\mathbf{A}^t\,\mathbf{Y}_m\,\mathbf{A})^{-1}\,\mathbf{A}^t[\Delta\mathbf{Y}_m\,(\mathbf{E}+\mathbf{A}\mathbf{e}')] \tag{8.19}$$

Combining Eqs. (8.17) and (8.19) gives the total variation $\Delta \mathbf{e}'$ in $\mathbf{e}'$ as,

$$\Delta \mathbf{e}' = (\mathbf{A}^t \mathbf{Y}_m \mathbf{A})^{-1} \mathbf{A}^t[\Delta \mathbf{I} - \mathbf{Y}_m \Delta \mathbf{E} - \Delta \mathbf{Y}_m (\mathbf{E} + \mathbf{A}\mathbf{e}')] , \quad (8.20)$$

when admittances in $\mathbf{Y}_m$, current sources $\mathbf{I}$, and voltage sources $\mathbf{E}$ all vary simultaneously.

Eq. (8.20) can be rewritten in the nodal admittance matrix form,

$$\Delta \mathbf{e}' = \mathbf{Z}[\Delta \mathbf{I}' - \Delta \mathbf{Y} \mathbf{e}'] \quad (8.21)$$

where $\mathbf{Y} = (\mathbf{A}^t \mathbf{Y}_m \mathbf{A})$, $\mathbf{Z} = \mathbf{Y}^{-1}$, and $\Delta \mathbf{Y} = (\mathbf{A}^t \Delta \mathbf{Y}_m \mathbf{A})$ and $\Delta \mathbf{I}'$ are the source increments $\Delta \mathbf{I}$ referred to the nodes. The circuit is further assumed to contain current sources only, so that $\mathbf{E} = \Delta \mathbf{E} = \mathbf{0}$. Only minor modifications are required to circuit analysis programs based on inversion of the nodal admittance matrix $\mathbf{Y}$ to retain the elements of the inverse $\mathbf{Z}$. First-order sensitivity can then be obtained directly from $\mathbf{Z}$ without further analysis.

Before proceeding, the notation identifying a circuit admittance and its sensitivity and the rules for filling $\mathbf{Y}$ must be recalled from Chapter 4. The element $y_{kl}^{ij}$ of the nodal admittance matrix $\mathbf{Y}$ is the self-admittance connected between circuit nodes ⓘ and ⓙ when $i = k$ and $j = l$. When $i \neq k$ or $j \neq l$, $y_{kl}^{ij}$ is the mutual admittance connected between nodes ⓘ and ⓙ controlled by nodes ⓚ to ⓛ. The element appears in the $\mathbf{Y}$ matrix in positions $Y_{ik}$ and $Y_{ij}$ as $+y_{kl}^{ij}$ and in positions $Y_{il}$ and $Y_{jk}$ as $-y_{kl}^{ij}$, where the two suffixes to $\mathbf{Y}$ identify the row and column numbers respectively.

Sensitivity to an individual circuit admittance involves these four elements only in $\Delta \mathbf{Y}$, and all other elements are zero. Direct use of Eq. (8.20) is therefore extremely inefficient since almost all of the multiplications involved in the product $\mathbf{Z}\Delta\mathbf{Y}$ are zero. The non-zero terms required in this product must be selected to overcome the inefficiency of the basic matrix equation. The utility of Eq. (8.20) lies in the expression of total variation in all the node voltages with respect to increments in all sources and admittances. It will be used in the discussion on tolerances in the next chapter.

In an $n$-node corcuits, assume that the input is applied at node ① and the output is at node ⓡ. The transfer function $T$ is therefore given by

$$T = e_r'/I_1' ,$$

which may be identified with element $Z_{r1}$ in $\mathbf{Z}$. The absolute sensitivity $S_{kl}^{ij}$ to the circuit admittance $y_{kl}^{ij}$ is by definition,

$$S_{kl}^{ij} = \frac{\partial T}{\partial y_{kl}^{ij}} = \frac{\partial Z_{r1}}{\partial y_{kl}^{ij}} .$$

Expanding for $\Delta \mathbf{e}'_r$ from Eq. (8.20), noting that all elements in $\Delta \mathbf{Y}$ are zero except for the $\delta y_{kl}^{ij}$ terms identified above,

$$\Delta e'_r = -\ [Z_{r1}\, Z_{r2} \ldots Z_{rn}] \begin{array}{c} \\ \\ i \\ \\ \\ j \\ \\ \\ \text{Row} \\ \text{Column} \end{array} \begin{array}{c} \left[\begin{array}{cccccc} 0 & & \cdot & & \cdot & \\ & & \cdot & & \cdot & \\ \ldots & & \delta y_{kl}^{ij} & \ldots & -\delta y_{kl}^{ij} & \ldots \\ & & \cdot & & \cdot & \\ & & \cdot & & \cdot & \\ \ldots & & -\delta y_{kl}^{ij} & \ldots & \delta y_{kl}^{ij} & \ldots \\ & & \cdot & & \cdot & \\ & & \cdot & & \cdot & \\ & & \cdot & & \cdot & 0 \end{array}\right] \\ \quad k \qquad\qquad l \end{array} \left[\begin{array}{c} e'_1 \\ \cdot \\ \cdot \\ \cdot \\ \cdot \\ \cdot \\ \cdot \\ \cdot \\ e'_n \end{array}\right]$$

$$= -(Z_{ri} - Z_{rj})\,(e'_k - e'_l)\,\delta y_{kl}^{ij}\ .$$

Therefore, $$\frac{\Delta e'_r/I'_1}{\delta y_{kl}^{ij}} = -(Z_{ri} - Z_{rj})\left(\frac{e'_k}{I'_1} - \frac{e'_l}{I'_1}\right)\ .$$

Expanding Eq. (8.16) with excitation at node ① only,

$$\begin{bmatrix} e'_1 \\ e'_2 \\ \\ e'_n \end{bmatrix} = \begin{bmatrix} Z_{11} & \ldots & Z_{1n} \\ Z_{21} & \ldots & \cdot \\ & & \cdot \\ Z_{n1} & \ldots & Z_{nn} \end{bmatrix} \begin{bmatrix} I'_1 \\ 0 \\ \\ 0 \end{bmatrix}$$

Clearly $$\frac{e'_k}{I'_1} - \frac{e'_l}{I'_1} = Z_{k1} - Z_{l1}$$

and since $$S_{kl}^{ij} = \mathop{\text{Limit}}_{\delta y \to 0} \left( \frac{\Delta e'_n/I'_1}{\delta y_{kl}^{ij}} \right) ,$$

$$S_{kl}^{ij} = -(Z_{ri} - Z_{rj})\,(Z_{k1} - Z_{l1})\ . \qquad (8.22)$$

Sensitivity has therefore been found as a product of two factors obtained from the elements of the inverse $\mathbf{Z}$ of the nodal admittance matrix. No further analysis has been necessary, provided that the elements of $\mathbf{Z}$ are retained during the matrix inverse procedure. Clearly, transfer functions corresponding to elements of $\mathbf{Z}$ other than $\mathbf{Z}_{r1}$ could have been chosen, resulting in similar expressions for all node voltage sensitivities with respect to any admittance variation in the network for any excited input node. The method can be extended to second-order sensitivity determination.

It must be emphasized that the value $S_{kl}^{ij}$ is complex and represents sensitivity to change in admittance. In most cases sensitivity to component variation is more relevant, and $S_i$ is generally assumed to be component sensitivity either in terms of admittance ($R^{-1}$, $C$, $L^{-1}$) or impedance ($R$, $C^{-1}$, $L$) units.

Since $\delta y$ equals $\delta(R^{-1})$. $\delta(\mathrm{j}\omega C)$ or $\delta(L^{-1}/\mathrm{j}\omega)$,

$$S_i = S_{kl}^{ij} \text{ for resistive admittance,}$$

$$S_i = \mathrm{j}\omega S_{kl}^{ij} \text{ for capacitive admittance, and}$$

$$S_i = (1/\mathrm{j}\omega)S_{kl}^{ij} \text{ for inductive admittance.}$$

**Adjoint Networks**

When analysis programs already in use do not provide the elements of the inverse $Z$ of the nodal admittance matrix, sensitivity can still be efficiently obtained, but by two analyses instead of one.

The factor $(Z_{k1} - Z_{l1})$ in Eq. (8.22) was first expressed in terms of the network voltages and input excitation as

$$Z_{k1} - Z_{l1} = (e'_k - e'_l)/I'_1 \ . \tag{8.23}$$

Similarly, $$Z_{ri} - Z_{rj} = \frac{e'_{ri}}{I'_i} - \frac{e'_{rj}}{I'_j} \ . \tag{8.24}$$

Unfortunately expression (8.24) implies two separate analyses, one to determine the value $e'_{ri}$ when the network is excited at node ⓘ and a second to determine $e'_{rj}$ with the network excited at node ⓙ. A total of $n$ analyses are therefore required to determine $e'_{ri}/I'_i$ for $i = 1, \ldots n$. All sensitivities can then be found.

The number of extra analyses can be reduced to one by inventing a new network, called the **adjoint** network which can be analysed to obtain all the factors required. The adjoint circuit has the property that its admittance matrix $\mathbf{Y}^*$ is the transpose of the admittance matrix $\mathbf{Y}$ of the original circuit. The * refers to the adjoint circuit, that is,

$$\mathbf{Y}^* = \mathbf{Y}^t$$

$$\mathbf{Z}^* = \mathbf{Z}^t$$

$$\mathbf{Z}^{*t} = \mathbf{Z} \tag{8.25}$$

$$Z_{ri} - Z_{nj} = Z_{ir}^* - Z_{jr}^* \ . \tag{8.26}$$

In the adjoint circuit

$$Z^*_{ir} - Z^*_{jr} = \frac{e^*_i - e^*_j}{I^*_r},$$

which requires excitation at node ⓡ only to determine $Z^*_{ir} - Z^*_{jr}$ for all values of $i$ and $j$. To avoid a confusion of notation the dash has been omitted from the adjoint circuit variables $e^*$ and $I^*$. By substitution in Eq. (8.22)

$$S^{ij}_{kl} = -\frac{(e'_k - e'_l)(e^*_i - e^*_j)}{I'_1 I^*_r}. \qquad (8.27)$$

All sensitivities can therefore be obtained from two separate analyses, one of the original network and one of the adjoint network. The link with the nodal analysis method is shown by Eq. (8.25). If the elements of **Z** are available the inverse **Z*** need not be computed, and the extra analysis to determine the factor $(e^*_i - e^*_j)/I^*_r$ is avoided. However, the adjoint method is important because rules have been formulated for generating adjoint networks which can then be analysed by any method. Extension to second-order sensitivity is also possible.

*Example* 8.2

Determine the sensitivity of gain $G$ with respect to the admittance $1/R_1$ in the attenuator circuit of Fig. 8.2.

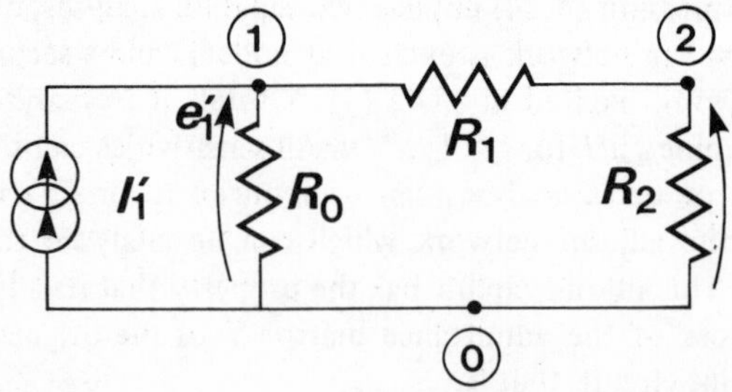


Fig. 8.2 – Circuit for Example 8.2.

By analysis,

$$G = \frac{e'_2}{I'_1} = \frac{R_0 R_2}{R_0 + R_1 + R_2}.$$

*Differentiation Method*

$$\frac{\partial G}{\partial(1/R_1)} = -R_1^2 \frac{\partial G}{\partial R_1}$$

$$= \frac{R_0 R_2 R_1^2}{(R_0 + R_1 + R_2)^2} \, .$$

*Differences*

Let $y_1 = 1/R_1$. The expression for gain becomes,

$$G = \frac{R_0 R_2 y_1}{(R_0 + R_2) y_1 + 1} \, .$$

Add a small increment $\delta y_1$ to $y_1$ and obtain increment $\delta G$ in $G$. Subtracting $G$ to obtain the difference yields

$$G + \delta G - G = \frac{R_0 R_2 (y_1 + \delta y_1)}{(R_0+R_2)(y_1+\delta y_1)+1} - \frac{R_0 R_2 y_1}{(R_0+R_2)\, y_1+1} \, ,$$

Simplifying and taking the limit as $\delta y_1$ tends to zero,

$$\delta G = \frac{R_0 R_2 \, \delta y_1}{[(R_0+R_2)(y_1+\delta y_1)+1] \, [(R_0+R_2) y_1+1]} \, ,$$

$$\lim_{\delta y_1 \to 0} \frac{\delta G}{\delta y_1} = \frac{R_0 R_2}{[(R_0+R_2) y_1 + 1]^2} \, ,$$

$$\frac{\partial G}{\partial(1/R_1)} = \frac{R_0 R_2 R_1^2}{[R_0+R_2+R_1]^2} \, .$$

*Nodal method*

Filling the admittance matrix yields,

$$\mathbf{Y} = \begin{bmatrix} 1/R_0 + 1/R_1 & -1/R_1 \\ -1/R_1 & 1/R_1 + 1/R_2 \end{bmatrix} .$$

Inversion gives

$$\mathbf{Z} = \mathbf{Y}^{-1} = \frac{R_0 R_1 R_2}{R_1 + R_2 + R_0} \begin{bmatrix} 1/R_1 + 1/R_2 & 1/R_1 \\ 1/R_1 & 1/R_0 + 1/R_1 \end{bmatrix}.$$

Substituting for sensitivity to $y_{12}^{12}$.

$$S_{12}^{12} = -(Z_{21} - Z_{22})(Z_{11} - Z_{21}) .$$

Therefore,

$$\frac{\partial G}{\partial(1/R_1)} = S_{12}^{12} = \frac{R_0 R_2 R_1^2}{(R_1 + R_2 + R_0)^2} .$$

*Adjoint method*
With excitation $I_1'$ to node ①,

$$\mathbf{e}' = \mathbf{Z}\,\mathbf{I}' \quad \text{where } \mathbf{I}' = [I_1' \;\; 0]^{\mathrm{t}} .$$

Solving for $\mathbf{e}'$,

$$\mathrm{e}_1' = \frac{R_0 (R_1 + R_2)}{R_0 + R_1 + R_2} I_1' , \quad \mathrm{e}_2' = \frac{R_0 R_2 I_1'}{R_0 + R_1 + R_2} .$$

For the adjoint network, excitation is at node ②. Noting that

$$\mathbf{Y}^* = \mathbf{Y},$$

$$\mathbf{e}^* = \mathbf{Z}^* \mathbf{I}^*$$

where $\mathbf{I}^* = [0 \;\; I_2']^{\mathrm{t}}$ and $\mathbf{Z}^* = \mathbf{Z}^{\mathrm{t}} = \mathbf{Z}$ ,

therefore,

$$\mathbf{e}^* = \mathbf{Z}\mathbf{I}^* .$$

Solving for $\mathbf{e}^*$

$$e_1^* = \frac{R_0 R_2 I_2^*}{(R_0 + R_1 + R_2)} \quad e_2^* = \frac{R_2 (R_0 + R_1) I_2^*}{(R_0 + R_1 + R_2)} .$$

Substituting for sensitivity to $y_{12}^{12}$,

$$S_{12}^{12} = -(e_1' - e_2')(e_1^* - e_2^*)/I_2' I_2^* \ ;$$

therefore,

$$\frac{\partial G}{\partial(1/R_1)} = S_{12}^{12} = \frac{R_0 R_2 R_1^2}{(R_0 + R_1 + R_2)^2} \ .$$

The same value for sensitivity has been found by all four methods. The problems contain further circuits which should be worked to gain familiarity particularly with the nodal method.

## 8.3 FREQUENCY DOMAIN SENSITIVITY

Sensitivity methods based on the nodal and adjoint methods give sensitivity at a given frequency as a complex number, that is.

$$S_i = a + jb \ . \tag{8.28}$$

Computation of $S_i$ over a range of frequencies results in sensitivity as a complex frequency response. In practice other responses are more convenient, and the values of $S_i$ obtained must be converted by using a number of simple formulae which will now be derived. Let the complex gain $K = A + jB$, therefore $S_i = \partial K/\partial p_i$.

*A) Real and Imaginary Parts (A, B)*

Since components are assumed to have real values,

$$\Delta K = \Delta A + j\Delta B = (a + jb)\Delta p_i \ ;$$

therefore,

| | |
|---|---|
| $\dfrac{\partial A}{\partial p_i} = \text{Re}\,\{S_i\}$ | $\dfrac{\partial B}{\partial p_i} = \text{Im}\,\{S_i\}$ . |

*B) Gain (neper) and Phase (radians) (G, P)*

By definition

$$G = \log_n |K| \qquad\qquad P = \tan^{-1}(B/A)$$

Differentiating,

$$\frac{\partial G}{\partial p_i} = \frac{\partial(\log_n|K|)}{\partial p_i} = \frac{1}{|K|}\frac{\partial(A^2+B^2)^{\frac{1}{2}}}{\partial p_i} = \frac{1}{|K|^2}\left(\frac{A\partial A}{\partial p_i} + \frac{B\partial B}{\partial p_i}\right)$$

$$\frac{\partial P}{\partial p_i} = \frac{\partial(\tan^{-1}B/A)}{\partial p_i} = \frac{A^2}{A^2+B^2}\frac{\partial(B/A)}{\partial p_i} = \frac{1}{|K|^2}\left(\frac{A\partial B}{\partial p_i} - \frac{B\partial A}{\partial p_i}\right).$$

These sensitivities can be developed further by recognizing them to be the real and imaginary parts of the same complex expression, that is,

$$\frac{\partial G}{\partial p_i} = \frac{1}{|K|^2} \text{Re}\left\{(A - \text{j}B)\left(\frac{\partial A}{\partial p_i} + \text{j}\frac{\partial B}{\partial p_i}\right)\right\}$$

$$\frac{\partial P}{\partial p_i} = \frac{1}{|K|^2} \text{Im}\left\{(A - \text{j}B)\left(\frac{\partial A}{\partial p_i} + \text{j}\frac{\partial B}{\partial p_i}\right)\right\}$$

Since $|K|^2 = (A + \text{j}B)(A - \text{j}B)$ ,

$$\frac{\partial G}{\partial p_i} = \text{Re}\left\{\frac{\partial A/\partial p_i + \text{j}\partial B/\partial p_i}{A + \text{j}B}\right\} \qquad \frac{\partial P}{\partial p_i} = \text{Im}\left\{\frac{\partial A/\partial p_i + \text{j}\partial B/\partial p_i}{A + \text{j}B}\right\}$$

Finally,

$$\frac{\partial G}{\partial p_i} = \text{Re}\left\{\frac{S_i}{K}\right\} \qquad \frac{\partial P}{\partial p_i} = \text{Im}\left\{\frac{S_i}{K}\right\} \tag{8.30}$$

*C) Gain (dB) and phase (degrees)* $(G_\text{D} P_\text{D}$

By definition,

$$G_\text{D} = (20/\log_n 10)G \qquad P_\text{D} = (180/\pi)P .$$

Sensitivities from $B$ are merely scaled to give

$$\frac{\partial G_\text{D}}{\partial p_i} = 8.68589\ \text{Re}\left\{\frac{S_i}{K}\right\} \qquad \frac{\partial P_\text{D}}{\partial p_i} = 57.29578\ \text{Im}\left\{\frac{S_i}{K}\right\}. \tag{8.31}$$

*D) Magnitude* $|K|$

With reference to the derivation for gain $G$ in section $B$ above,

$$\frac{\partial|K|}{\partial p_i} = \frac{|K|\,\partial(\log_n|K|)}{\partial p_i} ;$$

therefore

$$\boxed{\frac{\partial|K|}{\partial p_i} = |K|\,\mathrm{Re}\left\{\frac{S_i}{K}\right\}} \quad . \tag{8.32}$$

*E) Sensitivity $S_i$ from Gain and Phase Sensitivity*

As we saw in Chapter 6, poles and zeroes in factorized transfer functions make individual contributions towards gain and phase. Sensitivity $\partial G/\partial p_i$ and $\partial P/\partial p_i$, where $p_i$ is a pole or zero, is therefore rapidly obtainable, and it is convenient to be able to determine $S_i$ from them. Combining the two expressions (8.30) yields

$$\frac{S_i}{K} = \mathrm{Re}\left\{\frac{S_i}{K}\right\} + \mathrm{Im}\left\{\frac{S_i}{K}\right\}$$

$$= \partial G/\partial p_i + \mathrm{j}\partial P/\partial p_i \ ,$$

Hence,

$$\boxed{S_i = K\left(\frac{\partial G}{\partial p_i} + \mathrm{j}\frac{\partial P}{\partial p_i}\right)} \ . \tag{8.33}$$

*Example* 8.3

Obtain the sensitivity $S_i$ of the gain $K$ with respect to the capacitor $C$ of the circuit in Fig. 8.3 at $\omega = 10^6$ rad/sec. Derive the sensitivities of real part $A$, imaginary part $B$, magnitude $|K|$, gain $G$, and phase $P$, and verify Eq. (8.33).

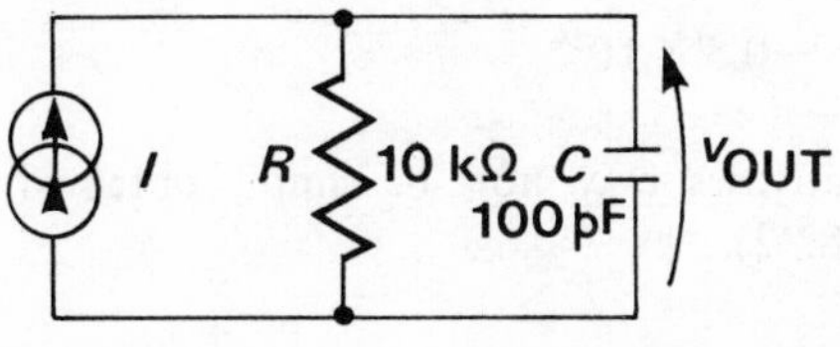


Fig. 8.3 – Circuit for Example 8.3.

By analysis for $K = v_{out}/I$,

$$K = \frac{R}{1 + j\omega CR}, \quad |K| = \frac{R}{\sqrt{(1 + \omega^2C^2R^2)}},$$

$$\text{Re}\,\{K\} = \frac{R}{1 + \omega^2C^2R^2}, \quad \text{Im}\,\{K\} = \frac{-\omega CR}{1 + \omega^2C^2R^2}.$$

Differentiating $K$ with respect to $C$,

$$S_i = \frac{\partial K}{\partial C} = \frac{-j\omega R^2}{(1 + j\omega CR)^2},$$

and $$\text{Re}\,\{S_i\} = \frac{-2\omega^2CR^3}{(1 + \omega^2C^2R^2)^2},$$

$$\text{Im}\,\{S_i\} = \frac{-\omega R^2(1 - \omega^2C^2R^2)}{(1 + \omega^2C^2R^2)^2}.$$

Substitution for $R = 10^4$. $C = 10^{-10}$, $\omega = 10^6$ yields

$$K = \frac{10^4}{1 + j} = 5000(1 - j)$$

$$|K| = 5000\sqrt{2}$$

$$\text{Re}\,\{K\} = 5000$$

$$\text{Im}\,\{K\} = -5000 .$$

Sensitivity $S_i$ is real in this circuit since

$$\text{Re}\,\{S_i\} = -0.5 \times 10^{14}, \quad \text{Im}\,\{S_i\} = 0 ,$$

and $$S_i = -0.5 \times 10^{14} .$$

The sensitivities may now be simply obtained by use of Eqs. (8.29), (8.30) and (8.32).

$$\frac{\partial A}{\partial C} = \text{Re}\,\{S_i\} = -0.5 \times 10^{14}$$

$$\frac{\partial B}{\partial C} = \text{Im}\,\{S_i\} = 0$$

$$\frac{\partial G}{\partial C} = \text{Re}\left\{\frac{S_i}{K}\right\} = \text{Re}\left\{-0.5 \times 10^{14}\frac{(1+\text{j})}{10^4}\right\}$$

$$= -0.5 \times 10^{10}$$

$$\frac{\partial P}{\partial C} = \text{Im}\left\{\frac{S_i}{K}\right\} = -0.5 \times 10^{10}$$

$$\frac{\partial |K|}{\partial C} = |K|\ \text{Re}\left\{\frac{S_i}{K}\right\} = 5000\sqrt{2}\,(-0.5 \times 10^{10})$$

$$= -3.54 \times 10^{13}\,.$$

Eqn. (8.33) may be verified by substitution to obtain $S_i$, that is,

$$S_i = K\left(\frac{\partial G}{\partial p_i} + \frac{\partial P}{\partial p_i}\right)$$

$$= 50000(1-\text{j})\,(-0.5 \times 10^{10} - \text{j}\;0.5 \times 10^{10})$$

$$= -5000(1-\text{j})\,(1+\text{j}) \times 0.5 \times 10^{10}$$

$$= -0.5 \times 10^{14}\,.$$

Eqs. (8.29), (8.30), (8.31), and (8.32) may be verified for all frequencies in this example by symbolic analysis and differentiation.

## 8.4 TRANSFER FUNCTION SENSITIVITIES

The general form of the factorized transfer function was given in Eq. (6.31) as

$$F(s) = \frac{K\Pi_{i=1}^{n_z}\,(s - z_i)^{m_i}}{\Pi_{j=1}^{n_p}\,(s - p_j)^{m_j}}\,, \tag{8.34}$$

where the poles and zeroes are either real or occur in complex pairs. The alternative form of transfer function in terms of the ratio of polynomials in $s$ is

$$F(s) = \frac{a_m s^m + a_{m-1}s^{m-1} + \ldots. + a_0}{b_l s^l + b_{l-1}s^{l-1} + \ldots + b_0} \tag{8.35}$$

where the coefficients $a_i$, $i = 0, 1, \ldots, m$, and $b_i$, $i = 0, 1, \ldots l$, are linear functions of the component values in most practical cases. The process of taking roots to obtain Eq. (8.34) from Eq. (8.35) destroys this useful property, and roots are in general only available in numerical form from computer programs.

Transfer or immittance functions in the $s$-domain are intermediate between components and responses. They are important in analysis because they may be evaluated at any frequency by the substitution $s = j\omega$, or they may be evaluated through the inverse Laplace transform to obtain transient response. Sensitivity analysis proceeds by differentiation of the transfer function, which need only be performed once, followed by evaluation as in analysis giving similar advantages.

Transfer functions are concerned with the inputs and outputs of linear circuits and systems, and sensitivity analysis through them is limited in the same way. Sensitivity determination through the transfer function makes use of symbolic analysis, even though this imposes limits on the maximum size of network which can be handled. The coefficients of $F(s)$ in symbolic form in both the numerator polynomial $N(s)$ and denominator polynomial $D(s)$ are easily evaluated, or first differentiared then evaluated giving polynomials in $s$. As we saw in Chapter 6 the listing by computer of coefficients in symbolic form is meaningless for circuits with more than a few nodes and branches. However, evaluation as polynomials in $s$ is practical, possibly for 20-node 50-branch circuits maximum. Important sensitivity applications lie in the analysis and design of RC active networks and filters, and in the approximation of required responses by transfer function. In these areas the electronic detail of amplifiers etc. is omitted by the use of models, and the resulting circuits are either small or they can be divided into sections.

A number of different sensitivity coefficients arise from the use of transfer functions. These are first introduced in Fig. 8.4 which shows a flow chart of transfer function and sensitivity analysis methods. In the remainder of this section all the sensitivities shown will be derived from the symbolic transfer function and properly defined.

**Numerator and Denominator Sensitivity**

The transfer function $F(s,\mathbf{p})$ obtained by symbolic analysis is in the form of two separate polynomials $N(s,\mathbf{p})$ and $D(s,\mathbf{p})$ where

$$F(s,\mathbf{p}) = N(s,\mathbf{p})/D(s,\mathbf{p}) \ . \qquad (8.36)$$

Recall that $\mathbf{p}$ is the vector of $n$ component values $[p_1, p_2 \ldots p_n]$. Substitution of component values yields

$$F(s) = N(s)/D(s) \ . \qquad (8.37)$$

which can also be obtained directly from $s$-domain analysis.

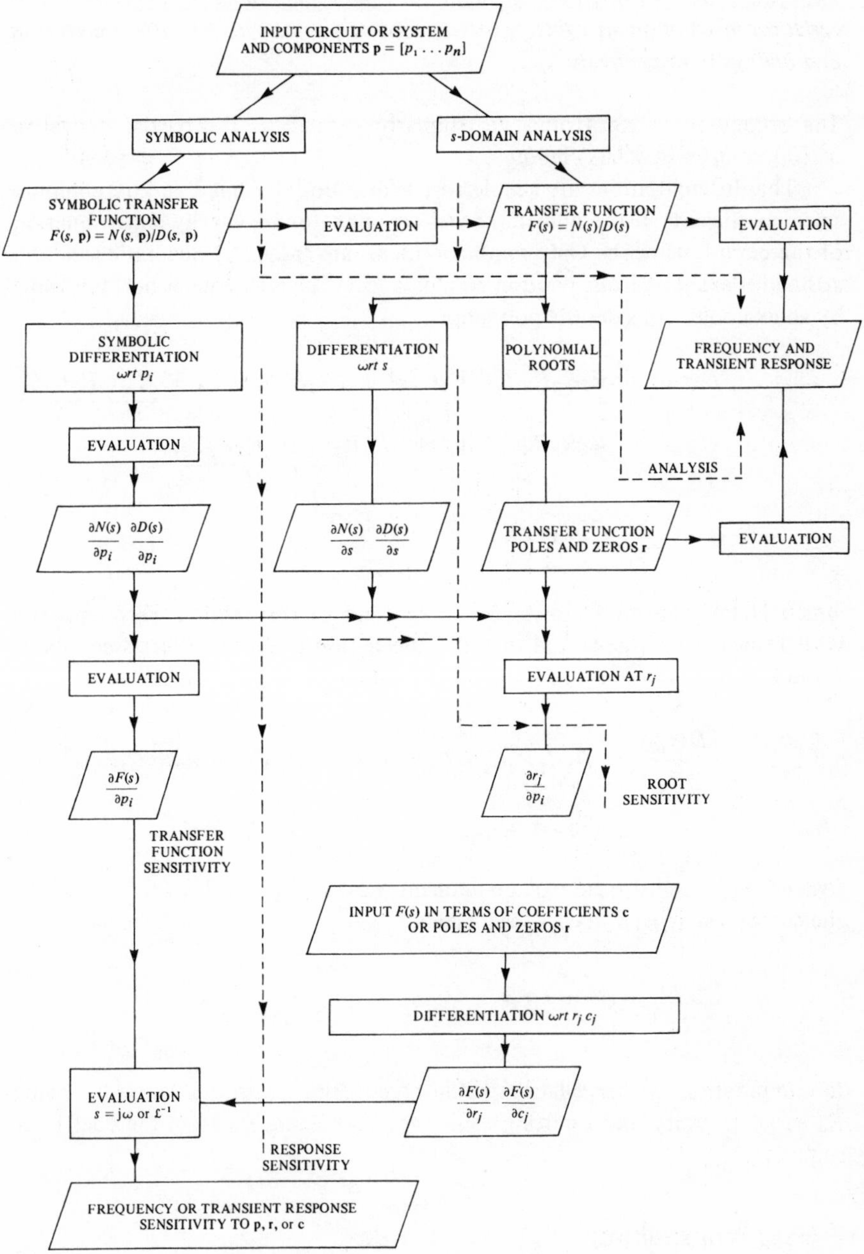


**Fig. 8.4 – Transfer function sensitivity analysis flowchart.**

*The first-order sensitivities of numerator and denominator polynomials to variation in component value $p_i$ are defined as the partial derivatives $\partial N(s)/\partial p_i$ and $\partial D(s)/\partial p_i$ respectively.*

The second-order sensitivities are similarly defined as the partial derivatives $\partial^2 N(s)/\partial p_i \partial p_j$ and $\partial^2 D(s)/\partial p_i \partial p_j$.

The differentiations are simple to perform both by hand and by computer since coefficients in the numerator and denominator are usually linear functions of the circuit elements. Only minor problems are caused by quadratic functions arising for example from gyration conductances. The technique is best illustrated by an example. Consider the polynomial

$$D(s,\mathbf{p}) = R_1R_2R_3R_4C_1C_2C_3 s^3 + (R_1R_3R_4C_1C_3 + \underline{R_1R_2R_3C_1C_3} + R_2R_3R_4C_2C_3)s^2 + (\underline{R_1R_2C_1} + \underline{R_2R_3C_3} + R_3R_4C_3)s + \underline{R_2}$$

which is taken from Example 8.4 at the end of this section. Differentiation with respect to $R_4$ is carried out by cancellation of $R_4$ from terms containing $R_4$ and by omission of terms (underlined) not containing $R_4$. Therefore

$$\frac{\partial D(s,\mathbf{p})}{\partial R_4} = R_1R_2R_3C_1C_2C_3s^3 + (R_1R_3C_1C_3 + R_2R_3C_2C_3)s^2 + R_3C_3s\ .$$

Evaluation by substitution of component values [1.0 2.0 0.5 1.0 1.0 1.0] for the components $[R_1R_2R_3C_1C_2C_3]$ yields

$$\frac{\partial D(s)}{\partial R_4} = s^3 + 1.5s^2 + 0.5s\ .$$

In computation the cancellation would be combined with evaluation by setting $R_4$ equal to unity and omitting terms not containing $R_4$ from the coefficient summing.

**Pole and Zero Sensitivity**

The poles and zeroes of the transfer function are the roots of the denominator and numerator polynomials $D(s)$ and $N(s)$ respectively. Let the poles and zeroes collectively be called the roots $\mathbf{r} = [r_1 r_2 \ldots.]$.

*The sensitivity of the pole or zero $r_j$ to variation in component value $p_i$ is defined as the partial derivative $\partial r_j/\partial p_i$.*

The derivative $\partial r_j/\partial p_i$ is equal to the value of $\partial s/\partial p_i$ evaluated at the point $s = r_j$ in the $s$-plane.

The numerator and denominator sensitivities derived from symbolic analysis above are converted to the required form $\partial s/\partial p_i$ on division by the factors $\partial N(s)/\partial s$ or $\partial D(s)/\partial s$.

For numerator roots, i.e. zeroes,

$$\frac{\partial r_j}{\partial p_i} = \left(\frac{\partial N(s)}{\partial p_i} \bigg/ \frac{\partial N(s)}{\partial s}\right)_{s=r_j} \tag{8.38}$$

$$\frac{\partial r_j}{\partial p_i} = \left(\frac{\partial D(s)}{\partial p_i} \bigg/ \frac{\partial D(s)}{\partial s}\right)_{s=r_j} . \tag{8.39}$$

Differentiation of numerator or denominator polynomials with respect to $s$ presents no problems, and $\partial r/\partial p_i$ is obtained as a ratio of two polynomials in $s$. This rational function in $s$ is then evaluated at each root in turn to give all the root sensitivities.

In computation this procedure suffers from the problem of common factors in the numerator and denominator of $\partial r_j/\partial p_i$ caused by multiple poles or zeroes. Without cancellation, evaluation of $\partial r_j/\partial p_i$ results in the indefinite answer zero/zero. Fortunately a high proportion of circuits do not contain multiple roots.

The **root locus** technique from automatic control theory is another approach to the same problem. It gives a graphical presentation of results involving large changes in system component values. The pole and zero sensitivity method given here is limited to small changes in component value since it is first-order only. Within this limitation the two methods give identical results, and first-order sensitivity is often used to assist in the construction of the root locus.

**Transfer Function Sensitivity to Component Variation**

*The first-order sensitivity of the transfer function $F(s)$ to variation in the component $p_i$ is defined as the partial derivative $\partial F(s)/\partial p_i$.*

Differentiation of Eq. (8.37) with respect to $p_i$ yields

$$\frac{\partial F(s)}{\partial p_i} = \left(D(s)\,\frac{\partial N(s)}{\partial p_i} - N(s)\,\frac{\partial D(s)}{\partial p_i}\right) \bigg/ D^2(s) \tag{8.40}$$

The terms in Eq. (8.40) are all available from evaluations of the transfer function and the numerator and denominator sensitivities as functions of $s$. The transfer function sensitivity is therefore easily obtained as a rational function. It should be noted that common factors in the numerator and denominator of $\partial F(s)/\partial p_i$ have not been cancelled, and problems of evaluation can arise particularly if inverse Laplace transformation is necessary.

An expression for second-order sensitivity $\partial^2 F(s)/\partial p_i \partial p_j$ can be obtained by differentiation of Eq. (8.40). This is left as an exercise.

**Transfer Function Sensitivity to Pole, Zero, and Coefficient Variation**

In designing a linear circuit for which the frequency or transient response is specified, the circuit transfer function is first obtained by a process called **approximation.** The transfer function is then **realized** as a circuit or system. Approximation procedures often involve iterative computation of the sensitivity of the transfer function, and through it the response, to variation in the poles and zeroes or the coefficients. These sensitivities are introduced here to complete and unify the treatment for transfer function sensitivity. Applications lie in the area of computer-aided circuit and control system design.

*The sensitivity of the transfer function $F(s)$ to variation in the pole or zero $r_j$ is defined as the partial derivative $\partial F(s)/\partial r_j$.*

Let the numerator and denominator polynomial coefficients $a_0\, a_1 \ldots a_m$ and $b_0\, b_1 \ldots b_l$ be collectively referred to as the coefficients $\mathbf{c} = [c_1\, c_2 \ldots]$.

*The sensitivity of the transfer function $F(s)$ to variation in the coefficient $c_j$ is defined as the partial derivative $\partial F(s)/\partial c_j$.*

Since poles and zeroes must occur in quadratic pairs it is incorrect to consider variation of one half of the pair alone. It is of more practical use to consider variation in complex pairs of roots in the form of separate sensitivity figures for the real part of the root and for the imaginary part of the root. The alternative form of factorized transfer function from Chapter 6 will therefore be used in place of that given in Eq. (8.34). In this section the transfer function

$$F(s) = \frac{K(s-z_1)^k\, \{(s-z_2)^2 + z_3^2\}^l}{(s-p_1)^m\, \{(s-p_2)^2 + p_3^2\}^n} \tag{8.41}$$

will be used. It omits the $\Pi$ which indicated the multiplication of factors and introduce numerous suffixes which would merely obscure the essential mathematics. In the quadratic factors the real numbers, $z_2$ and $z_3$ for example, are equal to the real and imaginary parts of the conjugate pair of roots, located at $z_2 \pm jz_3$ in this case. The roots are therefore given by

$$\mathbf{r} = [z_1\, z_2\, z_3\, p_1\, p_2\, p_3] \ .$$

Differentiating with respect to each root yields

$$\frac{\partial F(s)}{\partial z_1} = \frac{-kF(s)}{(s-z_i)} \qquad \frac{\partial F(s)}{\partial p_1} = \frac{mF(s)}{(s-p_1)}$$

$$\frac{\partial F(s)}{\partial z_2} = \frac{-2l(s-z_2)\,F(s)}{\{(s-z_2)^2+z_3^2\}} \qquad \frac{\partial F(s)}{\partial p_2} = \frac{2n(s-p_2)\,F(s)}{\{(s-p_2)^2+p_3^2\}}$$

$$\frac{\partial F(s)}{\partial z_3} = \frac{-2lz_3F(s)}{\{(s-z_2)^2+z_3^2\}} \qquad \frac{\partial F(s)}{\partial p_3} = \frac{2np_3}{\{(s-p_2)^2+p_3^2\}}\,F(s) \ . \tag{8.42}$$

Poles and zeroes are seen to be identical in effect but of opposite sign. In each case the power $k$, $l$, $m$, or $n$ of the factor involved in $F(s)$ is either reduced or increased by one, and for quadratic roots extra zero roots are inserted into the numerator. In computation these similarities are easily exploited for efficient evaluation of the sensitivities.

Numerator and denominator polynomials $N(s)$ and $D(s)$ are simply differentiated with respect to coefficients $a_i$ and $b_j$ to give

$$\frac{\partial N(s)}{\partial a_i} = s^i \qquad \frac{\partial D(s)}{\partial a_i} = 0$$

$$\frac{\partial N(s)}{\partial b_j} = 0 \qquad \frac{\partial D(s)}{\partial b_j} = s^j \ .$$

Use of Eq. (8.40) with components identified in this case with the coefficients yields

$$\frac{\partial F(s)}{\partial a_i} = \frac{s^i}{D(s)} \qquad \frac{\partial F(s)}{\partial b_j} = \frac{-N(s)}{D^2(s)}\,s^j \ . \tag{8.43}$$

The sensitivity expressions in Eqs. (8.42) for poles and zeroes and Eqs. (8.43) for coefficients are all transfer functions expressed as functions of $s$. As before in Eq. (8.40) cancellation of common factors cannot be assumed. However, it can usually be arranged in the pole, zero, and coefficient cases.

**Frequency Domain Sensitivity**

Evaluation of the transfer function partial derivatives is carried out as in analysis by the substitution $s = j\omega$. The frequency domain sensitivity $S_i$ is then obtained as a complex number, and the whole of Section 8.3 applies. Real part, imaginary part, gain, phase, and magnitude sensitivities may all be obtained by use of Eqs. (8.29) to (8.32).

It is worth recognizing the fact that gain and phase sensitivity to pole and zero variation evaluated through Eqs. (8.42) and (8.30) or (8.31), is inefficient. This is because poles and zeroes contribut individually to gain and phase, and direct differentiation gives a direct result.

Considering the zeroes in Eq. (8.41) the gain $G$(nepers) and phase $P$(radians) may be written

$$G = \frac{k}{2}\log(\omega^2+z_1^2) + \frac{l}{2}\log(\{z_2^2 + z_3^2 - \omega^2\}^2 + 4z_2^2\,\omega^2)$$

$$P = k\tan^{-1}(-\omega/z_1) + l\tan^{-1}(-2z_2\omega/\{z_2^2 + z_3^2 - \omega^2\})\ .$$

Additional terms due to the poles are identical except for a minus sign and have been omitted. Differentiating with respect to $z_1$, $z_2$ and $z_3$.

$$\frac{\partial G}{\partial z_1} = \frac{kz_1}{(\omega^2+z_1^2)}\ , \qquad \frac{\partial P}{\partial z_1} = \frac{k\omega}{(\omega^2+z_1^2)}\ ,$$

$$\frac{\partial G}{\partial z_2} = \frac{2F_3lz_2}{F_2}\ , \qquad \frac{\partial P}{\partial z_2} = \frac{2F_4l\omega}{F_2}\ ,$$

$$\frac{\partial G}{\partial z_3} = \frac{2F_1lz_3}{F_2}\ , \qquad \frac{\partial P}{\partial z_3} = \frac{4l\omega z_2z_3}{F_2}\ ,$$

where
$$F_1 = z_2^2 + z_3^2 - \omega^2$$

$$F_2 = F_1^2 + 4z_2^2\omega^2$$

$$F_3 = z_2^2 + z_3^2 + \omega^2$$

$$F_4 = z_2^2 - z_3^2 + \omega^2\ .$$

The value for complex frequency domain sensitivity $S_i$ for pole and zero variation is evaluated using Eq. (8.33), and all other frequency domain sensitivities can be obtained as before. Evaluation of sensitivity through Eqs. (8.44), with negative signs for poles, is recommended for frequency response sensitivity to pole and zero variations.

**Time Domain Sensitivity**

A standard theorem from the theory of Laplace transformation from Chapter 6 states that

$$\frac{\partial(\mathcal{L}^{-1}F(s))}{\partial p} = \frac{\mathcal{L}^{-1}\partial F(s)}{\partial p}$$

where $p$ is a parameter in $F(s)$ which we have identified as a component, a pole, a zero, or a coefficient in this section. Since $f(t) = \mathcal{L}^{-1}F(s)$

$$\frac{\partial f(t)}{\partial p} = \frac{\mathcal{L}^{-1}\partial F(s)}{\partial p} \; . \qquad (8.45)$$

Sensitivity in the time domain is therefore available for linear systems by inverse Laplace transformation of sensitivity derived as a function of $s$. This may be carried out using numerical inversion methods or, for sensitivity to poles and zeroes, partial fractions or convolution. Once the sensitivity has been obtained as a rational function, any inversion method may be used, and the reader is referred back to Chapter 6.

*Example* 8.4

Determine the transfer function and pole sensitivities to variation in the resistor $R_4$ for the filter shown in Fig. 8.5.

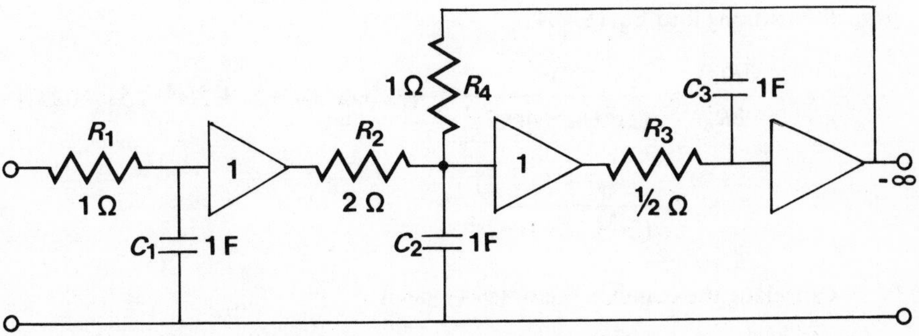


Fig. 8.5 – Low-pass filter for Example 8.4.

Assuming ideal amplifiers, symbolic analysis yields the transfer function

$$F(s) = -R_4 / \{R_1R_2R_3R_4C_1C_2C_3s^3 + s^2(R_1R_3R_4C_1C_3 + R_1R_2R_3C_1C_3 + R_2R_3R_4C_2C_3) + s(R_1R_2C_1 + R_2R_3C_3 + R_3R_4C_3) + R_2\}$$

$$= -R_4 / \{(1+sC_1R_1)\{s^2R_2R_3R_4C_2C_3 + s(C_3R_3R_4 + C_3R_2R_3) + R_2\}$$

$$= -2 \;/\; \{(s+1)(2s^2 + 2s + 2)\}$$

The filter is therefore shown to be a normalized, third-order, low-pass Butterworth with poles $-1, -1/2. \pm j\sqrt{3}/2$, and unity d.c. gain.

*Numerator and denominator sensitivity*
Differentiate $N(s)$ and $D(s)$ with respect to $R_4$ to obtain

$$\frac{\partial N(s)}{\partial R_4} = -1 \,,$$

$$\frac{\partial D(s)}{\partial R_4} = R_1R_2R_3C_1C_2C_3s^3 + (R_1R_3C_1C_3 + R_2R_3C_2C_3)s^2 + R_3C_3s$$

$$= s^3 + 1.5s^2 + 0.5s \,.$$

*Transfer function sensitivity*
Substituting into Eq. (8.40)

$$\frac{\partial F(s)}{\partial R_4} = \frac{1}{(2s^3+4s^2+4s+2)^2} \{-(2s^3+4s^2+4s+2) + 2(s^3+1.5s^2+0.5s)\}$$

$$= \frac{-(s^2 + 3s + 2)}{(2s^3+4s^2+4s+2)^2} \,.$$

Cancelling the common factor $(s+1)$ yields

$$\frac{\partial F(s)}{\partial R_4} = \frac{-(s + 2)}{(s+1)(s^2+s+1)^2} \,.$$

*Pole sensitivity*
Differentiation of $D(s)$ with respect to $s$ gives

$$\frac{\partial D(s)}{\partial s} = 6s^2 + 8s + 4 \ .$$

Substitute into Eq. (8.39) to obtain

$$\frac{\partial s}{\partial R_4} = \frac{s^3 + 1.5s^2 + 0.5s}{6s^2 + 8s + 4} \ .$$

At $\quad s = -1 \qquad \dfrac{\partial s'}{\partial R_4} = 0 \ ,$

$$s = -1/2 + j\sqrt{3}/2 \qquad \frac{\partial s}{\partial R_4} = 1/8 - j\sqrt{3}/8 \ ,$$

$$s = -1/2 - j\sqrt{3}/2 \qquad \frac{\partial s}{\partial R_4} = 1/8 + j\sqrt{3}/8 \ ,$$

The real root is therefore seen to be insensitive to $R_4$. The real part of the quadratic root has sensitivity 1/8 and the imaginary part $\mp\sqrt{3}/8$. Identical results can be obtained by solving the denominator cubic for the quadratic root in terms of $R_4$, that is

$$-\left(\frac{1}{4} + \frac{1}{2R_4}\right) \pm j \left(\frac{7}{4R_4} - \frac{1}{16} - \frac{1}{4R_4^2}\right) ,$$

and differentiating with respect to $R_4$. Evaluation gives the same sensitivity as above. This is one of the few practical examples which can be solved for roots in symbolic form and results compared.

## 8.5 LINEARIZED D.C. SENSITIVITY

Linear d.c. sensitivity can of course be determined by any of the general methods already described in this chapter. As in d.c. analysis, the solution obtained and all sensitivities are real numbers. The stability of operating point at all circuit nodes is of major importance in d.c. circuit design. It depends not only on component tolerances but also on power supply stability and independent parameters such as temperature. Node voltage sensitivity to the fixed sources is therefore required in addition to admittance sensitivity.

In non-linear circuits the process of linearization for analysis produces models involving incremental resistances and fixed sources. Voltage increments in the circuit must be assumed sufficiently small not to significantly change the admittances or sources in the model. Sensitivity through Eqs. (8.20) and (8.21) or the simplified versions (8.22) and (8.27) is then valid, and linear methods may be applied. It is only necessary to retain the inverse admittance matrix and the source vectors at the last iteration in the solution of linearized circuits, and all sensitivities can be obtained.

When linearization models depend on an independent parameter, usually temperature in transistor circuits, sensitivity analysis may be extended to give node voltage sensitivity to the independent parameter. This process will be followed with reference to temperature sensitivity in diode-resistor networks. Through the Ebers-Moll transistor model and similar models for field effect transistors, the technique will determine bias sensitivity to temperature or other variation which have environmental effects on circuit operation.

**Temperature Sensitivity**

In Chapter 3 non-linearities were linearized at a given operating point by a linear circuit model which incorporated an incremental resistance and a current or voltage source. The process for a diode is shown in Fig. 8.6.

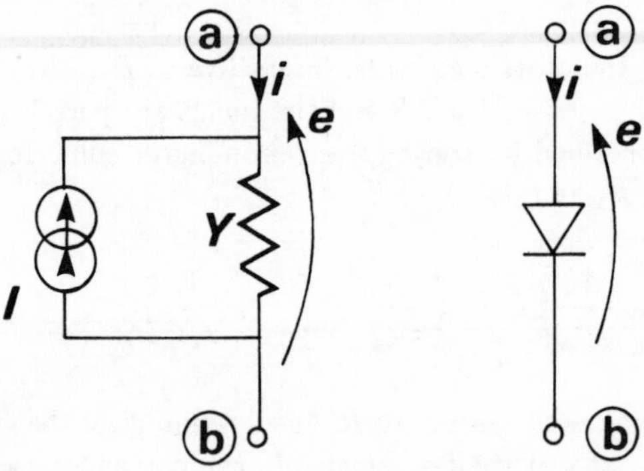


Fig. 8.6 – Diode linearization.

If the non-linearities are governed by the functional relationship

$$i = f(e) \tag{8.46}$$

$$Y = \frac{\partial i}{\partial e} = \frac{\partial f(e)}{\partial e} \tag{8.47}$$

the source current $I$ is obtained from KCL at node ⓐ as

$$I = Ye - i \ . \tag{8.48}$$

When the solution for d.c. voltages has converged, the model values $I$ and $Y$ will be fully defined. Other variables in the model, for example $E$ and $Z$, may be used in other methods of analysis, but for nodal analysis $I$ and $Y$ are most convenient.

Sensitivity to independent parameters, of which temperature is the most important, implies that the functional relationship should be replaced by

$$i = f(e,T) \; . \tag{8.49}$$

The sensitivity of admittance $Y$ to temperature $T$ is given by the partial derivative $\partial Y/\partial T$. Since $Y = \partial f(e,T)/\partial e$, from Eq. (8.47),

$$\frac{\partial Y}{\partial T} = \frac{\partial^2 f(e,T)}{\partial e \partial T} \; . \tag{8.50}$$

Differentiating Eq. (8.48) with respect to $T$ yields

$$\frac{\partial I}{\partial T} - \frac{e \partial Y}{\partial T} = - \frac{\partial f(e,T)}{\partial T} \; . \tag{8.51}$$

Eq. (8.51) relates to a single circuit branch. In matrix form for all $n_B$ branches.

$$\frac{\partial \mathbf{I}}{\partial T} - \frac{\partial \mathbf{Y}_m}{\partial T}\mathbf{e} = - \frac{\partial \mathbf{f}(\mathbf{e},T)}{\partial T} \tag{8.52}$$

where $$\frac{\partial \mathbf{I}}{\partial T} = \begin{bmatrix} \frac{\partial I_1}{\partial T} \frac{\partial I_2}{\partial T} & \cdots & \frac{\partial I_{n_B}}{\partial T} \end{bmatrix}^t$$

$$\frac{\partial \mathbf{Y}_m}{\partial T}\mathbf{e} = \begin{bmatrix} \frac{\partial Y_1}{\partial T} & & & 0 \\ & \frac{\partial Y_2}{\partial T} & & \\ & & \ddots & \\ 0 & & & \frac{\partial Y_{n_B}}{\partial T} \end{bmatrix} \begin{bmatrix} e_1 \\ \vdots \\ e_{n_B} \end{bmatrix}$$

$$\frac{\partial \mathbf{f}(\mathbf{e},T)}{\partial T} = \begin{bmatrix} \frac{f_1(e_1,T)}{\partial T} & \cdots & \frac{\partial f_{n_B}(e_{n_B},T)}{\partial T} \end{bmatrix}^t \; .$$

From Eq. (8.21)

$$\Delta \mathbf{e}' = \mathbf{Z}[\Delta \mathbf{I}' - \Delta \mathbf{Y}\, \mathbf{e}']$$

$$= \mathbf{Z}\, \mathbf{A}^t \left[ \frac{\partial \mathbf{I}}{\partial T} - \frac{\partial \mathbf{Y}_m}{\partial T} \mathbf{e} \right] \Delta T \ , \tag{8.53}$$

remembering that $\mathbf{I}' = \mathbf{A}^t\mathbf{I}$, $\mathbf{A}^t\ \Delta\mathbf{Y}_m \mathbf{A} = \Delta\mathbf{Y}$ and $\mathbf{e} = \mathbf{A}\mathbf{e}'$. Therefore

$$\Delta \mathbf{e}' = -\mathbf{Z}\, \mathbf{A}^t \, \frac{\partial \mathbf{f}(\mathbf{e},T)}{\partial T} \Delta T$$

The term $\frac{\partial \mathbf{f}(\mathbf{e},T)}{\partial T}$ represents the current sensitivities with respect to temperature $T$ in a two-terminal temperaturs sensitive non-linear resistance. The vector form collectively refers to all temperature sensitive branches. Pre-multiplication by the matrix $\mathbf{A}^t$ transforms and branch current sensitivities to an equivalent set of node-driving current sensitivities. Let these be denoted by $\partial\mathbf{f}'/\partial T$ and take the limit $\Delta T \to 0$ in Eq. (8.54) giving finally

$$\frac{\partial \mathbf{e}'}{\partial T} = -\ \mathbf{Z} \frac{\partial \mathbf{f}'}{\partial T} \tag{8.55}$$

Eq. (8.55) gives the sensitivity of all node voltages to changes in all non-linear elements with dependence on a common independent parameter, in this case temperature. The sensitivity is determined when the linearized solution has converged, by one matrix operation. In general there are a number of components in a circuit which depend on temperature, and the full matrix multiplication with the column vector $\partial\mathbf{f}'/\partial T$ may not be over-wasteful of computational effort. However, individual component sensitivities may be obtained in terms of the adjoint network or elements of the **Z** matrix as before.

Two-terminal self-impedances have been assumed, and the element $f_{ij}$ refers to the impedance connected between node ⓘ and node ⓙ in the circuit. In $\partial\mathbf{f}'/\partial T$ element $f_{ij}$ occurs with positive sign in row $i$ and with negative sign in row $j$. Considering node ⓡ only.

$$\frac{\partial e_r'}{\partial T} = -\ (Z_{ri} - Z_{rj})\ \frac{\partial f_{ij}}{\partial T}\ . \tag{8.56}$$

This equation may be interpested in terms of analysis of the adjoint network as

$$\frac{\partial e_r'}{\partial T} = -\ \frac{(e_i^* - e_j^*)}{I_r^*} \frac{\partial f_{ij}}{\partial T} \tag{8.57}$$

provided that the adjoint network is adjoint to the linearized circuit at the final iteration in the solution for node voltages. The total sensitivity to temperature at node ⓡ using Eqs. (8.56) of (8.57) is obtained by summing terms for all temperature-dependent branches.

*Example* 8.5

Determine the node voltage sensitivities at 290°K to temperature variation for the diode circuit shown in Fig. 8.7.

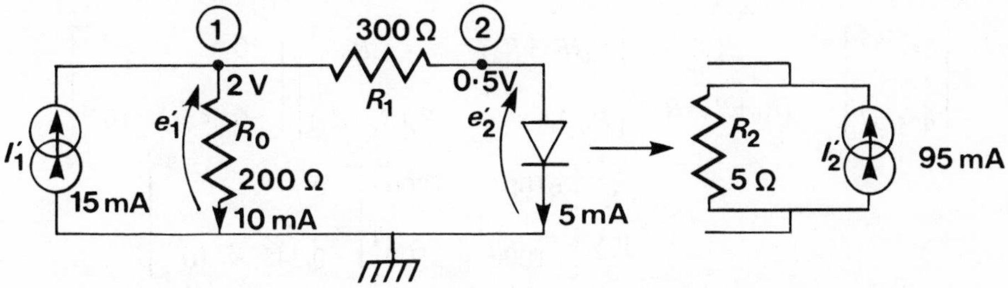


(a) Diode circuit (b) Diode linearization

Fig. 8.7 – Circuit for Example 8.5.

Assuming the diode relationship $i = 1.03 \times 10^{-11} \exp\left(\dfrac{11600}{T} e\right)$ between diode current $i$ and voltage $e$, the nominal solution given in Fig. 8.7 was obtained. Since the linearised circuit is identical to that analysed in Example 8.2, the same **Z** matrix is applicable. Therefore,

$$\mathbf{Z} = \frac{1}{R_0+R_1+R_2}\begin{bmatrix} R_0(R_1+R_2) & R_0R_2 \\ R_0R_2 & R_2(R_0+R_1) \end{bmatrix}.$$

Differentiating the diode relationship

$$\frac{\partial i}{\partial T} = -\frac{11600}{T^2} e \times 1.03 \times 10^{-11} \exp\left(\frac{11600}{T} e\right)$$

$$= -\frac{11600}{T^2} e\, i\ .$$

At $e = 0.5$ V, $i = 5$ mA, $T = 290$°K,

$$\frac{\partial i}{\partial T} = -\ 0.345 \times 10^{-3}\ .$$

The diode is connected from node ② to the reference node. $\partial i/\partial T$ can therefore be identified as $\partial f_{20}/\partial T$. It occurs in row 2 of the column matrix $\partial \mathbf{f}'/\partial \mathbf{T}$ which is therefore given by

$$\frac{\partial \mathbf{f}'}{\partial T} = \begin{bmatrix} 0 \\ -0.345 \times 10^{-3} \end{bmatrix} .$$

Substitution into Eq. (8.55) yields

$$\begin{bmatrix} \partial e_1'/\partial T \\ \partial e_2'/\partial T \end{bmatrix} = \frac{-1}{(R_0+R_1+R_2)} \begin{bmatrix} R_0(R_1+R_2) & R_0R_2 \\ R_0R_2 & R_2(R_0+R_1) \end{bmatrix} \begin{bmatrix} 0 \\ -0.345 \times 10^{-3} \end{bmatrix}$$

$$= -\frac{1}{505} \begin{bmatrix} 61000 & 1000 \\ 1000 & 2500 \end{bmatrix} \begin{bmatrix} 0 \\ -0.345 \times 10^{-3} \end{bmatrix}$$

$$= \begin{bmatrix} 0.683 \times 10^{-3} \\ 1.708 \times 10^{-3} \end{bmatrix} .$$

The sensitivity of nodes ① and ② to temperature is 0.683 mV/°K and 1.708 mV/°K respectively.

## 8.6 COMPUTER PROGAMMING

The BASIC computer program RNODE from Chapter 2 and ACNLU from Chapter 4 each contain subroutines for sensitivity analysis based on the nodal admittance matrix method of this chapter. The subroutines utilize the inverse NAM which is provided by the main circuit analysis program, and compute either the node voltage or voltage gain sensitivities to listed components. Thus these subroutines are fast and compact, and contribute little to storage or processing time in comparison with the main program.

### Sensitivity using RNODE Subroutine 2000

The computer program RNODE was fully described in Chapter 2, Section 2.5. At the end of an analysis run or non-linear iteration, the command SE calls the d.c. sensitivity subroutine 2000. This subroutine is described here since it is based on the theory from this chapter, and the subroutine list is given in the Appendix to this chapter.

D.C. sensitivity in RNODE is implemented directly from Eq. (8.20) which gives the total variation in node voltages as

$$\Delta \mathbf{e}' = (\mathbf{A}^t \mathbf{Y}_m \mathbf{A})^{-1} \mathbf{A}^t [\Delta \mathbf{I} - \mathbf{Y}_m \Delta \mathbf{E} - \Delta \mathbf{Y}_m (\mathbf{E} + \mathbf{A}\mathbf{e}')] \quad ,$$

noting that the terms $\mathbf{Y_m \Delta E}$ and $\mathbf{\Delta Y_m E}$ are changed to $\mathbf{Y \Delta E}$ and $\mathbf{\Delta YE}$ for branch-dependent VCCS. Separating out the terms of this equations we obtain the following sensitivities where $\mathbf{Z} = (\mathbf{A^t Y_m A})^{-1}$.

$$\Delta \mathbf{e}' = \mathbf{ZA^t \Delta I} \tag{8.58}$$

$$\Delta \mathbf{e}' = -\mathbf{ZA^t Y_m \Delta E} \tag{8.59}$$

$$\Delta \mathbf{e}' = -\mathbf{ZA^t}[\mathbf{\Delta Y_m E} + \mathbf{\Delta Y_m e}] \; . \tag{8.60}$$

As before $\mathbf{Y_m \Delta E}$ and $\mathbf{\Delta Y_m E}$ are modified for branch-dependent VCCS. Since admittance is either equal to the VCCS conductance $g_{jk}$ or equal to the inverse of resistance $r_k$,

$$\frac{\partial e'}{\partial r_k} = \{(e_k + E_k)/r_k^2\} \, [\mathbf{ZA^t}]_{\text{column } k} \tag{8.61}$$

$$\frac{\partial e'}{\partial g_{jk}} = -(e_k + E_k) \, [\mathbf{ZA^t}]_{\text{column } j} \; . \tag{8.62}$$

The term $E_k$ is omitted from Eq. (8.62) for branch-dependent VCCS.

Eqs. (8.58) to (8.62) are directly implemented in Subroutine 2000 while recalling the following solution storage locations in RNODE.

| | |
|---|---|
| Z(N,N) | contains the inverse NAM **Z**. |
| L(N,1) | contains the node voltages **e**′. |
| K(B,1) | contains the branch voltages **e**. |
| M(B,1) | contains the branch currents **i**. |
| J(B,1) | contains the resistor voltages **v**. |
| N | equals the number of nodes. |
| B | equals the number of branches. |
| S | equals the number of VCCS. |

Statements numbered 2000 to 2060 enter the matrix $\mathbf{ZA^t}$ into matrix **A** and print the sensitivity to all branch current sources. Statements 2090 and 2180 assemble the appropriate version of $-\mathbf{ZA^t Y_m}$ from Eq. (8.59) into matrix **Z** and print the sensitivity to all branch voltage sources. The inverse NAM in **Z** is therefore destroyed at this point in the subroutine but could be retained at the expense of more storage. Statement 2210 and 2260 assemble the matrix columns generated by Eq. (8.61) into matrix **Z** and print the sensitivity to branch resistors. Statements 2290 to 2400 assemble the columns generated by Eq. (8.62) into matrix **Z** and print the sensitivity to VCCS. Owing to the original maximum dimensions of matrix **Z** only the first B VCCS sensitivities are generated. However, it is unusual for the number of VCCS to exceed the number of branches, and this limitation remains. Statements 2420 and 2430 regenerate the original matrices **A** and **J** but matrix **Z** remains destroyed.

*Example* 2.5 *continued*

The computer run reproduced in Fig. 8.8 is a continuation of that given in Fig. 2.20 and completes Example 2.5 by calling the sensitivity subroutine.

```
NODE VOLTAGE SENSITIVITY
TO CURRENT SOURCES

-628.067         626.813         -1.2531          42.5555          41.3024
 1.12135         .131758          1.2531
-47.536          46.4743         -1.06174         520.672          519.61
 .950099         .111637          1.06174
-100287          98046.6         -2239.94         -82911.3         -85151.2
-2201.39         4441.34          2239.94
-353.125         345.238         -7.88718          267.85           259.962
 7.05788         .829301          7.88718

TO VOLTAGE SOURCES

 .923627        -7.63727E-02     3.13276E-05     -.075992          7.59921E-02
-2.80336E-05    -2.80336E-05    -3.21309E-04
 6.99059E-02     6.99059E-02     2.65434E-05     -.929772         -7.02284E-02
-2.37525E-05    -2.37525E-05    -2.7224E-04
 147.48          147.48          5.59985E-02      148.056         -148.056
 5.50349E-02    -.944965        -.574344
 .519301         .519301         1.9718E-04      -.478303          .478303
-1.76447E-04    -1.76447E-04    -2.02236E-03

TO RESISTORS

-6.11569E-06     1.15291E-06    -1.21538E-08    -3.35027E-07      5.05806E-08
-6.08337E-09    -4.07874E-08     1.1931E-06
-4.62874E-07     8.54814E-08    -1.02977E-08    -4.0991E-06       6.36335E-07
-5.15435E-09    -3.45585E-08     1.01089E-06
-9.76524E-04     1.8034E-04     -2.17251E-05     6.52737E-04     -1.0428E-04
 1.19427E-05    -1.37487E-03     2.13268E-03
-3.43849E-06     6.35004E-07    -7.64975E-08    -2.1087E-06       3.1836E-07
-3.82895E-08    -2.56721E-07     7.5095E-06

TO GM 1 TO  2

 2.98711E-02    -1.77973E-02
 2.53093E-02    -1.50794E-02
 53.395          34.9391
 .188012        -.112018

NEXT ? ER

STOP AT 1160
*
```

Fig. 8.8 – Computer run continued from Fig. 2.20 for Example 2.5.

One feature of the output is that sensitivity is obtained to all current and voltage sources even if they are not present in the original circuit. Voltage gain from input to output in the small-signal mode is equal to node ③ sensitivity to the branch 1 voltage source, that is 147.48, which is the figure previously obtained in Example 2.5.

**Node Voltage Sensitivity using ACNLU Subroutine 6000**

The computer program ACNLU was fully described in Chapter 4, Section 4.4. ACNLU contains provision for a sensitivity analysis run at the command SE. Node voltages and sensitivities to listed components are then printed at each analysis frequency with the limitation that a maximum of 9 sensitivities per run can be requested. The modification facility also applies to the sensitivity component list, and all required sensitivities for a circuit can be generated. The node voltage sensitivity subroutine 6000 is listed in the Appendix to this chapter.

Node voltage absolute sensitivity is given from Eq. (8.21) by

$$\Delta \mathbf{e}' = -\mathbf{Z}\Delta\mathbf{Y}\mathbf{e}' \,.$$

By substitution in $\Delta\mathbf{Y}$ of the four terms $\delta y_{kl}^{ij}$, we may obtain the intermediate result from the development of Eq (8.22) for node voltage sensitivity to admittance $y_{kl}^{ij}$ as

$$\frac{\partial e_r'}{\partial y_{kl}^{ij}} = -\,(Z_{ri} - Z_{rj})\,(e_k' - e_l') \,. \tag{8.63}$$

Subroutine 6000 directly programs this equation, and is called after the inverse NAM has been generated and node voltages computed. At this point matrices D(N,N) and F(N,N) contain the real and imaginary parts respectively of the inverse NAM **Z**. Matrices E(N,1) and V(N,1) similarly contain the real and imaginary parts of the node voltages.

Components for which sensitivities are required are originally listed for example as R,1 or GM,3. Matrix U(T9,6) contains this list of T9 (⩽9) components in columns 0 and 1 where column 0 is coded for component type by 1,2, or 3 for R, L or C respectively, or −1, −2, or −3 for various types of VCCS. Column 1 contains the component number. At the beginning of each analysis, the remaining columns 2 to 6 of U are filled with the component node connections *i, j, k,* and *l* as required by Eq. (8.63), and the component value respectively.

At each frequency W, subroutine 6000 fills matrices P(1,T9 + 1) and Q(1,T9 + 1) with real and imaginary parts respectively of the node voltages in column 1 and node voltage sensitivities to T9 components in columns 2 to T9 + 1. Sensitivity is computed at each node in turn from 1 to N. Statements 6080 to 6150 directly utilize Eq. (8.63) to give node voltage sensitivity to each component in turn as R8 + jL8. Statements 6160 to 6310 implement the following table which converts admittance sensitivity R8 + jL8 to absolute component sensitivity dependent on the admittance type from U(S8,0), S8 = 1 to T9.

| U(S8,0) TYPE | ADMITTANCE R,L,C,GM | REAL PART | IMAGINARY PART |
|---|---|---|---|
| | | COMPONENT SENSITIVITY | |
| −3 | jωGM | −ωL8 | ωR8 |
| −2 | GM/jω | L8/ω | −R8/ω |
| −1 | GM | R8 | L8 |
| 1 | 1/R | $-R8/R^2$ | $-L8/R^2$ |
| 2 | 1/jωL | $-L8/\omega L^2$ | $R8/\omega L^2$ |
| 3 | jωC | −ωL8 | ωR8 |

Statements 6330 to 6370 add the node voltage to matrices **P** and **Q** and use MAT PRINT to print the node voltage and sensitivities at node S9.

*Example* 8.6

Determine the node voltages and node voltage sensitivities of the bandpass filter in Fig. 8.9.

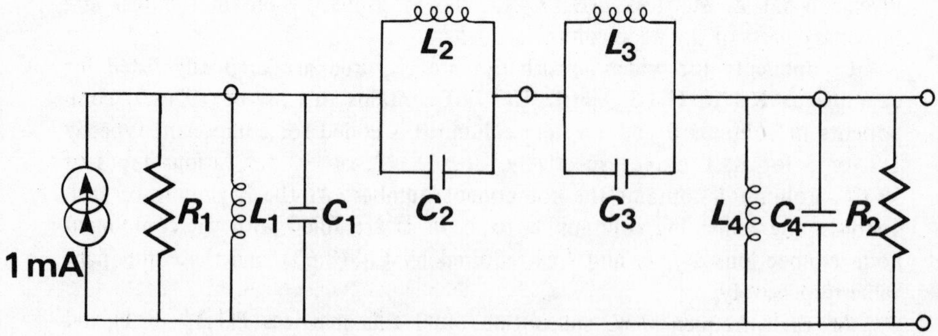


**Fig. 8.9 – Bandpass filter for Example 8.6**
$L_1 = L_4 = 1.919$ mH, $L_2 = 7.5964$ mH, $L_3 = 8.3438$ mH., $R_1 = R_2 = 1$ kΩ, $C_1 = C_4 = 1.677$ μF, $C_2 = 0.3857$ μF, $C_3 = 0.4236$ μF.

Sufficient information is given in Fig. 8.9 for direct input to the computer run reproduced in Fig. 8.10. The input voltage source has been changed to an equivalent current source of 1 mA in parallel with the 1 kΩ source resistence. Sensitivities to all inductors and capacitators were computed, and owing to the amount of output produced, two frequencies only are shown in Fig. 8.10. One of these frequencies is in the filter stop-band, and the second is just outside the edge of the pass-band which from 2761 Hz to 2850 Hz gives attenuation within the range −6 dB to −6.1 dB. However, it is worthwhile completing this run and repeating the analysis in terms of voltage gain because narrow bandpass filters with resonant circuits at nearby frequencies severely test the accuracy of a computer analysis.

This filter will be re-analysed in Chapter 9 by means of a specially written subroutine intended for random simulation. Nominal voltage gain

```
AC NAM ANALYSIS
RLC INPUTS: COMPONENT NUMBER,VALUE,2 NODE CONNECTIONS
GM  INPUTS: GM NUMBER,VALUE,TYPE,4 NODE CONNECTIONS
      TYPE: 1 FOR GM, 2 FOR GM/JW, 3 FOR JWGM
     NODES: SEND +, - , RECEIVE +, -
I   INPUTS: NODE,VALUE,PHASE DEGREES
NO.NODES         ? 3
NO. R,L,C,GM     ? 2,4,4,0
R ? 1,1000,0,1
R ? 2,1000,3,0
L ? 1,1.919E-3,1,0
L ? 2,7.5964E-3,1,2
L ? 3,8.3438E-3,2,3
L ? 4,1.919E-3,3,0
C ? 1,1.677E-6,1,0
C ? 2,.3857E-6,1,2
C ? 3,.4236E-6,2,3
C ? 4,1.677E-6,3,0
F-LOW,F+INC,F*MULT,NO.F
 ? 2760,90,1,2
OUTPUTS: NV NODE VOLTS      SE SENSITIVITY+NV       YM NAM
         DB GAIN ZIN ZOUT   SN SENSITIVITY+GAIN     ZM INV NAM
OUTPUT ? SE
NO. I SOURCES  ? 1
I ? 1,1E-3,0
SENSITIVITY COMPONENT LIST
INPUT R L C OR GM, COMPONENT NUMBER
NO.SENSITIVITIES  ? 8
CPT.TYPE, NO.
 ? L,1
 ? C,1
 ? L,2
 ? C,2
 ? L,3
 ? C,3
 ? L,4
 ? C,4
NEXT COMMAND: ER END RUN        NS NEW SENSITIVITY LIST
MODIFICATION: NC COMPONENTS     NF FREQUENCIES   NI SOURCES
            : NN IN/OUT NODES   NO OUTPUTS
      OUTPUT: YM 1/R WC 1/WL    ZM INV(NAM)      GO ANALYSE
NEXT ? GO
```

Fig. 8.10 – Computer analysis of band-pass filter from Example 8.6.

Fig. 8.10 (*continued*)

```
NODE VOLTAGES AND SENSITIVITIES GIVEN AS
REAL PARTS NV L 1            C 1           L 2            C 2
L 3            C 3           L 4           C 4
THEN IMAG.PARTS SIMILARLY
 2760  HZ.
NODE  1

 .590456        254.52        281871        -542.15        -9.40834E+06
-1935.64       -4.05257E+07  -1125.56       -1.24652E+06

 1.37639E-02   -5456.32      -6.04266E+06   -188.771       -3.2759E+06
-673.964       -1.41105E+07   3612.33        4.00052E+06
NODE  2

 1.20827       -3671.95      -4.06654E+06   -230.096       -3.99304E+06
-3264.11       -6.83394E+07  -6335.31       -7.0161E+06

-.42531        -11263.2      -1.24735E+07   -1478.15       -2.56516E+07
-3541.52       -7.41474E+07   6238.05        6.9084E+06
NODE  3

-7.39565E-02    4477.21       4.95833E+06    542.147        9.4083E+06
 1935.63        4.05256E+07   4477.19        4.95831E+06

 .485961        788.531       873267         188.768        3.27584E+06
 673.96         1.41104E+07   788.562        873302
 2850  HZ.
NODE  1

 .560802       -51.0409      -60272.3        2215.26        4.09912E+07
 455.537        1.01696E+07   367.278        433705

-3.00091E-03   -4769.05      -5.6316E+06    -383.245       -7.09157E+06
-78.8071       -1.75931E+06   3716.8         4.38903E+06
NODE  2

-.605201       -8345.75      -9.85519E+06   -4093.35       -7.57434E+07
-319.827       -7.13992E+06  -8760.89       -1.03454E+07

-.984604        5191.57       6.13054E+06    3264.26        6.04019E+07
 1110.62        2.47939E+07  -3229.69       -3.81383E+06
NODE  3

-2.44309E-02   -4214.25      -4.97646E+06   -2215.28       -4.09915E+07
-455.541       -1.01696E+07  -4214.25       -4.97646E+06

-.495679        230.322       271979         383.239        7.09144E+06
 78.8076        1.75932E+06   230.26         271906
NEXT ?
```

in dB is also obtained and can be compared from Fig. 9.11 with results here. Comparison must ultimately be made against analysis, using a large computer. This is perhaps an example where relative sensitivities would be useful since some idea of the individual component contributions to the tolerances found in Example 9.4 can be obtained. However, this becomes difficult when components vary in a related fashion as in an integrated circuit.

**Voltage Gain Sensitivity using ACNLU Subroutine 8000**

The computer program ACNLU from Chapter 4 contains provision for a second sensitivity analysis subroutine at the command SN. The open-circuit voltage gain sensitivities to listed components are then output together with gain, phase, real and imaginary parts at each frequency. As in subroutine 6000, a maximum of 9 sensitivities are allowed although the sensitivity list can be modified and the analysis repeated if necessary. The voltage gain sensitivity subroutine 8000 is listed in the Appendix to this chapter.

The opne-circuit voltage gain from input node ⓐ to output node ⓑ is given by $Z_{ba}/Z_{aa}$ where **Z** is the inverse NAM. The transfer function sensitivities $S_a$ and $S_b$ at nodes ⓐ and ⓑ respectively, are given from Eq. (8.22) by

$$S_b = \frac{\Delta e_b/I_a}{\Delta y_{kl}^{ij}} = -(Z_{bi} - Z_{bj})(Z_{ka} - Z_{la})$$

$$S_b = \frac{\Delta e_a/I_a}{\Delta y_{kl}^{ij}} = -(Z_{ai} - Z_{aj})(Z_{ka} - Z_{la}) \; .$$

Since voltage gain $K = (e_b/I_a)/(e_a/I_a)$,

$$\frac{\Delta K}{\Delta y_{kl}^{ij}} = \frac{S_b Z_{aa} - S_a Z_{ba}}{Z_{aa}^2} \; . \tag{8.64}$$

Subroutine 8000 directly programs Eq. (8.64) and prints voltage gain and sensitivities to listed components. The subroutine is called after analysis at each frequency W when the inverse NAM **Z** is available with real parts stored in matrix D(N,N) and imaginary parts in matrix F(N,N). Matrix U(T9,6) contains the list of T9 components for which sensitivity is required, in the format described for subroutine 6000. N, N1 and N2 contain the number of nodes, the input node, and the output node respectively.

Program statements 8010 to 8280 fill R5 and L5 with the real and imaginary parts of gain sensitivity to admittance and directly use Eq. (8.64). Row 0 and column 0 of matrices **D** and **F** are initially set to zero since the admittance $y_{kl}^{ij}$ may involve reference to elements in those locations which must be zero. On entry to the subroutine the variables S6, S7, S8, S9, C8, G8, L8, R8 contain respectively D(N2,N1), F(N2,N1), D(N1,N1), F(N1,N1), Real Part of Gain, Imaginary Part of Gain, Phase, and Gain dB.

Statements 8290 to 8440 convert admittance sensitivities to absolute component sensitivities by implementing the table of factors from Subroutine 6000 which depend on the component type held in U(15,0). (Note that R5 and L5 are used in place of R8 and L8 and I5 in place of S8). The real and imaginary parts of absolute component sensitivity are then obtained in R5 and L5. These are converted to gain and phase sensitivity using Eq. (8.31), where complex gain K = C8 + jL8. This occurs at statements 8450 to 8480 and results are finally printed.

*Example* 4.3 *continued*

The computer run reproduced in Fig. 8.11 is a continuation of that given in Fig. 4.24 and completes Example 4.3 by calling for open-circuit voltage

```
OUTPUT ? SN
SENSITIVITY COMPONENT LIST
INPUT R L C OR GM, COMPONENT NUMBER
NO.SENSITIVITIES  ? 4
CPT.TYPE, NO.
 ? R,1
 ? R,2
 ? C,1
 ? C,2
NEXT ? GO
INPUT NODE  1   OUTPUT NODE  6
HZ.           GAIN DB       PHASE DEG     RP            IP
CPT.NO.       DB SEN        DEG SEN       RP SEN        IP SEN
 1000          8.22209      -9.00016       2.54521      -.403128
R 1            4.35661E-04   2.55647E-03   1.45648E-04   9.33442E-05
R 2            1.00766E-04  -.015126      -7.68978E-05  -6.76607E-04
C 1            2.44429E+07   5.38229E+08   1.09494E+07   2.27749E+07
C 2           -6.65703E+06  -6.66912E+08  -6.64303E+06  -2.93168E+07
 2000          9.00947      -19.9854       2.65155      -.964318
R 1            1.86133E-03   2.51282E-03   6.105E-04    -9.03579E-05
R 2            2.05759E-04  -3.50072E-02  -5.26377E-04  -1.64291E-03
C 1            1.10084E+08   1.03493E+09   5.10239E+07   3.56733E+07
C 2           -3.74167E+07  -1.45992E+09  -3.59934E+07  -6.34084E+07
 3000          10.2244      -35.97         2.62627      -1.906
R 1            4.40986E-03  -6.97354E-03   1.10139E-03  -1.28733E-03
R 2           -7.64397E-04  -6.52932E-02  -2.40316E-03  -2.82511E-03
C 1            2.92288E+08   1.2083E+09    1.28572E+08  -8.7534E+06
C 2           -1.38633E+08  -2.43684E+09  -1.22981E+08  -8.12765E+07
 4000          11.2442      -61.1504       1.76083      -3.19639
R 1            6.35665E-03  -3.99928E-02  -9.42459E-04  -3.56831E-03
R 2           -6.05426E-03  -.100245      -6.81979E-03  -8.52816E-04
C 1            5.45828E+08   4.13968E+07   1.12961E+08  -1.99591E+08
C 2           -3.9758E+08   -3.02307E+09  -2.49249E+08   5.34025E+07
 5000          10.5032       86.4441      -.207831      -3.34444
R 1            2.34356E-03  -8.04067E-02  -4.74954E-03  -6.10712E-04
R 2           -.015324      -9.22716E-02  -5.01937E-03   6.23512E-03
C 1            5.3032E+08   -2.66302E+09  -1.68133E+08  -1.94537E+08
C 2           -6.69328E+08  -1.71393E+09  -8.40294E+07   2.63937E+08
 6000          7.7837        60.2018      -1.21757      -2.12615
R 1           -3.87464E-03  -8.15829E-02  -2.48427E-03   2.68213E-03
R 2           -1.87167E-02  -4.86961E-02   8.16634E-04   5.61634E-03
C 1            2.30476E+08  -3.83285E+09  -1.74538E+08   2.50338E+07
C 2           -6.52354E+08   5.49511E+07   9.34847E+07   1.58517E+08
 7000          4.6339        44.1604      -1.22307      -1.18774
R 1           -6.98066E-03  -6.61128E-02  -3.87562E-04   2.36584E-03
R 2           -1.77294E-02  -.020503       2.07147E-03   2.86205E-03
C 1            1.32388E+07  -3.58539E+09  -7.61891E+07   7.47254E+07
C 2           -5.36788E+08   8.0169E+08    9.22045E+07   5.6289E+07
 8000          1.77491       34.4206      -1.01193      -.693419
R 1           -8.17836E-03  -.052791       3.13904E-04   1.58527E-03
R 2           -1.62023E-02  -7.36424E-03   1.79849E-03   1.42354E-03
C 1           -9.93968E+07  -3.09293E+09  -2.5852E+07    6.25611E+07
C 2           -4.46451E+08   1.00018E+09   6.41175E+07   1.79767E+07
 9000         -.711004       28.0417      -.813235      -.433164
R 1           -8.64117E-03  -4.33406E-02   4.81387E-04   1.04609E-03
R 2           -1.49825E-02  -1.26886E-03   1.39317E-03   7.65184E-04
C 1           -1.58799E+08  -2.66367E+09  -5.26979E+06   4.57264E+07
C 2           -3.86402E+08   1.01256E+09   4.38328E+07   4.89797E+06
 10000        -2.87978       23.5402      -.658076      -.286689
R 1           -8.82388E-03  -3.66387E-02   4.85203E-04   7.12059E-04
R 2           -1.40884E-02   1.67944E-03   1.0758E-03    4.45717E-04
C 1           -1.92673E+08  -2.32608E+09   2.95871E+06   3.30758E+07
C 2           -3.46228E+08   9.66884E+08   3.10695E+07   322445
NEXT ? ER
```

Fig. 8.11 – Computer run continued from Fig. 4.24 for Example 4.3.

gain sensitivities to the main filter components. Results can be compared to those obtained by differentiation of the ideal transfer function. However, exact agreement will not be obtained since the circuit was analysed with imperfect amplifiers.

The wide spread of numerical values again suggests the use of relative sensitivity for comparison. However, use of the results for tolerance analysis or in computer-aided design may then become difficult. No doubt ACNLU will shortly have the optional choice between absolute and relative sensitivity since the modifications required to the two sensitivity subroutines are very minor.

## 8.7 CONCLUSION

Sensitivity is the vital link between analysis and design. Any circuit can be analysed but, if it does not meet the design specification, more information is necessary. This can come from alteration of a prototype or 'breadboard' circuit, by knowledge of the circuit behaviour from experience or insight into the design, or by sensitivity analysis. All of these methods are essentially the same. Each predicts the effect of a modification to the circuit.

Applications of sensitivity begin in the next chapter with tolerance analysis. Much more important applications under the general heading of *computer-aided circuit design* arise in approximation, optimization, and tolerance design of both linear and non-linear circuits or systems from d.c. to microwave frequencies. Since the basic techniques of sensitivity analysis given in this chapter, or improved and extended versions of them, are becoming widely available, an appreciation of, and ability to use, sensitivity methods is important to circuit design. This applies even when the circuit designer is not particularly comitted to computer-aided design techniques in general.

## FURTHER READING

Adby, P. R. and Dempster, M. A. H., (1974), *Introduction to Optimisation Methods,* Chapman and Hall, London.

Calahan, D. A., (1972), *Computer-aided Network Design,* revised ed., Chs. 5, 6, and 11, McGraw-Hill, New York.

Geher, K., (1971), Theory of Network Tolerances, Chs. 1 to 3, Academiai Kiado, Budapest.

Spence, R., *Sensitivity Analysis for Electric Circuits,* Imperial College, University of London, 1977-(to be published by Scientific Elsevier, Amsterdam).

Szentirmai, G., Editor, (1973), *Computer Aided Filter Design,* pp. 291-382, Institute of Electrical and Electronics Engineers, New York.

## PROBLEMS

8.1 Determine the absolute sensitivity of the open-circuit voltage gain to admittance, and convert to resistor, inductor, capacitor, and VCCS sensitivities.

8.2 Determine the gain (dB) and phase (degrees) sensitivities from the answers to Problem 8.1. Convert all absolute sensitivities to relative sensitivities.

8.3 Show that the real part of the relative complex gain sensitivity is not equal to the relative real part sensitivity.

8.4 Analyse the filter circuit in Fig. 8.12 by symbolic analysis and differentiate to obtain first order sensitivities to the passive components. Assume the operational amplifier to have infinite input impedance, 100 Ω output resitance, and $A = 1000$.

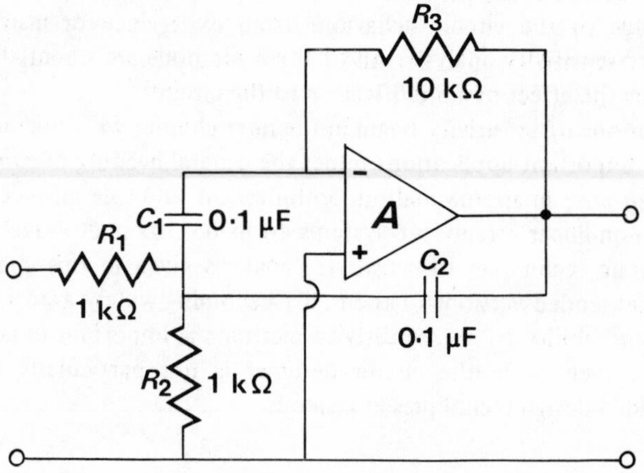


Fig. 8.12 – Problems 8.4 to 8.7.

8.5 Obtain the sensitivities to passive component variation in the filter circuit of Fig. 8.12 by differences at the frequency of maximum gain and at 1 kHz. Compare the sensitivities computed from 1% and 10% differences to those obtained in answer to Problem 8.4.

8.6 Fill the NAM for the filter in Fig. 8.12 at the two frequencies used for Problem 8.5 and invert the matrix. Derive the voltage gain sensitivities and compare to those obtained by differences and differentiation in Problems 8.4 and 8.5.

8.7 Determine the poles and zeroes of the voltage gain transfer function of the filter circuit in Fig. 8.12. Derive the root sensitivities to component variation using the method summarized in Fig. 8.4.

8.8 Extend the nodal analysis and adjoint network methods for sensitivity analysis to yield second order sensitivities.

8.9 Derive an expression for second order transfer function sensitivities $\partial^2 F(s)/\partial p_i\, \partial p_j$ by differentiation of Eq. (8.40).

*Computing*

8.10 Investigate the sensitivity of a number of familiar circuits, using programs RNODE, ACNLU, and TOPSEN. Compare results from ACNLU and TOPSEN which should agree, and for simple circuits at least, compare results with directly computed values.

8.11 Investigate the sensitivity of the bandpass filter in Fig. 4.29 to resistors $R_1$ to $R_7$ and capacitors $C_1$ to $C_4$. Compare computed results to those obtained by differences, using 1% and 10% changes in component values.

8.12 The filter in Fig. 4.29 is in fact integrated, and resistors $R_1$ to $R_7$ are constructed from etched rectangular areas of semiconductors doped to a resistance $r$ of 200 Ω/square. Resistors $R_1$, $R_3$, $R_4$ and $R_7$ are 200 μm wide and respectively 184, 327, 184, and 144 μm long. Resistors $R_5$ and $R_6$ are 50 μm wide and respectively 250 and 245 μm long. Resistor values are given by $rl/w$ where $l$ and $w$ are length and width. Determine the filter sensitivity to $r$ $l$ and $w$. Determine also the sensitivity to $c$ where all four capacitors are equal to $c$ and vary together. (Maximum expected errors in $r$, $l$, $w$, and $c$ are ± 20 Ω, ± 10 μm, ± 5 μm, and ± 2 pF respectively).

8.13 Add the option to output relative sensitivity to programs RNODE, ACNLU, and TOPSEN.

8.14 Add a subroutine to program RNODE which can be called when a non-linear iteration is completed, for the computation of temperature sensitivity.

8.15 Write a computer program to evaluate root sensitivity following the flow chart in Fig. 8.4. Program TOPSEN can be used for analysis and component sensitivity evaluation.

8.16 Investigate the possibility of creating an iterative optimization loop using program ACNLU for sensitivity analysis. The objective is to modify circuit component values so that a specified frequency response is approximated with least error. (See *Further Reading* on Computer-Aided Design and Optimization).

# APPENDIX

```
2000 REM SUBROUTINE - DC SENSITIVITY
2010 PRINT "NODE VOLTAGE SENSITIVITY"
2020 PRINT "TO CURRENT SOURCES"
2030 MAT D=TRN(A)
2040 MAT A=Z*D
2050 MAT PRINT A
2060 PRINT
2070 PRINT "TO VOLTAGE SOURCES"
2080 DIM Z[N,B]
2090 ON I9 THEN GOTO 2100, 2160
2100 FOR I1=1 TO N
2110   FOR I2=1 TO B
2120     LET Z[I1,I2]=-A[I1,I2]*Y[I2,I2]
2130   NEXT I2
2140 NEXT I1
2150 GOTO 2180
2160 MAT Z=A*Y
2170 MAT Z=(-1)*Z
2180 MAT PRINT Z
2190 PRINT
2200 PRINT "TO RESISTORS"
2210 FOR I1=1 TO N
2220   FOR I2=1 TO B
2230     LET Z[I1,I2]=J[I2,1]*A[I1,I2]*Y[I2,I2]^2
2240   NEXT I2
2250 NEXT I1
2260 MAT PRINT Z
2270 PRINT
2280 IF S=0 THEN GOTO 2420
2290 LET I3=S
2300 IF S>B THEN LET I3=B
2310 DIM Z[N,I3]
2320 PRINT "TO GM 1 TO ";I3
2330 ON I9 THEN GOTO 2340, 2350
2340 MAT J=K
2350 FOR I1=1 TO N
2360   FOR I2=1 TO I3
2370     LET Z[I1,I2]=-J[G[I2,2],1]*A[I1,G[I2,1]]
2380   NEXT I2
2390 NEXT I1
2400 MAT PRINT Z
2410 PRINT
2420 MAT A=TRN(D)
2430 MAT J=K+E
2440 RETURN

*
```

Fig. 8.13 – Sensitivity subroutine for program RNODE.

```
6000 REM SUBROUTINE - NODE VOLTAGE SENSITIVITIES
6010 DIM P[1,T9+1],Q[1,T9+1]
6020 LET E[0,1]=0
6030 LET V[0,1]=0
6040 FOR S9=1 TO N
6050   LET D[S9,0]=0
6060   LET F[S9,0]=0
6070   FOR S8=1 TO T9
6080     LET R8=D[S9,U[S8,3]]-D[S9,U[S8,2]]
6090     LET L8=F[S9,U[S8,3]]-F[S9,U[S8,2]]
6100     LET C8=E[U[S8,4],1]-E[U[S8,5],1]
6110     LET G8=V[U[S8,4],1]-V[U[S8,5],1]
6120     LET S6=R8*C8-L8*G8
6130     LET S7=R8*G8+C8*L8
6140     LET R8=S6
6150     LET L8=S7
6160     LET I5=U[S8,0]
6170     LET S7=U[S8,6]
6180     LET C8=SGN(I5)
6190     IF ABS(I5)=1 THEN GOTO 6230
6200     LET S6=R8
6210     LET R8=L8
6220     LET L8=S6
6230     ON ABS(I5) THEN GOTO 6240, 6270, 6300
6240     LET P[1,S8+1]=-I5*R8*S7^(-I5-1)
6250     LET Q[1,S8+1]=-I5*L8*S7^(-I5-1)
6260     GOTO 6320
6270     LET P[1,S8+1]=-C8/W*R8*S7^(-C8-1)
6280     LET Q[1,S8+1]=C8/W*L8*S7^(-C8-1)
6290     GOTO 6320
6300     LET P[1,S8+1]=-R8*W
6310     LET Q[1,S8+1]=L8*W
6320   NEXT S8
6330   LET P[1,1]=E[S9,1]
6340   LET Q[1,1]=V[S9,1]
6350   PRINT "NODE ";S9
6360   MAT PRINT P
6370   MAT PRINT Q
6380 NEXT S9
6390 RETURN

*
```

Fig. 8.14 – Sensitivity subroutines for program ACNLU.

Fig. 8.4 – *continued.*

```
8000 REM SUBROUTINE - O/C VOLTAGE GAIN SENSITIVITIES
8010 FOR I5=1 TO N
8020   LET D[0,I5]=0
8030   LET F[0,I5]=0
8040   LET D[I5,0]=0
8050   LET F[I5,0]=0
8060 NEXT I5
8070 FOR I5=1 TO T9
8080   LET R5=D[N1,U[I5,2]]-D[N1,U[I5,3]]
8090   LET L5=F[N1,U[I5,2]]-F[N1,U[I5,3]]
8100   LET R7=R5*S6-L5*S7
8110   LET L7=R5*S7+L5*S6
8120   LET R5=D[N2,U[I5,2]]-D[N2,U[I5,3]]
8130   LET L5=F[N2,U[I5,2]]-F[N2,U[I5,3]]
8140   LET R6=R5*S8-L5*S9
8150   LET L6=R5*S9+L5*S8
8160   LET R5=D[U[I5,4],N1]-D[U[I5,5],N1]
8170   LET L5=F[U[I5,4],N1]-F[U[I5,5],N1]
8180   LET R7=R7-R6
8190   LET L7=L7-L6
8200   LET R6=R7*R5-L7*L5
8210   LET L6=R7*L5+R5*L7
8220   LET R7=S8*S8-S9*S9
8230   LET L7=-2*S8*S9
8240   LET R5=R7*R7+L7*L7
8250   LET R7=R7/R5
8260   LET L7=L7/R5
8270   LET R5=R6*R7-L6*L7
8280   LET L5=R6*L7+R7*L6
8290   IF ABS(U[I5,0])=1 THEN GOTO 8330
8300   LET R7=R5
8310   LET R5=L5
8320   LET L5=R7
8330   LET R6=U[I5,0]
8340   LET L6=U[I5,6]
8350   LET R7=SGN(R6)
8360   ON ABS(R6) THEN GOTO 8370, 8400, 8430
8370   LET R5=-R6*R5*L6^(-R6-1)
8380   LET L5=-R6*L5*L6^(-R6-1)
8390   GOTO 8450
8400   LET R5=-R7/W*R5*L6^(-R7-1)
8410   LET L5=R7/W*L5*L6^(-R7-1)
8420   GOTO 8450
8430   LET R5=-R5*W
8440   LET L5=L5*W
8450   LET L6=C8*C8+G8*G8
8460   LET R7=8.68589*(R5*C8+L5*G8)/L6
8470   LET L7=57.2958*(L5*C8-R5*G8)/L6
8480   IF R6<0 THEN GOTO 8560
8490   ON R6 THEN GOTO 8500, 8520, 8540
8500   PRINT 'R';U[I5,1],R7,L7,R5,L5
8510   GOTO 8570
8520   PRINT 'L';U[I5,1],R7,L7,R5,L5
8530   GOTO 8570
8540   PRINT 'C';U[I5,1],R7,L7,R5,L5
8550   GOTO 8570
8560   PRINT 'GM';U[I5,1],R7,L7,R5,L5
8570 NEXT I5
8580 RETURN
```

*

CHAPTER 9

# Tolerance Analysis

Sensitivity was defined in the previous chapter as a measure of the variation of circuit performance due to changes in the values of individual circuit components. Thus the designer is able to predict the effect of component variation on his design, and in general to select a circuit of low sensitivity in preference to one with higher sensitivity. In practice all circuit components vary together, either during manufacture owing to manufacturing tolerances, or subsequently owing to ageing and variation in the environment. Both effects occur during the lifetime of a circuit, and it is the analysis of the combined effect of the sensitivity of all components with which we are concerned in this chapter.

The tolerance was defined in Eq. (8.1) by the Taylor series expansion for the variation $\Delta y$ of a performance function $y$ in terms of variations $\Delta p_1$, $\Delta p_2$, ... $\Delta p_n$ in circuit parameters $p_1$, $p_2$, ... $p_n$. Neglecting third and higher order derivatives the tolerance is given by

$$\Delta j = \Sigma_{i=1}^{n} \frac{\partial y}{\partial p_i} \Delta p_i + \frac{1}{2} \Sigma_{i=1}^{n} \Sigma_{j=1}^{n} \frac{\partial^2 y}{\partial p_i \partial p_j} \Delta p_i \, \Delta p_j \; . \tag{9.1}$$

In the terminology of Eq. (8.7) this may be expressed in matrix form by

$$\Delta y = \mathbf{G}^{t} \, \Delta \mathbf{p} + \tfrac{1}{2} \Delta \mathbf{p}^{t} \, \mathbf{H} \, \Delta \mathbf{p} + \ldots . \tag{9.2}$$

Additionally when $\mathbf{y}$ represents a vector output function and using first order terms only, we may write

$$\Delta \mathbf{y} = \mathbf{S} \, \Delta \mathbf{p} \; , \tag{9.3}$$

where $\mathbf{S}$ is defined in Eq. (8.5) as a matrix of first-order sensitivities.

Tolerance analysis methods based on finite increments in paramerer values are broadly labelled *worst case* methods. When variations are assumed to have normal (or *Gaussian*) distributions and analysis is through the standard equations for transmission of variance, tolerance analysis is by the *moment* method. *Random sampling* is used to simulate the construction of a large number of circuits with component variation defined in terms of a known distribution or by measurement. In general the three methods, which are in order of increasing computational effort, give correspondingly improved estimates of variation in circuit performance.

Before we discuss any method in detail, we must first establish the expected form of variation in component values for both discrete and integrated circuits. We will also take the opportunity to recall some fundamental probability theory.

## 9.1 COMPONENT VARIATION

### Discrete Components

Consider the example of a circuit which contains a resistor equal to 100 Ω, and the manufacture of one thousand circuits. At some stage in the design a decision is made that the resistor tolerance is to be ±5%, and eventually the resistors are obtained from a component manufacturer. The designer can expect no more than a guarantee that all resistor values lie between 95 Ω and 105 Ω. However, the extent that further assumptions can be made concerning the variation of the resistor values will have an important effect on the number of circuits manufactured which are within specification. Alternatively, if all the circuits produced are to be within a specification, then the reduced variation in component value will increase the component cost. Whichever approach to tolerance design is used, knowledge of the variation in component values will assist in minimizing production costs.

The distribution of resistance values in our sample of 1000 resistors with nominal value equal to 100 Ω may be visualized by means of a histogram. The range of possible resistance values is divided into a number of *cells*. for example from 95 to 96 Ω, 96 to 97 Ω, and so on to 104 to 105 Ω The resistor values are then measured and the number of resistors in each cell tabulated. The histogram is then plotted as shown in Fig. 9.1(a) which is typical for resistors which have not been subject to selection.

Fig. 9.1(b) shows the envelope of the histogram plotted in terms of the relative number of resistors compared to the total sample. The graph then shows the *probability density* and has the property that the area under the graph, is unity.

Several other distributions typical of those obtained for discrete components are shown in Fig. 9.2. These include symmetric distributions (graphs 1, 2, and 4), a skew distribution (graph 3), a uniform distribution (graph 4), and a double peaked distribution (graph 2) arising, for example, from the partial removal of 1% resistors from a production run of 5% resistors. All distributions have unit area.

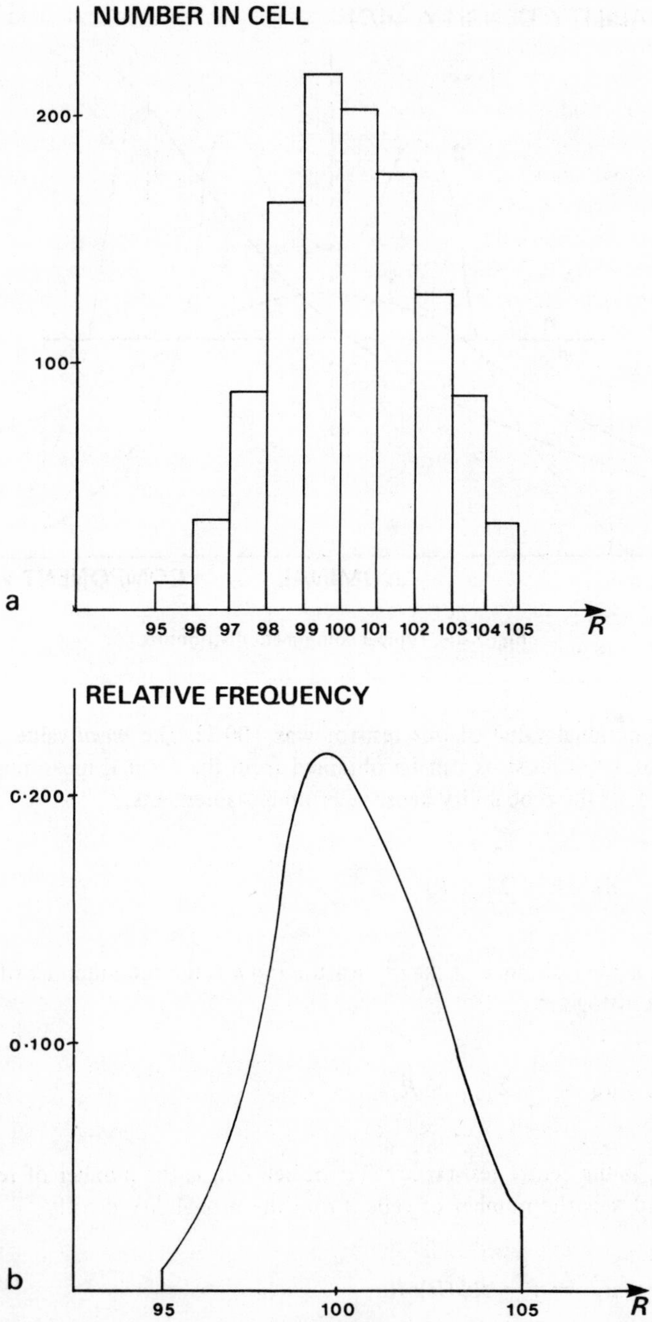


Fig. 9.1 – Resistor distribution.

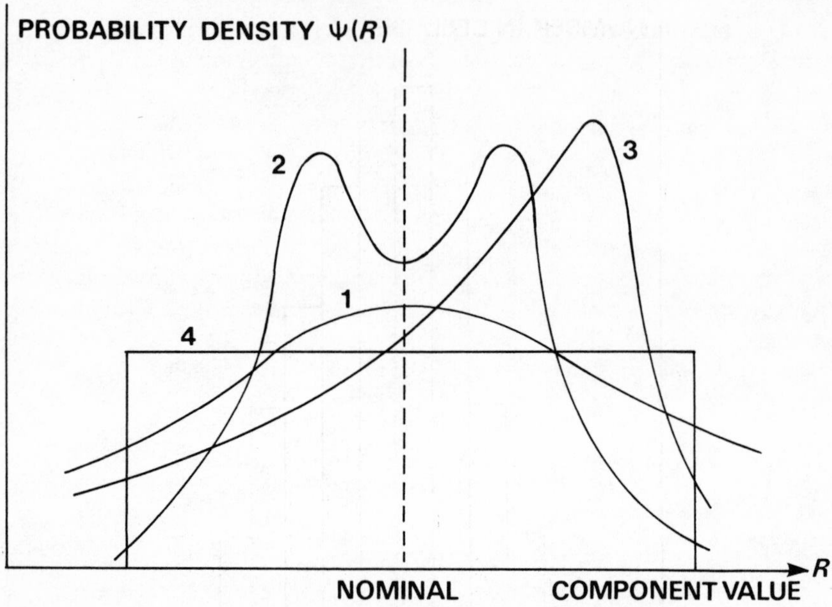


Fig. 9.2 – Typical component distributors.

The nominal value of our resistor was 100 Ω. The *mean* value $\mu_R$ of our sample of 1000 resistors can be obtained from the original measurements, the histogram, or the probability density. From measurements,

$$\mu_R = \frac{1}{n} \Sigma_{i=1}^{n} R_i \ , \tag{9.4}$$

where $R_i$ is the resistance of the $i^{th}$ resistor and $n$ is the total number of resistors. From the histogram,

$$\mu_R = \frac{1}{n} \Sigma_{i=1}^{k} m_i \bar{R}_i \ ,$$

where $\bar{R}_i$ is the centre resistance value of cell $i, m_i$ is the number of resistors in cell $i$, and $k$ is the number of cells. From the probability density,

$$\mu_R = \int_{-\infty}^{\infty} R\psi(R)\mathrm{d}R \ ,$$

where $\psi(R)$ is the probability density as a function of resistance $R$.

The spread or width of a distribution is measured by its **variance** $\sigma_R^2$, or **second moment**, defined for continuous distributions by

$$\sigma_R^2 = \int_{-\infty}^{\infty} (R - \mu_R)^2 \, \psi(R) \mathrm{d}R \ .$$

The square root of variance, $\delta_R$, is called the **standard deviation.** Higher moments, using the third and higher powers of $(R - \mu_R)$ in the integral, are sometimes used to describe the skewness and other properties of unusually shaped distributions, but they are of only limited use in circuit applications. From measurements,

$$\sigma_R^2 = \frac{1}{n} \Sigma_{i=1}^{n} (R_i - \mu_R)^2 = \frac{1}{n} (\Sigma_{i=1}^{n} R_i^2) - \mu_R^2 \ , \tag{9.5}$$

and from the histogram,

$$\sigma_R^2 = \frac{1}{n} \Sigma_{i=1}^{k} m_i (\bar{R}_i - \mu_R)^2 = \frac{1}{n} (\Sigma_{i=1}^{k} m_i \bar{R}_i) - \mu_R^2$$

All variables were defined in the previous paragraph.

**Integrated Circuit Components**

Integrated circuits (ICs) are originally processed in the form of a wafer (or slice) containing a few hundred identical IC chips. Circuit parameters vary continuously over the slice within the limits set by the process specification. However the variation within a single chip can be considerably less than that over a slice. Fig. 9.3 shows typical variation of sheet resistance in the form of contours of constant resistance over a complete slice. The overall tolerance limit of the resistance on the slice is ±20% but within the limited area of a single chip the spread is no more than ±5%. At two nearby points matching could be expected within ±1%.

Measurements of this type are available to integrated circuit designers, and variations in IC components are often more fully defined than for discrete components. Variations can usually be modelled by Gaussian distributions since there are no arbitrary selection procedures which interfere with the naturally occurring component values. However, the matching of component values in an IC chip, which is of extreme importance to IC circuit design and arises because distributions for different components are not independent, must also be modelled.

All electronic components in an integrated circuit are dependent on a limited number of *process parameters* and are therefore closely related. Consider two

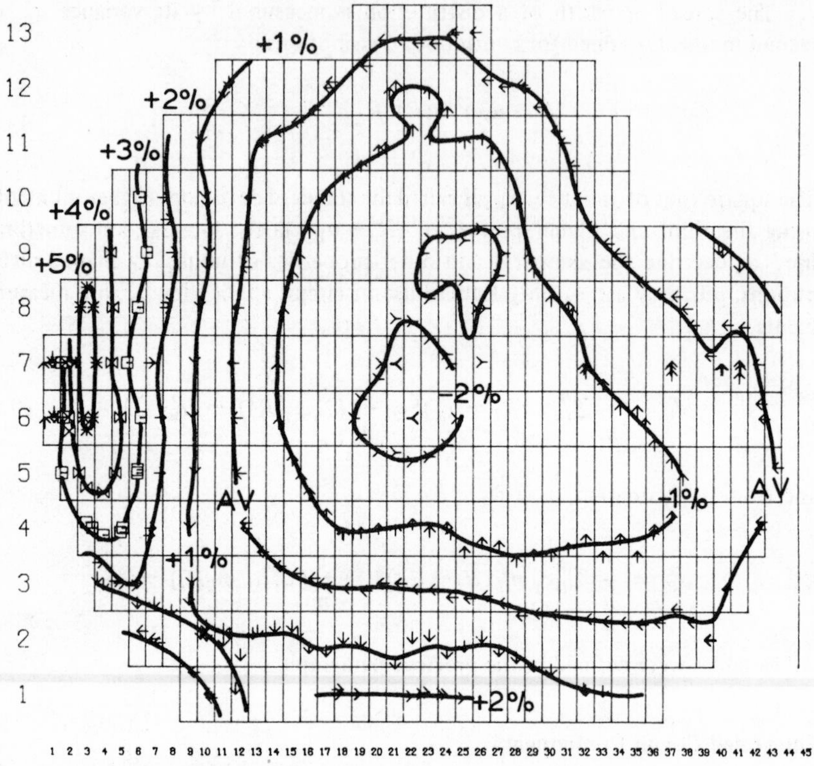


Fig. 9.3 – Resistance contours on integrated circuit wafer.
(Courtesy Phillips Research Laboratories)

resistors in the base diffusion layer of a monolithic IC. Each resistor value is given in terms of its width $w$, length $l_G$, equivalent length of connection resistance $l_C$, and sheet sensitivity $\rho$ $\Omega$/square by

$$\frac{(l_G + l_C)}{w}\rho = \frac{l\rho}{w} \tag{9.6}$$

Variation in the connection resistance equivalent length $l_C$ swamps all variation in the geometric length $l_G$, and the distributions of the process parameters in the equation for resistance can be considered independent. However, the two resistor distributions are closely related as they both depend on three common parameters which have almost matching values when the resistors are located alongside each other.

This behaviour is modelled in probability theory by the **correlation coefficient** which is a measure of the relationship between two distributions. If the variances of our two resistors $R_A$ and $R_B$ with mean values $\mu_A$ and $\mu_B$ are $\delta_A^2$ and $\delta_B^2$, then the **covariance** is defined by the expression

$$\sigma_A \, \sigma_B \, \rho_{AB} = \frac{1}{n} \, \Sigma_{i=1}^{n} \, (R_{A,i} - \mu_A) \, (R_{B,i} - \mu_B) \; . \tag{9.7}$$

The correlation coefficient $\rho_{AB}$ is obtained from the convariance by dividing out the two standard deviations. The definition of covariance is similar to that of variance but takes sample values from each distribution simultaneously. Therefore any tendancy for them to be related will result in a predominance of terms in the summation of a particular sign, and $\rho_{AB}$ will be non-zero. In the extreme case when both resistors are equal to all times the summation for covariance is the same as that for variance, and $\sigma_A \, \sigma_B \, \rho_{AB} = \sigma_A^2 = \sigma_B^2$ and $\rho_{AB} = 1$. The value of correlation coefficient is therefore limited by $-1 \leqslant \rho \leqslant 1$.

When the two distributions are independent the sign of the product term in Eq. (9.7) will be random and the covariance, and therefore the correlation coefficient, will be zero. If, however, the correlation coefficient equals zero, we cannot conclude that the distributions are independent but only that they are uncorrelated.

Later, under the heading of the moment method (Section 9.3), we will discuss the transmission of variance. In Example 9.3 we will obtain typical values for correlation coefficients in integrated circuits.

**The Gaussian Distribution**

The definitions of standard deviation, correlation coefficient, mean value, and other characteristic parameters of the distribution are independent of the shape of the distribution. However, it is often sufficiently accurate in tolerance analysis to assume that a component distribution is Gaussian (or *Normal*) with mean value and standard deviation equal to that of the actual distribution. This assumption may be justified first by the observation that components in practice do have distributions of the general form of a Gaussian distribution. Secondly, the combined effect of the distributions of a number of different components on a network response more closely approximates a Gaussian distribution as the number of components increases. The properties of the Gaussian distribution must therefore be established in relation to tolerance analysis and design.

The Gaussian **distribution function** is defined by the equation

$$\phi(x) = \int_{-\infty}^{x} \frac{1}{\sigma\sqrt{2\pi}} \, e^{-(x-\mu)^2/2\sigma^2} \, dx \; , \tag{9.8}$$

where $\phi(x)$ is the probability that the sample value lies between $-\infty$ and $x$. The probability density, which was the function plotted in Figs. 9.1(b) and 9.2, is obtained by differentiation of $\phi(x)$ yielding

$$\psi(x) = \frac{d\phi(x)}{dx} = \frac{1}{\sigma\sqrt{2\pi}} e^{-(x-\mu)^2/2\sigma^2} . \tag{9.9}$$

Interpretation of the graph of probability density is through the incremental area $\psi(x)dx$ which gives the probability of a sample assuming a value in the interval $x$ to $x + dx$. Graphs of $\psi(x)$ for various values of standard deviation $\sigma$ are shown in Fig. 9.4. The densities are symmetric about the value $\mu$ since $\psi(x-\mu) = \psi(x+\mu)$ and the use of the symbol $\mu$ for mean value is justified.

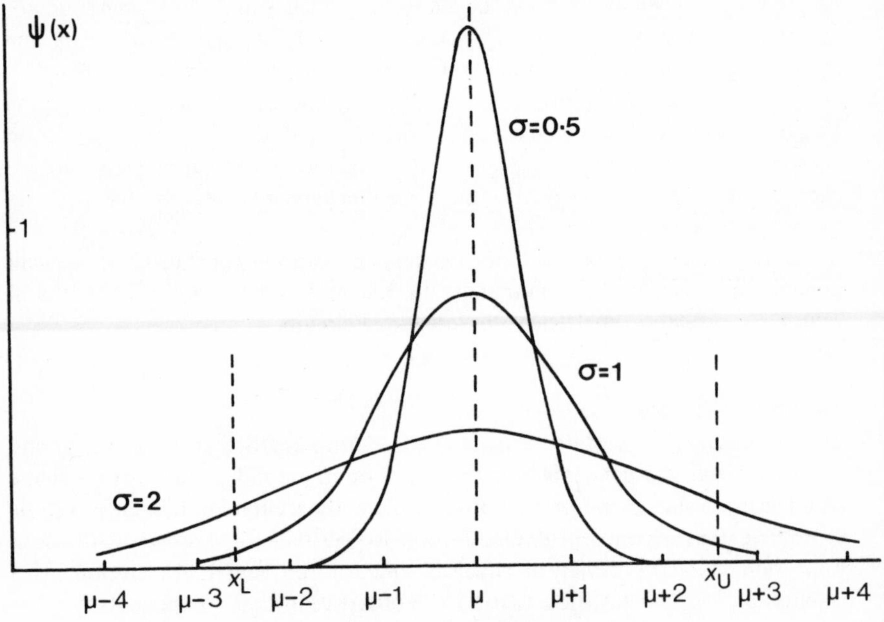


Fig. 9.4 – Gaussian probability density.

If we assign upper and lower tolerance limits $x_U$ and $x_L$ such that the region $x_L < x < x_U$ is acceptable, then the probability of a sample assuming an acceptable value is given by

$$\int_{x_L}^{x} \psi(x)dx = \int_{x_L}^{x_U} \frac{1}{\sigma\sqrt{2\pi}} e^{-(x-\mu)^2/2\sigma^2} dx \tag{9.10}$$

and is called the **yield**. Since the total area under the probability density graph is unity, the probability of a sample assuming an unacceptable value is the **rejection ratio** = (1-yield). Both figures are usually quoted as a percentage. When tolerance limits are equally distant from the mean value, the yield is available from published tables in terms of a parameter $K$ where

$$K\sigma = \mu - x_L = x_U - \mu \ .$$

Letting $t = (x - \mu)/\sigma$ in Eq. (9.10) and taking advantage of symmetry,

$$\text{Yield} = 2 \int_0^K \frac{1}{\sqrt{2\pi}} e^{-t^2/2} \, dt \ .$$

This integral may be evaluated to obtain the following table.

**Table 9.1**

| Yield % | 68 | 90 | 95 | 98 | 99 | 99.5 | 99.8 | 99.9 | 99.99 |
|---|---|---|---|---|---|---|---|---|---|
| Rejection rate % | 32 | 10 | 5 | 2 | 1 | 0.5 | 0.2 | 0.1 | 0.01 |
| $K$ | 1 | 1.64 | 1.96 | 2.33 | 2.58 | 2.81 | 3.09 | 3.29 | 3.89 |

**The Uniform Distribution**

The probability density of a component with uniform distribution between upper and lower tolerance limits $x_U$ and $x_L$ is defined by

$$\psi(x) = 1/(x_U - x_L) \ , \qquad x_L \leqslant x \leqslant x_U$$

and

$$\psi(x) = 0 \ , \qquad x < x_L \text{ or } x > x_U \ .$$

The density is shown in Fig. 9.2, graph 4. The mean value $\mu$ is of course $(x_U + x_L)/2$, and the variance $\sigma^2$ is given by $(x_U - x_L)^2/12$. Therefore,

$$\sigma^2 = (x_U - \mu)^2/3 = (\mu - x_L)^2/3 \ .$$

The tolerance limit corresponds to $\sqrt{3}\ \sigma$, and from Table 9.1 for $K = \sqrt{3}$, the rejection ratio is about 7% for the corresponding Gaussian distribution used in the moment method of tolerance analysis.

## 9.2 WORST CASE ANALYSIS

Worst case tolerance analysis is a non-statistical method based on direct use of the tolerance equation (9.1). Taking the first-order term only, a tolerance limit $\Delta y$ of an output or performance function $y$ is given by

$$\Delta y = \Sigma_{i=1}^{n} \frac{\partial y}{\partial p_i} \Delta p_i \ , \qquad (9.11)$$

where $p_i$, $i$=1, 2, . . . $n$ are $n$ nominal circuit component values subject to tolerance limits $\Delta p_i$. The maximum, or worst case, upper tolerance limit $\bar{y}$ of the output $y$ occurs when all terms within the summation in Eq. (9.11) are positive. If the sensitivity $\partial y/\partial p_i$ is positive, the upper tolerance limit $\bar{p}_i$ of $p_i$ is used so that the product $(\bar{p}_i - p_i)\,(\partial y/\partial\, p_i)$ is positive. When the sensitivity is negative the lower tolerance limit $\underline{p}_i$ must be chosen so that the product remains positive. The worst case lower tolerance limit $\underline{y}$ will be obtained if all selected component tolerance limits are opposite to those selected for $\bar{y}$ since the product terms in the summation then become negative.

This simple scheme for worst case tolerance analysis is summarized by the following steps.

STEP 1. Use the nominal component values to evaluate the nominal output $y$.
2. Evaluate the differential sensitivity $\partial y/\partial p_i$ for all components.
3. Select component tolerance limits of the same sign as sensitivity and use Eq. (9.11) to determine $\bar{y}$.
4. Select component tolerance limits of the opposite sign to those selected in STEP 3 and use Eq. (9.11) to evaluate $\underline{y}$.

The same procedure applies if $y$ represents an output vector; for example, several d.c. voltages in a circuit, or the voltage gain at various frequencies. It is preferable in practice to reanalyse the circuit, using the predicted extrenum values instead of using Eq. (9.11) in STEPS 3 and 4.

*Example* 9.1

Determine the worst case tolerance limits for the amplifier gain of the circuit in Fig. 9.5. Assume resistor tolerances to be ±2% and the operational amplifier gain tolerance to be +50% and −10%.

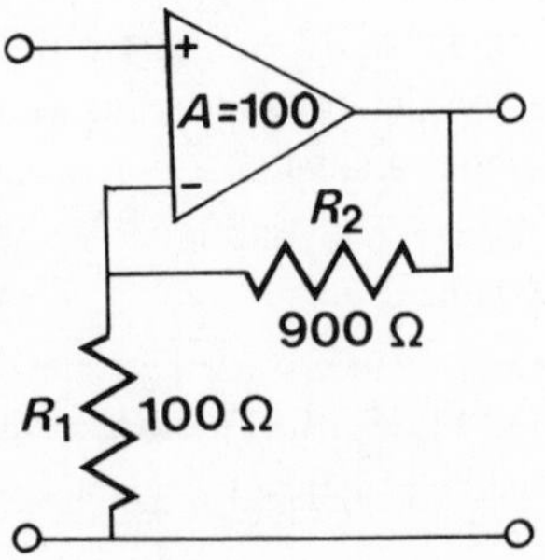


Fig. 9.5 – Amplifier for Example 9.1.

Symbolic analysis followed by differentiation and substitution yields,

$$G = 1 \Bigg/ \left\{ \frac{1}{A} + \frac{R_1}{R_1+R_2} \right\} = 9.1$$

$$\frac{\partial G}{\partial A} = -G^2 \left\{ \frac{-1}{A^2} \right\} = 83 \times 10^{-4}$$

$$\frac{\partial G}{\partial R_1} = -G^2 \left\{ \frac{R_2}{(R_1+R_2)^2} \right\} = -747 \times 10^{-4}$$

$$\frac{\partial G}{\partial R_2} = -G^2 \left\{ \frac{-R_1}{(R_1+R_2)^2} \right\} = 83 \times 10^{-4} .$$

Absolute component tolerances are

$\Delta A$ +50 −10

$\Delta R_1$ +2 −2 Ω

$\Delta R_2$ +18 −18 Ω .

Tolerance limits are selected for a positive product in the summation for the upper tolerance limit $\overline{G}$ given by

$$\overline{G} = G + \frac{\partial G}{\partial A}\Delta A + \frac{\partial G}{\partial R_1}\Delta R_1 + \frac{\partial G}{\partial R_2}\Delta R_2$$

$$= 9.1 + \{83 \times 50 + (-747)(-2) + 83 \times 18\} \times 10^{-4}$$

$$= 9.1 + 0.71 .$$

The lower tolerance limit $\underline{G}$ is obtained by selecting limits for negative product terms, that is,

$$\underline{G} = G + \{83 \times (-10) + (-747)(+2) + 83 \times (-18)\} \times 10^{-4}$$

$$= 9.1 - 0.38 .$$

The tolerance limits for amplifier gain are therefore $9.1^{+0.71}_{-0.38}$.

It is interesting to compare this result, achieved using first order sensitivity, with that obtained by direct substitution of the extreme component values into the gain expression. The exact tolerance limit is then $9.1^{+0.606}_{-0.376}$.

The tolerance design of this amplifier may be improved by *centreing* the component values within their tolerance limits. Consider the operational amplifier gain to have a nominal value of 116 with tolerance limits $^{+34}_{-26}$. The operational amplifier gain spread is identical to that used previously. However, a new value of $R_2$ is required, equal to 886 Ω, which gives the same overall gain as previously. At ±2% the new tolerance limits of $R_2$ are ±17.7 Ω. Recalculation of the exact tolerance limits for gain gives $G = 9.1^{+1.481}_{-0.485}$. Thus the maximum deviation of gain from nominal has been reduced by 1½% to just over 5%. Use of the numerical centre of the tolerance limits for $A$, that is, 120, does not give the best centering of $G$, because of non-linearity of the gain $G$ as a function of $A$.

**Large Change Sensitivity**

In practice, tolerance limits are of the order ±1% to ±20% with some circuit components subject to much larger variation. Very few components are accurate to limits less than ±0.1%, because of their expense. Variation of a component within its tolerance limit frequently exceeds the range over which first-order differential sensitivity may be regarded as accurate and is therefore considered to be a *large change*. The **incremental** or **large change sensitivity** of an output $y = f(p)$, where $p$ is a component value, is defined by the ratio of increments $\Delta y/\Delta p$ and is valid only for the increment $\Delta p$ used. In general $\Delta y/\Delta p$ will vary as $\Delta p$ is changed, and in the limit as $\Delta p \to 0$ it will of course equal the differential sensitivity $\partial y/\partial p$.

The relationship between a circuit output and a component value may be so non-linear that the differential sensitivity and the incremental sensitivity may even be of opposite sign. This is particularly true if the variation of other components is also considered as in worst case analysis.. One important result of these large change effects is that the worst case output may occur at component values intermediate between the nominal values and tolerance limits. We will therefore take an introductory look at the problem and leave advanced solution algorithms to texts on computer-aided design.

The source of many difficulties with worst case analysis using sensitivity lies in the *bilinear* relationship between output transfer functions and an individual component. As we noted in Section 6.1 of Chapter 6, the numerator $N(s)$ and denominator $D(s)$ of a transfer function $F(s)$ are both linear in a component variable. Therefore we may write

$$F(s) = \frac{N(s)}{D(s)} = \frac{N_1(s) + pN_2(s)}{D_1(s) + pD_2(s)}, \tag{9.12}$$

where $p$ is a component variable. More complicated expressions also occur, for example a gyrator leads to a biquadratic expression in the gyration conductance. Other circuits may have automatic tuning or gain control leading to very odd behaviour sometimes difficult to describe by sensitivity.

Differentiating Eq. (9.12) with respect to $p$ and omitting the function of $s$ for clarity,

$$\frac{\partial F}{\partial p} = \frac{D_1N_2 - N_1D_2}{(D_1 + pD_2)^2} \; . \tag{9.13}$$

Differential sensitivity from this equation can be seen to be dependent on the component value $p$ and therefore changes as $p$ varies within its tolerance limit. Sensitivity also depends on the remaining component value through $N_1$,$N_2$,$D_1$ and $D_2$, and it changes if other components change. Differentiating again,

$$\frac{\partial^2 F}{\partial p^2} = \frac{-(D_1N_2 - N_1D_2)}{(D_1 + p\,D_2)^3} \times 2D_2 \; , \tag{9.14}$$

and once more,

$$\frac{\partial^3 F}{\partial p^3} = \frac{(D_1N_2 - N_1D_2)}{(D_1 + p\,D_2)^4} \times 2\,D_2 \times 3\,D_2 \; . \tag{9.15}$$

These expressions are easily found, remembering that $N_1(s)$, $N_2(s)$, $D_1(s)$, and $D_2(s)$ are independent of $p$. The general form for the $n^{\text{th}}$ derivative is clearly,

$$\frac{\partial^n F}{\partial p^n} = \left(\frac{\partial F}{\partial p}\right)\left(\frac{-D_2}{D}\right)^{n-1} n! \; . \tag{9.16}$$

The incremental change $\Delta F$ in $F$ can now be predicted by substitution of derivatives in the Taylor expansion for $\Delta F$ given by

$$\Delta F = \frac{\partial F}{\partial p}\Delta p + \frac{1}{2!}\frac{\partial^2 F}{\partial p^2}\Delta p^2 + \frac{1}{3!}\frac{\partial^3 F}{\partial p_3}\Delta p^3 \; \ldots \; .$$

Therefore, $\Delta F = \dfrac{\partial F}{\partial p}\Delta p - \dfrac{D_2}{D}\dfrac{\partial F}{\partial p}\Delta p^2 + \dfrac{D_2^2}{D^2}\dfrac{\partial F}{\partial p}\Delta p^3 \; \ldots$

and summation of the geometric series yields

$$\Delta F = \frac{\partial F}{\partial p}\Delta p \Bigg/ \left\{1 + \frac{D_2}{D}\Delta p\right\}$$

that is, $$\frac{\Delta F}{\Delta p} = \frac{\partial F}{\partial p} \Bigg/ \left\{1 + \frac{D_2}{D}\Delta p\right\}. \tag{9.17}$$

This equation relates the large change sensitivity and differential sensitivity for network functions with bilinear dependence on a single component variable.

One unexpected feature of this result is that $\Delta F/\Delta p=0$ when $\partial F/\partial p=0$, since all higher partial derivatives are also zero. The component involved is therefore redundant to the circuit and could be removed. However, zero sensitivity is unlikely to occur simultaneously at all frequencies. It is clear from Eq. (9.17) that large change sensitivity may be of any sign or go to infinity dependent on the sign and value of $D_2\Delta p/D$ in comparison with unity. Worst case analysis is therefore prone to inaccuracy, and quite sophisticated computer programs are required to make a useful prediction for a large circuit.

The most straightforward modification of the worst case analysis procedure given at the beginning of this section assumes that the worst case output values are determined by re-analysis. In STEP 3 the computation of sensitivity is then only used to indicate which tolerance limit should be selected in the analysis for $\bar{y}$. Using the methods of Chapter 8, sensitivity is obtained at the chosen tolerance limits as an extension of the analysis. If any value for sensitivity has changed sign compared to that obtained at the nominal component values, then the opposite tolerance limit for that component should have been chosen. The chosen tolerance limits are therefore corrected and the worst case analysis repeated. STEP 4 of the procedure should be similarly checked. The modified STEPS 3 and 4 for worst case analysis are summarized as follows.

STEP 3
- a Select tolerance limits of the same sign as sensitivity.
- b Analyse to obtain worst case $\bar{y}$ and re-compute sensitivity at the tolerance limits.
- c Check and correct the selection of tolerance limits and re-analyse to obtain the corrected worst case $\bar{y}$.

STEP 4
- a Select the opposite tolerance limits to those used in STEP 3c.
- b Analyse to obtain worst case $\underline{y}$ and recompute sensitivity at the tolerance limits.
- c Check and correct the selection of tolerance limits and re-analyse to obtain the corrected worst case $\underline{y}$.

*Example* 9.2

Determine the worst case limits for output voltage $v$ of the circuit in Fig. 9.6.

Analysis yields

$$v = \frac{30\,R_2R_3 + 16\,R_1R_2}{R_1R_2 + R_2R_3 + R_1R_3} = 15.33\text{ V (nominal)}.$$

The worst case limits to $v$ are obtained

when $R_1 = 900\ \Omega$, $R_2 = 1100\ \Omega$, $R_3 = 1500\ \Omega$ and $\bar{v} = 16.38$ V,

and $R_1 = 1100\ \Omega$, $R_2 = 900\ \Omega$, $R_3 = 1500\ \Omega$ gives $\underline{v} = 14.72$ V.

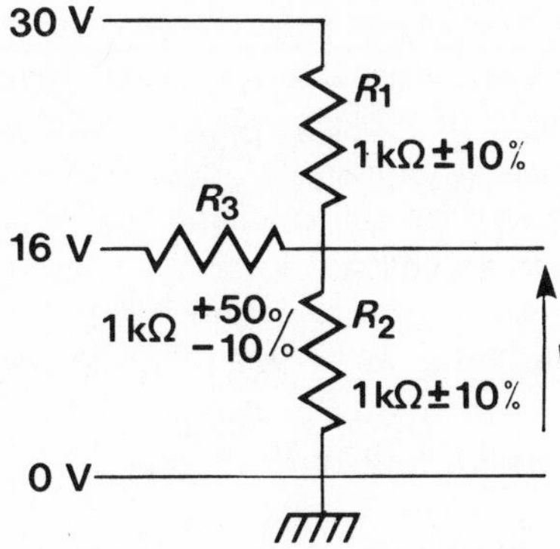


Fig. 9.6 – Circuit for Example 9.2.

The worst case for both upper and lower limits requires the use of the maximum value of $R_3$. The selection of the appropriate component tolerance limits is made from sensitivity computations as follows.

STEP 2 Sensitivity ($\times 10^3$ throughout) at nominal component values.

$$\frac{\partial v}{\partial R_1} = -4.89 \qquad \frac{\partial v}{\partial R_2} = 5.1 \qquad \frac{\partial v}{\partial R_3} = -0.2$$

STEP 3a Select tolerance limits such that $(\partial v/\partial p)\Delta p$ is positive.

$$R_1 = 900\ \Omega \qquad R_2 = 1100\ \Omega \qquad R_3 = 900\ \Omega\ .$$

3b Recompute sensitivity.

$$\frac{\partial v}{\partial R_1} = -5.39 \qquad \frac{\partial v}{\partial R_2} = 13.11 \qquad \frac{\partial v}{\partial R_3} = +0.13$$

3c Corrected tolerance limits.

$$R_1 = 900\ \Omega \qquad R_2 = 1100\ \Omega \qquad R_3 = 1500\ \Omega\ .$$

These values yield $\overline{v} = 16.36$ V.

4a Select opposite tolerance limits.

$$R_1 = 1100\ \Omega \quad R_2 = 900\ \Omega \qquad R_3 = 900\ \Omega$$

4b Recompute sensitivity.

$$\frac{\partial v}{\partial R_1} = -4.12 \qquad \frac{\partial v}{\partial R_2} = 12.0 \qquad \frac{\partial v}{\partial R_3} = -0.64$$

4c Corrected tolerance limits.

$$R_1 = 1100\,\Omega \quad R_2 = 900\,\Omega \qquad R_3 = 1500\,\Omega\ .$$

These values yield $\underline{v} = 14.72$ V.

In both steps the modified procedure has selected the appropriate extreme value of $R_3$ after recomputing sensitivities.

## 9.3 THE MOMENT METHOD

Tolerance analysis by the moment method makes use of the distribution functions from Section 9.1 which describe the statistical variation of circuit components. All variations are assumed to be Gaussian, and any that are not are replaced by Gaussian distribution with the same mean value and variance. Analysis by the moment method can be expressed in a convenient matrix form for computation by definition of the covariance matrix which contains both variances and correlations of $n$ variables in an $n \times n$ matrix.

The covariances, defined by Eq. (9.7) for each pair of components, may be assembled with an $n \times n$ matrix describing the joint variation of $n$ components. Each component $p_1, p_2, \ldots p_n$ varies with standard deviations $\sigma_{p_1}, \sigma_{p_2}, \ldots \sigma_{p_n}$. The correlation coefficients $\rho_{p_{i,j}}$, $i=1,2, \ldots n, j=1,2, \ldots n$, specify the relationships between the component distribution of $p_i$ and $p_j$, noting that $\rho_{p_{i,i}}=1$ and $\rho_{p_{i,j}} = \rho_{p_{j,i}}$. Standard deviations and correlation can be assembled into matrices $\sigma_p$ and $\rho_p$ defined by

$$\sigma_p = \begin{bmatrix} \sigma_{p_1} & & & 0 \\ & \ddots & & \\ & & \ddots & \\ 0 & & & \sigma_{p_n} \end{bmatrix} \quad \text{and } \rho_p = \begin{bmatrix} 1 & \rho_{p_{1,2}} & \cdots & \rho_{p_{1,n}} \\ & 1 & & \\ & & \ddots & \\ \rho_{p_{n,1}} & & & 1 \end{bmatrix}$$

The component convariance matrix **P** is then defined by

$$\mathbf{P} = \sigma_p\, \rho_p\, \sigma_p^t \qquad (9.18)$$

The general term $P(i,j)$ of **P** is equal to $\sigma_{p_i}\,\rho_{p_{i,j}}\sigma_{p_j}$ and may be determined by measurements as in Eq. (9.7). Since the diagonal element $P(i,i)$ contains $\rho_{p_{i,i}}$ which equals unity, the standard deviation $\sigma_{p_i}$ may be extracted as the square root of $P(i,i)$. Standard deviation may then be divided out from the off-diagonal terms to yield the correlation coefficients.

The linearized relationship between variation $\Delta\mathbf{p}$ in component values and variation $\Delta\mathbf{y}$ in the output vector was given in Eq. (9.3) as

$$\Delta\mathbf{y} = \mathbf{S}\Delta\mathbf{P}\ .$$

The transmission of variance through this linear set of equations to yield the output convariance matrix **Q** is given by

$$\mathbf{Q} \cong \mathbf{SPS}^t \tag{9.19}$$

This equation is approximate because of our assumption that the variation of the output is adequately described by first-order sensitivity over the range of component variation. More familiar forms of Eq. (9.19), known as the **general variance law**, give variance in non-matrix form as

$$\sigma_{Q_i^2} = \Sigma_{k=1}^{n}\ \Sigma_{l=1}^{n}\ \frac{\partial Q_i}{\partial p_k}\,\frac{\partial Q_i}{\partial p_l}\,\sigma_{p_k}\,\sigma_{p_l}\,\rho_{p_{k,l}} \tag{9.20}$$

Similarly the correlation coefficient is given by

$$\rho_{Q_{i,j}} = \frac{1}{\sigma_{Q_i}\sigma_{Q_i}}\ \Sigma_{k=1}^{n}\ \Sigma_{l=1}^{n}\ \frac{\partial Q_i}{\partial p_k}\,\frac{\partial Q_j}{\partial p_l}\,\sigma_{p_k}\sigma_{p_l}\,\rho_{p_{k,l}}\ . \tag{9.21}$$

The moment method of tolerance analysis consists quite simply of the application of Eq. (9.19) or Eqs. (9.20) and (9.21). The procedure may be summarized by the following steps.

STEP 1. Use nominal component values to evaluate the nominal output vector **y**.

2. Evaluate the differential sensitivity matrix **S** for all components.
3. Evaluate the component covariance matrix **P** from specified component standard deviations and correlation coefficients.
4. Use Eq. (9.19) to evaluate the output covariance **Q** and extract the output standard deviation and correlation coefficients.

*Example* 9.3

Determine the standard deviations and correlation coefficient of two integrated circuit resistors defined by the following table. Examine their application in the amplifier used in Example 9.1.

**Table 9.2**

| | Nominal value $R_1$ | $R_2$ | Tolerance Absolute | Matching | Units |
|---|---|---|---|---|---|
| Sheet resistance $r$ | 200 | 200 | ±20 | ±5 | Ω/square |
| Width $w$ | 45 | 15 | ±1.5 | ±0.2 | μm |
| Length $l$ | 45 | 135 | ±5 | ±1 | μm |
| Value $lr/w$ | 200 | 1800 | – | – | Ω |

We must first determine the correlation coefficients implied by the specified matching tolerances of process variables. The absolute tolerance of sheet resistance for example, is ±20 Ω/square or ±10%, and the matching tolerance is ±5 Ω/square or $\pm 2\frac{1}{2}\%$ of nominal. Matching is usually exploited as means of maintaining the accuracy of a ratio $t = r_1/r_2$ where $r_1$ and $r_2$ are here equal to the two values of sheet resistance at nearby points on a chip. Application of Eq. (9.20) yields

$$\sigma_t^2 = \left(\frac{\partial t}{\partial r_1}\sigma_{r_1}\right)^2 + \left(\frac{\partial t}{\partial r_2}\sigma_{r_2}\right)^2 + 2\,\frac{\partial t}{\partial r_1}\,\frac{\partial t}{\partial r_2}\,\sigma_{r_1}\,\sigma_{r_2}\,\rho\ ,$$

where $\rho$ is the unknown correlation. Since $r_1{=}r_2$, the sensitivities may be obtained as

$$\frac{\partial t}{\partial r_1} = \frac{1}{r_1} \text{ and } \frac{\partial t}{\partial r_2} = \frac{-1}{r_2} = \frac{-1}{r_1}\ ,$$

and $\quad \sigma_t^2 = (\sigma_{r_1}^2 + \sigma_{r_2}^2 - 2\sigma_{r_1}\,\sigma_{r_2}\,\rho)/r_1^2\ .$

Using the equality $\sigma_{r_2} = \sigma_{r_1}$ ,

$$\sigma_{r_1}/r_1\sigma_t = 1/\sqrt{2-2\rho}\ . \tag{9.22}$$

Standard deviations in Eq. (9.22) may be replaced by tolerance limits from Table 9.1 using the same value $K$ for both deviations. The absolute tolerance limit therefore equals $K\sigma_{r_1}$ and the matching tolerance limit equals $K\sigma_t r_1$, where the factor $r_1$ arises because $\sigma_t$ refers to the unit nominal value of $t$. We may therefore obtain the general expression

$$\frac{\text{Absolute Tolerance}}{\text{Matching Tolerance}} = \frac{1}{\sqrt{2-2\rho}}\,, \tag{9.23}$$

which applies to tolerance limits and standard deviations of equal-valued matching variables. The following table is easily evaluated, and the correlation implied by the ratio of tolerances of 4:1 is found to be 0.07. Note that zero correlation yields a matching tolerance equal to $\sqrt{2}$ times the absolute tolerance, and even worse figures are obtained for negative correlation. We can now proceed with the example.

**Table 9.3**

| $\frac{\text{Absolute tolerance}}{\text{Matching tolerance}}$ | 0.7 | 1 | $\sqrt{2}$ | 2 | 4 | 7 | 22 |
|---|---|---|---|---|---|---|---|
| correlation $\rho$ | 0 | 0.5 | 0.75 | 0.875 | 0.97 | 0.99 | 0.999 |

The covariance matrix **P** of parameters $l_1, l_2, w_1, w_2, r_1$, and $r_2$ is determined as in Eq. (9.18) by the product $\sigma_p \rho_p \sigma_p{}^t$. Using $K$=4 from Table 9.1 and $\rho_p$ from Table 9.3, **P** is given by

$$\begin{bmatrix} 1.25 & & & & & 0 \\ & 1.25 & & & & \\ & & 0.375 & & & \\ & & & 0.375 & & \\ & & & & 5 & \\ 0 & & & & & 5 \end{bmatrix} \begin{bmatrix} 1 & 0.98 & & & & 0 \\ 0.98 & 1 & & & & \\ & & 1 & 0.99 & & \\ & & 0.99 & 1 & & \\ & & & & 1 & 0.97 \\ 0 & & & & 0.97 & 1 \end{bmatrix} \begin{bmatrix} 1.25 & & & & & 0 \\ & 1.25 & & & & \\ & & 0.375 & & & \\ & & & 0.375 & & \\ & & & & 5 & \\ 0 & & & & & 5 \end{bmatrix}$$

The sensitivity matrix **S** is derived by differentiation of the equations for $R_1$ and $R_2$ given by

$$r_1 = l_1 r_1/w_1 \text{ and } R_2 = l_2 r_2/w_2\ .$$

Thus the sensitivity matrix is arranged as

$$\mathbf{S} = \begin{bmatrix} \dfrac{\partial R_1}{\partial l_1} & \dfrac{\partial R_1}{\partial l_2} & \dfrac{\partial R_1}{\partial w_1} & \dfrac{\partial R_1}{\partial w_2} & \dfrac{\partial R_1}{\partial r_1} & \dfrac{\partial R_1}{\partial r_2} \\ \dfrac{\partial R_2}{\partial l_1} & \dfrac{\partial R_2}{\partial l_2} & \dfrac{\partial R_2}{\partial w_1} & \dfrac{\partial R_2}{\partial w_2} & \dfrac{\partial R_2}{\partial r_1} & \dfrac{\partial R_2}{\partial r_2} \end{bmatrix}$$

$$= \begin{bmatrix} 4.44 & 0 & -4.44 & 0 & 1 & 0 \\ 0 & 13.33 & 0 & -120 & 0 & 9 \end{bmatrix}.$$

Evaluation of $\mathbf{Q} = \mathbf{SPS}^t$ yields the output covariance

$$\mathbf{Q} = \begin{bmatrix} 58.6 & 383 \\ 383 & 4328 \end{bmatrix}.$$

Our output is in fact the two resistor values $R_1$ and $R_2$. Extracting standard deviations and the correlation coefficient from $\mathbf{Q}$ yields

$$\sigma_{R_1} = \sqrt{58.6} = 7.65 \quad \text{or } 3.8\% \text{ of nominal}$$

$$\sigma_{R_1} = \sqrt{4328} = 65.8 \quad \text{or } 3.7\% \text{ of nominal}$$

and $\rho_{R_1 R_2} = 0.765$.

Consider now the two resistors used in the amplifier circuit of Fig. 9.5 from Example 9.1 (but with gain $A$ fixed at 100). The standard deviation of the amplifier gain, nominally 9.1, is evaluated from Eq. (9.19) or (9.20), using sensitivities recomputed for the increased resistor values. Substitution into Eq. (9.19) of sensitivities $\partial G/\partial R_1 = -0.037$ and $\partial G/\partial R_2 = 0.0042$ and $\mathbf{Q}$ yields

$$\sigma_G^2 = [-0.037 \quad 0.0042] \begin{bmatrix} 58.6 & 383 \\ 383 & 4328 \end{bmatrix} \begin{bmatrix} -0.037 \\ 0.0042 \end{bmatrix},$$

and $\sigma_G = 0.193$ or 2.1% of nominal. This is not a particularly low figure since it represents a tolerance limit of about ±8%. However, matching tolerance limits for adjacent resistors can be significantly less in practice than the values used in this example. We might note finally that the contribution of correlation due to resistor matching in the standard deviation arises from the off-diagonal terms in **Q**. If these are set to zero, $\sigma_G = 0.394$ or 4.3% of nominal. Matching has therefore reduced the standard deviation $\sigma_G$ by one half.

It is worthwhile at this point to compare the information obtained from worst case analysis with that from the moment method.

*Worst Case Analysis*
Assume for comparison that discrete component resistor tolerances in the amplifier from Example 9.1 are ±5% and that $A$ is fixed at 100. Worst case analysis yields $G = 9.1 \pm 0.74$ or ±8.1% of nominal.

*The worst case prediction is that 5% discrete resistors achieve 8.1% gain accuracy with 100% yields of working circuits.*

*Moment Method*
In Example 9.3 absolute tolerance limits of the integrated circuit resistors are approximately ±25%. Two statistical predictions were made.

I Using the matching tolerances due to the integrated circuit the standard deviation of amplifier gain $\sigma_G$ was 0.193 or 2.1% of nominal. We can therefore infer from Table 9.1 that one circuit in ten thousand will exceed a tolerance limit of ±8.1%.

*The moment method prediction I is that integrated circuit 25% resistors achieve 8.1% gain accuracy with 99.99% yield.*

II Without using the matching correlation the standard deviation of amplifier gain was 0.394 or 4.3% of nominal. We can infer that six circuits in one hundred will exceed the ±8.1% tolerance limit.

*The moment method prediction II is that uncorrelated (that is, discrete) 25% resistors achieve 8.1% gain accuracy with 94% yield.*

Comparison of the two discrete resistor predictions suggests that the use of 5% resistors to achieve 8.1% gain tolerance is likely to be unnecessarily expensive when 25% resistors achieve the same with 94% yield. The least manufacturing cost of the amplifier will lie somewhaere between these two extremes. The integrated circuit has a much higher yield, and it may well be conomic to use the design with a lower gain tolerance. For example, yields are 99.9% at 7% tolerance and 99.5% at 6% tolerance. Detailed consideration of these tolerances design decisions is left to specialist texts on computer-aided tolerance design.

**Tolerance Fields**

When a circuit has $m$ separate outputs, $\mathbf{y}$ is a vector and the $m \times m$ output covariance matrix $\mathbf{Q}$ contains the joint variation of all outputs. The inevitable presence of correlation in $\mathbf{Q}$ makes it very difficult to estimate yield without random sampling because the yield required equals the probability that the outputs are simultaneously within all tolerance limits. When the output is a frequency response, for example, there is a high probability due to correlation that if the output is within the tolerance limit at one frequency then it will also be within limits at nearby frequencies. Yield is therefore likely to be increased by response correlation at successive data points.

The moment method of tolerance analysis is at best approximate owing to the use of first-order sensitivity, and the evaluation of yield for vector output is best deferred since it involves excessive computational effort. The moment method therefore remains fast and easily programmed. It can be widely applied during circuit design to assist in predicting tolerance performance so that the best of several circuit alternatives can be selected. Once the design has been finalized and tolerances assigned, random sampling can be used to obtain accurate measures of the output covariances, worst case limits, and yield.

We are left with the problem of interpreting the output covariance matrix so that comparisons between circuits can be made. Fortunately the **tolerance field**, which is a graph of the response standard deviation against frequency or time, provides a valuable indication of overall response variations even though correlation is not specifically represented. Tolerance fields for a second-order RC-active low-pass filter are shown in Fig. 9.7 (see Fig. 9.13 in Example 9.5 for the circuit). Resistor and capacitor standard deviations are 1.67% of nominal, and correlations for graphs (a), (b), and (c) between component variations $R_1$, $R_2$, $C_1$ and $C_2$ are given by

$$\rho_a = \begin{bmatrix} 1 & 0 & 0 & 0 \\ 0 & 1 & 0 & 0 \\ 0 & 0 & 1 & 0 \\ 0 & 0 & 0 & 1 \end{bmatrix}, \quad \rho_b = \begin{bmatrix} 1 & 0.95 & 0 & 0 \\ 0.95 & 1 & 0 & 0 \\ 0 & 0 & 1 & 0.95 \\ 0 & 0 & 0.95 & 1 \end{bmatrix}$$

and

$$\rho_c = \begin{bmatrix} 1 & 0.95 & -0.9 & -0.9 \\ 0.95 & 1 & -0.9 & -0.9 \\ -0.9 & -0.9 & 1 & 0.95 \\ -0.9 & -0.9 & 0.95 & 1 \end{bmatrix}$$

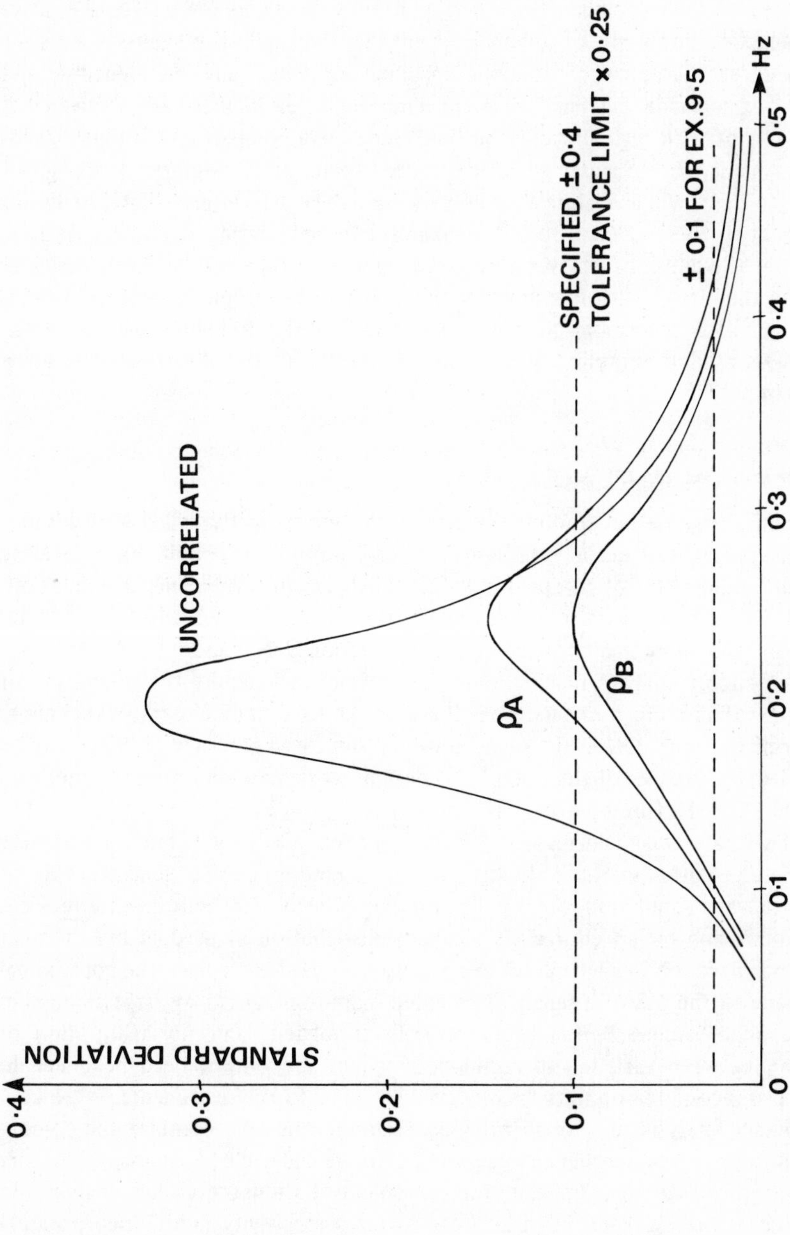


Fig. 9.7 – Tolerance fields for second-order filter.

The worst case tolerance limit may be assessed as $\pm 4\sigma$ for the purpose of comparison with specified tolerances. The specified tolerance limit is therefore indicated on the tolerance field as one quarter of the actual limit. (Where positive and negative limits are of unequal magnitude, the smallest magnitude is used). Areas of each graph exceeding the specified limit may then be identified and comparisons made between different circuits, or as in this case between the same circuit with different correlations. Qualitative comparisons of likely yields can be made, and in this case with specified limits of ±0.4, graph (a) will give a poor yield, (c) will give almost 100% yield, and (b) will be intermediate. Attempts to quantify these comparisons, for example by integrating the tolerance field graph or the graph of $\sigma^2$ over a range of frequencies, do not always successfully predict the circuit with maximum yield. Computed examples of tolerance fields together with corresponding yields are given in the following section which considers random sampling. Full numerical results for this illustration are given in Example 9.5.

## 9.4 RANDOM SAMPLING

Tolerance analysis by random sampling (also less correctly called Monte Carlo analysis) simulates the manufacture of a large number of circuits by generating pseudo-random sets of component values. Each circuit is then analysed and both worst case and statistical measures of variation in circuit performance are evaluated by the basic methods described in Section 9.1. Random sampling may involve circuit analysis many thousands of times and require the generation of millions of random numbers. It is therefore an extremely expensive technique in terms of computation time, and until the improved accuracy obtained can be justified by analysis of the completed design, worst case and moment methods should be used in preference.

There are two main reasons for the improved accuracy of random sampling compared to other methods. Firstly, the use of random number generators allows the accurate simulation of any known component distribution and replaces approximation by an equivalent Gaussian distribution as used in the moment method. Secondly, exact circuit analysis can be used to replace the approximation due to the use of a sensitivity model. Variation caused by large changes in component value are therefore correctly simulated, and no assumption of linarity between variation in component values and output need be made. In fact, this second reason must be qualified by a few further comments.

Exact analysis of a small integrated circuit can take considerable computation time when frequency response is to be determined. It may therefore prove unpractical to carry out, for example, one thousand circuit analyses. In such cases the designer may be forced to use a sensitivity model incorporating first-order and preferable second-order sensitivities even though some accuracy will be lost. An intermediate situation can be achieved for active networks and filter circuits by the use of symbolic transfer function analysis. Thus much of the labour of repeated analysis is avoided by the initial transfer function analysis.

The random sampling method can be summarized by the following steps.

STEP 1 Generate a random set of circuit component values with specified distributions including matching or correlation.

2 Analyse the circuit to obtain the output performance vector.

3a Update the summations in Eqs. (9.5) and (9.7) to obtain output variances and covariances.

b Retain the maximum and minimum values of each output for worst case tolerance limits.

c If tolerance field limits are specified, count the number of circuit simulations with output entirely within the tolerance limits.

4 Repeat STEPS 1 to 3 until a sufficiently large number of sets of component values have been generated to give the required statistical accuracy.

5 Determine the output variances and correlation coefficient matrix together with worst case tolerance field and yield.

**Generation of Random Numbers**

Most computers provide a standard function for the selection of a random number from a uniform distribution between zero and unity. Random numbers from almost any arbitrary probability density function can be generated by modification of those selected from this uniform distribution.

Various probability density functions $\psi(x)$ were shown in Fig. 9.2. The distribution function $\phi(x)$ equals the integral of the probability density and is given by

$$\phi(x) = \int_{-\infty}^{x} \psi(x)\mathrm{d}x \ .$$

$\phi(x)$ strictly increases from 0 to 1 as $x$ increases from $-\infty$ to $+\infty$ since the probability density is assumed to be non-negative and have unit area. Fig.9.8(a) for example, shows a triangular function $\psi(x)$ defined by

$$\psi(x) = 1 + x \qquad x \leqslant 1$$

$$\psi(x) = 1 - x \qquad x > 1$$

The corresponding distribution function $\phi(x)$ can be obtained from the two linear segments of the density by integration.

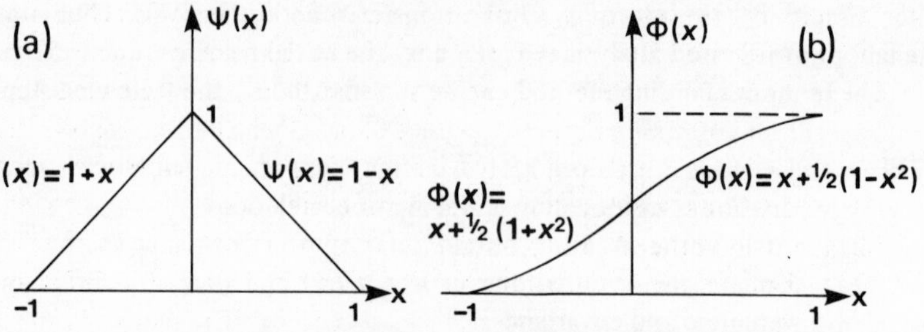


Fig. 9.8 – (a) Triangular probability density function; (b) Corresponding distribution function.

For $\quad x \leqslant 0 \quad \phi(x) = \int_{-1}^{x} (1+x)\mathrm{d}x = \dfrac{x^2}{2} + x + \dfrac{1}{2}\;.$

For $\quad x > 0 \quad \phi(x) = \dfrac{1}{2} + \int_{0}^{x} (1-x)\mathrm{d}x = \dfrac{-x^2}{2} + x + \dfrac{1}{2}\;.$

The graph of $\phi(x)$ is shown in Fig. 9.8(b).

If we now generate random numbers $y = \phi(x)$ uniformly in the range 0 to 1, the solution for $x$ will yield random numbers distributed with probability density function $\psi(x)$. For the triangular density function,

$$y = \frac{x^2}{2} + x + \frac{1}{2} \qquad \text{for } x \leqslant 0\;,$$

$$\text{therefore,}\quad x = -1 + \sqrt{2y} \qquad \text{for } y \leqslant \frac{1}{2}\;, \qquad (9.24)$$

$$\text{and,}\quad y = \frac{-x^2}{2} + x + \frac{1}{2} \qquad \text{for } x > 0\;,$$

$$\text{therefore,}\quad x = 1 - \sqrt{2-2y} \qquad \text{for } y > \frac{1}{2}\;. \qquad (9.25)$$

The procedure for generating a random number with the triangular density function initially involves a selection of $y$ from a uniform distribution between 0. and 1. This is followed by substitution into either Eq. (9.24) or (9.25) to obtain the required random number.

The general procedure (the *inverse* method) for any specified distribution is similarly carried out in two steps. For the process to be efficient, the equations for $x$ in terms of $y$ must be as simple and as few as possible. This usually involves provision of an approximation to the function $\psi(x)$ which can be integrated and solved to find the required function $x = f(y)$. Piecewise linear approximations, of which the triangular distribution could be an example, are a last resort for discontinuous and other difficult distributions.

The Gaussian distribution has been approximated by many different functions for random number generation by the inverse method. The required probability density function is given by

$$\psi(x) = \frac{1}{\sqrt{2\pi}} e^{-x^2/2}$$

for mean value equal to zero and standard deviation equal to unity. Kahn's approximation, for example, applies to the distribution $2\psi(x)$ for $x \geqslant 0$, thus generating the magnitude of the required Gaussian distribution. The sign is then randomly generated. Kahn's approximation is given by

$$\frac{2}{\sqrt{2\pi}} e^{-x^2/2} \approx \frac{2ke^{-kx}}{(1+e^{-kx})^2} ,$$

where $k = \sqrt{8/\pi}$. Therefore,

$$\int_0^x \frac{2ke^{-kx}}{(1+e^{-kx})^2} = \frac{1-e^{-kx}}{1+e^{-kx}}$$

and

$$x = 0.626657 \log_e \left\{\frac{1+y}{1-y}\right\}. \tag{9.26}$$

Random numbers from a Gaussian distribution are specified by mean value $\mu$ and standard deviation $\sigma$. They are therefore generated by substitution of uniformly distributed random numbers $y$ between 0 and 1 in the expression

$$\mu \pm 0.626657\, \sigma \log_e \left\{\frac{1+y}{1-y}\right\}, \tag{9.27}$$

where the sign is also generated randomly.

An expression similar to Eq. (9.26) but exact, which uses a trigonometric combination of two random numbers, is given by

$$x = \left\{-2 \log_e y_1\right\}^{\frac{1}{2}} \sin (2\pi y_2) , \tag{9.28}$$

where $y_1$ and $y_2$ are two independent random numbers from the uniform distribution between 0 and 1. A second independent Gaussian random number can be generated from the original $y_1$ and $y_2$ by using $\cos(2\pi y_2)$ in place of $\sin(2\pi y_2)$. Efficiency is therefore improved at the expense of extra program control logic.

**Correlated Random Numbers**

Correlated random numbers with Gaussian distribution fully defined by the covariance matrix **P** can be generated provided that **P** is positive definite. Mean values are assumed equal to zero and added later.

Using the definitions for $\sigma_p$ and $\rho_p$ in Eq. (9.18), the required covariance matrix for the components $\mathbf{p} = [p_1\, p_2 \ldots p_n]^t$ is given by

$$\mathbf{P} = \sigma_p \rho_p \sigma_p{}^t \; .$$

We are able to generate uncorrelated sets of random numbers $\mathbf{y} = [y_1 y_2 \ldots y_n]^t$ with unit standard deviations by the inverse method of the previous section. Their covariance matrix **Y** therefore equals the unit matrix **I**. Assume that the linear relationship

$$\mathbf{p} = \mathbf{S}\,\mathbf{y}$$

relates the random numbers **y** to the required random component values **p**. From the transmission of variance Eq. (9.19) we obtain

$$\begin{aligned} \mathbf{P} &= \mathbf{S}\,\mathbf{Y}\,\mathbf{S}^t \\ &= \mathbf{S}\,\mathbf{I}\,\mathbf{S}^t \end{aligned} \tag{9.29}$$

The problem therefore is to determine the matrix **S** given the component covariance matrix **P**.

If the matrix of correlation coefficients $\rho_p$ is assumed to be positive definite it may be diagonalized to yield

$$\rho_p = \mathbf{Q}\,\Lambda\,\mathbf{Q}^{-1} \; ,$$

where $\Lambda$ is the diagonal matrix of *real* eigenvalues and **Q** is an *orthogonal* matrix of normalized eigenvectors. Since **Q** is orthogonal $\mathbf{Q}^{-1} = \mathbf{Q}^t$ and

$$\begin{aligned} \mathbf{P} &= \sigma_p \rho_p \sigma_p \\ &= \sigma_p\, \mathbf{Q}\,\Lambda\,\mathbf{Q}^t\, \sigma_p{}^t \\ &= (\sigma_p\, \mathbf{Q}\,\Lambda^{\frac{1}{2}})\,\mathbf{I}(\sigma_p\, \mathbf{Q}\,\Lambda^{\frac{1}{2}})^t \; . \end{aligned} \tag{9.30}$$

By comparison of Eqs. (9.29) and (9.30)

$$\mathbf{S} = \sigma_p \, \mathbf{Q} \, \mathbf{\Lambda}^{\frac{1}{2}}$$

and

$$\mathbf{p} = (\sigma_p \, \mathbf{Q} \, \mathbf{\Lambda}^{\frac{1}{2}})\mathbf{y} \; . \tag{9.31}$$

Generation of correlated Gaussian random numbers with specified covariance matrix **P** proceeds as follows.

STEP 1 On initial entry only, determine **Q** and $\mathbf{\Lambda}^{\frac{1}{2}}$ and generate **S**.

2 Generate $n$ independent Gaussian random numbers **y** with unit standard deviations and zero mean values.

3 Use $\mathbf{p} = \mathbf{Sy}$ to obtain the correlated random component values.

4 Add mean values.

In practice there are faster methods of finding a suitable matrix **S** which do not use eigenvalues and eigenvectors. For example, direct solution of Eq. (9.29) is feasible if **S** is assumed to be an upper triangle matrix. Also the method known as *conditional decomposition* carries out STEP 1 about ten times faster than the eigenvalue method. However, the bulk of the computation time is occupied in the repeated use of STEPS 2 to 4, and the method selected for finding **S** is a matter of convenience.

**Computer Program**

RANSIM is a computer program for random simulation written in BASIC for on-line computation from time-shared terminals. The program is admittedly rather slow for use from a terminal, as it carries out a considerable amount of computation between data input and the output of results. A clock is in fact provided to indicate that the program is operative since users become uneasy when terminals go quiet. However, the intended application of the program for exploratory design and teaching can utilize a relatively small number of random samples, typically from 100 to 500. Accurate results are then obtained from a batch processed version of RANSIM which operates with the data and circuit analysis subroutine previously checked on-line. The program list is given in the Appendix to this chapter. Approximately 7 kbytes of memory are required to run Example 9.4 which has ten components and ten response frequencies. Execution times of the order of one to ten minutes have been experienced using Data General BASIC 3.8 on the NOVA 1220 computer.

Program RANSIM closely follows the five steps given earlier in this section for the random sampling method.

*Data input*

Program statements 10 to 640 are primarily used for data input. The following data inputs are required:

| | |
|---|---|
| Number of components and responses | N9 and M9 |
| Nominal component values | Row 0 of matrix P(N9, N9) |
| Component standard deviations | Entered into matrix T(1, N9) then held in R(N9) |
| Component correlation matrix (only if correlated) | Entered into matrix P(N9,N9). The covariance matrix is then generated in **P**. |
| Response tolerance limits (only if yield is to be computed) | Upper limits are entered in row 1 of matrix L(2,M9). Lower limits are in row 2. Nominal response is generated in row 0. |
| Number of random simulations | T9. |

*Random number generation (Step 1)*

Independent random numbers with Gaussian distribution are generated by the defined function FNR(S), statement 60, where S is the required standard deviation. This function is directly programmed from Eq. (9.28). Random numbers could be generated more efficiently with a subroutine written for speed, possibly best in machine code, but this has not been done in this version of RANSIM since the programs in this book are intended to be easily adapted to other computers.

Correlated random numbers are generated in a separate subroutine with statement numbers 9900 to 9929. This subroutine is a direct translation of the published program, CACM algorithm 425. On first entry the component covariance matrix is conditionally decomposed into the form required for Eq. (9.31). Matrix R(N9) retains the component standard directions. N9 independent random numbers are then generated in calculation space Q(N9) using function FNR(S). Program statements 9921 to 9928 produce the output matrix of correlated random numbers X(N9) using a version of Eq. (9.31). At all subsequent entries only statement numbers 9921 to 9929 need to be executed.

*Circuit analysis subroutine 2000 (Step 2)*

Circuit analysis by any of the computer programs in this book is inordinately slow. This is not a problem for a single analysis although for large circuits a speed increase of at least 10:1 is a necessity and involves more sophisticated numerical techniques. However, a typical random simulation may require 10 000 repeated analyses for frequency response with somewhat fewer analyses on-line. The problem is simply stated:

Given an input vector of N9 component values in X(N9),
generate a response vector of M9 values in Y(M9).

Usually in active network design this implies a frequency response, but any other measure of circuit performance can be programmed, for example d.c. node voltages, or transient response. The overriding requirement is for analysis speed, and almost any effort by the designer or by the computer to reduce the analysis to simple evaluation is justified.

In RANSIM, subroutine 2000 must be separately written for each circuit analysed. Typical examples given for small filters in Examples 9.4 and 9.5 show what can be achieved using symbolic analysis. By programming for maximum speed in a specially written subroutine, analysis time can be reduced by a factor of 10 to 100 times.

The only practical restriction on writing subroutine 2000 is that variable names must be chosen to avoid those already in use in the main program. The analysis must terminate in a RETURN statement without running into the random number generator beginning at statement 9900. The final output of RANSIM is obtained in columns. The first of these can be labelled in the analysis subroutine using M$(10), for example M$ = "HZ." and row 0 of matrix S(1,M9) which is printed under the heading M$ can be filled, for example with the response frequencies.

*Statistical analysis (Steps 3 and 4)*

Statements 650 to 990 are concerned with the statistical and worst case analysis of the random simulation. After each call to subroutine 2000 the randomly generated response is obtained in matrix Y(M9). The response covariance matrix A(M9,M9) is progressively updated using Eqs. (9.5) and (9.7), assuming T9 simulations. Additionally row 0 of **A** is updated to contain the worst case upper tolerance limit of the response, and column 0 is similarly updated for the lower limit. The yield of circuits with response entirely within specified tolerance limits is contained in A(0,0).

The statistical analysis in RANSIM assumes that the nominal response is equal to the mean response, and excessive derivation from that assumption will result in errors. The extent of these errors was checked for Example 9.4 which exhibits this effect because it contains resonant circuits. Errors were introduced but were not considered sufficiently large to warrant changes in the on-line program. The minor re-programming necessary to first compute mean values of response and then to remove the mean from the variances and co-variances is a worthwhile problem as minor difficulties arising from statistical inaccuracy are introduced.

If response correlation is never required, it is probably worthwhile to modify statement numbers 950 to 970 and omit statements 1110 to 1200 so that execution is faster. Storage can also be saved by reducing matrix **A** to three columns.

*Output (Step 5)*

Statements 1000 to 1200 are concerned with the output of results. The response covariance matrix A(M9, M9) is first reduced to a matrix of correlation coefficients while extracting response standard deviation into matrix S(1,M9). Outputs are then printed in five columns containing,

1. Response identification, for example frequencies, from row 0 of S(1,M9),
2. Nominal response from row 0 of L(1,M9),
3. response upper tolerance limit from row 0 of A(M9,M9),
4. response lower of tolerance limit from column 0 of A(M9,M9), and
5. response standard deviation from row 1 of S(1,M9).

The yield is printed if computed, and this is followed by the response correlation matrix if required.

The program finally enables the user to call for further simulations using the same data. This allows the program, particularly the analysis subroutine, to be checked with a few simulations only before a large amount of computation time is expended and possibly wasted.

*Other program variables*

Control variables I6, I7, I8, L8, L9 .
Calculation space I9, J9, K9, N$(3), Q(N9), T(1,N9).

*Example* 9.4

Deremine the tolerance fields, worst case tolerance limits, and yield of the bandpass filter redrawn in Fig. 9.9, from Example 8.6 where the filter was first analysed. Use ten frequencies in the pass- and stop-bands with abosolute tolerance limits specified by the following table.

| Hz | 2650 | 2695 | 2760 | 2775 | 2795 | 2815 | 2835 | 2850 | 2920 | 2965 |
|---|---|---|---|---|---|---|---|---|---|---|
| Upper Limit dB | 4 | 4 | 0.1 | 0.1 | 0.1 | 0.1 | 0.1 | 0.1 | 4 | 4 |
| Lower Limit dB | −50 | −50 | −1 | −0.2 | −0.2 | −0.2 | −0.2 | −1 | −50 | −50 |

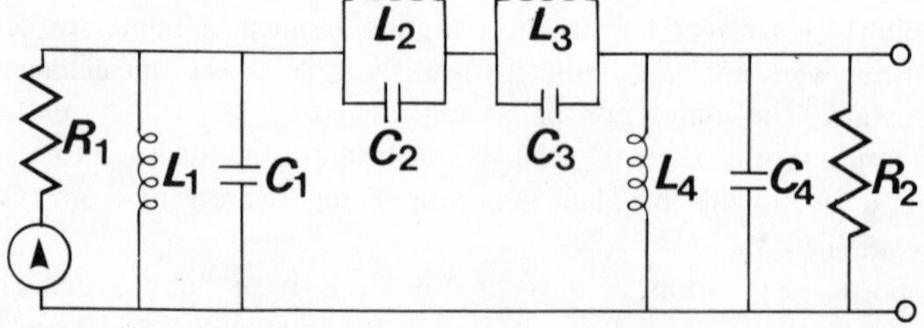


Fig. 9. 9 – Bandpass filter for Example 9.4.
R1 = R2 = 1k L1 = L4 = 1.919 mH L2 = 7.5964 mH L3 = 8.3438 mH C1 = C4 = 1.677 μF C2 = 0.3859 μF C3 = 0.4236 μF.

These tolerances are superimposed on the nominal response which can be added in RANSIM. Assume components to be uncorrelated with standard deviation equal to 0.33% of nominal; that is, component tolerances are about ±1%.

The analysis subroutine written for this filter is given in Fig. 9.10. The seven statements 2050 to 2110 which carry out the analysis are remarkably compact considering the complexity so far encountered in analysing a general circuit. The designer has of course done all the hard work in the interest of rapid evaluation. The analysis is symbolic but proceeds through a succession of circuit reductions instead of through the transfer function. Component values in X(N9) produced by the random number generator are in the order, R1, R2, L1, L2, L3, L4, C1, C2, C3, C4.

```
2000 REM SUBROUTINE - ELLIPTIC BANDPASS FILTER
2010 RESTORE
2015 LET M$="HZ."
2020 FOR F=1 TO 10
2030   READ W
2035   LET S[0,F]=W
2040   LET W=6.28319*W
2050   LET Z1=W*X[3]/(1-W^2*X[3]*X[7])
2060   LET Z2=W*X[4]/(1-W^2*X[4]*X[8])
2070   LET Z2=Z2+W*X[5]/(1-W^2*X[5]*X[9])
2080   LET Y3=W*X[10]-1/W/X[6]
2090   LET Y1=X[1]-Z1*Z2/X[2]-Y3*X[1]*(Z1+Z2)
2100   LET Y2=Z1-Z1*Z2*Y3+X[1]/X[2]*(Z1+Z2)
2110   LET Y[F]=4.34295*LOG(Z1^2/(Y1^2+Y2^2))
2120 NEXT F
2130 RETURN
2140 DATA 2650,2695,2760,2775,2795,2815,2835,2850,2920,2965

*
```

Fig. 9.10 – Analysis subroutine for bandpass filter.

The computer run is reproduced in Fig. 9.11, and results in the passband are interpreted graphically in Fig. 9.12. The predicted yield of circuit responses entirely within the specified tolerance limits, is 46%. From the computer output and graphs in Figs. 9.11 and 9.12 it is clear that the critical areas of response where failures occur lie at the edges of the passband. The predicted yield of 46% is perhaps surprising in view of the remaining results. Even so, a much higher yield would be expected of a passive circuit, and this emphasizes the accuracy required for the components of narrow passband filters, for this filter was analysed with 1% components.

The graph for response mean value was obtained from a modified version of RANSIM so that the effect of differences between nominal and mean could be assessed. The worst deviation just outside the passband was 1 dB. This led to errors of just over 0.1 dB in standard deviation and different though similar correlations. Worst case limits and yield are of course not affected by these errors.

```
RANSIM  RANDOM SIMULATION
NO. COMPONENTS, NO. RESPONSES ? 10,10
NOMINAL COMPONENT VALUES
? 1E3,1E3,1.919E-3,7.5964E-3,8.3438E-3,1.919E-3,1.677E-6,.3859E-6,.4236E-6?
 1.677E-6
COMPONENT STANDARD DEVIATIONS
? 3.33,3.33,.064E-4,.253E-4,.278E-4,.064E-4,.056E-7,.013E-7,.014E-7,.056E-7
CORRELATED YES OR NO ? NO
YIELD TEST LIMITS YES OR NO ? YES
ALL RESPONSE UPPER TOLERANCE LIMITS, THEN ALL LOWER LIMITS
? 4,4,.1,.1,.1,.1,.1,.1,4,4,-50,-50,-1,-.2,-.2,-.2,-.2,-1,-50,-50
ADD NOMINAL RESPONSE TO TOLERANCE LIMITS YES OR NO ? YES
NUMBER OF SIMULATIONS ? 500
.....................................
             RESPONSE       UPPER          LOWER          STANDARD
HZ.          NOMINAL        LIMIT          LIMIT          DEVIATION
 2650        -40.0544       -36.0726       -47.3051       1.79845
 2695        -35.5394       -28.4461       -73.2727       5.26802
 2760        -6.16208       -6.00098       -8.23046       .472108
 2775        -6.08824       -5.98842       -6.76089       .130072
 2795        -6.06238       -5.98189       -6.56213       .10442
 2815        -6.05026       -5.97847       -6.89208       .125695
 2835        -6.08785       -5.9894        -7.09029       .165411
 2850        -6.10452       -5.98828       -8.05932       .366348
 2920        -35.4684       -26.3316       -78.4702       5.17194
 2965        -40.8028       -36.5336       -61.2836       2.84531
PERCENTAGE YIELD  45.9995
RESPONSE CORRELATION MATRIX YES OR NO ? YES
 1            -.761824      -.598545       .623752        .448229
-.305684      -.277484       8.56483E-02   -3.00118E-02   -4.55759E-02

-.761824       1             .881166       -.179272       -8.70363E-02
 .20094        .233464      -2.9056E-03    1.70097E-02    1.54567E-02

-.598545       .881166       1             2.45521E-02    3.66274E-02
 .378401       .483735       9.32827E-02   -.102542       .235063

 .623752      -.179272       2.45521E-02   1              .817452
 2.15461E-02   5.21972E-02   .483576       .221806        -.142848

 .448229      -8.70363E-02   3.66274E-02   .817452        1
 4.90054E-02  -3.97626E-03   .32905        .139495        -.148183

-.305684       .20094        .378401       2.15461E-02    4.90054E-02
 1             .880093       .131891       -.116035       .508959

-.277484       .233464       .483735       5.21972E-02    -3.97626E-03
 .880093       1             7.17889E-02   -.297118       .741134

 8.56483E-02  -2.9056E-03    9.32827E-02   .483576        .32905
 .131891       7.17889E-02   1             .800507        -.325478

-3.00118E-02   1.70097E-02  -.102542       .221806        .139495
-.116035      -.297118       .800507       1              -.687696

-4.55759E-02   1.54567E-02   .235063       -.142848       -.148183
 .508958       .741134      -.325478       -.687696       1
FURTHER SIMULATION YES OR NO ? NO

END AT 1250
*
```

Fig. 9.11 – Computer run for Example 9.4.

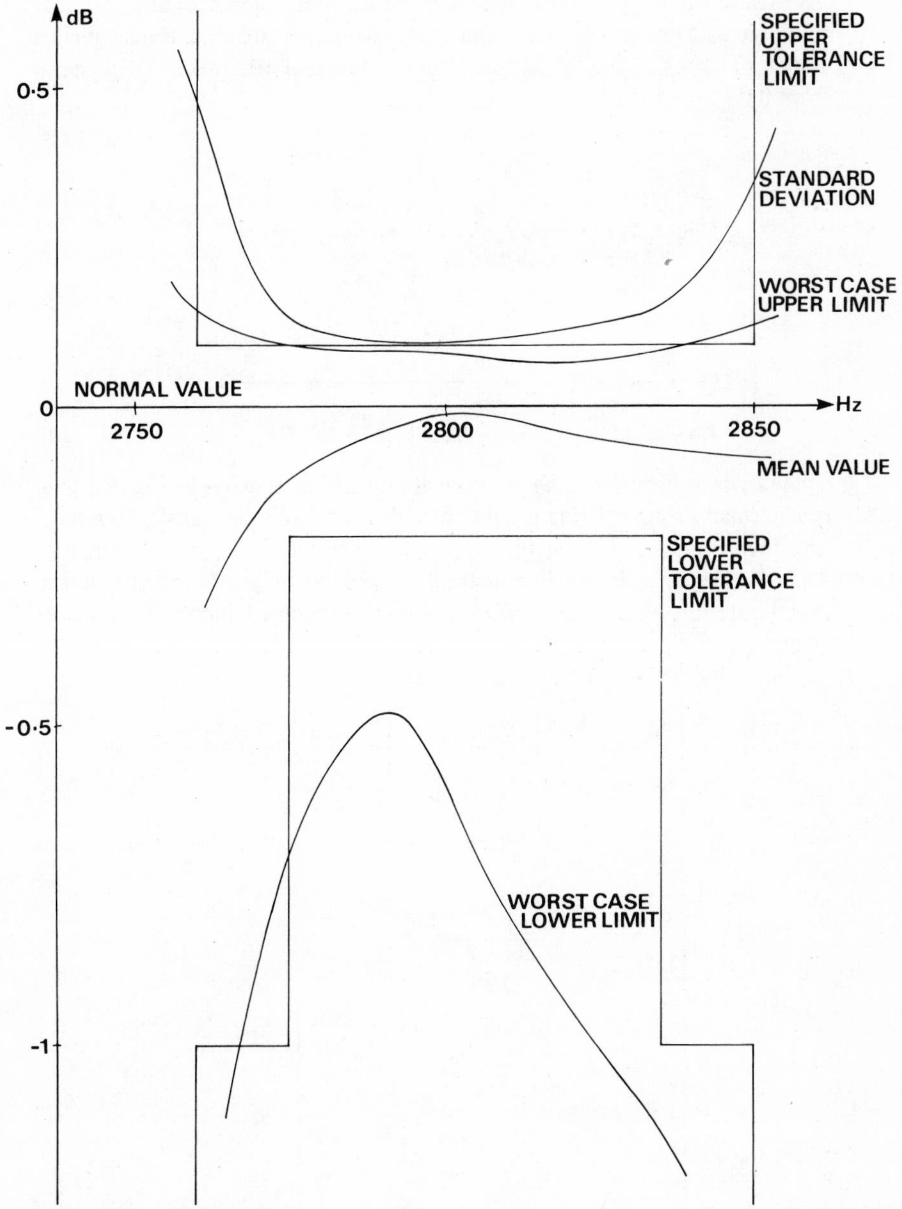


Fig. 9.12 – Computed results for bandpass filter. Passband 2760 to 2850 Hz.

*Example* 9.5

Investigate the effect of component correlation on tolerances and yield for the lowpass filter in Fig. 9.13. Use component standard deviations equal to 1.67% of nominal, and yield test limits of nominal ±0.1 dB at frequencies from 0.05 to 0.5 Hz.

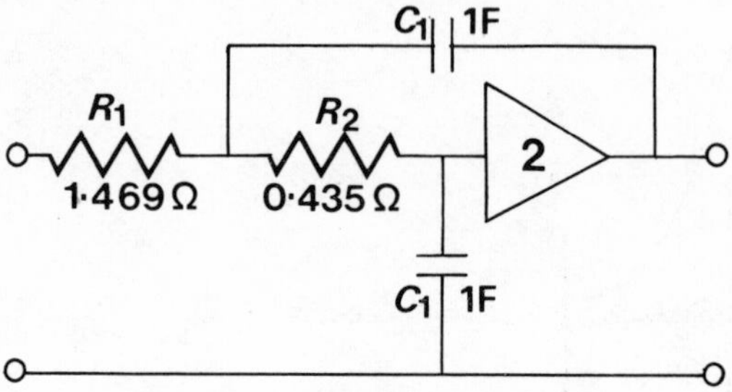


Fig. 9.13 – Filter for Example 9.5.
$F(s) = 2/\{1 + s(R_1C_2 + R_2C_2 - R_1C_1) + s^2R_1R_2C_1C_2\}$

Analysis subroutine 2000 written for this filter is given in Fig. 9.14. It is particularly simple, being programmed directly from the transfer function. The first computer run for uncorrelated components is reproduced in Fig. 9.15. Performance is not impressive with yield at only 12% because worst case limits are considerably in excess of the specified limits. Two subsequent sets of computed results are given in Fig. 9.16. These correspond to the following component correlation matrices:

$$\rho_b = \begin{bmatrix} R_1 & R_2 & C_1 & C_2 \\ 1 & 0.95 & 0 & 0 \\ 0.95 & 1 & 0 & 0 \\ 0 & 0 & 1 & 0.95 \\ 0 & 0 & 0.95 & 1 \end{bmatrix}$$

$$\rho_c = \begin{bmatrix} R_1 & R_2 & C_1 & C_2 \\ 1 & 0.95 & -0.9 & -0.9 \\ 0.95 & 1 & -0.9 & -0.9 \\ -0.9 & -0.9 & 1 & 0.95 \\ -0.9 & -0.9 & 0.95 & 1 \end{bmatrix}$$

Both of these matrices could represent IC forms of the filter with an actual realization somewhere between the two. The best yield obtained at 51% is a considerable improvement in performance compared to the uncorrelated version. Results for standard deviation were used to illustrate tolerance fields, and graphs of standard deviation plotted against frequency for this filter were given in Fig. 9.7. The graphs show a peak at the filter cut-off frequency of 1 rad/sec, which successively reduces as correlation is introduced. Computed worst case limits are generally in approximate agreement with tolerance fields obtained from standard deviation with a factor about equal to three between two values.

```
2000 REM ANALYSIS SUBROUTINE - LOWPASS FILTER
2010 IF I8>0 THEN GOTO 2030
2020 LET M$="HZ."
2030 FOR I=1 TO M9
2040   IF I8>0 THEN GOTO 2060
2050   LET S[0,I]=I/20
2060   LET W=.31416*I
2070   LET C3=X[1]*X[3]
2080   LET C4=X[2]*X[4]
2090   LET C5=X[1]*X[4]
2100   LET Y[I]=2/SQR((1-W^2*C3*C4)^2+W^2*(C5+C4-C3)^2)
2110 NEXT I
2120 RETURN

*
```

Fig. 9.14 – Analysis subroutine for Example 9.5. Component order in X(4) is $R_1$, $R_2$, $C_1$, $C_2$.

```
RANSIM  RANDOM SIMULATION
NO. COMPONENTS, NO. RESPONSES ? 4,10
NOMINAL COMPONENT VALUES
? 1.469,.435,1,1
COMPONENT STANDARD DEVIATIONS
? .0245,.00726,.0167,.0167
CORRELATED YES OR NO ? NO
YIELD TEST LIMITS YES OR NO ? YES
ALL RESPONSE UPPER TOLERANCE LIMITS, THEN ALL LOWER LIMITS
? .1,.1,.1,.1,.1,.1,.1,.1,.1,.1,-.1,-.1,-.1,-.1,-.1,-.1,-.1,-.1,-.1,-.1
ADD NOMINAL RESPONSE TO TOLERANCE LIMITS YES OR NO ? YES
NUMBER OF SIMULATIONS ? 250
...................
          RESPONSE       UPPER          LOWER          STANDARD
HZ.       NOMINAL        LIMIT          LIMIT          DEVIATION
 .05        2.11228        2.12731        2.09619        5.55418E-03
 .1         2.5122         2.60183        2.41202        3.31611E-02
 .15        3.35654        3.75456        2.91631        .1505
 .2         3.65821        4.8501         2.85766        .337775
 .25        2.23677        2.8035         1.81171        .15774
 .3         1.32267        1.52901        1.13441        .070986
 .35        .87            .973832        .769547        4.02626E-02
 .4         .61973         .68968         .557632        2.63028E-02
 .45        .466342        .516797        .423989        1.87298E-02
 .5         .364964        .403227        .334074        1.41176E-02
PERCENTAGE YIELD  12
RESPONSE CORRELATION MATRIX YES OR NO ? NO
FURTHER SIMULATION YES OR NO ? NO

END AT 1250
*
```

Fig. 9.15 – Computer run for Example 9.5 with uncorrelated components.

```
....................
              RESPONSE         UPPER          LOWER          STANDARD
HZ.           NOMINAL          LIMIT          LIMIT          DEVIATION
 .05            2.11228          2.12789        2.09902      5.47922E-03
 .1             2.5122           2.59531        2.44424      2.82975E-02
 .15            3.35654          3.5804         3.16595      7.73184E-02
 .2             3.65821          3.92752        3.34781       .116669
 .25            2.23677          2.6387         1.90994       .151793
 .3             1.32267          1.54538        1.1311       8.78858E-02
 .35             .87             1.00805         .751042     5.35167E-02
 .4              .61973           .71264         .538885     3.58611E-02
 .45             .466342          .533294        .407583     2.58134E-02
 .5              .364964          .415697        .320143     1.95575E-02
PERCENTAGE YIELD  36.3997
RESPONSE CORRELATION MATRIX YES OR NO ? NO
FURTHER SIMULATION YES OR NO ? NO

END AT 1250
*
```

```
....................
              RESPONSE         UPPER          LOWER          STANDARD
HZ.           NOMINAL          LIMIT          LIMIT          DEVIATION
 .05            2.11228          2.12284        2.10185      3.82427E-03
 .1             2.5122           2.56877        2.45829      1.99148E-02
 .15            3.35654          3.51967        3.20293      5.80359E-02
 .2             3.65821          3.90394        3.39148      9.78139E-02
 .25            2.23677          2.56096        2.02347       .107548
 .3             1.32267          1.49961        1.19413      6.12274E-02
 .35             .87              .977029        .789645     3.71424E-02
 .4              .61973           .690988        .564996     2.48565E-02
 .45             .466342          .51739         .426522     1.78817E-02
 .5              .364964          .403506        .334574     1.35439E-02
PERCENTAGE YIELD  51.1994
RESPONSE CORRELATION MATRIX YES OR NO ? NO
FURTHER SIMULATION YES OR NO ? NO

END AT 1250
*
```

Fig. 9.16 – Correlated component results for Example 9.5.
Upper set – Correlation matrix $\rho_b$; Lower set – Correlation matrix $\rho_c$.

## 9.5 CONCLUSION

When the nominal circuit design has been completed and a final full analysis carried out, the designer should have all possible information about the behaviour of his circuit. However, circuit design is far from complete. A design may appear to be perfectly acceptable on the basis of nominal behaviour and sensitivity yet be quite useless in practice. Indeed, many of the published RC active circuit filter realizations belong to this category. Design will be completed only when tolerances have been assigned, and a full random simulation carried out to assess yield, production costs and behaviour as components vary together.

This chapter does little more than introduce tolerance analysis which is a most important area of circuit design. The major problem of tolerance assignment has not even been considered.

## FURTHER READING

Arley, N. and Buch, K. Randar., (1966). *Introduction to the Theory of Probability and Statistics,* Wiley, New York.

Calahan, D. A., (1972), *Computer-aided Network Design,* Ch. 7, McGraw-Hill, New York.
Geher, K., (1971), *Theory of Network Tolerances,* Akademiai Kiado, Budapest.
Tocher, K. D., (1963), *The Art of Simulation,* English Universities Press, London.

## PROBLEMS

9.1 Analyse the circuit in Fig. 9.17 by the worst case method assuming $K$, $C_1$, $C_2$ to vary by ±0.1, ±0.1, and ±0.02 respectively. The required output voltage across $C_2$ is the real part of $F(s)$ at 0.2, 0.4, . . . 1 rad/sec. If $C_1$ and $5C_2$ match to within ±5% what are the new worst case limits?

9.2 Analyse the circuit in Fig. 9.17 by the moment method assuming component standard deviations for $K$, $C_1$, and $C_2$ equal to 0.03, 0.03, and 0.006 respectively and correlation between $C_1$ and $C_2$ variations equal to 0.95.

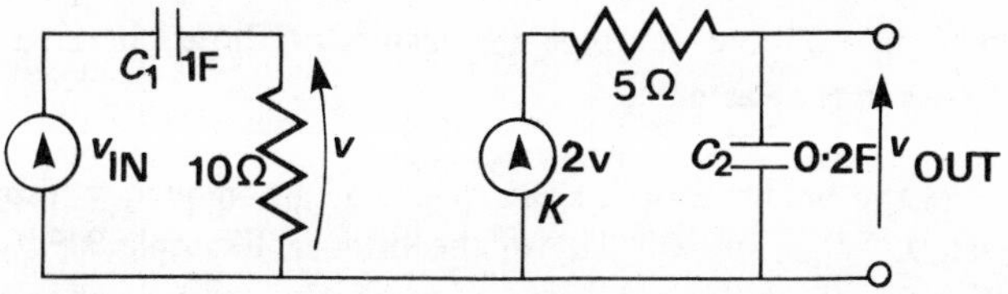


Fig. 9.17 – Problems 9.1 to 9.3.

9.3 Obtain the unit impulse response of the circuit in Fig. 9.17 at $t$=1, 2, and 3 secs, and determine the transient response tolerance field by the moment method.

9.4 $E(i,n)$ is the frequency response absolute sensitivity matrix to $n$ component values, $F(j,n)$ is the transient response sensitivity matrix, $G(n,n)$ is the component covariance matrix, $V(i,i)$ is the frequency response covariance matrix, and $W(j,j)$ is the transient response covariance matrix. Use the two basic relationships $\mathbf{W} = \mathbf{F\ G\ F^t}$ and $\mathbf{V} = \mathbf{E\ G\ E^t}$, and the definition of the generalized inverse, for example when $i > n$ $\mathbf{E^{-1}} = \mathbf{(E^tE)^{-1}E^t}$, and show that $\mathbf{W} = \mathbf{(FE^{-1})V(FE^{-1})^t}$ and $\mathbf{V} = \mathbf{(EF^{-1})W(EF^{-1})^t}$.

9.5 Check the accuracy of the expressions in Problem 9.4 for tolerance field transfer between frequency and time domains. using the results from Problems 9.1 to 9.3.

9.6 Determine the worst case limits of the output voltage of the zener diode regulator in Fig. 9.18.

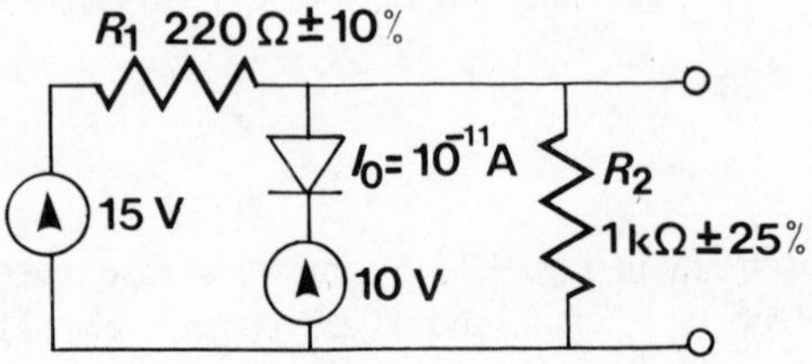


Fig. 9.18 – Problems 9.6 and 9.7.

9.7 Determine the worst case limits in Problem 9.6 if temperature also varies by ±25°C.

9.8 Obtain the standard deviations and correlation coefficients of the frequency response magnitude for the filter circuit in Example 9.5. Use the moment method and compare the results of those obtained by random simulation in the example.

9.9 What are the predicted worst case values of the frequency response magnitude at 0.1, 0.2, and 0.3 Hz for the filter in Example 9.5. Compare the results with the tolerance limits obtained by random simulation in the example.

*Computing*

9.10 Check the performance of the random number generator in program RANSIM by writing a suitable replacement for the normal circuit analysis subroutine.

9.11 Examine tolerance fields produced by generation of uniform random numbers within component tolerance limits. Compare response limits against computed worst case limits for the circuits in Figs. 9.5 and 9.13.

9.12 Modify the program RANSIM so that response mean values are also determined and taken into account in the simulation. Check the effect of this correction for the bandpass filter from Example 9.4.

9.13 Generate tolerance fields for the filter in Problem 4.13. Assume uncorrelated component values with 1% tolerances. A suitable yield test for the filter is based on the required gain dB from 5.8 Hz to 6.2 MHz which must be within the limits −8 dB to −12.5 dB. Also gain must be less than −65 dB, −45 dB and −40 dB at 2.5 MHz, 4.43 MHz, and 7.5 MHz respectively. Assume the gain at 6 MHz to be adjusted to nominal by an automatic gain control.

9.14 Refer to Problem 8.12. Assume amplifier gains fixed and determine tolerance fields for the filter due to the specified tolerances for resistivity $r$, length $l$, width $w$, and basic capacitance $c$.

9.15 Write a program for worst case analysis. This is probably a worthwhile extension to program RNODE.

9.16 Write a program for moment method analysis. This could possibly be an extension to progran ACNLU or TOPSEN since they both generate the necessary sensitivities.

## APPENDIX

```
0010 PRINT 'RANSIM  RANDOM SIMULATION'
0020 PRINT 'NO. COMPONENTS, NO. RESPONSES';
0030 INPUT N9,M9
0040 DIM A[M9,M9],L[2,M9],P[N9,N9],S[1,M9],T[1;N9]
0050 DIM Q[N9],R[N9],X[N9],Y[M9],N$[3],M$[10]
0060 DEF FNR(S)=SQR(-2*LOG(RND(1)))*SIN(6.28318*RND(1))*S
0070 LET M$=' '
0080 PRINT 'NOMINAL COMPONENT VALUES'
0090 MAT INPUT T
0100 PRINT
0110 LET I8=0
0120 LET I7=0
0130 FOR I9=1 TO N9
0140   LET X[I9]=T[1,I9]
0150   LET P[0,I9]=X[I9]
0160 NEXT I9
0170 GOSUB 2000
0180 FOR I9=1 TO M9
0190   LET L[0,I9]=Y[I9]
0200   LET S[0,I9]=I9
0210 NEXT I9
0220 PRINT 'COMPONENT STANDARD DEVIATIONS'
0230 MAT INPUT T
0240 PRINT
0250 PRINT 'CORRELATED YES OR NO';
0260 INPUT N$
0270 LET L8=1
0280 IF N$='NO' THEN GOTO 0340
0290 LET L8=2
0300 PRINT 'COMPONENT CORRELATION MATRIX'
0310 MAT INPUT P
0320 PRINT
0330 GOTO 0350
0340 MAT P=IDN
0350 FOR I9=1 TO N9
0360   ON L8 THEN GOTO 0370, 0390
0370   LET P[I9,I9]=T[1,I9]
0380   GOTO 0420
0390   FOR J9=1 TO N9
0400     LET P[I9,J9]=P[I9,J9]*T[1,I9]*T[1,J9]
0410   NEXT J9
0420 NEXT I9
0430 PRINT 'YIELD TEST LIMITS YES OR NO';
0440 INPUT N$
0450 LET L9=1
0460 IF N$='NO' THEN GOTO 0570
0470 LET L9=2
0480 PRINT 'ALL RESPONSE UPPER TOLERANCE LIMITS, THEN ALL LOWER LIMITS'
0490 MAT INPUT L
0500 PRINT 'ADD NOMINAL RESPONSE TO TOLERANCE LIMITS YES OR NO';
0510 INPUT N$
0520 IF N$='NO' THEN GOTO 0570
0530 FOR I9=1 TO M9
0540   LET L[1,I9]=L[1,I9]+L[0,I9]
0550   LET L[2,I9]=L[2,I9]+L[0,I9]
0560 NEXT I9
0570 PRINT 'NUMBER OF SIMULATIONS';
0580 INPUT T9
0590 MAT A=ZER
0600 LET A[0,0]=0
0610 FOR I9=1 TO M9
0620   LET A[0,I9]=L[0,I9]
0630   LET A[I9,0]=L[0,I9]
0640 NEXT I9
```

*

Fig. 9.19 – Program RANSIM.

Fig. 9.19 – *continued*

```
0650 LET I6=0
0660 FOR I8=1+I7 TO T9+I7
0670   LET I6=I6+.08
0680   IF I6<1 THEN GOTO 0710
0690   PRINT ".";
0700   LET I6=0
0710   ON L8 THEN GOTO 0720, 0760
0720   FOR J9=1 TO N9
0730     LET X[J9]=FNR(P[J9,J9])+P[0,J9]
0740   NEXT J9
0750   GOTO 0800
0760   GOSUB 9900
0770   FOR J9=1 TO N9
0780     LET X[J9]=X[J9]+P[0,J9]
0790   NEXT J9
0800   GOSUB 2000
0810   ON L9 THEN GOTO 0870, 0820
0820   FOR J9=1 TO M9
0830     IF Y[J9]>L[1,J9] THEN GOTO 0870
0840     IF Y[J9]<L[2,J9] THEN GOTO 0870
0850   NEXT J9
0860   LET A[0,0]=A[0,0]+100/T9
0870   FOR J9=1 TO M9
0880     IF Y[J9]<=A[0,J9] THEN GOTO 0910
0890     LET A[0,J9]=Y[J9]
0900     GOTO 0930
0910     IF Y[J9]>=A[J9,0] THEN GOTO 0930
0920     LET A[J9,0]=Y[J9]
0930   NEXT J9
0940   FOR J9=1 TO M9
0950     FOR K9=1 TO M9
0960       LET A[J9,K9]=A[J9,K9]+(Y[J9]-L[0,J9])*(Y[K9]-L[0,K9])/T9
0970     NEXT K9
0980   NEXT J9
0990 NEXT I8
1000 PRINT
1010 PRINT " ","RESPONSE","UPPER","LOWER","STANDARD"
1020 PRINT M$,"NOMINAL","LIMIT","LIMIT","DEVIATION"
1030 FOR I9=1 TO M9
1040   LET S[1,I9]=SQR(A[I9,I9])
1050 NEXT I9
1060 FOR I9=1 TO M9
1070   PRINT S[0,I9],L[0,I9],A[0,I9],A[I9,0],S[1,I9]
1080 NEXT I9
1090 IF L9=1 THEN GOTO 1110
1100 PRINT "PERCENTAGE YIELD ";A[0,0]
1110 PRINT "RESPONSE CORRELATION MATRIX YES OR NO";
1120 INPUT N$
1130 IF N$="NO" THEN GOTO 1210
1140 FOR I9=1 TO M9
1150   FOR J9=1 TO M9
1160     LET A[I9,J9]=A[I9,J9]/S[1,I9]/S[1,J9]
1170     PRINT A[I9,J9],
1180   NEXT J9
1190   PRINT
1200 NEXT I9
1210 PRINT "FURTHER SIMULATION YES OR NO";
1220 LET I7=1
1230 INPUT N$
1240 IF N$="YES" THEN GOTO 0570
1250 END
```

*

Fig. 9.19 – *continued*

```
9900 REM SUBROUTINE - CORRELATED RANDOM NUMBER GENERATOR
9901 DIM P[N9,N9],R[N9],Q[N9],X[N9]
9902 IF I8>1 THEN GOTO 9921
9903 REM  CONDITIONAL DECOMPOSITION
9904 FOR K9=1 TO N9-1
9905   IF P[K9,K9]<=0 THEN GOTO 9917
9906   LET R[K9]=SQR(P[K9,K9])
9907   FOR I9=K9+1 TO N9
9908     LET P[I9,K9]=P[K9,I9]/P[K9,K9]
9909   NEXT I9
9910   FOR I9=K9+1 TO N9
9911     FOR J9=I9 TO N9
9912       LET P[I9,J9]=P[I9,J9]-P[I9,K9]*P[K9,J9]
9913     NEXT J9
9914   NEXT I9
9915 NEXT K9
9916 IF P[N9,N9]>0 THEN GOTO 9919
9917 PRINT 'NON-POSITIVE DEFINITE COVARIANCE MATRIX'
9918 STOP
9919 LET R[N9]=SQR(P[N9,N9])
9920 REM RANDOM NUMBER GENERATOR
9921 FOR I9=1 TO N9
9922   LET Q[I9]=FNR(R[I9])
9923   LET X[I9]=Q[I9]
9924   IF I9=1 THEN GOTO 9928
9925   FOR J9=1 TO I9-1
9926     LET X[I9]=X[I9]+P[I9,J9]*Q[J9]
9927   NEXT J9
9928 NEXT I9
9929 RETURN

*
```

# Index